NORTHERN HORIZON

URSA MAJOR "BIG DIPPER"

DRACO

URSA MINOR "LITTLE DIPPER"

POLARIS "NORTH STAR"

POLLUX
CASTOR

GEMINI

CEPHEUS

CYGNUS "NORTHERN CROSS"

VEGA
LYRA

CAPELLA

DENEB

AURIGA

CASSIOPEIA

SAGITTA

PERSEUS

DELPHINUS

ANDROMEDA

ALTAIR

EASTERN HORIZON

BETELGEUSE

TAURUS

ALDEBARAN

PLEIADES

TRIANGULUM

ARIES

AQUILA

WESTERN HORIZON

ORION

PEGASUS

RIGEL

PISCES

AQUARIUS

CAPRICORNUS

CETUS

FOMALHAUT

GRUS

SOUTHERN HORIZON

# THE NIGHT SKY IN NOVEMBER

Chart time (Local Standard Time):

10 pm...First of November
9 pm...Middle of November
8 pm...Last of November

THE NIGHT SKY IN JANUARY

Chart time (Local Standard Time):

10 pm...First of January
9 pm...Middle of January
8 pm...Last of January

# DISCOVERING THE UNIVERSE

## Fifth Edition

## Neil F. Comins
University of Maine

## William J. Kaufmann III
Late of San Diego State University

W. H. Freeman and Company
New York

*Dedicated to my mother, Pearl Mirianne Comins
and to the memory of my father, Francis Malcolm Comins*

*Acquisitions Editor and Marketing Manager:* Kimberly Manzi
*Developmental Editor:* Barbara Brooks
*Project Editor:* Diane Cimino Maass
*Director of Media:* Patrick Shriner
*Media Editor:* Matthew Fitzpatrick
*Cover and Text Designer:* Cambraia (Magalhaes) Fernandes
*Illustration Coordinator:* Bill Page
*Illustrations:* Fine Line Illustrations
*Photo Researcher:* Jennifer MacMillan
*Production Coordinator:* Susan Wein
*Composition:* Bonnie Stuart and Sheridan Sellers
*Manufacturing:* RR Donnelley & Company

*Cover Image and Frontispiece:* An artist's perception of our Milky Way galaxy and the smaller Sagittarius dwarf spheroidal. The Sagittarius dwarf, which looks here like an explosion of stars, is orbiting the Milky Way. Our Galaxy's gravitational forces are pulling the dwarf apart and assimilating its stars. The perspective shown here is that of a viewer 500 light years (4750 trillion kilometers or 3000 trillion miles) above the plane of the galactic disk.

Library of Congress Cataloging-in-Publication Data
Comins, Neil F., 1951–
Discovering the universe / by Neil F. Cumins & William J. Kaufmann
III. —5th ed.
p.      cm.
Kaufmann's name appears first on the previous ed.
Includes bibliographical references and index.
ISBN: 0-7167-3732-9, 0-7167-3637-3 (pbk.)
1. Astronomy.   2. Cosmology.   I. Kaufmann, William J.   II. Title.
QB43.2.K376     1999
520—dc21

Printed in the United States of America
\ First printing, 1999

# TEXT OVERVIEW

# CONTENTS

**viii**      Contents

# To Students

You've come to astronomy at the right time. Discoveries are skyrocketing, and this book conveys the adventure and excitement of astronomy at the beginning of a new century. To make the most of it, take a few minutes now to preview its learning tools. As you read you will...

## BATTLE COMMON MISCONCEPTIONS

Many of our commonly held ideas about astronomy are misconceptions. We may misinterpret what our senses tell us. Some laws of nature may defy common sense, or our sources of information may be incomplete or inaccurate. Questioning our beliefs is difficult, even disconcerting, but keeping an open mind prepares us to examine the natural world rationally. "What Do You Think?" questions at the beginning of each chapter address common misconceptions shared by most nonastronomers. As you read, check your understanding against the correct science as presented in the text and keyed by number to each question. Questions and answers appear together at the end of the chapter in a section called "What Did You Think?"

## WITNESS RECENT BREAKTHROUGHS

You'll see science as it's happening, from the latest discoveries of planets outside the solar system to tantalizing evidence of water ice trapped on the Moon to how stars form, evolve and burn out or blow up. Updates are available on our Web site.

## CONDUCT YOUR OWN OBSERVATIONS

Throughout the book, a series of Discover for Yourself boxes and end-chapter Observing Projects offer hands-on experience in observational astronomy. Many take advantage of the award-winning Starry Night planetarium software included in the text's CD. You will find tips on buying your first telescope and orienting yourself to star charts and planetarium software.

## EXERCISE CRITICAL THINKING

"What if the Moon didn't exist?" Questions like this challenge you to look at the natural world from new perspectives and help you gain insight into how the universe works. A series of eight "What If..." essays appear at the end of selected chapters. Your instructor may also assign chapter-ending "What If..." questions as part of your course work. How do astronomers know that their theories are correct? How does astronomy connect to your own life experience? Brief asides within the text called "Insights into Science" relate what you're learning to the way scientists think and work.

## BUILD YOUR FOUNDATION IN ASTRONOMY

Each of the book's four parts opens with a Build Your Foundation minichapter that sets the stage for the chapters that follow. To help you place the science in context, a series of four timelines called "An Astronomer's Almanac," one for each text part, gives you a bird's-eye view of discoveries in astronomy relative to other events in science and in history. To clarify your learning goals, chapters open with an overview list of upcoming topics. To guide you through some essential equations used in astronomy—whatever your math background—additional explanations, examples, and practice items are set off in a series of numbered Astronomer's Toolboxes.

## THRILL TO SPECTACULAR IMAGES

Illustrations designed to reinforce what the words say help you see processes step by step, and deepen your understanding. Astronomers view the universe through instruments that detect light at wavelengths our eyes alone cannot see. Wavelength tabs show the portion of the electromagnetic spectrum used to create the photos that illustrate each topic.

## EXPLORE INTERACTIVE TECHNOLOGY

The CD and Web site that accompany this text offer you a connection between the fundamentals of astronomy and what astronomers do. The CD contains study features including interactive quizzes and exercises that will help you better understand the ideas in the book. It also gives you a chance to see over 75 splendid astronomical animations and videos from NASA and other sources—often the best way to see the results of the work that many astronomers do. The student edition of the outstanding Starry Night planetarium software on the CD may engage you the most. Use Starry Night to view accurate depictions of the skies from anywhere on Earth, learn about over 14,000 stars, and trace the paths of planets and comets. Use Starry Night to complete selected Observing Projects and to enhance your understanding of text concepts throughout the course. The book's Web site offers additional Starry Night activities, additional study guide features, and links to the vast astronomical resources on the Internet. The Web site address is:

**www.whfreeman.com/astronomy**

For a demonstration of all of these learning tools, just turn the page.

# Discover
## What Astronomy's All About

**BUILD YOUR FOUNDATION SECTIONS** preview the upcoming chapters and introduce important background. Like each chapter, they begin with a stunning photograph. Their careful overview sets the stage for each of the book's four parts.

### II
## BUILD YOUR FOUNDATION
### The Solar System

*In the chapters ahead you will discover*

* how the solar system formed from a cloud of interstellar gas and dust

* why the environment of the early solar system was much more violent than it is today

* that the planets fall into groups with similar properties

* how astronomers characterize each planet's "personality" by its size, chemical composition, rotation rate, atmosphere, and features

* that the moons throughout the solar system are extremely varied

* that the solar system also contains debris—asteroids, meteoroids, and comets—left over from its birth

* why the sun shines

R I V U X G
**The Planets of Our Solar System**
This photomontage of photographs shows the nine planets. At the top (from left to right) are Mercury, Venus, Earth, and Mars. At the bottom (from right to left) are Jupiter, Saturn, Uranus, Neptune, and Pluto. (The images are not reproduced to the same scale. See Figure II-9 for a scale drawing.)

---

**AN ASTRONOMER'S ALMANAC,** a dynamic timeline, offers students a perspective in astronomy relative to other historical events. An Astronomer's Almanac is located after each Foundation.

**Canyon Fog**
Delicate fog created by the scant water vapor in Mars's atmosphere fills canyons of Labyrinthus Noctis at sunrise and sunset.

## The Other Terrestrial Planets

*In this chapter you will discover*

* Mercury, a Sun-scorched planet with a heavily cratered, Moonlike surface and an Earthlike, iron core

* Venus, perpetually shrouded in thick, poisonous clouds and mostly covered by gently rolling hills

* Mars, the inspiration for scientific and popular speculation about extraterrestrial life

**CHAPTER OPENERS** include clear section-by-section summaries of the chapter material and feature striking opening images.

## An Astronomer's Almanac

**586 B.C.**
Thales of Miletus predicts solar eclipse.

**1512-1543**
Nicolas Copernicus proposes heliocentric cosmology in his Commentariolus and De Revolutionibus Orbium Coelestium.

**1715**
Edmund Halley calculates shadow path of a solar eclipse over Earth's surface.

**ca. 270 B.C.**
Aristarchus of Samos proposes heliocentric cosmology.

**1589-1609**
Galileo Galilei proposes that all objects fall with the same acceleration, independent of their masses; builds his first telescope, a refractor.

*Greek Golden Age*

*European Renaissance*

**350 B.C.**
Aristotle proposes spherical Earth, geocentric cosmology.

**1609-1619**
Johannes Kepler publishes his three laws of planetary motion.

**1766**
Henry Cavendish discovers hydrogen.

**2136 B.C.**
Chinese astronomers record solar eclipse.

**1576-1601**
Tycho Brahe makes precise observations of stars and planets.

**1800-1803**
William Herschel discovers infrared radiation from the Sun. Thomas Young demonstrates wave nature of light. John Dalton proposes that matter is composed of atoms of different weights.

**1665-1704**
Isaac Newton deduces gravitational force from the orbit of the Moon; builds first reflecting telescope; proves that planets moving under influence of gravitational force obey Kepler's laws; publishes compendium on light, Optics.

**ca. A.D. 125**
Claudius Ptolemy refines and details geocentric cosmology in his Almagest.

## WHAT DO YOU THINK?

1. How many stars are there in the solar system?

2. Was the solar system created as a direct result of the birth of the universe?

3. How long has the Earth existed?

4. Is Pluto the always the farthest planet from the sun?

5. What shapes the moons have?

## WHAT DID YOU THINK?

1. *Is the temperature on Mercury, the closest planet to the Sun, higher than the temperature on Earth?* The temperature on the daytime side of Mercury is much higher than on Earth, but the temperature on the nighttime side of Mercury is much lower than on Earth, because Mercury rotates so slowly and has little atmosphere to retain heat.

2. *What planet is most similar to Earth?* Venus is most similar to Earth in size, chemistry, and distance from the Sun. Mars, more commonly believed to be Earth's "sister plan-

et," is more similar to Earth (than Venus) only in its rotation rate and seasons.

3. *What is the composition of the clouds surrounding Venus?* The clouds are composed primarily of sulfuric acid.

4. *Does Mars have surface liquid water today?* No, but there are strong indications that it had liquid water in the distant past.

5. *Is life known to exist on Mars today?* No current life has yet been discovered on Mars.

---

## DISCOVER FOR YOURSELF    The Inner Solar System    ✦ Starry Night.

Let's get some practice observing the inner planets and some insight into their motions in the night sky. Planetary software such as Starry Night can help. So, start up your Starry Night program and put on the equatorial (Eq) and ecliptic (Ec) grids and the planet list.

**Mercury** In the planet list, double click on Mercury. Your view will change until Mercury is in the sky. Mercury should lock in the center of your screen. Where is the Sun? If Mercury is west of the Sun, run time backward until before sunrise, stopping before Mercury sets in the east. This shows Mercury as the "morning star." Now, run time forward at 3 minutes per time step and you will see the white dot of Mercury fade. Conversely, if Mercury is east of the Sun, run the time forward until the Sun has set but Mercury is still up. Mercury is now the "evening star." In either case, note how close Mercury always is to the horizon in the night sky.

Making sure that you are "locked" on Mercury, hide the daylight and the horizon. Start time running at a rate of 1 sidereal day per time step. Describe the motion of the Sun relative to Mercury.

You can learn more about Mercury as seen from Earth by trying Observing Projects 38 at the end of this chapter.

**Venus** Repeat the same procedures for Venus as we have described for Mercury. What similarities and differences do you see?

**Mars** Making sure that the horizon and daylight are on, lock on Mars. If it is "up" during the day, run the program forward at a rate of 1 sidereal day per time step or more until Mars is up at night. Turn off the horizon and daylight and determine how far (in angle) the Sun is away from Mars. You can make this measurement by "dragging" the celestial sphere around and measuring angles from the grid. What is Mars's phase? Does the Sun ever get this far from Mercury or Venus?

Relock on Mars and run the program at a rate of 1 sidereal day until the Sun is very close to Mars. What is Mars's phase now? You can learn more about Mars's unusual motion relative to Earth's orbit by trying the Observing Projects 39 at the end

---

## DISCOVERING ASTRONOMY

**1871-1873** Dimitri Mendeleyev develops periodic table of the elements. Henry Draper develops spectroscopy. James Clerk Maxwell asserts that light is an electromagnetic phenomenon.

**1885-1888** Johann Balmer expresses spectral lines of hydrogen mathematically. Heinrich Hertz detects radio waves.

**1900** Max Planck explains blackbody radiation. Paul Villard discovers gamma rays.

**1930-1934** Karl Jansky builds first radio telescope. James Chadwick discovers the neutron. Bernhard Schmidt builds his Schmidt optical reflecting telescope.

**1942-1949** J.S. Hey detects radio waves from the Sun. First astronomical telescope launched into space. Herbert Friedman detects X-rays from the Sun. 200-in. optical reflecting telescope begins operation on Mt. Palomar, California.

**1963-1967** Largest single-dish radio telescope, 300 meters across, begins operation at Arecibo, Puerto Rico. First VLBI images.

**1990-1996** Hubble Space Telescope launched. Keck 10-meter optical/infrared telescopes begin operation at Mauna Kea, Hawaii. SOHO solar observatory launched.

Industrial Revolution

Information Age

**1840-1849** J. W. Draper invents astronomical photography; first photographs of the Moon. Christian Doppler proposes that wavelength is affected by motion. Lord Rosse completes 60-in. reflecting telescope at Birr Castle in Ireland. Armand Fizeau and Jean-Bernard Foucault measure speed of light accurately.

**1895-1897** Wilhelm Roentgen discovers X-rays. Joseph Thomson detects the electron. Yerkes 40-in. optical refracting telescope completed.

**1913** Niels Bohr proposes quantum theory of the atom.

**1975** First CCD astronomical observations.

**1980** VLA radio observatory completed, Socorro, New Mexico.

---

## WHAT IF...

### Earth's Axis Tilted on the Ecliptic?

Imagine the Earth tilted so that its axis of rotation lies in the plane of its orbit about the Sun (see the accompanying figure). New Earth still rotates once every 24 hours. We arbitrarily fix New Earth's north pole to point toward the star Tau Tauri, just north of the bright star Aldebaran in the constellation Taurus (see Figure II-14.)

**New Seasons** Let's see how the seasons unfold on New Earth. It is March 21, the date of the spring equinox. The Sun is directly over the equator. For the next three months, the Sun rises higher in the northern hemisphere. Unlike on our Earth, the Sun does not stop moving northward when it is over 23H° north latitude. Rather, the Sun rises farther and farther north day by day until it appears over the north pole of New Earth at the summer solstice, around June 22. Three months later, at the autumnal equinox, the Sun again rises over the equator, and day and night have the same length everywhere.

The Sun appears over the south pole of New Earth at the winter solstice, around December 22. The Sun's apparent motion through New Earth's sky completes the seasonal cycle by moving north, appearing over the equator once again around March 21.

**New Climate** Day and night take on new meanings for inhabitants of New Earth. On our Earth, the regions above the Arctic Circle and below the Antarctic Circle have days or weeks of continual light in summer and days or weeks of constant darkness in winter. But on New Earth,

every place has extended winter periods of constant darkness followed by extended summer periods of constant daylight. Spring and fall on New Earth have daily cycles of daylight and darkness, which separate the periods of continual daylight and darkness.

**A New Day** At the latitude of Atlanta, GA, (33°46¢) on New Earth, the day-night cycle occurs for only seven and a half months out of the year. During the other four and a half months, there is continuous day or continuous night, coupled with harsh summers and winters. With variations, this sequence of events occurs everywhere on New Earth.

The seasonal cycle on New Earth prevents the formation of permanent polar ice caps. Polar regions experience the same tropical heating and high temperatures as the equatorial regions of our Earth. The polar regions of New Earth in winter are exceptionally cold, so seasonal polar ice caps may form. Because the southern polar cap resting on Antarctica is not permanent, the oceans, and the shorelines on the continents, are higher than those on our Earth.

If seasonal polar ice caps form, the dominant force controlling weather may shift from the jet streams that circle our Earth along lines of latitude to a pole-to-pole flow. This is the condensation flows seen on Mars. Thermal flows created by intense heating at one location and cooling at others may replace our Earth's trade winds and other east-west winds. How else on New Earth compare to our world?

## AN ASTRONOMER'S TOOLBOX 3-1

### Photon Energies

The energies of all photons with the same wavelength are the same, and a photon's energy (or, equivalently, its wavelength) is the only property that differs from photon to photon. We conclude that all the photons with the same energy are identical.

The energy $E$ of any given photon can be calculated from the equation

$$E = \frac{hc}{\lambda}$$

where Planck's constant $h$ is $6.6 \times 10^{-34}$ kg m²/s, the speed of light $c$ is 300,000 km/s, and the photon's wavelength is $\lambda$.

*Example:* A photon of red light with wavelength 700 nm has what energy?

$$E_{red} = \frac{(6.6 \times 10^{-34} \ kg \cdot m^2/s) \ (300{,}000 \ km/s)}{700 \ nm}$$

$$E_{red} = 2.8 \times 10^{-19} J$$

(A joule, abbreviated J, is a unit of energy.) Each photon of red light with wavelength 700 nm has an energy of $2.8 \times 10^{-19}$ J.

*Compare!* In one second a 25-watt lightbulb emits 25 J of energy.

*Self-test:* A photon has energy $4.9 \times 10^{-19}$ J. Calculate its wavelength. Looking ahead to Figure 4-10, what is the photon's color?

---

**ASTRONOMER'S TOOLBOXES** help you with mathematics by highlighting key formulas. Selected toolboxes include self-tests (answers appear at the end of the book).

---

**INSIGHTS INTO SCIENCE**, brief asides within the text, relate text topics to the nature of scientific inquiry.

**Insight into science  Fact and fiction**
Science fiction does foreshadow real scientific discoveries. (Can you think of some examples?) But science also reins in the fantasies of the science fiction writer. In the first decades of the twentieth century, millions of people, including top scientists, believed that technologically advanced life existed on Mars.

While the inner regions of the solar system were heating up, temperatures in the outer regions of the solar nebula remained quite cool. Ice and ice-coated dust, along with hydrogen and helium gas, were able to survive in these cooler regions. The large outer planets—

---

### Key Words

caldera
retrograde rotation

greenhouse effect
scarp

hot-spot volcanism

### Key Ideas

✳  All four inner planets are primarily composed of rock and metal, and thus they are classified as terrestrial.

**Mercury**

✳  At its greatest elongations, Mercury can be seen only briefly after sunset or before sunrise.

✳  The *Mariner 10* spacecraft passed near Mercury in the mid-1970s, providing pictures of its surface. The Mercurian surface is pocked with craters like those of the Moon, [...] smooth plains exist between these craters. Lon[...] cliffs meander across the surface of Mercury. T[...] ably formed as the planet cooled, solidified, an[...]

✳  The impact of a large object long ago form[...] Caloris Basin on Mercury and shoved up jumb[...] the opposite side of the planet.

✳  Mercury has an iron core much like that o[...]

**Venus**

✳  Venus is similar to the Earth in size, mass[...] density, but it is covered by unbroken, highly r[...] that conceal its other features from Earth-base[...]

✳  Venus's clouds consist of droplets of concentrated sulfuric acid along with substantial amounts of yellowish sulfur dust. Active volcanoes on Venus may be a constant source of this sulfurous material.

✳  The surface pressure on Venus is 98 atm, and the surface temperature is 750 K. Both temperature and pressure decrease as altitude increases. Venus's exceptionally high temperature is caused by the greenhouse effect, as the dense carbon dioxide atmosphere traps and retains heat emitted by the planet.

### Review Questions

1  Why is Mercury so difficult to observe? Wh[...] time to see the planet? (*Tip:* The "Discover for [...] this chapter can help.)

2  Compare the surfaces of Mercury and our M[...] they similar? How are they different?

3  Compare the interiors of Mercury and Earth[...] similar? How are they different?

4  What kind of tectonic features are found on M[...] are they probably much older than tectonic featur[...]

5  Briefly describe at least one theory explainin[...] has a large iron core.

6  Astronomers often refer to Venus as the Eart[...] physical properties do the two planets have in c[...] ways are the two planets dissimilar?

7  Why is it hotter on Venus than on Mercury?

---

### Advanced Questions

19  What evidence do we have that the surface features on Mercury were not formed during recent geologic history?

20  Venus takes 440 days to move from greatest western elongation to greatest eastern elongation, but it needs only 144 days to go from greatest eastern elongation to greatest western elongation. With the aid of a diagram, explain why.

21  As seen from Earth, the brightness of Venus changes as it moves along its orbit. Describe the main factors that determine Venus's variations in brightness as seen from Earth. (*Tip:* See the discussion of Venus in Chapter 2.)

22  How might Venus's cloud cover change if all of Venus's volcanic activity suddenly stopped? How might these changes affect the overall Venusian environment?

23  Compare Venus's continents with those on the Earth. What do they have in common? How are they different?

24  Explain why Mars has the longest synodic period of all the planets, although its sidereal period is only 687 days.

25  With carbon dioxide accounting for about 95% of the atmospheres of both Mars and Venus, why do you suppose there is little greenhouse effect on Mars today?

26  Could the polar regions of Mars reasonably be expected to harbor life-forms, even though the Martian regolith is sterile at the *Viking* lander sites?

### Discussion Questions

27  If you were planning a return mission to Mercury, what features and observations would be of particular interest to you?

28  If you were designing a space vehicle to land on Venus, what special features would you think necessary? In what ways would this mission and landing craft differ from a spacecraft designed for a similar mission to Mercury?

29  Suppose someone told you that the *Viking* mission failed to detect life on Mars simply because the tests were designed to

detect terrestrial life-forms, not Martian life-forms. How would you respond?

30  Compare the scientific opportunities for long-term exploration offered by the Moon and Mars. What difficulties would there be in establishing a permanent base or colony on each of these two worlds?

31  Imagine you are an astronaut living at a base on Mars. Describe your day's activities, what you see, the weather, the spacesuit you are wearing, and so on.

### What If …

32  Mars rotated once every 20 days rather than once every 1.026 days? What would be different?

33  The carbon dioxide content of the Earth's atmosphere grew? What would happen to the Earth? This is not a completely

hypothetical question, because carbon dioxide levels ar[...] increasing today. (Do you know why?)

### Observing Projects

34  Refer to the table below to determine the dates of the next two or three greatest elongations of Mercury. Consult such magazines as *Sky & Telescope* and *Astronomy*, or use your *Starry Night* planetarium software, to see if any of these greatest elongations is going to be especially favorable for viewing the planet. If so, make plans to be one of those rare individuals who has actually seen the innermost planet of the solar system. Set aside several evenings (or mornings) around the date of the favorable elongation to reduce the chances of being "clouded out." Select an observing site that has a clear, unobstructed view of the horizon where the Sun sets (or rises). Make arrangements to have a telescope at your disposal. Search for the planet on the

dates you hav[...]
through your[...]

---

**END-OF-CHAPTER PEDAGOGY** includes recaps of Key Words and Ideas, Review Questions, Advanced Questions, Discussion Questions, "What If?" Questions, and Observing Projects—many utilizing Starry Night from the CD-ROM.

39  The "Discover for Yourself" box will introduce you to Mars's motion. With Starry Night, notice that the equatorial coordinate system usually flows smoothly past Mars. Run the program until the coordinate grid behind Mars begins to slow down. Unlock Mars by moving it to the left quarter of the screen.

# Discover
## Out-of-This-World Technology

★ **Web Resources**
(for students and instructors)

★ **CD-ROM**
(packaged with every text)

**DISCOVERING THE UNIVERSE 5.0**
with Starry Night™ Astronomy Software
Neil F. Comins
William J. Kaufmann III

**DISCOVERING THE UNIVERSE**
COMINS · KAUFMANN
**ONLINE COMPANION**

For the Student

For the Instructor

**Starry Night™ Observation Exercises**

FREEMAN ASTRONOMY

© 2000 W. H. FREEMAN AND COMPANY    WEB DESIGN: SUMANAS, INC.

**MEDIA SUPPLEMENTS** bring technology to the forefront of learning astronomy. The book's Web site offers an Online Study Guide and links to the treasures of astronomy on the Internet. The accompanying CD-ROM contains first-rate planetarium software and popular features such as animations and study support. An Instructor's CD facilitates displaying images and animations during lectures.

## Kepler's First and Second Laws

According to Kepler's first law, every planet travels around the Sun along a(n) __?__ orbit with the __?__ at one focus. __?__ is at the other focus. According to Kepler's second law, the line joining the planet and the Sun sweeps out __?__ areas in equal intervals of time. This means that the shaded segment C-D is __?__ the shaded segment A-B.

*Diagram labels: B, A, C, Sun, Perihelion, D, Aphelion*

Drag the correct word or phrase from the list below to fill in the blanks in the paragraph above:

- (?) An asteroid
- (?) circular
- (?) Earth
- (?) elliptical
- (?) equal
- (?) equal to
- (?) hyperbolic
- (?) large
- (?) larger than
- (?) No object
- (?) planet
- (?) small
- (?) smaller then
- (?) Sun
- (?) The planet

**ONLINE COMPANION WEB SITE** offers study guide features including chapter self-quizzes, interactive exercises, flash cards to learn key words, and chapter objectives.

## Web Links
**Chapter 6: The Other Inner Planets**

Website Notes: A web link will open in an external window. To return to this page, click the close box in the upper left hand corner of the external window (e.g., do not use your browser's "Back" button).

**The Nine Planets: Mercury.** The Nine Planets: A Multimedia Tour of the Solar System is an overview of the history, mythology, and current scientific knowledge of each of the planets and moons in our solar system. The Mercury page includes text and images and provides links to additional Mercury-related web sites.

**Views of the Solar System: Mercury.** Another good overview of current knowledge about Mercury. Includes text, photos, diagrams, animations.

**WEB LINKS** for each book chapter on our Web site set you on a path to explore the wealth of Internet sites focusing on all aspects of astronomy.

Starry Night™

**Pisces Info**

**Pisces**
The Fish

Possessive form: Piscium

Pisces the Fish is a zodiacal constellation that contains the Vernal Equinox; the position of the Sun on the first day of spring. Due to precession, the Sun's position on this day moves slowly through the zodiac.

**STARRY NIGHT™ PLANETARIUM SOFTWARE** is offered on our CD in a student version. Use it to view the skies from anywhere on Earth, learn about over 14,000 stars, and trace the paths of planet and comets. This awarding-winning, easy-to-use software actively engages users in learning astronomy. Exercises in the book and on its Web site tie the software to specific book topics.

Discovering the Universe 5/e                    Animations & Videos

Chapter Browser

Animations & Videos

Interactive Exercises

Q & A

Astronomer's Toolbox

CD/Web Updates

About the CD

Planetary Orbits (Fdn II)
The Birth of the Solar System (Fdn II)
Mercury's Cratered Surface (Chap 6)
Formation of a Scarp (Chap 6)
Mercury's Magnetic Field (Chap 6)
*Magellan* Maps a Planet (Chap 6)
Volcanoes on Venus (Chap 6)
Craters, Volcanoes, and Highlands on
    Venus (Chap 6)
An Immense Chasm on Venus (Chap 6)
Impact Craters and Volcanoes on
    Venus (Chap 6)
Valles Marineris and the Giant Volancoes of Mars
    (Chap 6)
A Martian Chasm (Chap 6)
Surface of Mars as Seen by the *Viking* I Lander
    (Chap 6)

**DISCOVERING THE UNIVERSE 5.0 CD-ROM** provides a variety of study features as well as stunning and informative animations and videos from NASA and other sources—all in offered in the context of specific areas in the book.

# Discover
## The Process of Science

You'll gain the most from reading science if you know what to expect.

★ **Look for patterns** Start by examining the big picture. For example, gravitational attraction explains why the Earth holds us down, why the planets orbit the Sun, why the moons orbit the planets, why the Sun orbits with the myriad other stars of our Milky Way galaxy, and why clusters of galaxies traverse the universe together.

★ **Remember that a picture is worth a thousand words.** Text illustrations visually reinforce what the words say, and the CD offers even more visual reinforcement.

★ **Summarize sections and chapters in your own words.** This helps you to understand and to organize what you learn. The detailed Contents provides an outline of the headings in each chapter. Use it to preview the material you are about to read and to see how it fits into the chapter and the course. Taking a minute to do this primes you to learn.

★ **Check your understanding.** Do you understand concepts well enough to explain them on homework and tests? For example, we all substitute the words "rotation" and "revolution" in everyday conversation, but these words have completely different meanings to astronomers. Test yourself as you go along. The self-tests on the CD are especially useful for this, as are old tests (with answers) that your instructor may have put on reserve in the library or on the Internet.

★ **Learn to use all the navigating tools in the book.** Among these are chapter headings, glossary, and index. The better you are at finding information, the easier it will be for you to check on concepts you have seen before but may have forgotten or need to learn but cannot immediately locate.

★ **Look for connections between what you read and your own life experience.** For example, even if you have never used a telescope, you may have used binoculars, which work in exactly the same way. Comparing how you think and interact with the world to how scientists do these things, as discussed in the Insights Into Science, connects you more deeply to the material in this book.

★ **Use software resources.** The CD-ROM accompanying this text features an easy-to-use student version of Starry Night, actual planetarium software that lets you experience astronomy hands-on. The CD's learning tools—animations, videos, interactive exercises, and self-quizzes—contain valuable astronomical information to help you learn astronomy chapter by chapter.

---

**DISCOVER FOR YOURSELF**  Starry Night

### Astronomy on the Internet

The Internet offers an ever-expanding choice of sites covering an enormous array of topics in astronomy. The first site you should bookmark is the one developed specifically to accompany this text:

http://www.whfreeman.com/astronomy

The text's Web site features an Online Companion designed to further your understanding of astronomy and help you excel in the course. It includes interactive exercises, quizzes, help in mastering the text's key words, and chapter objectives for every chapter of the text. It also includes an extensive, annotated list of outside Web sites for each chapter. These listings will be helpful to you to sort through the many thousands of astronomy sites in the world, but don't forget to also discover sites on your own. Below is my list of the five most exciting astronomy sites on the Web. They are great starting points in your discovery of astronomy on the Internet.

★ Space Telescope Electronic Information Service (Hubble Space Telescope)
http://www.stsci.edu/top.html

★ NASA Information Services
http://www.nasa.gov/NASA_homepage.html

★ Students for the Exploration and Development of Space (SEDS)
http://seds.lpl.arizona.edu/AstroWeb
http://www.stsci.edu/astroweb/astronomy/html

★ The Astronomical Society of the Pacific (a major organization of amateur and professional astronomers)
http://www.aspsky.org

# PREFACE

This new edition of *Discovering the Universe* clarifies the big picture for students by enhancing the distinction between theory and observation in astronomy. The CD-ROM has been greatly enhanced, and I am especially excited about the Starry Night planetarium software included on the CD. I used it extensively in the year preceding the release of the book and found it excellent both for instruction and for hands-on student activities. That is why I have written several questions based on this software for the book, and many more are posted on the book's homepage.

Several features are new to this edition:

★ "An Astronomer's Almanac" timelines, one for each of the four main sections of the book, offer students a birds-eye view of the progress of discovery in astronomy relative to other events in science and in history.

★ New "Insights into Science" give students an understanding of how science and scientists work, explaining, for example, using Occam's razor as a guiding principle.

★ A total of eight "What If..." essays. twice as many as in the previous edition, stimulate students to think about the natural world from new perspectives. Brief, new end-chapter "What If..." questions likewise stimulate critical thinking, reinforce students' understanding, and broaden their view of the universe.

★ Chapter-opening surveys appear in bullet-list format to help students focus on their learning goals efficiently.

I wrote *Discovering the Universe* with special attention to the way people learn. It evolved from my perceptions of the backgrounds, needs, and expectations of over 7500 students to whom I have taught introductory astronomy over the last twenty years.

*Discovering the Universe* can be covered in a single semester or quarter course. Instructors are often torn between a desire to cover a broad range of topics and the restrictions of time. The four parts of this edition will make it easier for you to choose among topics when needed, and I have tried to ensure that concepts and relationships do not get lost in a staggering amount of detail. Most paperback introductory textbooks are just pared down versions of an author's longer text, but not this one. This is the only astronomy text I write, and it's designed especially for the course.

## It's easier to study related topics

In the realm of the solar system, for example, each planet is explored separately, followed by a discussion of its moons. Students will study each planet from its atmosphere to the properties of its interior—from what we can see to what our theories tell us. Within the two chapters on the planets, comparative planetology tables summarize physical properties to help students grasp the overall structure of the solar system.

This edition again includes some essential equations used in astronomy. Additional explanations and practice with unfamiliar equations are set off in numbered boxes, each called An Astronomer's Toolbox. Worked examples help those who are uncomfortable with mathematics follow the reasoning step by step, whatever their math background. Self-tests reinforce this practice and answers appear at the end of the book.

## Examining misconceptions promotes real understanding

The previous edition of *Discovering the Universe* pioneered the practice of directly addressing common misconceptions about astronomy. That effort expands in this edition. I have written this book expressly to help students identify their own misconceptions and ponder alternatives. Throughout the text I explain why common, incorrect beliefs are wrong.

I want students to understand and retain what they learn in this course. Memorizing for an exam is one thing; true understanding is something else altogether. Extensive research shows that it is virtually impossible for anyone to accept new information that conflicts with their established beliefs unless they first understand why their old ideas are wrong. Nearly everyone develops ideas about astronomy prior to taking a course in this subject. Misconceptions and deep-seated beliefs that are inconsistent with accepted scientific knowledge are inevitable as students struggle to understand our complex environment. Such incorrect beliefs do not imply that students are either stupid or incapable of learning; rather, they show how fertile and active the human mind is. My research into misconceptions about astronomy has led me to identify over 1400 of them, along with their origins and common reasoning that leads to incorrect beliefs.

To help students become aware of their beliefs, I present a series of numbered questions at the beginning of each chapter under the heading What Do You Think? Each question addresses a common misconception. Urge students to honestly consider their own answers before reading the text. When the correct science is presented in the chapter, an icon with the corresponding question number appears in the margin. The same questions and their answers are listed at the end of each chapter under the heading What Did You Think?

## Each part builds a foundation for understanding

Presenting facts isn't enough. A book has to guide students through the facts to the ideas behind them. There-fore, each of this book's four parts begins with a "Build Your Foundation" minichapter that overviews upcoming chapters and presents essential physical concepts. Presenting this basic material just when it is needed reinforces students' sense of discovery.

The book begins the study of astronomy with what people have seen for millennia. As a result, students start out with familiar sights and concepts, then move into realms visible only through telescopes and which they may know less well. *Discovering the Universe* has been written so that stars can be studied before or after planets, as instructors may prefer.

★  Part I, Discovering Astronomy, begins with naked-eye astronomy, the nature of gravity, and its effects on the planets and other bodies. Then we turn to the nature of light and how telescopes work, along with how stars and other astronomical bodies emit light.

★  Part II, The Solar System, begins with the Earth and Moon. We explore each of the planets and their moons, the smaller bodies in the solar system, and finally the Sun.

★  Part III, The Stars, describes the nature and evolutionary paths of stars, concluding with a discussion of black holes.

★  Part IV, The Universe, explores the Milky Way Galaxy, other galaxies, quasars, and the entire universe, including information about the possible acceleration of the universe. The book ends by exploring the age-old questions of whether other life exists in the universe.

## Students learn how to observe

Practical, hands-on information about astronomy as well as connections among the book, the accompanying CD-ROM, and the Internet appear in Discover for Yourself... boxes throughout the book. One helps students buy their first telescope, another orients them to the unfamiliar names on star charts, a third walks them through their first on-line session as they explore the power of the CD-ROM accompanying this text.

## Chapters preview, summarize, and review the big picture

People learn in different ways. Some are visual learners; others rely more heavily on text. Bill Kaufmann recognized that when he introduced large, full-color illustrations to the first edition of this book. The text itself is designed to help students more easily understand and retain information:

★ Full-sentence headings introduce each topic. Grouped beneath a broader main heading of only a word or two, they also provide an outline for review.

★ Important astronomical terms appear in **boldface** type where they are introduced and also at the end of the chapter in a Key Terms list followed by a full summary list of Key Ideas. Key terms are again defined in the Glossary.

★ Review Questions and Advanced Questions at the end of each chapter stimulate students to explore the ideas in the text. Discussion Questions require even more thought, more physics and chemistry, and sometimes even outside reading. Any question relying on mathematics is flagged with an asterisk and answered briefly at the back of the book. As mentioned above, "What If…" questions will stimulate viewing the universe from new perspectives. The question sets also include Observing Projects, many for use in conjunction with Starry Night and the star charts located inside the book's covers.

## Supplements support students and instructors

Students can experience the strong connection between astronomy and technology through the electronic media supplements that support this text:

★ The CD-ROM that accompanies the text features a student version of the award winning Starry Night planetarium software from Sienna Software. It enables students to learn astronomy by getting directly involved with astronomy—both through exploring the intuitive software on their own as well as by following the numerous text-specific activities found in the text and on its web site Starry Night. In addition to Starry Night, the CD contains revised and expanded versions of all of the features that made the CD for the previous edition of the text the most successful textbook CD in the college sciences. These returning features include over 75 animations and videos from NASA and other sources, interactive exercises, self-quizzes, and the illustration program from the text.

★ Rounding out this unbeatable combination of electronic-based supplements for students is the new Online Companion web site for *Discovering the Universe*. It can be found at

http://www.whfreeman.com/astronomy

The Online Companion web site is an electronic chapter-by-chapter study guide for students. It features interactive exercises and quizzes to help students prepare for tests and quizzes; flash card exercises to

aid in the mastery of the text's hundreds of key words; chapter objectives for each chapter and foundation to help organize study; and an extensive, annotated list of outside web sites for each chapter. It also offers Starry Night activities specifically developed for use with many chapters of the text.

The explosion of electronic media options in astronomy education is also reflected in the supplements for instructors.

★ The printed Instructor's Resource Guide contains solutions to all text computational problems and other end-of-chapter questions, additional questions, teaching strategies, suggestions for class discussions and projects, and a comprehensive list of outside resources.

★ The Online Instructor's Guide for this text can be found at:

http://www.whfreeman.com/astronomy

It offers specific ideas for using the electronic supplements for this book when teaching from it, as well as the password-protected answers to special quizzes covering the materials found on these electronic supplements.

★ Printed Windows and Macintosh versions of our Test Bank contain over 1500 multiple-choice test questions revised and expanded by T. Alan Clark and William J. F. Wilson of the University of Calgary.

★ For the first time, we are also offering secure Online Testing/Quizzing capability to instructors who want to assign questions using the web.

★ A set of 100 full-color overhead transparencies represents key diagrams from the text.

★ With our Instructor's Resource CD instructors with projection units can display all of the hundreds of diagrams and photographs in the textbook during lectures—using either our free Presentation Manager Pro software or commercially-available presentation software and JPEG files that we provide.

## Acknowledgments

I am deeply grateful to the astronomers and teachers who reviewed the manuscript of this edition:

Gordon Baird, *University of Mississippi*
Henry E. Bass, *University of Mississippi*
Michael Bennett, *DeAnza College*
John W. Burns, *Mt. San Antonio College*
Gerald Cecil, *University of North Carolina*
Chris Clemens, *University of North Carolina*
Antoinette Cowie, *University of Hawaii*
David Hedin, *Northern Illinois University*
Kenneth Janes, *Boston University*

William C. Keel, *University of Alabama*
William, Keller, *St. Petersburg Junior College*
Marvin D. Kemple, *IUPUI*
R.M. MacQueen, *Rhodes College*
Rahul Mehta, *University of Central Arkansas*
J. Ward Moody, *Brigham Young University*
Gerald H. Newsom, *Ohio State University*
William C. Oelfke, *Valencia Community College*
John P. Oliver, *University of Florida*
Melvyn Oremland, *Pace University*
Sidney Perkowitz, *Emory University*
Adam W. Rengstorf, *Indiana University*
James A. Roberts, *University of North Texas*
Dwight P. Russell, *University of Texas in El Paso*
John D. Silva, *University of Massachusetts at Dartmouth*
Michael L. Sitko, *University of Cincinnati*
Alex G. Smith, *University of Florida*
John Wallin, *George Mason University*
Jeff S. Wright, *Elon College*

Extra thanks are due John Burns, Chris Clemens, and Gerald H. Newsom for especially deep reviews.

I wish to express continuing thanks to John Haber, development editor, and Holly Hodder, acquisitions editor, who helped me learn the ropes of textbook writing. Special thanks go to Barbara Brooks, development editor, who brought an exciting new perspective to this edition; to Patrick Shriner and Matthew Fitzpatrick for coordinating the Supplements and CD components; and to Kimberly Manzi and Charlie Van Wagner for seeing the entire project through to publication.

Thanks are due the talented book team: Diane Cimino Maass, project editor; Susan Wein, production coordinator; Cambraia (Magalhaes) Fernandes, designer; Jennifer MacMillan, photo editor; Sheridan Sellers and Bonnie Stuart, layout. Thanks also to Bill Page, Maria Eppes, and to Fine Line Illustrations.

Finally, special thanks to my wife, Suzanne, and to my sons, James (who helped with preliminary computer sketches and some figures) and Joshua. All put up with Daddy hard at work on "the book."

Every effort has been made to make this book as error-free as possible. Nevertheless, some inaccuracies have inevitably crept in. I would appreciate hearing from you if you find an error or wish to comment on the text.

Neil F. Comins
neil.comins@umit.maine.edu

# BUILD YOUR FOUNDATION

II

## Discovering Astronomy

*In the chapters ahead you will discover*

✳ the scientific process astronomers use to study the universe

✳ how astronomers organize the night sky

✳ how the Earth's motion on its axis and its motion around the Sun cause changes on Earth and in what we see in the sky at night

✳ that the Moon's orbit causes eclipses and its changing phases

✳ Kepler's orbital laws and Newton's gravitational law

✳ the nature of light and how telescopes work

✳ the structure of atoms and what observing light from them reveals about the universe

R I V U X G
**The Hubble Space Telescope**
Our quest for knowledge about the universe is epitomized by the building of such technological marvels as the Hubble Space Telescope (HST), shown here in 1990 as it was being inserted into orbit from the Space Shuttle. Since the installation of corrective optics in 1993, this telescope, now fully functional, is providing new insights into the nature of everything in space.

## WHAT DO YOU THINK?

1  What makes a theory scientific?

2  What do astronomers do?

3  Why is astronomy worth learning?

Look for answers beside the boxed numbers in the text margins.

Figure I-1  **The Starry Sky**  Daylight hides all signs of the stars and other wonders shining above the sky's azure curtain. Our thoughts about the universe overhead focus on the Sun, the Moon, and our ever-changing weather. But, ah, the night! No light show, artist's brush, or poet's words can truly capture the beauty of this breathtaking panorama. This photograph was taken in the Saguaro National Park, Arizona.

For thousands of years people have looked up at the sky and found themselves inspired to contemplate the nature of the universe. How was it created? Where did the Earth, Moon, and Sun come from? What are the planets and stars made of? What are our place and role in the cosmic scope of space and time?

The beauty of the star-filled night sky or the drama of an eclipse alone makes astronomy fascinating (Figure I-1). But there are also practical reasons for an interest in the universe. The ancient Greeks knew the connection between the changing height of the noontime Sun and the different patterns of stars in the night throughout the year. This information enabled them to predict the seasons, a useful skill for farming. Many early seafaring cultures were also aware that the positions of the Moon and Sun influence the tides; this knowledge helped these people plan and navigate sailing voyages.

## I-1  The universe is comprehensible

What causes these relationships between the Earth and heavenly bodies? Astronomical phenomena were first explained as the result of supernatural forces and divine intervention. The heavens were thought to be populated by demons and heroes, gods and goddesses. Yet, despite superstitious beliefs, some people have always realized that the universe is logical and comprehensible. Astronomical cycles, such as the seasons, the reappearance of stars, the tides, and day and night, led many early societies to study the patterns and motions found in the night sky. Progress in science is embodied in improved technology, such as the invention of the telescope. Improvements in technology have in turn led to further discoveries about the fundamental laws of **physics**, the science that investigates the nature of matter and energy and the relationships between them.

This section begins with the "everyday" aspects of astronomy, from naked-eye observations of the sky to the motion of the planets and the use of telescopes to widen our vistas. You will discover the process scientists use to explore natural phenomena through systematic observations and refined theories, and you will see how the knowledge gained by the scientific study of astronomy has led to an understanding of events and phenomena that our ancestors never could have imagined.

## I-2  Science demands systematic observations

Astronomers gain knowledge by using the **scientific method** to observe, predict, and explain physical reality. Observations lead scientists to create a **scientific theory**, an idea or collection of ideas that proposes to explain the observed phenomenon. Scientific theories are expressed mathematically as **models**. For example, Newton's law (or theory, as such ideas are now called) of gravity is written as an equation that predicts how bodies attract each other. This model predicts that the Sun's gravitational force makes the planets move in elliptical orbits.

A scientific theory can be independently tested and potentially disproved. Newton's ideas can be tested and potentially disproved by observations and thus qualify as scientific theory. The idea that God created the Earth in six days cannot be tested, much less disproved. It is not a scientific theory but rather a matter of faith.

Scientific theories also make testable *predictions* that can be verified using new observations and experiments.

Testing is a crucial aspect of the scientific method, which requires that the theory accurately forecast the results of new observations in its realm of validity.

If a theory proves inconsistent with observations, then the theory is either modified, applied only in limited circumstances, or discarded in favor of a more accurate explanation. For example, Newton's law of gravitation is entirely adequate for describing the motion of the Space Shuttle around the Earth or the Earth around the Sun, but it is inaccurate in the vicinity of a black hole, where matter is especially dense. In this latter realm, Newton's law of gravitation is supplanted by Einstein's theory of general relativity, which accurately describes gravitational behavior in a much wider range of conditions than Newton's, at the cost of greater mathematical complexity. When applied to the motion of the Earth around the Sun, general relativity gives the same results as Newton's law of gravitation.

 **Insight into science   Theories and beliefs**  New theories are personal creations, but science is not a personal belief system. Scientific theories make predictions that can be tested independently. If everyone who performs tests of the theory's predictions gets results consistent with the theory, the theory is considered valid. In comparison, belief systems such as which sports team or political system is best are personal matters. People will always hold differing opinions about such issues.

Astronomy is the most engaging of the physical sciences because so much that we observe through telescopes is surrounded with mystery. We see so many objects, but we see them all from one vantage point in space and time, and, therefore, we cannot always be certain of their true shape and form or of their history. In applying the scientific method to astronomy, we necessarily form theories based on this limited observational data and hope that in time additional observations will substantiate these theories.

Many scientific theories discussed in the following chapters are supported by a broad range of observations while others are still largely unsubstantiated. These latter theories await confirming observations that may be made in the next few years or may prove to remain beyond our physical limits. Thus, the scientific method can be summarized in five words: observe, theorize, predict, test, modify. I urge you to watch for applications of the scientific method in action throughout this book.

The power of the scientific method was demonstrated centuries ago, during the late Renaissance, when a few courageous scientists proposed that the Earth orbits the Sun. The prevailing belief system of the sixteenth century held that the Earth was the center of everything. This

**Figure I-2**   **Galileo's Telescope**  When Galileo turned his telescope toward the sky in the early 1600s, he discovered craters on the Moon, spots on the Sun, the phases of Venus, and satellites orbiting Jupiter. These controversial discoveries flew in the face of conventional wisdom and threatened to undermine the teachings of organized religion of the time.

makes sense: To the untrained eye, all the astronomical objects appear to orbit our planet!

This Earth-centered theory ran into trouble when observations of the motions of the stars, planets, Moon, and Sun were shown to be inconsistent with the predictions of the theory. Centuries of modifications to the Earth-centered theory made it truly unwieldy, and even those changes could not keep it consistent with increasingly accurate observations. In the mid-1500s, the Polish mathematician Nicolaus Copernicus resurrected a theory first proposed nearly 15 centuries earlier—that the Earth orbits the Sun. He was motivated by an effort to simplify the celestial scheme. This Sun-centered theory of the known universe gained strength when Italian scientist Galileo Galilei, the first person to point a telescope toward the sky, saw the moons of Jupiter orbit that planet. This discovery flew in the face of the Earth-centered theory and fueled the search to discover the relationship between the Earth and the rest of the cosmos.

 **Insight into science   Shedding misconceptions**  Ideally, a scientist must be willing to discard even the most cherished theories if they fail to agree with observation and experiment. However, scientists are human; they have beliefs that they are loath to let go. Often a disproved theory, such as the belief that the Earth is at the center of the universe, dies only with its advocates.

# AN ASTRONOMER'S TOOLBOX I-1

## Powers-of-Ten Notation

Astronomy is a science of extremes. As we examine various cosmic environments, we find an astonishing range of conditions, from the incredibly hot, dense centers of stars to the frigid, near-perfect vacuum of interstellar space. To describe such divergent conditions accurately, we need a wide range of both large and small numbers. Astronomers avoid such confusing terms as "a million billion billion" (1,000,000,000,000,000,000,000,000) by using a standard shorthand system. All the cumbersome zeros that accompany such a large number are consolidated into one term consisting of 10 followed by an *exponent,* which is written as a superscript and called the **power of ten.** The exponent merely indicates how many zeros you would need to write out the long form of the number. Thus,

$$10^0 = 1$$
$$10^1 = 10$$
$$10^2 = 100$$
$$10^3 = 1000$$
$$10^4 = 10,000$$

and so forth. The exponent tells you how many tens must be multiplied together to yield the desired number. For example, ten thousand can be written as $10^4$ ("ten to the fourth") because $10^4 = 10 \times 10 \times 10 \times 10 = 10,000$. Similarly, 273,000 can be written $2.73 \times 10^5$.

In scientific notation, numbers are written as a figure between 1 and 10 multiplied by the appropriate power of 10. The distance between the Earth and the Sun, for example, can be written as $1.5 \times 10^8$ km. Once you get used to it, you will find this notation more convenient than writing "150,000,000 kilometers" or "one hundred and fifty million kilometers."

This powers-of-ten system can also be applied to numbers that are less than 1 by using a minus sign in front of the exponent. A negative exponent tells you that the location of the decimal point is as follows:

$$10^0 = 1$$
$$10^{-1} = 0.1$$
$$10^{-2} = 0.01$$
$$10^{-3} = 0.001$$
$$10^{-4} = 0.0001$$

and so forth. For example, the diameter of a hydrogen atom is $1.1 \times 10^{-8}$ cm. That is more convenient than saying "0.000000011 centimeter" or "11 billionths of a centimeter." Similarly, .000728 equals $7.28 \times 10^{-4}$.

Using the powers-of-ten shorthand, one can write large or small numbers like these compactly:

$$3,416,000 = 3.416 \times 10^6$$
$$0.000000807 = 8.07 \times 10^{-7}$$

Because powers-of-ten notation bypasses all the awkward zeros, a wide range of circumstances can be numerically described conveniently:

$$\text{one thousand} = 10^3$$
$$\text{one million} = 10^6$$
$$\text{one billion} = 10^9$$
$$\text{one trillion} = 10^{12}$$

and also

$$\text{one thousandth} = 10^{-3} = 0.001$$
$$\text{one millionth} = 10^{-6} = 0.000001$$
$$\text{one billionth} = 10^{-9} = 0.000000001$$
$$\text{one trillionth} = 10^{-12} = 0.000000000001$$

The accompanying figure shows how clearly the powers-of-ten notation expresses the scale of objects, ranging from subatomic particles like the proton to the size of the observable universe.

Astronomy is a quantitative science; its discoveries are based on facts and expressed in terms of numbers and associated units, like 1800 seconds or $8.3 \times 10^{12}$ kilograms. The incredible ranges of distances, sizes, and masses in astronomy require a shorthand for large and small numbers called "powers of ten" or **scientific notation.** (Please read An Astronomer's Toolbox I-1 if you are not familiar with scientific notation.) We will also use the

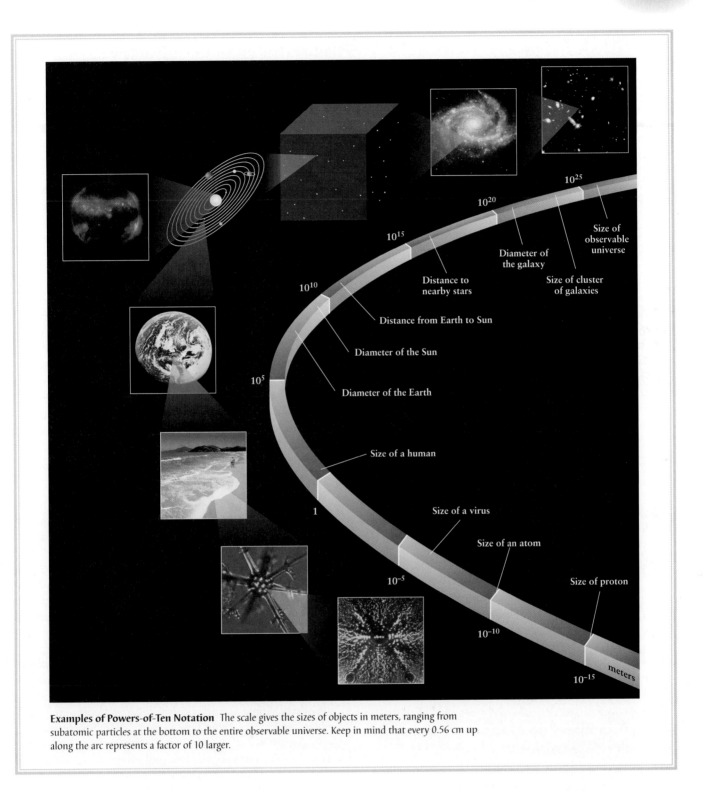

**Examples of Powers-of-Ten Notation** The scale gives the sizes of objects in meters, ranging from subatomic particles at the bottom to the entire observable universe. Keep in mind that every 0.56 cm up along the arc represents a factor of 10 larger.

metric system of units, which is now standard in science. Table I-1 lists some comparisons and conversions between metric and the traditional British units of measure. Even the metric system is too cumbersome for studying such remote objects as stars, galaxies, and the farthest realms of the universe. An Astronomer's Toolbox I-2 thus introduces other important astronomical distance units.

## DISCOVER FOR YOURSELF    The Earth-Centered Universe

At the dawn of the twenty-first century, most of us find it hard to understand why anyone would believe that the Sun, planets, and stars orbit the Earth. After all, we *know* that the Earth spins on its axis; we *know* that the gravitational force from the Sun holds the planets in orbit, just as the Earth's gravitational force holds the Moon in orbit. And we *know* that the stars in the night sky all lie far past the boundaries of our solar system. These facts have become part of our understanding of the motions of the heavenly bodies.

Psychologists call this background information that we use to help explain things a conceptual framework. Any conceptual framework contains all the information we take for granted. For example, when the Sun rises, moves across the sky, and sets today, we take for granted that it is the Earth's rotation that causes the Sun's apparent motion.

Our ancestors possessed a different conceptual framework. They didn't know that the Earth rotates. They didn't know that the then-mysterious force that held them to the ground is the same force that attracts the Earth to the Sun and the Moon to the Earth. They didn't know that the Sun is a star, just like the fixed points of light in the sky. And they didn't know any of the other laws of physics related to motion that we take for granted.

Because they did not feel the Earth move under their feet, nor see any other indication that the Earth is in motion, our forebears sensed nothing to support the belief that the Earth moves. (We now know that one indication of planetary rotation is the wind patterns, but they didn't know the cause of the winds either.) The obvious conclusion for one who has a prescientific conceptual framework is that the Earth stays put while objects in the heavens move around it.

This prescientific conceptual framework for understanding the motions of the heavenly bodies was strictly based on the senses, that is, people observed motions and drew "obvious," commonsense conclusions. Today, we incorporate the known and tested laws of physics in our understanding of the natural world. Many of these concepts are utterly counterintuitive, and, therefore, the conceptual frameworks we possess are less consistent with common sense than those held in the past. Studying science helps us develop intuition that is consistent with the actual workings of nature.

**Table I-1**
**Common Conversions Between British and Metric Units**

| 1 inch | = | 2.54 centimeter (cm) |
|--------|---|----------------------|
| 1 cm | = | .394 inches |
| 1 yard | = | .914 meters (m) |
| 1 meter | = | 1.09 yards = 39.37 cm |
| 1 mile | = | 1.61 kilometer (km) |
| 1 km | = | .621 miles |

## I-3 Astronomers do much more than just make observations

The process of astronomical discovery involves much more than observing and recording the heavens. The research activities of astronomers today fall into three rough categories: observing, recording, and analyzing observations; theorizing; and computer modeling. Most people think that an astronomer spends his or her time observing the sky, spending long nights directing the most powerful eyes on Earth to reveal the secrets of space. In reality, research telescopes are constantly booked, and most astronomers who get observing time—even a few weeks each year—consider themselves very lucky.

Planning observations and analyzing data takes up the majority of an observational astronomer's time. Most observational astronomers use data-collecting equipment provided by research observatories, but some design and build their own specialized apparatus, which they then connect to existing telescopes.

Other astronomers, theoreticians called *astrophysicists*, never use telescopes. Rather, they hypothesize or expand on theories in an effort to explain the observations of others. Some astrophysicists construct computer models to predict outcomes based upon existing theories.

Astronomical discovery is always exciting and sometimes totally unexpected. Unusual observations often lead astronomers and astrophysicists in new directions of research as they try to explain what they see or reconcile apparent contradictions between theories and observations. For example, astronomers recently observed evidence that neutrinos, particles created inside stars and previously believed to be massless, actually have mass. If confirmed, this result will have a major impact on theories about the evolution and fate of the entire universe.

# AN ASTRONOMER'S TOOLBOX 1-2

## Astronomical Distances

Throughout this book we will find that some of our traditional units of measure become cumbersome. It is fine to use kilometers to measure the diameters of craters on the Moon or the heights of volcanoes on Mars. However, it is as awkward to use kilometers to express distances to planets, stars, or galaxies as it is talk about the distance from New York City to San Francisco in millimeters. Astronomers have therefore devised new units of measure.

When discussing distances across the solar system, astronomers use a unit of length called the **astronomical unit (AU)**, which is the average distance between the Earth and the Sun:

$$1 \text{ AU} \approx 1.5 \times 10^8 \text{ km} \approx 9.3 \times 10^7 \text{ miles}$$

Jupiter, for example, is an average of 5.2 times farther from the Sun than is the Earth. Thus, the distance between the Sun and Jupiter can be conveniently stated as 5.2 AU. This can be converted into kilometers or miles using the relationship above.

When talking about distances to the stars, astronomers choose between two different units of length. One is the **light-year (ly)**, which is the distance that light travels in a vacuum (in the absence of air) in one year:

$$1 \text{ ly} \approx 9.46 \times 10^{12} \text{ km} \approx 63{,}000 \text{ AU}$$

One light-year is roughly equal to six trillion miles. Proxima Centauri, the star nearest to our solar system, is just over 4.2 ly from Earth.

The second commonly used unit of length is the **parsec (pc)** (the distance at which two objects separated by 1 AU make an angle of 1 arcsecond). Imagine taking a journey far into space, beyond the orbits of the outer planets. Watching the solar system as you move away, the angle between the Sun and the Earth becomes smaller and smaller. When the Sun and Earth are side by side and you measure the angle between them as $1/3600°$ (called 1 arcsecond), you have reached a distance astronomers call 1 parsec, as shown in the figure below. The parsec turns out to be longer than the light-year. Specifically,

$$1 \text{ pc} \approx 3.09 \times 10^{13} \text{ km} \approx 3.26 \text{ ly}$$

Thus, the distance to the nearest star can be stated as 1.3 pc as well as 4.2 ly. Whether one uses light-years or parsecs is a matter of personal taste.

For even greater distances, astronomers commonly use *kiloparsecs* (kpc) and *megaparsecs* (Mpc), in which the prefixes simply mean "thousand" and "million," respectively:

$$1 \text{ kpc} = 10^3 \text{ pc}$$
$$1 \text{ Mpc} = 10^6 \text{ pc}$$

For example, the distance from Earth to the center of our Milky Way Galaxy is about 8.6 kpc, and the rich cluster of galaxies in the direction of the constellation Virgo is 20 Mpc away.

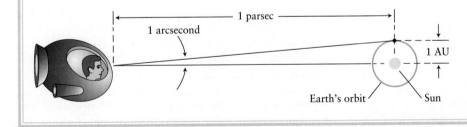

1 parsec · 1 arcsecond · 1 AU · Earth's orbit · Sun

**A Parsec** The parsec, a unit of length commonly used by astronomers, is equal to 3.26 ly. The parsec is defined as the distance at which 1 AU perpendicular to the observer's line of sight makes an angle of 1 arcsecond.

**3** In turn, the theories developed to explain the universe have led to a much deeper and richer understanding of the Earth. Many people think that astronomy deals only with the faraway and is of no significance to everyday life. But consider, as one example among many, how Newton's law of gravitation explains both the motions of the planets and also why we and every-

thing around us is held to the Earth's surface. Gravity also explains the time it takes an egg falling off a table to hit the floor and the hang time for basketball players under the net. By understanding the law of gravitation, engineers can control the friction between your car's tires and the road or design an airplane wing to lift a jumbo jet.

RIVUXG

RIVUXG

a

b

**Figure I-3**    **Visible and Nonvisible Radiation**  **(a)** This X-ray telescope was carried aloft in 1994 by the Space Shuttle. The inset shows an X-ray image of the Sun.  **(b)** This solar observatory in Arizona takes visible light photographs of the Sun, such as the one shown in the inset. Comparing these images reveals how important observing nonvisible radiation from astronomical phenomena is to furthering our understanding of how the universe operates.

## I-4  Telescopes enlarge our vision of the universe

How do astronomers gather information? Beginning with Galileo, astronomers peered directly through telescopes to observe visible starlight. Just over a hundred years ago, however, scientists began recording astronomical images on film, and they also began discovering forms of radiation invisible to our unaided eyes. Along with visible light, these radio waves, infrared rays, ultraviolet rays, X rays, and gamma rays are called **electromagnetic radiation**. During the past half century, astronomers have constructed telescopes that can detect nonvisible forms of electromagnetic radiation emitted by objects in space (Figure I-3). And every form of electro-

magnetic radiation has revealed new information about these objects.

Whether located on Earth or in orbit above the obscuring effects of our atmosphere, telescopes give us views vastly superior to anything our eyes can see. They are crucial to our understanding of familiar objects like the Sun, and they also give important clues about the nature of such exotic objects as neutron stars, pulsars, quasars, and black holes.

Most of our information about the universe still comes from imaging the various forms of electromagnetic radiation. But today astronomers are gathering an ever-growing body of firsthand cosmic knowledge from space debris found on Earth and the Moon, as well as from exotic particles that pepper the Earth from space.

Planets and Moons          Comets and other space debris

Stars

Interstellar gas and dust          Galaxies

Exotic objects

**Figure I-4**  **Inventory of the Universe**  Pictured here are examples of the major categories of objects that have been found throughout the universe. You will discover more about each type in the chapters that follow. All these photographs are at visible wavelengths, except the last, exotic objects, which is a radio image.

Examples include neutrinos and so-called cosmic rays (misnamed, because cosmic rays are actually particles). Astronomers are also on the brink of directly detecting the fluctuations in the fabric of space called gravitational radiation predicted by Einstein's theory of general relativity. Using these nonelectromagnetic sources of information, we may soon be able to see directly into the hearts of stars and further back toward the beginning of time.

What, then, have astronomers seen of the universe? Figure I-4 inventories some of the features we will explore in this text, including planets, stars, black holes, galaxies, and quasars. Each object is constantly changing; each had an origin and each will have an end. We will study these processes, too.

The dreams of science fiction writers, such as Jules Verne and H. G. Wells, pale when compared to current reality. Ours is an age of exploration and discovery more profound than any since the voyages of Columbus and Magellan revealed that we live on a spherical Earth. We have walked on the Moon and dug into Martian soil. Our probes have discovered active volcanoes and barren ice fields on the satellites of Jupiter. We have visited the shimmering rings of Saturn. We have seen evidence of planets orbiting other stars.

Never before has so much been revealed in so short a span of time. As you proceed through this book, you will gain a new appreciation of the awesome power of the human mind to reach out, to explore, to observe, and to comprehend. One of the great lessons of modern astronomy is that by gaining, sharing, and passing on knowledge, we can transcend the limitations of our bodies and the brevity of human life.

## WHAT DID YOU THINK?

1 *What makes a theory scientific?* A theory is an idea or set of ideas proposed to explain something about the natural world. A theory is scientific if it makes objective predictions that can be objectively tested and potentially disproved.

2 *What do astronomers do?* Most astronomers spend their time analyzing observational data, building equipment to make better observations, creating theories, and carrying out computer simulations. Observational astronomers spend only a few days or weeks each year actually using research telescopes.

3 *Why is astronomy worth learning?* It provides us with much practical information, such as the nature of the gravitational force, as well as an understanding of the origin and evolution of the universe and our place in it.

## Key Words

| | | | | |
|---|---|---|---|---|
| astronomical unit (AU), 7 | light-year (ly), 7 | parsec (pc), 7 | power of ten, 4 | scientific notation, 4 |
| electromagnetic radiation, 8 | model, 3 | physics, 2 | scientific method, 2 | scientific theory, 2 |

## Key Ideas

* The universe is comprehensible.

* The scientific method is a procedure for formulating theories that correctly predict how the universe behaves.

* A scientific theory must be testable, that is, capable of being disproved.

* Theories are tested and verified by observation or experimentation and result in a process that often leads to their refinement or replacement and to the progress of science.

* Observations of the heavens have led astronomers to discover some fundamental physical laws of the universe.

* Space debris collected on Earth, as well as particles from space, such as neutrinos and cosmic rays, and an exotic energy called gravitational radiation are providing astronomers with new insights into the universe.

## Review Questions

The answers to all computational problems, which are preceded by an asterisk (*), appear at the end of the book.

1 Compile a list of five examples of counterintuitive astronomy ideas, such as "the Earth rotates," that we now accept as correct.

*2 Convert the following distances into scientific notation: (a) Earth's mass: 5,974,000,000,000,000,000,000,000 kg;

(b) Sun's mass: 1,989,999,000,000,000,000,000,000,000,000 kg; (c) Sun's radius: 695,990 km; (d) one year: 31,558,000 s.

*3 Convert: (a) 8.3 pc into light-years; (b) 6.52 ly into parsecs; (c) 8450 AU into kilometers; (d) $2.7 \times 10^3$ Mpc into kiloparsecs.

## Advanced Questions

4 The dictionary defines *astrology* as "the study that assumes and attempts to interpret the influence of the heavenly bodies on human affairs." Based on what you know about scientific theory, is astrology a science? Why or why not? Feel free to explore astrology further if you wish before answering this question.

## What If ...

5 The laws of physics changed from time to time? What coping skills would life require, if it could exist at all? What else besides the laws of physics would scientists need to comprehend in order to understand and make predictions about the universe?

## Observing Projects

6 Install the Starry Night planetarium software from the CD-ROM in your book into your computer. Determine whether and, if so, when the Moon is visible during the day today and when it is visible during the night. Determine which, if any, of the following naked-eye planets are observable in the sky tonight: Mercury, Venus, Mars, Jupiter, and Saturn.

*Hints*: (1) If you are just learning to use Starry Night, experiment with settings as much as necessary. You can always return to your starting screen by clicking on *Return Home* in the Settings menu or by clicking on the house icon in the window normally on the left of your screen. (2) You can move the sky around by moving your mouse up until a little hand appears. Hold the left mouse button as you move the mouse and you will move the sky. (3) You can change the time of day and change the rate at which time is changed. (4) Toggling the planets on and off will let you identify planets presently in the sky.

# An Astronomer's Almanac

**586 B.C.**
Thales of Miletus predicts solar eclipse.

**1512-1543**
Nicolas Copernicus proposes heliocentric cosmology in his Commentariolus and De Revolutionibus Orbium Coelestium.

**ca. 270 B.C.**
Aristarchus of Samos proposes heliocentric cosmology.

**1589-1609**
Galileo Galilei proposes that all objects fall with the same acceleration, independent of their masses; builds his first telescope, a refractor.

**1715**
Edmund Halley calculates shadow path of a solar eclipse over Earth's surface.

**350 B.C.**
Aristotle proposes spherical Earth, geocentric cosmology.

Greek Golden Age

European Renaissance

**1609-1619**
Johannes Kepler publishes his three laws of planetary motion.

**1766**
Henry Cavendish discovers hydrogen.

**2136 B.C.**
Chinese astronomers record solar eclipse.

**1576-1601**
Tycho Brahe makes precise observations of stars and planets.

**1800-1803**
William Herschel discovers infrared radiation from the Sun. Thomas Young demonstrates wave nature of light. John Dalton proposes that matter is composed of atoms of different weights.

**1665-1704**
Isaac Newton deduces gravitational force from the orbit of the Moon; builds first reflecting telescope; proves that planets moving under influence of gravitational force obey Kepler's laws; publishes compendium on light, Optics.

**ca. A.D. 125**
Claudius Ptolemy refines and details geocentric cosmology in his Almagest.

Moon

Sun

Earth

# DISCOVERING ASTRONOMY

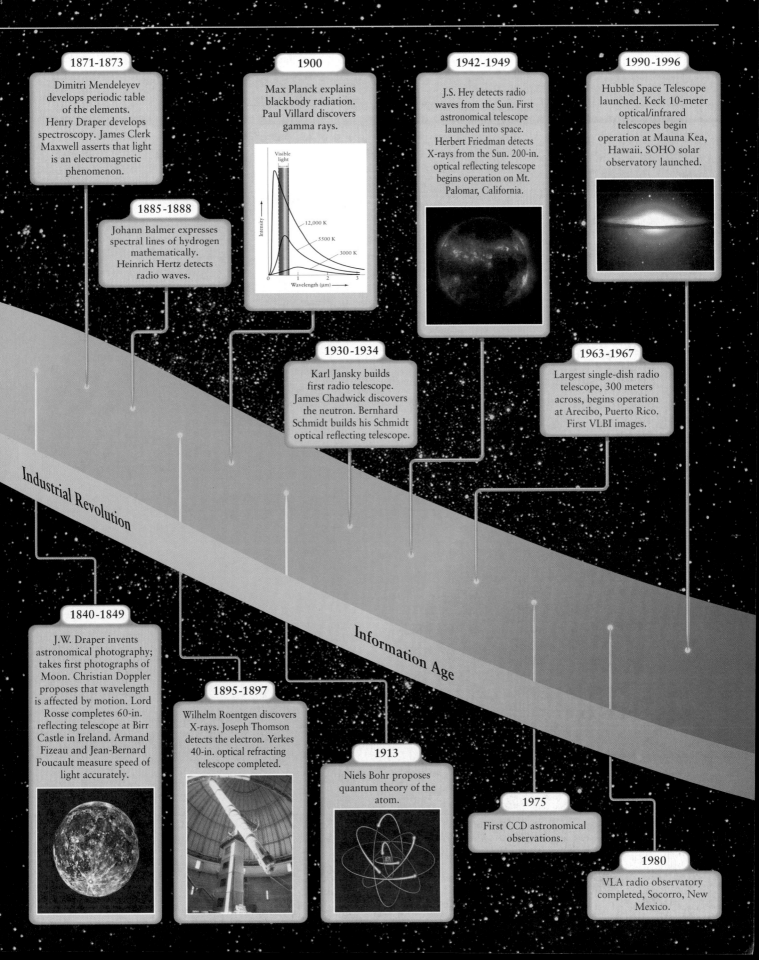

**1871-1873**

Dimitri Mendeleyev develops periodic table of the elements. Henry Draper develops spectroscopy. James Clerk Maxwell asserts that light is an electromagnetic phenomenon.

**1885-1888**

Johann Balmer expresses spectral lines of hydrogen mathematically. Heinrich Hertz detects radio waves.

**1900**

Max Planck explains blackbody radiation. Paul Villard discovers gamma rays.

**1942-1949**

J.S. Hey detects radio waves from the Sun. First astronomical telescope launched into space. Herbert Friedman detects X-rays from the Sun. 200-in. optical reflecting telescope begins operation on Mt. Palomar, California.

**1990-1996**

Hubble Space Telescope launched. Keck 10-meter optical/infrared telescopes begin operation at Mauna Kea, Hawaii. SOHO solar observatory launched.

**1930-1934**

Karl Jansky builds first radio telescope. James Chadwick discovers the neutron. Bernhard Schmidt builds his Schmidt optical reflecting telescope.

**1963-1967**

Largest single-dish radio telescope, 300 meters across, begins operation at Arecibo, Puerto Rico. First VLBI images.

Industrial Revolution

Information Age

**1840-1849**

J.W. Draper invents astronomical photography; takes first photographs of Moon. Christian Doppler proposes that wavelength is affected by motion. Lord Rosse completes 60-in. reflecting telescope at Birr Castle in Ireland. Armand Fizeau and Jean-Bernard Foucault measure speed of light accurately.

**1895-1897**

Wilhelm Roentgen discovers X-rays. Joseph Thomson detects the electron. Yerkes 40-in. optical refracting telescope completed.

**1913**

Niels Bohr proposes quantum theory of the atom.

**1975**

First CCD astronomical observations.

**1980**

VLA radio observatory completed, Socorro, New Mexico.

# Discovering the Night Sky

## *In this chapter you will discover*

* ✳ how astronomers map the night sky to help them locate objects in it

* ✳ that the Earth's spin on its axis causes day and night

* ✳ how the tilt of the Earth's axis of rotation and the Earth's motion around the Sun combine to create the seasons

* ✳ that the Moon's orbit around the Earth creates the phases of the Moon and lunar and solar eclipses

* ✳ how the year is defined and how the calendar was developed

**Circumpolar Star Trails**
This long exposure, taken from Australia's Siding Spring Mountain and aimed at the south celestial pole, shows the rotation of the sky. The building in the foreground houses the Anglo-Australian Telescope, one of the largest telescopes in the southern hemisphere. During the exposure, someone carrying a flashlight walked along the catwalk on the outside of the telescope dome. Another flashlight made the wavy trail at ground level.

R I V U X G

## WHAT DO YOU THINK?

1. Is the North Star—Polaris—the brightest star in the night sky?

2. Do astronomers regard constellations as simply the familiar patterns of stars in the sky first identified by ancient stargazers?

3. What causes the seasons?

4. How many zodiac constellations are there?

5. Does the Moon have a dark side that we never see from Earth?

6. Is the Moon ever visible during the daytime?

 hen you gaze at the sky on a clear, dark night, there seem to be millions of stars twinkling overhead. In reality, the unaided human eye can detect only about 6000 stars over the entire sky. At any one time, you can see roughly 3000 stars in dark skies, because only half of the stars are above the *horizon*, the boundary between the Earth and the sky.

You probably have noticed patterns formed by bright stars and are probably familiar with some common names for these patterns, such as the bowl-shaped Big Dipper and broad-shouldered Orion. These recognizable patterns of stars, which we call constellations in everyday conversation, have names derived from ancient legends (Figure 1-1a).

a

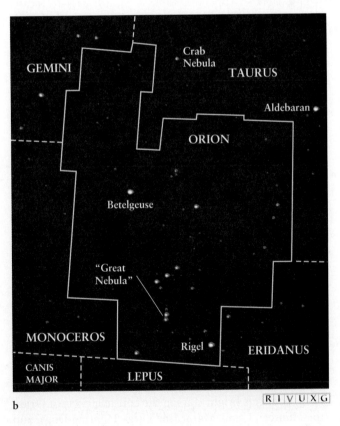

b

## Figure 1-1   The Constellation Orion   (a) The pattern of stars called Orion is a prominent winter constellation. From the United States, it is easily seen high above the southern horizon from December through March. You can see in this photograph that the various stars have different colors, something to watch for when you observe the night sky. (b) The region of the sky called Orion and parts of other nearby constellations are depicted on this photograph. *All* the stars inside the boundary of Orion are members of that constellation. The celestial sphere is covered by 88 constellations of differing sizes and shapes.

# PATTERNS OF STARS

## 1-1 Constellations make locating stars easy

You can orient yourself on Earth with the help of easily recognized constellations. For instance, if you live in the northern hemisphere, you can use the Big Dipper to find the direction north. To do this, locate the Big Dipper and imagine that its bowl is resting on a table. If you see the Dipper upside down in the sky, as you frequently will, imagine the dipper resting on an upside-down table above it. Locate the two stars of the bowl farthest from the Big Dipper's handle. These are called the *pointer stars*. Draw a mental line through these stars leading away from the table, as shown in Figure 1-2. The first *moderately bright* star you then encounter is Polaris, also called the North Star because it is located almost directly over the Earth's north pole. So, while Polaris is not even among the 20 brightest stars (Appendix Table A-5), it *is* easy to locate. Whenever you face Polaris, you are facing north. East is then on your right, south is behind you, and west is on your left.

The Big Dipper example illustrates the fact that easily recognized constellations make it easy to locate other stars. The most effective way to do this is to use vivid visual connections, especially those of your own devising. For example, imagine gripping the handle of the Big Dipper and slamming its bowl through the imaginary table and onto the head of Leo (the Lion). Leo comprises the first group of bright stars your dipper encounters. As shown in Figure 1-2, the brightest star in this group is Regulus, the dot of the backward question mark that traces the lion's mane. As another example, put the Big Dipper back on its table and then follow the arc of its handle away from its bowl. The first bright star you encounter along that arc beyond the handle is Arcturus in Boötes (the Shepherd). Follow the same arc further to the prominent bluish star Spica in Virgo (the Virgin). Spotting these stars is easy if you remember the saying "Arc to Arcturus and speed on to Spica."

During the winter months in the northern hemisphere, you can see some of the brightest stars in the sky. Many of them are in the vicinity of the "winter triangle," which connects bright stars in the constellations of Orion (the Hunter), Canis Major (the Large Hunting Dog), and Canis Minor (the Small Hunting Dog), as shown in Figure 1-3. The winter triangle is nearly overhead during the middle of winter at midnight. It is easy to find Sirius, the brightest star in the night sky, by locating the belt of Orion and following a straight mental line from it to the left (as you face Orion). The first bright star you encounter is Sirius.

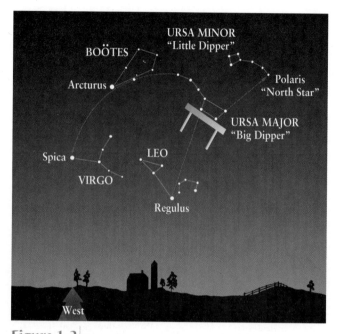

**Figure 1-2** **The Big Dipper as a Guide** In the northern hemisphere, the Big Dipper is an easily recognized pattern of seven bright stars. This star chart shows how the Big Dipper can be used to point out the North Star as well as the brightest stars in three other constellations. Note that the Big Dipper appears right side up in this drawing, but at other times of the night it appears upside down. Why do you suppose this happens?

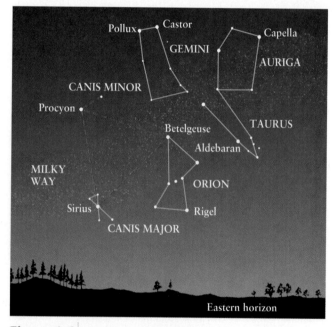

**Figure 1-3** **The Winter Triangle** This star chart shows the eastern sky as it appears during the evening in December. Three of the brightest stars in the sky make up the winter triangle. In addition to the constellations involved in the triangle, Gemini (the Twins), Auriga (the Charioteer), and Taurus (the Bull) are also shown.

## DISCOVER FOR YOURSELF | The Constellations

Many students take an introductory astronomy course expecting to be taught the familiar constellations, something most teachers just don't have time to cover. You can, however, learn them on your own using two helpful techniques for memorizing the night sky.

1. Observe easily identified constellations and use these to find nearby, lesser-known ones. For example, you may not remember where the star Aldebaran in Taurus is in the sky, but if you remember that Orion is fighting Taurus, you can easily locate Aldebaran by following a line defined by the belt of Orion to the right (away from Sirius). The first bright star you encounter is Aldebaran.

2. Make your own connections between the constellations. Have fun while you're doing it. Chances are that you will remember the phrase "slam the Big Dipper's bowl downward to hit Leo the Lion on the head" rather than "the first bright group of stars directly below the Big Dipper is Leo."

An extremely efficient way to learn the constellations is to use the board and card game *Stellar 28*. In addition, astronomy computer programs such as Starry Night, included with this text, are available. Starry Night shows many things besides what constellations are up at night, including the motion, location, and phases of the planets; the location of deep sky objects, such as nebulae; and the sky as seen from any location on Earth on any date.

Another good way to familiarize yourself with the night sky is to use star charts to see which constellations are up each night. You will find a set of star charts from the *Griffith Observer* magazine at the front and back of this book. To use the charts, first select the one that best corresponds to the date and time of your observation. Take the chart outside at night and compare it directly with the sky. Hold the chart vertically and turn it so that the direction you are facing shows at the bottom. Using a flashlight with a red plastic coating over the light will make it easier to read the chart without constricting your vision. You will find stargazing a surprisingly enjoyable experience.

---

**Insight into science** **Flexible thinking** Part of learning science is learning to look at things from different perspectives. For example, when learning to identify the prominent constellations, be sure to learn them from different orientations (that is, with the star chart rotated at different angles), so that you can find them at different times of the night and the year.

The "summer triangle," which graces the summer sky as shown in Figure 1-4, connects the bright stars Vega in Lyra (the Harp), Deneb in Cygnus (the Swan), and Altair in Aquila (the Eagle). A conspicuous portion of the Milky Way forms a beautiful background for these constellations, which are nearly overhead during the middle of summer at midnight.

Astronomers require more accuracy in locating dim objects than is possible simply by moving from constellation to constellation. They have therefore created a celestial map and applied a coordinate system to it, analogous to the coordinate system of north-south latitude and east-west longitude used to navigate on the Earth. If a star's celestial coordinates are known, it can then be quickly located. For such a sky map to be useful in finding stars,

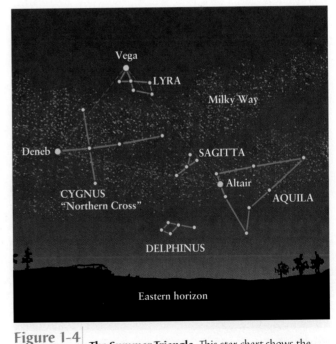

**Figure 1-4** **The Summer Triangle** This star chart shows the northeastern sky as it appears in the evening in June. In addition to the three constellations involved in the summer triangle, the faint stars in the constellations of Sagitta (the Arrow) and Delphinus (the Dolphin) are also shown.

the stars must be fixed on it as cities are fixed on maps of the Earth.

## 1-2 The celestial sphere aids in navigating the sky

If you look at the night sky year after year, you will see that the stars do indeed appear fixed relative to each other. Furthermore, throughout each night the entire pattern of stars appears to rigidly orbit the Earth. We employ this Earth-based view of the heavens to make celestial maps by pretending that the stars are attached to the inside of an enormous hollow shell, the **celestial sphere,** with the Earth at its center (Figure 1-5).

We discussed the common usage of the word *constellation* at the beginning of this chapter. Astronomers technically use the word to describe an entire region of the sky and all the objects in that region (see Figure 1-1b). The celestial sphere is thereby divided into 88 unequal regions, and these regions are what astronomers refer to when they use the term **constellations.** (Astronomers call the traditional star patterns *asterisms* rather than constellations when there is a danger of confusion.) Some constellations (like Ursa Major) are very large, while others (like Sagitta) are relatively small. To locate stars, we might

say "Albireo in the constellation Cygnus," much as we would refer to "Ithaca in New York State."

The stars seem fixed on the celestial sphere only because of their remoteness. In reality, they are at widely varying distances from the Earth, and they do move relative to each other. But we neither see their motion nor perceive their relative distances because the stars are so far from here. You can understand this by imagining a jet plane traveling at 1000 kilometers (620 miles) an hour roaring just overhead. Its motion is unmistakable. But the same plane, moving at the same speed, barely seems to budge when located along the distant horizon.

The stars (other than the Sun) are all more than 40 trillion kilometers (25 trillion miles) from us. Therefore, although the patterns of stars in the sky do change, their great distances prevent us from seeing those changes over the course of a human lifetime. Thus, as unrealistic as it is, the celestial sphere is so useful for navigating the heavens that it is used by astronomers even at the most sophisticated observatories around the world.

As shown in Figure 1-5, we can project key geographic features from Earth out into space to establish directions and bearings. If we expand the Earth's equator onto the celestial sphere, we obtain the **celestial equator.** The celestial equator divides the sky into northern and southern hemispheres, just as the Earth's equator divides the Earth into two hemispheres. We can also imagine extending the Earth's north and south poles out into space along the Earth's axis of rotation. Doing so gives us the **north celestial pole** and the **south celestial pole,** also shown in Figure 1-5. With the celestial equator and poles as reference features, astronomers denote the position of an object in the sky in much the same way that latitude and longitude are used to specify a location on Earth.

Just as we need two coordinates (latitude and longitude) to find any location on Earth, two coordinates are needed to locate any object on the celestial sphere. The equivalent to latitude on Earth is **declination** on the celestial sphere. It is measured north or south of the celestial equator. The equivalent of longitude on Earth is **right ascension** on the celestial sphere, measured around the celestial equator (Figure 1-5).

We will see later in this chapter that the Sun moves in a closed line around the celestial sphere during the course of a year. The celestial equator and the Sun's path intersect at two points. The equivalent on the celestial sphere of the Earth's prime meridian (from which degrees of longitude are measured on Earth) is where the Sun crosses the celestial equator moving northward. Angles of right ascension are measured from this point, called the *vernal equinox.*

In navigating on the celestial sphere, astronomers measure the distance between objects in terms of angles. If you are not familiar with measuring the distance between objects using this method, read An Astronomer's Toolbox 1-1.

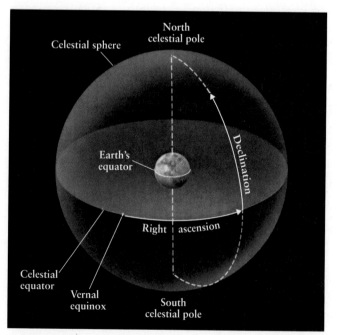

## Figure 1-5    **The Celestial Sphere**  The celestial sphere is the apparent "bowl" or hollow sphere of the sky. The celestial equator and poles are projections of the Earth's equator and axis of rotation out into space. The north celestial pole is therefore located directly over the Earth's north pole, while the south celestial pole is directly above the Earth's south pole.

# AN ASTRONOMER'S TOOLBOX 1-1

## Observational Measurements Using Angles

Astronomers have inherited many useful concepts from antiquity. For example, ancient mathematicians invented angles and a system of angular measure that is still used to denote the positions and apparent sizes of objects in the sky. For example, to locate stars, we don't need to know their distances from Earth (which are all different). All we need to know is the angle from one star to another in the sky, a property that remains fixed over our lifetimes.

An **arc angle**, often just called an **angle**, is the opening between two lines that meet at a point. Angular measure is a method of describing the size of an angle. The basic unit of angular measure is the **degree**, designated by the symbol °. A full circle is divided into 360°. A right angle measures 90°. As shown in the accompanying figure, the angle between the two "pointer stars" in the Big Dipper is about 5°.

Astronomers also use angular measure to describe the apparent sizes of celestial objects. For example, imagine looking up at the full Moon. The angle covered by the Moon's diameter is nearly ½°. We therefore say that the **angular diameter**, or angular size, of the Moon is ½°. Alternatively, astronomers say that the Moon "subtends" an angle of ½°. In this context, *subtend* means "to extend across."

The adult human hand held at arm's length provides a means of estimating angles. For example, your fist covers an angle of 10°, whereas the tip of

**Estimating Angles with the Human Hand**  Various parts of the adult human hand extended to arm's length can be used to estimate angular distances and sizes in the sky.

your finger is about 1° wide. Various segments of your index finger extended to arm's length can be similarly used to estimate angles a few degrees across.

To talk about smaller angles, we subdivide the degree into 60 arcminutes (abbreviated 60 arcmin or 60′). An arcminute is further subdivided into 60 arcseconds (abbreviated 60 arcsec or 60″). A dime viewed face on from a distance of 1 mile has an angular diameter of about 2 arcsec.

From everyday experience, we know that an object looks big when it is nearby but small when it is far away. The angular size of an object therefore does not necessarily tell you anything about its actual physical size. For example, the fact that the Moon's angular diameter is ½° does not tell you how big the Moon really is. But if you also happen to know the distance to the Moon, then you can calculate the Moon's physical diameter. In general, the physical diameter of an object can be calculated from the equation:

physical diameter = distance × tan (angular diameter)

where tan (angular diameter) means the tangent of the angle denoted "angular diameter". In the Moon's case, using a measured distance (Appendix Table A-3) of 384,400 km and an angular diameter of ½°, we find the diameter to be roughly 3350 km. The difference between this and the exact diameter of 3476 km is due primarily to the approximate value of ½° that we have used.

**The Big Dipper**  The angular distance between the two "pointer stars" at the front of the Big Dipper is about 5°. For comparison, the angular diameter of the Moon is about ½°.

## 1-3 Earth's rotation causes the stars to appear to move

The Earth spins on its axis. Such motion is called **rotation.** The Earth's 24-hour rotation causes the constellations—as well as the Sun, Moon, and planets—to appear to rise on the eastern horizon, move across the sky, and set on the western horizon. Bear in mind that the Sun and Moon rise due east and set due west only on certain days of the year. The **diurnal motion,** or daily motion, of the celestial bodies is apparent in time-exposure photographs, such as the one that opens this chapter. The Earth's rotation causes day and night because it makes the Sun appear to follow a diurnal path across the sky.

People who spend time outdoors at night are familiar with the diurnal motion of the stars. Take a friend outside on a clear, warm night to observe it for yourself. Soon after dark, find a spot away from bright lights and note the constellations in the sky relative to some prominent landmarks near you on Earth. A few hours later, check again from the same place. You will find that the entire pattern of stars (as well as the Moon, if it is visible) has shifted. New constellations will have risen above the eastern horizon, while other constellations will have disappeared below the western horizon. If you check again just before dawn, you will find the stars that were just rising in the east when the night began are now low in the western sky.

Different constellations are visible at night during different times of the year. This shift occurs because the Earth orbits, or *revolves,* around the Sun. **Revolution** is the motion of any astronomical object around another astronomical object. The Earth takes one year, or about 365¼ days, to go once around the Sun. As a result of this motion, the darkened, nighttime side of the Earth is turned toward different parts of the heavens at different times of the year. Another result of the Earth's motion around the Sun is that every star rises approximately 4 minutes earlier each night than it did the night before.

We spoke earlier of stars rising on the eastern horizon and setting on the western horizon. Depending on your latitude, some of the stars and constellations never disappear below the horizon. Instead, they trace complete circles in the sky over the course of each night. To understand why this happens, imagine that you are standing on the Earth's north pole at night. Looking straight up, you see Polaris. Because the Earth is spinning around its axis directly under your feet, all the stars appear to move from left to right in horizontal rings around you. The exception is Polaris, which always remains directly overhead. As seen from the north pole, no stars rise or set (Figure 1-6). They just seem to revolve around Polaris in horizontal circles. Stars and constellations that never go below the horizon are called **circumpolar.** All stars visible from the north or south pole are circumpolar.

**Figure 1-6** **Motion of Stars at the North Pole** Because the Earth rotates around the north pole, stars seen from there appear to move in huge, horizontal circles. This is the same effect you would get by standing up in a room and spinning around; everything would appear to move in circles around you.

**Figure 1-7** **Rising of Stars at the Equator** Standing on the equator, you would be perpendicular to the axis around which the Earth rotates. As seen from there, the stars rise straight up on the eastern horizon and set straight down on the western horizon. This is the same effect you get when driving straight over the crest of a hill; the objects on the other side of the hill appear to move straight upward as you descend.

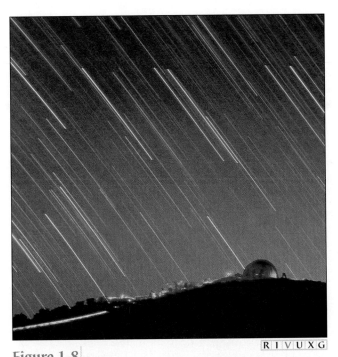

RIVUXG

**Figure 1-8**  **Rising and Setting of Stars at Middle North Latitudes** Unlike the motion of the stars at the poles (see Figure 1-6), the stars at all other latitudes do change angle above the ground throughout the night. This time-lapse photograph shows stars setting over Benson, Arizona. The latitude determines the angle at which the stars rise and set.

Now visualize yourself at the equator. All the stars appear to rise straight up in the eastern sky and set straight down in the western sky (Figure 1-7). Polaris is barely visible on the northern horizon. While Polaris never sets, all the other stars do, and therefore none of the stars are circumpolar as seen from the equator.

You may already have concluded from these two mental exercises that the angle at which the stars rise and set depends on your viewing latitude. Figure 1-8, for example, shows stars setting at 32° north latitude. In Orono, Maine (44°45′ north latitude), where this book was written, stars rise at an angle of nearly 45° to the eastern horizon and set at an angle of 45° to the western horizon. Polaris is fixed at 45° above the horizon in Orono's northern sky, not directly overhead, as it is at the pole or on the horizon as seen from the equator.

If you live in the northern hemisphere, Polaris is always located above your northern horizon at an angle equal to your latitude. Only the stars and constellations that pass between Polaris and the land directly below it are circumpolar. The farther north you go in the northern hemisphere, the greater the number of stars and constellations that are circumpolar. The opposite is true in the southern hemisphere.

# EARTHLY CYCLES

The everyday rhythms of the Earth and all life on it arise from three celestial motions: the Earth's rotation, which causes day and night; the Earth's revolution around the Sun, which creates the seasons and the year; and the Moon's revolution around the Earth, which creates the lunar phases, the cycle of tides, and the spectacular phenomena we call eclipses.

Contrary to common sense, the seasons are not caused by the change in the Earth's distance from the Sun. If the seasons were caused by the changing distance from the Earth to the Sun, all parts of the Earth should have the same seasons at the same time. In fact, the northern and southern hemispheres have exactly opposite seasons. Furthermore, the Earth is closest to the Sun on January 3 of each year—the dead of winter in the northern hemisphere! We will explore the seasons further in Section 1-6.

**Insight into science**  **Expect the unexpected** Many phenomena in the universe defy common-sense explanations. The process of science requires that we question the obvious, that is, what we think we know. The fact that the changing distance from the Earth to the Sun has a negligible effect on the seasons is an excellent example.

## 1-4 The speed of the Earth's rotation determines the length of the day

The Sun's daily motion through the sky provided our ancestors' earliest reference for time, because the Sun's location determines whether it is day or night, whether we are awake or asleep, and whether it is time for breakfast or dinner. In other words, the Sun determines the length of the **solar day**, upon which our 24-hour day is based. However, the length of the solar day varies throughout the year. The Sun is not a perfect timekeeper, because the Earth's orbit around it is not circular, as we will study in Chapter 3. Because the Earth moves slightly more rapidly along its orbit when it is near the Sun than when it is farther away, the speed of the Sun across the sky also varies throughout the year. Using the average time interval between consecutive noontimes to determine the solar day corrects this problem, and this average defines our 24-hour day.

Astronomically, noon is defined as the instant when the Sun is highest in the sky. However, at different longitudes the Sun is highest at different times. Thus, astronomical noon in New York City occurs slightly earlier than it does in Philadelphia. Fortunately, the concept of time zones removes this problem. In a **time zone**, everyone

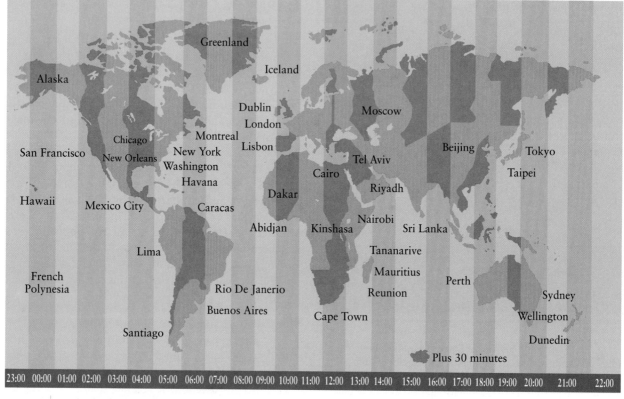

**Figure 1-9** **Time Zones of the World** For convenience, Earth's 360° circumference is divided into 24 time zones. Ideally, each time zone would run due north-south. However, political considerations make some of zones irregular. Indeed, there are even a few zones only a half hour wide.

agrees to set their clocks alike. Time zones, originally developed for scheduling rail transportation, are based on the time at 0° longitude in Greenwich, England, a location called the *prime meridian*. With some variation due to geopolitical boundaries, every 15° of longitude around the globe begins a new time zone. The resulting 24 time zones are shown in Figure 1-9. Going from one time zone to the next usually requires you to change the time on your wristwatch by exactly 1 hour.

## 1-5 The Earth's orbit around the Sun determines the length of the year

Just as the day originates in the Earth's rotation, the *year* is the unit of time based on the Earth's revolution about the Sun. The Earth does not take exactly 365 days to orbit the Sun, and so the year is not exactly 365 days long. Ancient astronomers realized that the length of a year is approximately 365¼ days. The Roman emperor Julius Caesar established the system of leap years to account for this extra quarter of a day. By adding an extra day to the

calendar every four years, Caesar hoped to ensure that seasonal astronomical events, such as the beginning of spring, would occur on the same date year after year.

Caesar's system would have been perfect if the year were exactly 365¼ days long and the Earth's rotation axis never changed direction. Unfortunately, this is not the case. Thus, over time, a discrepancy accumulated between Caesar's "Julian" system and actual time. Annual events began to fall on different dates each year. To straighten things out, Pope Gregory XIII reformed the Julian calendar in 1582. He began by dropping ten days (October 5, 1582, was proclaimed to be October 15, 1582), which brought the first day of spring back to March 21. Next, he modified Caesar's system of leap years.

Caesar had added February 29 to every calendar year that is evenly divisible by four. For example, 1988, 1992, and 1996 were all leap years with 366 days. But this system produces an error of about three days every four centuries. To solve the problem, Pope Gregory decreed that only the century years evenly divisible by 400 should be leap years. For example, the years 1700, 1800, and 1900 (which would have been leap years under Caesar's system) were not leap years under the improved Gregorian system. But the year 2000—which can be divided evenly by 400—is a leap year.

We use the Gregorian system today. It assumes that the year is 365.2425 mean solar days long, which is very close to the length of the *tropical year,* defined as the time interval from one vernal equinox to the next. In fact, the error is only one day in every 3300 years. That won't cause any problems for a long time. The major impact of our annual journey around the Sun is the unfolding of the seasons. While their influence on our lives is obvious enough, their cause is a fairly subtle combination of effects related to the Earth's orbit.

## 1-6 The seasons are caused by the tilt of the Earth's axis of rotation

The seasonal changes we experience actually result from the tilt of the Earth's axis of rotation relative to the plane of our orbit around the Sun. Imagine that you could see the stars even during the day, so that you could follow the Sun's apparent motion against the background constellations throughout the year. (The Sun appears to move among the stars, of course, because the Earth orbits around it.)

From day to day, the Sun would trace a straight path along the celestial sphere. You can do this with your Starry Night software by locking on the Sun, turning on the Ecliptic and equatorial coordinate grid, and then turning off the daylight. As you can see in Figure 1-10a, this path, called the **ecliptic**, makes a closed circle bisecting the celestial sphere.

The term *ecliptic* has another use in astronomy. The Earth orbits the Sun in a plane also called the ecliptic. You can see that the two ecliptics exactly coincide. Imagine yourself on the Sun watching the Earth move day by day. The path of the Earth on the celestial sphere as seen from the Sun is precisely the same as the path of the Sun as seen from the Earth (Figure 1-10b).

> **Insight into science  Define your terms** The words "ecliptic" and "constellation" have more than one scientific meaning. Most often, however, scientists restrict words to one well-defined meaning, because more than one meaning can lead to misunderstandings. Be sure that you understand each word in astronomy and how to use it.

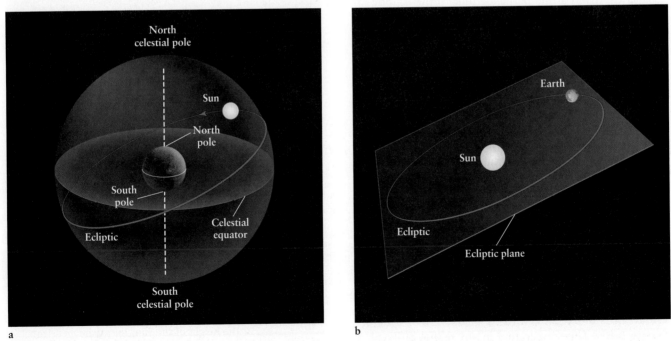

Ecliptic

a                                              b

**Figure 1-10**   **The Ecliptic  (a)** The ecliptic is the apparent annual path of the Sun on the celestial sphere. **(b)** The ecliptic is also the plane described by the Earth's path around the Sun. The planes created by the two ecliptics exactly coincide.

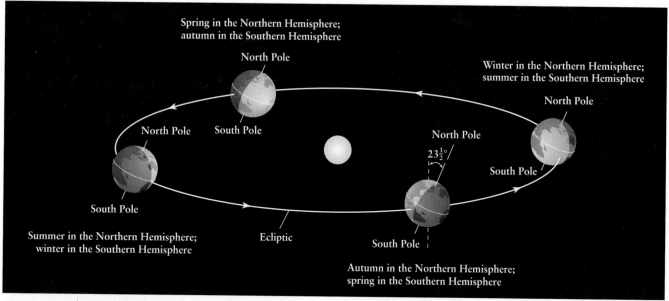

Spring in the Northern Hemisphere;
autumn in the Southern Hemisphere

North Pole

North Pole    South Pole

North Pole

South Pole

Summer in the Northern Hemisphere;
winter in the Southern Hemisphere

Ecliptic

Winter in the Northern Hemisphere;
summer in the Southern Hemisphere

North Pole

$23\frac{1}{2}°$

South Pole

North Pole

South Pole

Autumn in the Northern Hemisphere;
spring in the Southern Hemisphere

**Figure 1-11**   **The Tilt of the Earth's Axis** The Earth's axis of rotation is tilted 23½° from being perpendicular to the plane of the Earth's orbit. The Earth maintains this orientation (with its north pole aimed at the north celestial pole near the star Polaris) throughout the year as it orbits the Sun. Consequently, the amount of solar illumination and the number of daylight hours at any location on Earth varies in a regular fashion throughout the year.

The celestial equator and the ecliptic are different circles on the celestial sphere because the Earth's axis is tilted 23½° away from a line perpendicular to the ecliptic, as shown in Figure 1-11. Except for tiny changes each year, discussed below, the Earth maintains this tilted orientation as it orbits the Sun; Polaris is above the north pole throughout the year. Thus, during part of the year, the northern hemisphere is tilted toward the Sun and the southern hemisphere is tilted away. Half a year later, the situation is reversed. By the way, the seasons are *not* caused by the fact that one hemisphere is slightly closer to the Sun than the other—that difference in distance is much, much too small to cause any temperature difference.

When the north pole is tilted toward the Sun, the Sun rises higher in the northern sky than when that pole is pointing away from the Sun (Figure 1-12). The higher the Sun rises during the day, the more daylight hours. During these longer periods of daylight, more light and heat energy from the Sun strike that hemisphere than during the seasons when the north pole is pointed away from the Sun, and, therefore, the Sun does not rise nearly as high in the sky. Furthermore, when the Sun is higher in the sky, its energy is more concentrated on the Earth's surface (Figure 1-13). The temperature in any region on Earth is determined by the duration of daylight there and the height of the Sun in the sky.

Because of the tilt of the Earth's axis of rotation, the ecliptic and the celestial equator are inclined to each

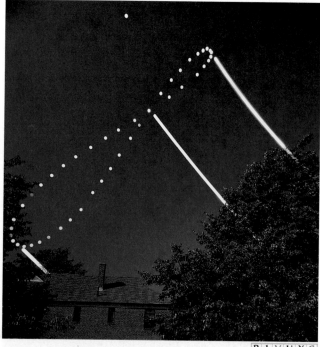

R I V U X G

**Figure 1-12**   **The Height of the Sun** The maximum height of the Sun in the sky varies throughout the year because of the 23½° tilt of the Earth's axis. This photograph shows the height of the Sun in the sky at the same clock time at ten-day intervals throughout the year, as well as its changing east-west location. The streaks show the path of the Sun on three different days.

a

b

**Figure 1-13** **The Energy Deposited by the Sun** In the northern hemisphere, the angle of the Sun above the southern horizon determines how much heat and light strike each square meter of ground. **(a)** During the summer at middle latitudes, a shaft of sunlight illuminates a nearly circular patch of ground at noon. **(b)** During the winter, the same shaft of sunlight at noon strikes the ground at a steeper angle, spreading the same amount of sunlight over a larger, oval shape. Because the sunlight's energy is diluted over a larger area, the ground receives less heat during the winter than during the summer.

other by 23½°, as shown in Figure 1-14. These two circles intersect at only two points, which are exactly opposite each other on the celestial sphere. Both points are called **equinoxes** (from the Latin words meaning "equal night"), because when the Sun appears at either point, it is directly over the Earth's equator, and there are 12 hours of daytime and 12 hours of nighttime everywhere on Earth on that day.

The **vernal equinox** occurs around March 21, when the Sun crosses the celestial equator heading northward; the **autumnal equinox** occurs six months later, around September 22, with the Sun heading southward. The vernal equinox is the "prime meridian" of the celestial sphere, as discussed earlier. The midpoints between the vernal and autumnal equinoxes along the ecliptic are significant as well. The point on the ecliptic farthest north of the celestial equator is the **summer solstice**. It signals the day of the year in the northern hemisphere with the largest number of daylight hours (and the fewest daylight hours in the southern hemisphere). The Sun reaches this point around June 21 each year. Six months later, around December 21, the Sun is farthest south of the celestial equator at the **winter solstice**, the day of the year in the northern hemisphere with the fewest daylight hours (and the most daylight hours in the southern hemisphere).

The Sun is highest in the northern sky on the summer solstice. This marks the beginning of summer in the northern hemisphere. As the Sun moves southward, the amount of daylight decreases. The autumnal equinox marks a midpoint in the amount of heat deposited by the Sun onto the northern hemisphere and is the beginning of autumn. When the Sun reaches the winter solstice, it is

lowest in the northern sky and is above the horizon for the shortest time of the year. This is the beginning of winter. Returning northward, the Sun crosses the celestial equator once again on the vernal equinox, the beginning of spring.

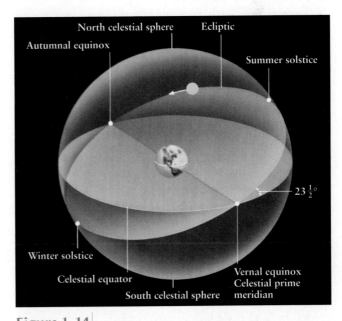

**Figure 1-14** **The Seasons Are Coupled to Equinoxes and Solstices** The ecliptic is inclined to the celestial equator by 23½° because of the tilt of the Earth's axis of rotation. The ecliptic and the celestial equator intersect at two points called the equinoxes. The northernmost point on the ecliptic is called the summer solstice; the southernmost point is called the winter solstice.

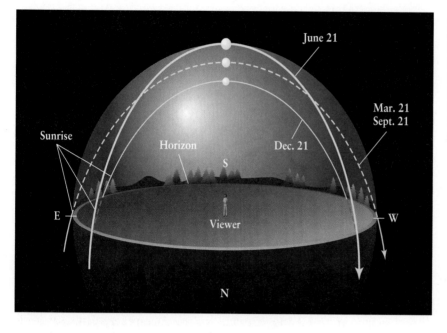

**Figure 1-15** **The Sun's Daily Path** On the first day of spring and the first day of fall, as seen from middle latitudes in the northern hemisphere, the Sun rises precisely in the east and sets precisely in the west. During summer in the northern hemisphere, the Sun rises in the northeast and sets in the northwest. In winter, the Sun rises in the southeast and sets in the southwest.

Figure 1-15 shows the seasonal changes in the Sun's daily path across the sky. On the first day of spring or autumn (when the Sun is at one of the equinoxes), the Sun rises directly in the east and sets directly in the west. Daytime and nighttime are of equal duration everywhere on Earth on those days only.

During the northern hemisphere's summer months, when the northern hemisphere is tilted toward the Sun, the Sun rises in the northeast and sets in the northwest. The Sun provides more than 12 hours of daylight in the northern hemisphere and passes high in the sky at noontime. At the summer solstice, the Sun is as far north as it gets, giving the greatest number of daylight hours to the northern hemisphere.

During the northern hemisphere's winter months, when the northern hemisphere is tilted away from the Sun, the Sun rises in the southeast. Daylight lasts for fewer than 12 hours, as the Sun skims low over the southern horizon and sets in the southwest. Night is longest in the northern hemisphere when the Sun is at the winter solstice.

People at different latitudes see the noontime Sun at different angles in the sky each day. The farther from the equator you are, the lower the Sun is in the sky at noon. The farther north you go, less of the Sun's heat and light energy are deposited (see Figure 1-13) and, therefore, the colder the land is. At latitudes above 66½° north latitude, the Sun does not rise at all during certain times of the year. Six months later these same regions of the Earth have continuous sunlight for weeks or months, hence the name "Land of the Midnight Sun." The same is true in reverse for lands below 66½° south latitude.

The Sun takes one year to complete a trip around the ecliptic. Since there are about 365¼ days in a year and 360° in a circle, the Sun appears to move along the ecliptic at a rate of slightly less than 1° per day. The constellations through which the Sun moves throughout the year are called **zodiac** constellations. We cannot see the stars of these constellations when the Sun is among them, of course, but we can plot the Sun's path on the celestial sphere to determine through which constellations it moves. One traditionally learns that there are 12 zodiac constellations, which are used by astrologers. These 12 are based on constellation boundaries used in antiquity. Those boundaries have since been redefined by astronomers, and the Sun moves through 13 modern constellations throughout the year. (The thirteenth zodiac constellation is Ophiuchus, the Serpent Holder. The Sun passes through Ophiuchus from December 1 to December 19 of each year.) Table 1-1 lists all the zodiac constellations and the dates the Sun passes through them.

## 1-7 Precession is a slow, circular motion of the Earth's axis of rotation

As noted above, the position of the Earth's axis of rotation changes slightly with respect to the celestial sphere (that is, it "points" in a slightly different direction) each year. This change in orientation is caused by gravitational forces from the Moon and Sun. Recall that gravity is the universal force of attraction between all matter. The Earth's rotation creates an "equatorial bulge"—our plan-

### Table 1-1
### The 13 Constellations of the Zodiac

| Constellation | Dates of Sun's Passage Through |
|---|---|
| Pisces | March 13–April 20 |
| Aries | April 20–May 13 |
| Taurus | May 13–June 21 |
| Gemini | June 21–July 20 |
| Cancer | July 20–August 11 |
| Leo | August 11–September 18 |
| Virgo | September 18–November 1 |
| Libra | November 1–November 22 |
| Scorpius | November 22–December 1 |
| Ophiuchus | December 1–December 19 |
| Sagittarius | December 19–January 19 |
| Capricorn | January 19–February 18 |
| Aquarius | February 18–March 13 |

et's diameter is about 43 km (27 mi) greater at the equator than from pole to pole.

Because of the Earth's tilted axis of rotation, the Sun and Moon are usually not located directly over the Earth's equator. As a result, their gravitational attraction on the Earth tries to force the equatorial bulge to be as close to them as possible. However, the Earth does not respond to these forces from the Sun and Moon by straightening up perpendicular to them on its axis. Instead, it changes the direction in which its axis of rotation points on the celes-

tial sphere—a motion called **precession**. This is exactly the same behavior that is exhibited by a spinning top (Figure 1-16). If the top were not spinning, gravity would pull it over on its side. But when it is spinning, the combined actions of gravity and rotation cause the top's axis of rotation to precess in a circular path. As with the toy top, the combined actions of gravity and rotation cause the Earth's axis to trace out a circle in the sky while remaining tilted about 23½° away from the perpendicular.

The Earth's rate of precession is slow compared to human time scales. It takes about 26,000 years for the north celestial pole to trace out a complete circle around the sky, as shown in Figure 1-17. (Of course, the south celestial pole executes a similar circle in the southern sky.) At the present time, the Earth's axis of rotation points within 1° of the star Polaris. In 3000 B.C., it was pointing near the star Thuban in the constellation of Draco (the Dragon). In A.D. 14,000, the pole star will be Vega in Lyra (the Harp).

As the Earth's axis of rotation precesses, its equatorial plane also moves. Because the Earth's equatorial plane defines the location of the celestial equator in the sky, the celestial equator also precesses. Because the intersections of the celestial equator and the ecliptic define the equinoxes, these key locations in the sky also shift slowly from year to year. This entire phenomenon is often called the **precession of the equinoxes**. This change was discovered by the great Greek astronomer Hipparchus in the second century B.C. Today, the vernal equinox is located in the constellation Pisces (the Fishes). Two thousand years ago, it was located in Aries (the Ram). Around the year A.D. 2600, the vernal equinox will move into Aquarius (the Water Bearer).

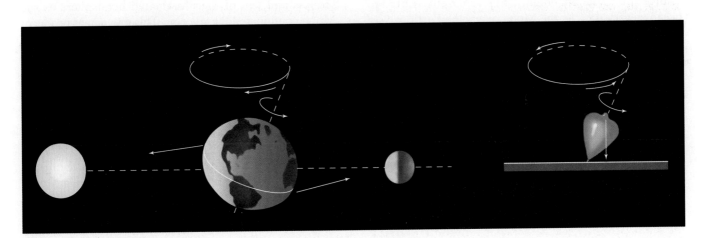

**Figure 1-16**  **Precession** The gravitational pulls of the Moon and the Sun on the Earth's equatorial bulge cause the Earth to precess. As the Earth precesses, its axis of rotation slowly traces out a circle in the sky. The situation is analogous to that of a spinning top. As the top spins, the Earth's gravitational pull causes the top's axis of rotation to move in a circle.

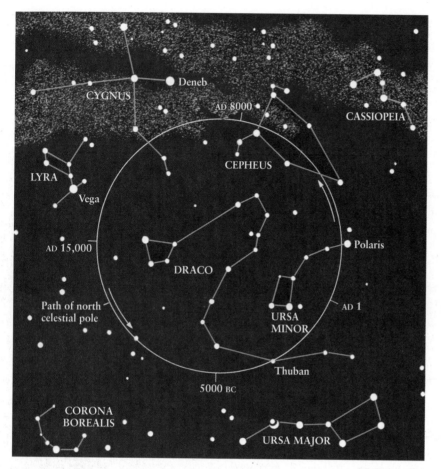

**Figure 1-17** **The Path of the North Celestial Pole** As the Earth precesses, the north celestial pole slowly traces out a circle among the northern constellations. At the present time, the north celestial pole is near the moderately bright star Polaris, which serves as the pole star.

## 1-8 The phases of the Moon originally inspired the concept of the month

The Moon's contribution to the Earth's precession causes changes that take millennia. Other lunar effects are noticeable every day. As the Moon orbits the Earth, we see different **lunar phases.** As with the Earth and other spherical bodies, the Sun illuminates half of the Moon at any time. The Moon's phase depends on how much of its sunlit hemisphere is exposed to our Earth-based view. When the Moon is closest to the Sun in the sky, its dark hemisphere faces the Earth. This phase, during which the Moon is at most a tiny crescent, is called *new Moon* (Figure 1-18).

During the seven days following the new phase, more of the Moon's illuminated hemisphere becomes exposed to our view, resulting in a phase called the *waxing crescent Moon.* At *first quarter Moon*, we see half of the illuminated hemisphere and half of the dark hemisphere. "Quarter Moon" refers to how far in its cycle the Moon has gone, rather than what fraction of the Moon appears lit by sunlight.

During the next week still more of the illuminated hemisphere can be seen from Earth, giving us the phase

called the *waxing gibbous Moon.* When the Moon arrives on the opposite side of the Earth from the Sun, we see almost all of the fully illuminated hemisphere, which is called the *full Moon.* Over the following two weeks, we see less and less of the illuminated hemisphere as the Moon continues along its orbit. This movement produces the phases called *waning gibbous Moon, third quarter Moon,* and *waning crescent Moon.*

The Moon completes a full cycle of phases in 29½ days. Thus, the side of the Moon facing away from Earth, its *far side,* is not always dark, and the Moon's far side is not necessarily its "dark side."

Figure 1-18 shows the Moon at various positions in its orbit. Remember that the bright side of the Moon is on the right (west) side of the waxing Moon, while the bright side is on the left (east) side of the waning Moon. When looking at the Moon through a telescope, the best place to see details is where the shadows are longest. This occurs at the boundary between the bright and dark regions, called the **terminator.**

Figure 1-18 also shows local time around the globe, from noon, when the Sun is highest in the sky, to midnight, when it is on the opposite side of the Earth. These time markings can be used to correlate the phase and

# Phases of the Moon

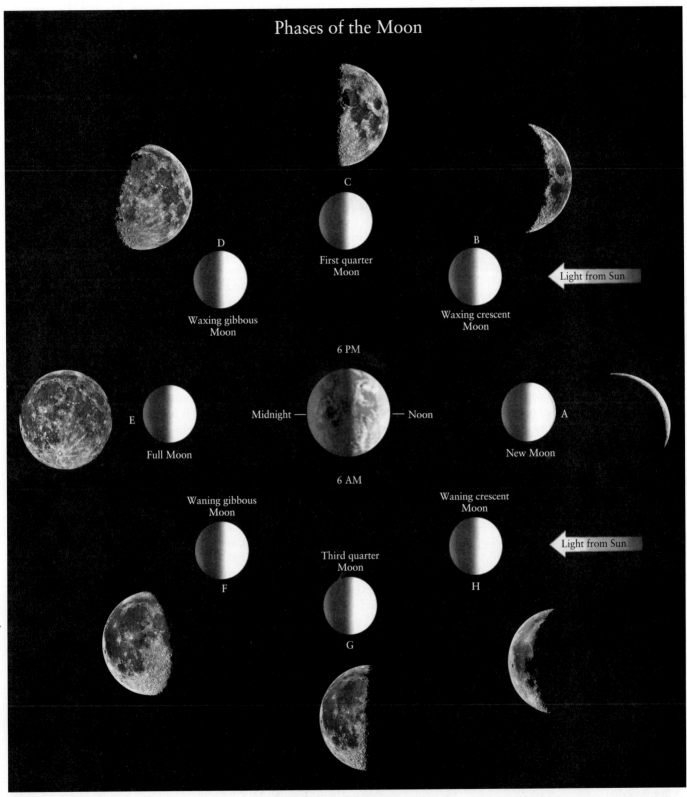

**Figure 1-18**  **The Phases of the Moon**  The diagram shows the Moon at eight locations on its orbit as viewed from far above the Earth's north pole. The corresponding photographs show the resulting lunar phases as seen from Earth. Light from the Sun illuminates one-half of the Moon at all times, while the other half is dark. It takes about 29½ days for the Moon to go through all its phases.

**Figure 1-19** **The Moon During the Day** The Moon is visible at some time during daylight hours virtually every day. The time of day or night it is up in our sky depends on its phase.

with celestial events. Thus, astronomers have traditionally been responsible for telling time. Indeed, of the four ways in which time cycles are set, three are astronomical in origin: Time is determined by the positions of the Moon, Sun, or stars or, in our own age, by technological means, such as atomic clocks.

The approximately four weeks that the Moon takes to complete one cycle of its phases inspired our ancestors to invent the concept of a month. Astronomers find it useful to define two types of months, depending on whether the Moon's motion is measured relative to the stars or to the Sun. Neither type corresponds exactly to the months of our usual calendar, which have different lengths.

The **sidereal month** is the time it takes the Moon to complete one full orbit of 360° around the Earth (Figure 1-20). This is also the time it takes the Moon to start at one place on the celestial sphere and return to exactly the same place again. Indeed, sidereal motion of any astronomical body is motion measured with respect to the distant stars. This true orbital period for the Moon is approximately 27.3 days. The **synodic month**, or **lunar month,** is the time it takes the Moon to complete one 29½-day cycle of phases (that is, from new Moon to new Moon or from full Moon to full Moon) and thus is measured with respect to the Sun rather than the distant stars.

The synodic month is longer than the sidereal month, because the Earth is orbiting the Sun while the Moon goes through its phases. As shown in Figure 1-20, the Moon must travel *more* than 360° along its orbit to complete a cycle of phases (for example, from one new Moon to the next), which takes about 2.2 days longer than the sidereal month.

Both the sidereal month and synodic month vary somewhat, because the gravitational pull of the Sun on the Moon affects the Moon's speed as it orbits the Earth. The sidereal month can vary by as much as 7 hours, while the synodic month can vary by as much as 12 hours.

position of the Moon with the time of day. For example, at first quarter, the Moon is 90° east of the Sun in the sky; hence, moonrise occurs approximately at noon. At full Moon, the Moon is opposite the Sun in the sky; thus, moonrise occurs at sunset. Using this information, you can see that the Moon is visible during the daytime (Figure 1-19) for a part of most days of the year.

Since the dawn of civilization, people have sought accurate timekeeping systems. Ancient Egyptians wanted to know when the Nile would flood, and farmers everywhere needed to know when to plant crops. Migratory tribes wanted to know when the weather would change. Religious leaders scheduled observances in accordance

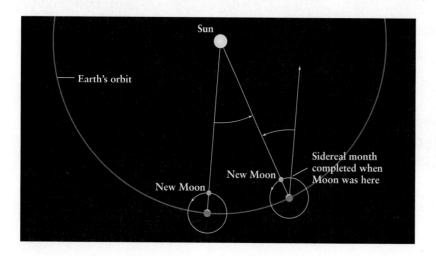

**Figure 1-20** **The Sidereal and Synodic Months** The sidereal month is the time it takes the Moon to complete one revolution with respect to the background stars. However, because the Earth is constantly moving in its orbit about the Sun, the Moon must travel through more than 360° to get from one new Moon to the next. The synodic month is the time between consecutive new moons or consecutive full moons. Thus, the synodic month is slightly longer than the sidereal month.

The terms *synodic* and *sidereal* are also used in discussing the motion of the other bodies in the solar system. The synodic period of a planet is the time between consecutive straight alignments between the Sun, Earth, and that planet (during which time the planet also goes through a cycle of phases, as seen from the Earth). On the other hand, any orbit measured with respect to the distant stars is called "sidereal," including orbits of the planets around the Sun, as well as orbits of moons around their planets. The Earth's sidereal year is 365.2564 days. The difference between the sidereal year and the tropical year is due primarily to the Earth's precession (see Section 1-7).

# ECLIPSES

## 1-9 Eclipses occur only when the Moon crosses the ecliptic during the new or full phase

Eclipses are among the most spectacular natural phenomena. During a lunar eclipse the brilliant full Moon often darkens to a deep red, while during a solar eclipse, broad daylight is transformed into an eerie twilight, as the Sun seems to be blotted from the sky. A **lunar eclipse** occurs when the Moon passes through the Earth's shadow. This can happen only when the Sun, Earth, and Moon are in a straight line at full Moon. A **solar eclipse** occurs when the Moon's shadow moves across the Earth's surface. As seen from Earth, the Moon moves in front of the Sun. This can happen only when the Sun, Moon, and Earth are aligned at new Moon.

At first glance, it would seem that eclipses should happen at every new and full Moon, but, in fact, they occur much less often. Eclipses occur infrequently because the Moon's orbit is tilted 5° out of the ecliptic, as shown in Figure 1-21. Because of this tilt, new Moon and full Moon usually occur when the Moon is either above or below the plane of the Earth's orbit. In such positions, a perfect alignment between the Sun, Moon, and Earth is not possible and an eclipse cannot occur.

When the Moon crosses the plane of the ecliptic during its new or full phase, an eclipse takes place. The Moon crosses the ecliptic at what is called the **line of**

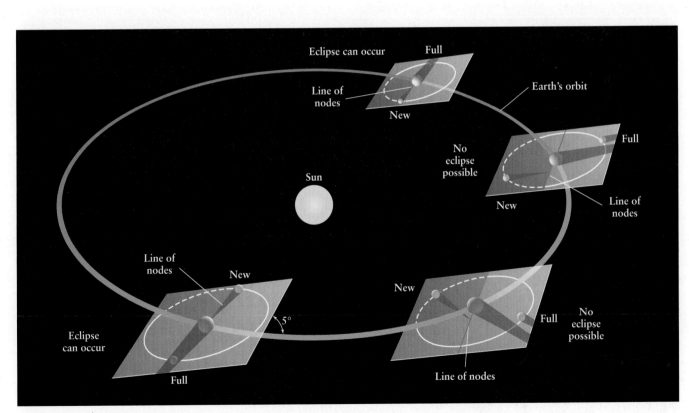

**Figure 1-21**  **Conditions for Eclipses** A solar eclipse occurs only if the Moon is very nearly on the ecliptic at new Moon. A lunar eclipse occurs only if the Moon is very nearly on the ecliptic at full Moon. When new Moon or full Moon phases occur away from the ecliptic, no eclipse is seen, because the Moon and the Earth do not pass through each other's shadows.

nodes (Figure 1-22). By calculating the number of times a new Moon takes place on the line of nodes, we find that at least two and no more than five solar eclipses occur each year. Lunar eclipses occur just about as frequently as solar eclipses, but the maximum number of eclipses (solar plus lunar) possible in a year is seven.

## 1-10 There are three types of lunar eclipse

Depending on how the Moon travels through the Earth's shadow, three kinds of lunar eclipse may occur. The Earth's shadow has two distinct parts, as shown in Figure 1-23. The **umbra** is the part of the shadow where all direct sunlight is blocked by the Earth; the **penumbra** of the shadow is where the Earth blocks only some of the sunlight. A **penumbral eclipse,** when the Moon passes through only the Earth's penumbra, is easy to miss. The Moon still looks full, just a little dimmer than usual or even reddish in color.

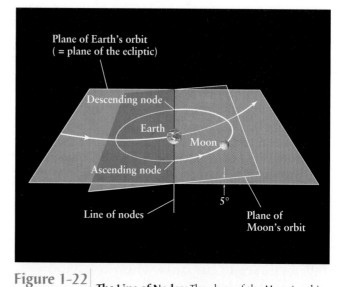

**Figure 1-22**    **The Line of Nodes** The plane of the Moon's orbit is tilted slightly with respect to the plane of the Earth's orbit. These two planes intersect along a line called the line of nodes.

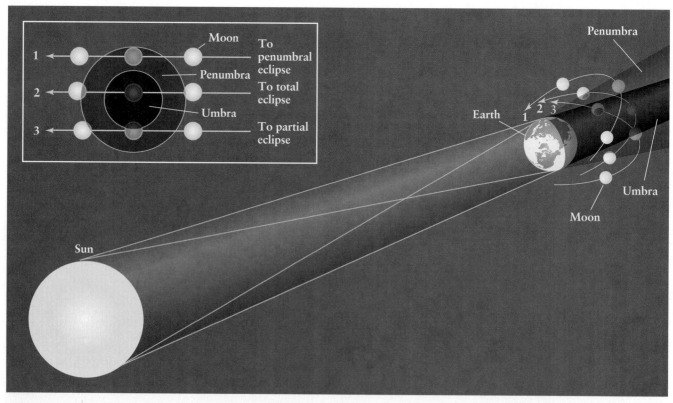

**Figure 1-23**    **Three Types of Lunar Eclipses** People on the nighttime side of the Earth see a lunar eclipse when the Moon moves through the Earth's shadow. The umbra is the darkest part of the shadow. In the penumbra, only part of the Sun is covered by the Earth. The inset shows the various lunar eclipses that occur, depending on the Moon's path through the Earth's shadow.

### Table 1-2
### Lunar Eclipses, 1999–2003

| Date | Visible from | Type | Duration of totality (h:min) |
|---|---|---|---|
| 1999 July 28 | Americas, Asia, Australia | Partial | |
| 2000 January 21 | America, Africa, Europe | Total | 1:16 |
| 2000 July 16 | Western Americas, Eastern Africa, Asia, Australia | Total | 1:46 |
| 2001 January 9 | Eastern America, Europe, Africa, Asia, Australia | Total | 1:03 |
| 2001 July 5 | Eastern Africa, Asia, Australia | Partial | |
| 2001 December 30 | Americas, Asia, Australia | Penumbral | |
| 2002 May 26 | Western Americas, Asia, Australia | Penumbral | |
| 2002 June 24 | South America, Europe, Africa, Western Asia | Penumbral | |
| 2002 November 20 | Americas, Europe, Africa, Western Asia | Penumbral | |
| 2003 May 16 | Americas, Europe, Africa | Total | 0:53 |
| 2003 November 9 | Americas, Europe, Africa, Asia | Total | 0:26 |

When only part of the lunar surface passes through the umbra, a bite seems to be taken out of the Moon, and we see a **partial eclipse**. When the Moon travels completely into the umbra, as sketched in Figure 1-23, we see a **total eclipse** of the Moon. Totality is the time when the entire Moon is in the Earth's umbra. Eclipses with the maximum duration of totality, lasting for up to 1 hour 47 minutes, occur when the Moon is closest to the Earth and travels directly through the center of the umbra. Table 1-2 lists all the total and partial lunar eclipses from late 1999 through 2003.

Even during a total eclipse, the Moon does not completely disappear. A small amount of sunlight passing through the Earth's atmosphere is bent into the Earth's umbra. The light deflected into the umbra is primarily red and orange, and thus the darkened Moon glows faintly in rust-colored hues during totality, as shown in Figure 1-24. At sunrise and sunset the Sun appears red or orange for the same reason, specifically, because at those times red and orange light are deflected from the Sun toward you. Other colors, especially violet, blue, and green, are actually deflected by the air so much more than red, orange, and yellow that the former colors from the Sun do not enter the Earth's umbra or reach us at sunrise and sunset.

Lunar eclipses are perfectly safe to watch with the naked eye. However, solar eclipses are *never* safe to view without suitable eye protection. *Viewing the Sun directly at any time without an approved filter causes permanent eye damage.*

## 1-11 There are also three types of solar eclipse

Because of their different distances from Earth, the Sun and the Moon have nearly the same angular diameter as seen from Earth—about ½°. When the Moon just "fits" over the Sun, the result is a *total solar eclipse*. You must

R I V U X G

**Figure 1-24** **A Total Eclipse of the Moon** Notice the distinctly reddish color of the Moon in this photograph, taken by an amateur astronomer during the lunar eclipse of September 6, 1979.

be at a location within the Moon's umbra in order to see a total solar eclipse.

During those few precious moments, hot gases (the **solar corona**) surrounding the Sun can be observed and photographed (Figure 1-25). Astronomers can then learn more about the Sun's temperature, chemistry, and atmospheric activity. You can see in Figure 1-26 that only the tip of the Moon's umbra ever reaches the Earth's surface. As the Earth turns and the Moon orbits, the tip traces an **eclipse path** across the Earth's surface. Only people within this path are treated to the spectacle of a total solar eclipse. Figure 1-26 shows the dark spot produced by the Moon's umbra on the Earth's surface during a total solar eclipse.

The Earth's rotation and the orbital motion of the Moon cause the umbra to race along the eclipse path at speeds in excess of 1700 km/h (1050 mph). Totality never lasts for more than 7½ minutes at any one location on the eclipse path, and it usually lasts for only a few moments.

The Moon's umbra is also surrounded by a penumbra. During a solar eclipse, the Moon's penumbra extends

**Figure 1-25**   **A Total Eclipse of the Sun**  During a total solar eclipse, the Moon completely covers the Sun's disk, and the solar corona can be photographed. This halo of hot gases extends for millions of kilometers into space. This photograph was taken by an amateur astronomer during the solar eclipse of July 11, 1991.

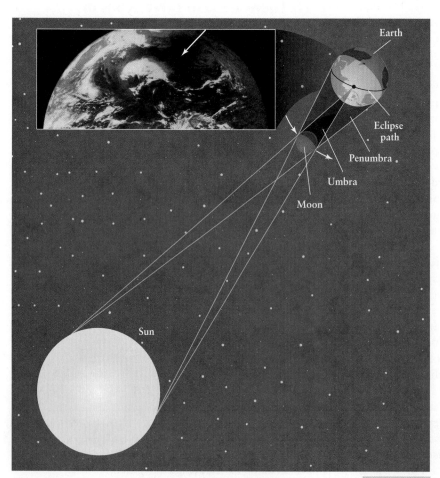

**Figure 1-26**   **The Geometry of a Total Solar Eclipse**  During a total solar eclipse, the tip of the Moon's umbra traces an eclipse path across the Earth's surface. People inside the eclipse path see a total solar eclipse, whereas people inside the penumbra see only a partial eclipse. The photograph in this figure shows the Moon's shadow on the Earth. It was taken from an Earth-orbiting satellite during the total eclipse of March 7, 1970. The Moon's umbra appears as a dark spot on the eastern coast of the United States.

**Table 1-3**
**Solar Eclipses, 1999–2003**

| Date | Type | Area | Notes |
|---|---|---|---|
| 1999 August 11 | Total | Atlantic, Europe, Middle East, India | Maximum length 2 min 23 s |
| 2000 February 5 | Partial | Antarctic, southern Indian Ocean | 58% eclipsed |
| 2000 July 1 | Partial | Southern Atlantic | 48% eclipsed |
| 2000 July 31 | Partial | Northern Europe, Greenland, Canada, western United States | 60% eclipsed |
| 2000 December 25 | Partial | Canada, United States, Mexico | 73% eclipsed |
| 2001 June 21 | Total | Atlantic, South Africa, Madagascar | Maximum length 4 min 55 s |
| 2001 December 14 | Annular | Pacific, Central America | Maximum length 3 min 52 s |
| 2002 June 10 | Annular | Pacific | Maximum length 23 s |
| 2002 December 4 | Total | South Africa, Indian Ocean, Australia | Maximum length 2 min 3 s |
| 2003 May 31 | Annular | Atlantic, Greenland | Maximum length 3 min 36 s |
| 2003 November 23 | Total | Antarctica | Maximum length 1 min 57 s |

over a large portion of the Earth's surface. When only the penumbra sweeps across the Earth's surface, the Sun is only partly covered by the Moon. This circumstance results in a *partial eclipse of the Sun*. Similarly, people in the penumbra of a total eclipse see a partial eclipse.

The Moon's orbit around the Earth is not quite a perfect circle. The distance between the Earth and the Moon, which averages 384,400 km (238,900 mi), varies by a few percent as the Moon goes around the Earth. The width of the eclipse path depends primarily on the Earth-Moon distance during an eclipse. The eclipse path is widest—up to 270 km (170 mi)—when the Moon happens to be at the point in its orbit nearest the Earth. Usually, however, the path is much narrower.

If a solar eclipse occurs when the Moon is farthest from the Earth, then the Moon's umbra falls short of the Earth and no one sees a total eclipse. From the Earth's surface, the Moon then appears too small to cover the Sun completely, and a thin ring or "annulus" of light is seen around the edge of the Moon at mid-eclipse. This type of eclipse is called an **annular eclipse** (Figure 1-27). The length of the Moon's umbra is nearly 5000 km (3100 mi) shorter than the average distance between the Moon and the Earth's surface. Thus, the Moon's shadow often fails to reach the Earth, making annular eclipses more common than total eclipses. Table 1-3 lists all the total, partial, and annular solar eclipses from late 1999 through 2003.

A total solar eclipse is a dramatic event. The sky begins to darken, the air temperature falls, and the winds

**Figure 1-27**   RIVUXG   **An Annular Eclipse of the Sun**  This composite of five exposures taken at sunrise in Costa Rica shows the progress of an annular eclipse of the Sun that occurred on December 24, 1974. Note that at mid-eclipse the edge of the Sun is visible around the Moon.

increase as the Moon's umbra races toward you. All nature responds: Birds go to roost, flowers close their petals, and crickets begin to chirp as if evening had arrived. As totality approaches, the landscape is bathed in shimmering bands of light and dark as the last few rays of sunlight peek out from behind the edge of the Moon. Finally, the corona blazes forth in a star-studded daytime sky. It is an awesome sight but should only be viewed through a suitable protective filter.

## WHAT DID YOU THINK?

1️⃣ *Is the North Star—Polaris—the brightest star in the night sky?* Polaris is a star of medium brightness compared with other stars visible to the naked eye.

2️⃣ *Do astronomers regard constellations as simply the familiar patterns of stars in the sky first identified by ancient stargazers?* Astronomers sometimes use the common definition of a constellation as a pattern of stars. Formally, however, a constellation is an entire region of the celestial sphere and all the stars and other objects in it.

Viewed from Earth, the entire sky is covered by 88 different-sized constellations. If there is any room for confusion, astronomers refer to the patterns as asterisms.

3️⃣ *What causes the seasons?* The tilt of the Earth's rotation axis with respect to the ecliptic causes the seasons. They are not caused by the changing distance from the Earth to the Sun that results from the shape of Earth's orbit.

4️⃣ *How many zodiac constellations are there?* There are 13 zodiac constellations, the lesser-known one being Ophiuchus.

5️⃣ *Does the Moon have a dark side that we never see from Earth?* Half of the Moon is always dark. Whenever we see less than a full moon, we are seeing part of the Moon's dark side. So, the dark side of the Moon is not the same as the far side of the Moon, which we never see from Earth.

6️⃣ *Is the Moon ever visible during the daytime?* The Moon is visible at some time during daylight hours almost every day of the year. Different phases are visible during different times of the day.

## Key Words

| | | | |
|---|---|---|---|
| angle, 19 | declination, 18 | penumbra, 32 | solar eclipse, 31 |
| angular diameter (angular size), 19 | degree (°), 19 | penumbral eclipse, 32 | summer solstice, 25 |
| annular eclipse, 35 | diurnal motion, 20 | precession, 27 | synodic month (lunar month), 30 |
| arc angle, 19 | eclipse path, 34 | precession of the equinoxes, 27 | terminator, 28 |
| autumnal equinox, 25 | ecliptic, 23 | revolution, 20 | time zone, 21 |
| celestial equator, 18 | equinox, 25 | right ascension, 18 | total eclipse, 33 |
| celestial pole, 18 | line of nodes, 31 | rotation, 20 | umbra, 32 |
| celestial sphere, 18 | lunar eclipse, 31 | sidereal month, 30 | vernal equinox, 25 |
| circumpolar stars, 20 | lunar phases, 28 | solar corona, 34 | winter solstice, 25 |
| constellation, 18 | partial eclipse, 33 | solar day, 21 | zodiac, 26 |

## Key Ideas

### Patterns of Stars

✱   The surface of the celestial sphere is divided into 88 unequal regions called constellations.

### Earthly Cycles

✱   The celestial sphere appears to rotate around the Earth once in each day-night cycle. In fact, of course, it is the Earth's rotation that causes that apparent motion.

✳ The poles and equator of the celestial sphere are determined by extending the axis of rotation and the equatorial plane of the Earth out onto the celestial sphere.

✳ Earth's axis of rotation is tilted at an angle of 23½° from the perpendicular to the plane of the Earth's orbit. This tilt causes the seasons.

✳ Equinoxes and solstices are significant points along the Earth's orbit that are determined by the relationship between the Sun's path on the celestial sphere (the ecliptic) and the celestial equator.

✳ The Earth's axis of rotation slowly changes direction relative to the stars over thousands of years, a phenomenon called precession. Precession is caused by the gravitational pull of the Sun and Moon on the Earth's equatorial bulge.

✳ The length of the day is based upon the average motion of the Sun along the celestial equator, which produces the 24-hour day upon which our clocks are based.

✳ The phases of the Moon are caused by the relative positions of the Earth, Moon, and Sun. The Moon completes one cycle of phases in a synodic month, which averages 29½ days.

✳ The Moon completes one orbit around the Earth with respect to the stars in a sidereal month, which averages 27.3 days.

## Eclipses

✳ The shadow of an object has two parts: the umbra, where direct light from the source is completely blocked; and the penumbra, where the light source is only partially obscured.

✳ A lunar eclipse occurs when the Moon moves through the Earth's shadow. During a lunar eclipse, the Sun, Earth, and Moon are in alignment, and the Moon is in the plane of the ecliptic.

✳ A solar eclipse occurs when the Earth passes through the Moon's shadow. During a solar eclipse, the Sun, Earth, and Moon are in alignment, and the Moon is in the plane of the ecliptic.

✳ Depending on the relative positions of the Sun, Moon, and Earth, lunar eclipses may be penumbral, partial, or total, and solar eclipses may be annular, partial, or total.

## Review Questions

1  How are constellations useful to astronomers?

2  What is the celestial sphere, and why is this ancient concept still useful today?

3  What is the celestial equator, and how is it related to the Earth's equator? How are the north and south celestial poles related to the Earth's axis of rotation?

4  What is the ecliptic, and why is it tilted with respect to the celestial equator?

5  By about how many degrees does the Sun move along the ecliptic each day?

6  Why does the tilt of the Earth's axis relative to its orbit cause the seasons as the Earth revolves around the Sun. Draw a diagram to illustrate your answer.

7  What are the vernal and autumnal equinoxes? What are the summer and winter solstices? How are these four points related to the ecliptic and the celestial equator?

8  What is precession, and how does it affect our view of the heavens?

9  How does the daily path of the Sun across the sky change with the seasons?

10  Why is it warmer in the summer than in the winter?

11  Why is it convenient to divide the Earth into time zones?

12  Why does the Moon exhibit phases?

13  What is the difference between a sidereal month and a synodic month? Which is longer? Why?

14  What is the line of nodes, and how is it related to solar and lunar eclipses?

15  What is the difference between the umbra and the penumbra of a shadow?

16  What is a penumbral eclipse of the Moon? Why do you suppose that it is easy to overlook such an eclipse?

17  Which type of eclipse—lunar or solar—do you think most people have seen? Why?

18  How is an annular eclipse of the Sun different from a total eclipse of the Sun? What causes this difference?

19  When is the next leap year?

20  At which phase(s) of the Moon does a solar eclipse occur? A lunar eclipse?

21  Is it safe to watch a solar eclipse without eye protection? A lunar eclipse?

22  During what phase is the Moon least visible in the daytime?

## Advanced Questions

The answers to all computational problems, which are preceded by an asterisk (*), appear at the end of the book.

23  During what phase(s) does the Moon rise after sunrise and before sunset? After sunset and before sunrise? At sunset? At sunrise?

24  Why can't a person in Australia use the Big Dipper to find north?

25  Are there any stars in the sky that are not members of a constellation?

26  At what location on Earth is the north celestial pole seen on the horizon?

27  Where do you have to be on the Earth in order to see the Sun at its zenith? If you stay at one such location for a full year, on how many days will the Sun pass through the zenith?

28  Where do you have to be on Earth in order to see the south celestial pole directly overhead? What is the maximum possible elevation of the Sun above the horizon at that location? On what date is this maximum elevation achieved?

29  At what point on the horizon does the Sun rise at the time of the vernal equinox?

30  Consult a star map of the southern hemisphere and determine which, if any, of the bright southern stars could someday become south celestial pole stars.

31  Are there stars in the sky that never set where you live? Are there stars that never rise where you live? Does your answer depend on your location on Earth? Why or why not?

32  Using a diagram, demonstrate that your latitude on Earth is equal to the altitude of the north celestial pole above your northern horizon.

33  Using a star map, determine which bright stars, if any, could someday mark the location of the vernal equinox. Give the approximate years when this should happen.

34  What is the phase of the Moon if it (a) rises at 3 A.M.? (b) sets at 9 P.M.? At what time does (c) the full Moon set? (d) the first quarter Moon rise?

35  What is the phase of the Moon if, on the first day of spring, the Moon is located at (a) the vernal equinox, (b) the summer solstice, (c) the autumnal equinox, (d) the winter solstice?

*36  How many more sidereal months than synodic months are there in a year? Why?

37  How do we know that the phases of the Moon are not due to the Moon moving in the Earth's shadow?

38  Why do the paths of total solar eclipses fall more frequently on oceans than on land?

39  Can one ever observe an annular eclipse of the Moon? Why or why not?

40  During a lunar eclipse, does the Moon enter the Earth's shadow from the east or the west? Explain your answer.

41  Do we see all of the Moon's surface from the Earth? *Hint:* Carefully examine the photographs in Figure 1-18.

42  What is the latitude of the place shown in Figure 1-12?

43  Why is a small crescent of light often observed on the Moon when it is exactly in the new phase?

44  Make a drawing of an annular solar eclipse as seen from space. Be sure to make clear how it differs from a total solar eclipse.

*45  Assuming that the Sun subtends an angle of ½° in our sky and is at a distance of $1.496 \times 10^{11}$ m, what is the Sun's diameter? Divide this by 2 to find the Sun's radius and explain why this result is slightly different from the value given in Appendix Table A-6.

## Discussion Questions

46  Examine a list of the 88 constellations. Are there any constellations whose names obviously date from modern times? Where are these constellations located? Why do you suppose they do not have archaic names?

47  In his novel *King Solomon's Mines*, H. Rider Haggard described a total solar eclipse that was seen in both South Africa and the British Isles. Is such an eclipse possible? Why or why not?

48  Describe how a lunar eclipse would look if the Earth had no atmosphere.

49  Examine a listing of total solar eclipses over the next several decades. What are the chances that you might be able to travel to one of the eclipse paths? Do you think you might go through your entire life without ever seeing a total eclipse of the Sun?

# What If ...

50   The Moon moved about the Earth in an orbit perpendicular to the plane of the Earth's orbit? What would the cycle of lunar phases be? Would solar and lunar eclipses be possible under these circumstances?

51   The Earth's axis of rotation were tilted at a different angle? What would the seasons be like where you are now if the axis of rotation is (a) 0° and (b) 45° to its orbital plane. What

would be different from today about the seasons and the day-night cycle if you lived at one of the Earth's poles?

52   You watched the Earth from the Moon? What would you see for Earth's (a) daily motion, (b) motion along the celestial sphere, and (c) cycle of phases.

# Observing Projects

53   On a clear, cloud-free night, use the star charts within the covers of this book to see how many constellations of the zodiac you can identify. Which ones are easy to find? Which are difficult?

54   Using Starry Night with all pattern lines off, see if you can identify the zodiac constellations that are visible in the sky this evening (set the time to sometime after sunset). Use the star charts within the covers of this book for guidance. If you have difficulty finding zodiac constellations, click on the ecliptic to highlight the part of the sky in which the zodiac is located. Watch out for planets, which are also found in the zodiac—you may have to ignore one of these bright "stars" (planets) in picking out the constellation patterns. Clicking on an object with the arrow cursor will tell you whether it is a star or a planet. Switching off daylight and hiding the horizon if necessary, determine which zodiac constellation the Sun is passing through today. You can check all your results by turning the constellation lines on.

55   Examine the star charts that are published monthly in such popular astronomy magazines as Sky & Telescope and Astronomy. How do they differ from the star charts within the covers of this book? On a clear, cloud-free night, use one of these star charts to locate the celestial equator and the ecliptic on the night sky. Note the inclination of the Milky Way to the ecliptic and celestial equator. What do your observations tell you about the orientation of the Earth and its orbit around the Sun relative to the rest of the Galaxy?

56   Observe the Moon on each clear night over the course of a month. Note the Moon's location among the constellations and

record that location on a star chart that also shows the ecliptic. After a few weeks, your observations will begin to trace the Moon's orbit. Identify the orientation of the line of nodes by marking the points where the Moon's orbit and the ecliptic intersect. On what dates is the Sun near the nodes marked on your star chart? Compare these dates with the dates of the next solar and lunar eclipses.

57   Use Starry Night with the daylight and horizon off and the ecliptic on to observe the location of the Moon on the celestial sphere. Center on the Moon and note whether it lies on the ecliptic. Set the motion to increment by 1 day per timestep. Keeping the Moon locked in the center of your screen, step ahead. Determine how many days elapse between successive times when the Moon is on the ecliptic. Reset to today and move forward until the Moon is either full or new *and* it is on the ecliptic. What type of eclipse will occur on that day? Confirm your observations by checking that date with the eclipses listed in this book (Tables 1-2 and 1-3) or elsewhere.

58   It is quite possible that a lunar eclipse will occur while you are taking this course. If so, look up the precise time that it will happen on the Internet; in the current issue of a reference from the U.S. Naval Observatory, such as the *Astronomical Almanac* or *Astronomical Phenomena*; or in such magazines as *Sky & Telescope* and *Astronomy*, which generally run articles about eclipses the month before they happen. Make arrangements to observe the next lunar eclipse. Note the times at which the Moon enters and exits the Earth's umbra.

# Earth's Axis Tilted on the Ecliptic?

Imagine the Earth tilted so that its axis of rotation lies in the plane of its orbit about the Sun (see the accompanying figure). New Earth still rotates once every 24 hours. We arbitrarily fix New Earth's north pole to point toward the star Tau Tauri, just north of the bright star Aldebaran in the constellation Taurus (see Figure II-14.)

**New Seasons** Let's see how the seasons unfold on New Earth. It is March 21, the date of the spring equinox. The Sun is directly over the equator. For the next three months, the Sun rises higher in the northern hemisphere. Unlike on our Earth, the Sun does not stop moving northward when it is over 23½° north latitude. Rather, the Sun rises farther and farther north day by day until it appears over the north pole of New Earth at the summer solstice, around June 22. Three months later, at the autumnal equinox, the Sun again rises over the equator, and day and night have the same length everywhere.

The Sun appears over the south pole of New Earth at the winter solstice, around December 22. The Sun's apparent motion through New Earth's sky completes the seasonal cycle by moving north, appearing over the equator once again around March 21.

**New Climate** Day and night take on new meanings for inhabitants of New Earth. On our Earth, the regions above the Arctic Circle and below the Antarctic Circle have days or weeks of continual light in summer and days or weeks of constant darkness in winter. But on New Earth, *every place* has extended winter periods of constant darkness followed by extended summer periods of constant daylight. Spring and fall on New Earth have daily cycles of daylight and darkness, which separate the periods of continual daylight and darkness.

**A New Day** At the latitude of Atlanta, GA, (33°46′) on New Earth, the day-night cycle occurs for only seven and a half months out of the year. During the other four and a half months, there is continuous day or continuous night, coupled with harsh summers and winters. With variations, this sequence of events occurs everywhere on New Earth.

The seasonal cycle on New Earth prevents the formation of permanent polar ice caps. Polar regions experience the same tropical heating and high temperatures as the equatorial regions of our Earth. The polar regions of New Earth in winter are exceptionally cold, so seasonal polar ice caps may form. Because the southern polar cap resting on Antarctica is not permanent, the oceans, and the shorelines on the continents, are higher than those on our Earth.

If seasonal polar ice caps form, the dominant force controlling weather may shift from the jet streams that circle our Earth along lines of latitude to a pole-to-pole flow. This is the condensation flows seen on Mars. Thermal flows created by intense heating at one location and cooling at others may replace our Earth's trade winds and other east-west winds. How else would New Earth compare to our world?

# Gravitation and the Waltz of the Planets

## In this chapter you will discover

✳ the scientific revolution that dethroned Earth from its location as the center of the universe

✳ Copernicus's argument: the planets orbit the Sun

✳ Kepler's demonstration of the shape of planetary orbits

✳ Galileo's first views of craters on the Moon and moons orbiting Jupiter

✳ how Isaac Newton used all these discoveries to formulate the basic laws of physics

✳ how the Sun holds the planets in their orbits

**Apollo 11 Leaving the Moon**
A lunar module returns from the Moon after completing a successful manned mission. This photograph was taken from the spacecraft in which the astronauts returned to Earth. All orbital maneuvers to the Moon and back were based on Newtonian physical principles. Astronomers use Newton's explanations to understand a wide range of phenomena, from the motions of double stars to the rotation of the Galaxy.

R I V U X G

## WHAT DO YOU THINK?

1 What is the shape of the Earth's orbit around the Sun?

2 Do the planets orbit the Sun at constant speeds?

3 Do the planets all orbit the Sun at the same speed?

4 How much force does it take to keep an object moving in a straight line at a constant speed?

5 How does an object's mass differ when measured on the Earth and on the Moon?

The groundwork for modern science was set down around 2500 years ago, when Pythagoras and his followers began using mathematics to describe natural phenomena. About 200 years later, Aristotle asserted that the universe is comprehensible: It is governed by regular laws. Just as important, the Greeks were also among the first to leave a written record of their ideas, so that succeeding generations could develop, criticize, and test their conclusions. This concept evolved into writing physical theories in mathematical terms so that we can test the theories by observing nature. This chapter explores that work.

Early Greek astronomers had already tried to explain the motion of the five then-known planets: Mercury, Venus, Mars, Jupiter, and Saturn. Most Greeks had a *geocentric* view of the universe: They assumed that the Sun, the Moon, the stars, and the planets revolve about the Earth. A theory of the overall structure and evolution of the universe is called a **cosmology**, so the Greeks held a geocentric cosmology. The geocentric cosmology is so intuitive that it held sway for nearly 2000 years. It was only in the face of more and more accurate observations of planet motions among the stars that its validity came into question.

# ORIGINS OF A SUN-CENTERED UNIVERSE

The Greeks knew that the positions of the planets slowly shift against the background of "fixed" stars in the constellations. In fact, the word *planet* comes from a Greek term meaning "wanderer." The Greeks observed that planets do not move at uniform rates through the constellations. From night to night, as viewed in the northern hemisphere, they usually move slowly to the left (eastward) relative to the background stars. This eastward movement is called **direct motion.** Occasionally, however, a planet seems to stop and then back up for several weeks or months. This westward movement is called **retrograde motion.**

These planetary motions are much slower than the apparent daily movement of the entire sky caused by the Earth's rotation, and so they are superimposed on it. Therefore, the planets always rise in the east and set in the west, as the stars do. Both direct and retrograde motion are best detected by mapping the nightly position of a planet against the background stars over a long period. For example, Figure 2-1 shows the path of the planet Mars from May 2003 through December 2003.

Explaining the motions of the five planets in a geocentric (Earth-centered) universe was one of the main challenges facing the astronomers of antiquity. The effort resulted in an increasingly contrived model, especially in explaining retrograde motion. The mechanical description of the geocentric model of the universe (An Astronomer's Toolbox 2-1) was complex, and as observations improved, the model increasingly failed to fit the data. Because simplicity and accuracy are hallmarks of science, the complex geocentric model had to give way to a simpler, more elegant one.

The ancient Greek astronomer Aristarchus had proposed a more straightforward explanation of planetary motion, namely, that all the planets, including the Earth, revolve about the Sun. The retrograde motion of Mars, for example, occurs when the Earth overtakes and passes it, as shown in Figure 2-2. The occasional retrograde movement of a planet is then merely the result of our changing viewpoint—an idea that is beautifully simple compared to a geocentric system with all its complex planetary motions.

**Insight into science   Keep it simple** When several competing theories describe the same concepts with the same accuracy, scientists choose the simplest one. That basic tenet, formally expressed by William of Occam in the fourteenth century, is known as **Occam's razor.** Indeed, the original form of the heliocentric cosmology was appealing not because it was more accurate, it wasn't, but because it made the same predictions within a simpler model than did the geocentric cosmology. *Remember Occam's razor.*

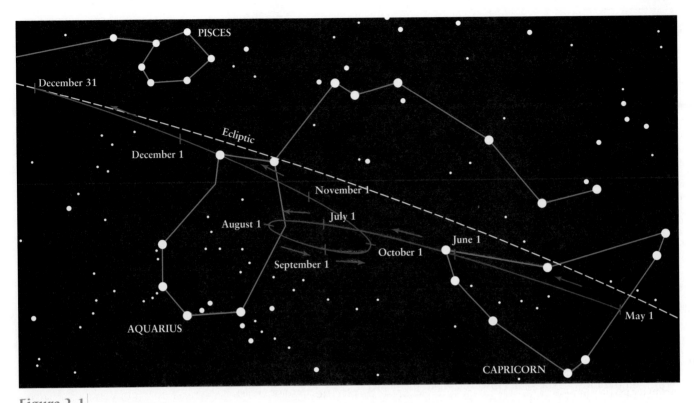

**Figure 2-1** **The Path of Mars in 2003** During the second half of 2003, Mars will move across the constellations of Capricornus, Aquarius, and Pisces. From August 1 through October 1, Mars's motion will be retrograde.

## 2-1 Copernicus devised the first comprehensive heliocentric cosmology

Dethroning Earth from its central role in the universe was difficult for a variety of reasons. Important among them is the fact that the Earth just doesn't seem to move! Combining this with the human desire to be at the center of everything and geocentric religious teaching led to a delay of more than 1300 years before someone had the

insight and determination to work out the details of Aristarchus's **heliocentric** (Sun-centered) **cosmology**. That person was the sixteenth-century Polish lawyer, physician, mathematician, economist, monk, and artist Nicolaus Copernicus.

Copernicus turned his attention to astronomy in the early 1500s. He found that by assuming that everything orbits the Sun rather than the Earth, he could determine which planets are closer to the Sun than the Earth and

**Figure 2-2** **A Heliocentric Explanation of Planetary Motion** The Earth travels around the Sun more rapidly than does Mars. Consequently, as the Earth overtakes and passes this slower-moving planet, Mars appears (from points 4 through 6) to move backward among the background stars for a few months.

### Geocentric Cosmology

The Greeks developed many theories to account for retrograde motion and the resulting loops that the planets trace out against the background stars. One of the most successful ideas was expounded by the last of the great ancient Greek astronomers, Ptolemy, who lived in Alexandria, Egypt, 1900 years ago. The basic concept is sketched in the accompanying figure. Each planet is assumed to move in a small circle called an **epicycle,** the center of which moves in a larger circle called a **deferent,** whose center is offset from the Earth. As viewed from Earth, the epicycle moves eastward along the deferent, and both circles rotate in the same direction (counterclockwise).

Most of the time, the motion of the planet on its epicycle adds to the eastward motion of the epicycle on the deferent. Thus, the planet is seen to be in direct (eastward) motion against the background stars throughout most of the year. However, when the planet is on the part of its epicycle nearest the Earth, its motion along the epicycle subtracts from the motion of the epicycle along the deferent. The planet thus appears to slow and then halt its usual eastward movement among the constellations, even seeming to go backward for a few weeks or months. Using this concept of epicycles and deferents, Greek astronomers were able to explain the retrograde loops of the planets.

Using the wealth of astronomical data in the library at Alexandria, including records of planetary positions covering hundreds of years, Ptolemy deduced the sizes of the epicycles and deferents and the rates of revolution needed to produce the recorded paths of the planets. After years of arduous work, Ptolemy assembled his calculations in the *Almagest,* in which the positions and paths of the Sun, Moon, and planets were described with unprecedented accuracy. In fact, the *Almagest* was so successful that it became the astronomer's bible. For more than 1000 years,

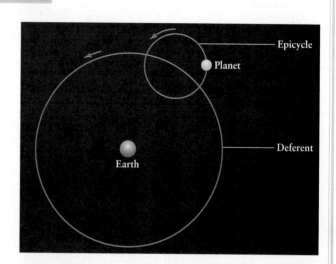

**A Geocentric Explanation of Planetary Motion** Each planet revolves about an epicycle, which in turn revolves about a deferent centered approximately on the Earth. As seen from Earth, the speed of the planet on the epicycle alternately adds to or subtracts from the speed of the epicycle on the deferent, thus producing alternating periods of direct and retrograde motion.

Ptolemy's cosmology endured as a useful description of the workings of the heavens.

Eventually, however, the commonsense explanation of the Earth-centered cosmology began to go awry. Errors and inaccuracies that were unnoticeable in Ptolemy's day compounded and multiplied over the years, especially errors due to precession, the slow change in the direction of the Earth's axis of rotation. Fifteenth-century astronomers made some cosmetic adjustments to the Ptolemaic system. However, the system became less and less satisfactory as more complicated and arbitrary details were added to keep it consistent with the observed motions of the planets. After Newton's time, scientists knew that circular motion required a force acting on the body. However, nothing in Ptolemy's epicycle theory produced such a force.

which are farther away. Because Mercury and Venus are always observed fairly near the Sun, Copernicus correctly concluded that their orbits must lie inside that of the Earth. The other visible planets—Mars, Jupiter, and Saturn—can be seen high in the sky in the middle of the night, when the Sun is far below the horizon. This can occur only if the Earth comes between the Sun and a planet. Copernicus

therefore concluded that the orbits of Mars, Jupiter, and Saturn lie outside the Earth's orbit. Three more distant planets (Uranus, Neptune, and Pluto) were discovered after the telescope was invented.

The geometrical arrangements among the Earth, another planet, and the Sun are called **configurations.** For example, when Mercury or Venus is directly between the

Earth and the Sun, as in Figure 2-3, we say the planet is in a configuration called an **inferior conjunction**; when a planet is on the opposite side of the Sun from the Earth, its configuration is called a **superior conjunction**.

The angle between the Sun and a planet as viewed from the Earth is called the planet's **elongation**. At **greatest elongation**, Mercury and Venus are as far from the Sun in angle as they can be. This is about 28° for Mercury and about 47° for Venus. Because Mercury and Venus can never be farther than their greatest elongations from the Sun, neither of these planets is ever seen very high in the night sky. When either Mercury or Venus rises before the Sun, it is visible in the eastern sky as a very bright "star" and is often called the "morning star." Similarly, when either of these two planets sets after the Sun, it is always low in the western sky and is then called the "evening star." Because they are so bright and sometimes appear to change color due to motion of the Earth's atmosphere, Venus and Mercury are often mistaken for UFOs. (The same motion of the air causes the road in front of your car to shimmer on a hot day.)

Planets whose orbits are larger than Earth's have different configurations. For example, when Mars is located beyond the Sun, as seen from Earth, it is said to be in **conjunction**. When it is opposite the Sun in the sky, the planet is at **opposition**. It is not difficult to determine when a planet happens to be located at one of the key positions in Figure 2-3. For example, when Mars is at opposition, it appears high in the sky at midnight.

It is easy to follow a planet as it moves from one configuration to another. However, these observations alone

| Table 2-1 | | |
|---|---|---|
| **Synodic and Sidereal Periods of the Planets** | | |
| | Synodic Period | Sidereal Period |
| Mercury | 116 d | 88 d |
| Venus | 584 d | 225 d |
| Earth | — | 1.0 yr |
| Mars | 780 d | 1.9 yr |
| Jupiter | 399 d | 11.9 yr |
| Saturn | 378 d | 29.5 yr |
| Uranus | 370 d | 84.0 yr |
| Neptune | 368 d | 164.8 yr |
| Pluto | 367 d | 248.5 yr |

do not tell us the planet's actual orbit, because the Earth, from which we make the observations, is also moving. Copernicus was therefore careful to distinguish between two characteristic time intervals, or *periods*, of each planet.

Recall from your study of the Moon in Chapter 1 that the **sidereal period** of orbit is the true orbital period. A planet's sidereal period is the time it takes the planet to complete one circuit around the Sun as would be measured by a fixed observer at the Sun's location watching each planet move through the background stars. Equivalently, by removing the effect of the Earth's motion, astronomers on Earth observing the motion of each planet among the background stars can calculate its sidereal period. The **synodic period** is the time that elapses between two successive identical configurations as seen from the Earth—from one opposition to the next, for example, or from one conjunction to the next.

Thus, nearly 500 years ago, Copernicus was able to obtain the first six entries shown in Table 2-1 (the others are contemporary results included for completeness). Copernicus was then able to devise a straightforward geometric method for determining the distances of the planets from the Sun. His answers turned out to be remarkably close to the modern values, as shown in Table 2-2. From these two tables it is apparent that the farther a planet is from the Sun, the longer it takes to complete its orbit.

> **Insight into science    Take a fresh look**  When the science is hard to visualize, try another perspective. For example, a planet's sidereal period of orbit is easy to understand from the perspective of the Sun but more complicated from the Earth. The synodic period of each planet, on the other hand, is easily determined from the Earth. As we will see later, especially when we study Einstein's theories of relativity, each of these perspectives is called a *frame of reference*.

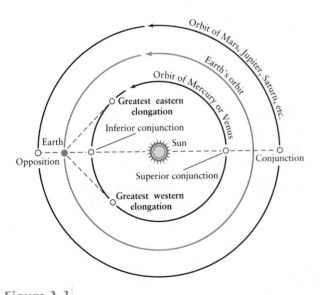

**Figure 2-3**  **Planetary Configurations** It is useful to specify key points along a planet's orbit, as shown in this figure. These points identify specific geometric arrangements between the Earth, another planet, and the Sun.

**Table 2-2**
**Average Distances of the Planets from the Sun**

| | Measurement (AU) | |
| --- | --- | --- |
| | By Copernicus | Modern |
| Mercury | 0.38 | 0.39 |
| Venus | 0.72 | 0.72 |
| Earth | 1.00 | 1.00 |
| Mars | 1.52 | 1.52 |
| Jupiter | 5.22 | 5.20 |
| Saturn | 9.07 | 9.54 |
| Uranus | Unknown | 19.19 |
| Neptune | Unknown | 30.06 |
| Pluto | Unknown | 39.53 |

Copernicus explained his heliocentric cosmology, including supporting observations and calculations, in a book entitled *De revolutionibus orbium coelestium* ("On the Revolutions of the Celestial Spheres"), which was published in 1543, the year of his death. Copernicus's great insight was the simplicity of a heliocentric cosmology compared to geocentric views. However, Copernicus incorrectly assumed that the Earth travels along a circular path around the Sun. As a result, his predictions were no more accurate than those of a geocentric theory!

## 2-2 Tycho Brahe made astronomical observations that disproved ancient ideas about the heavens

In November 1572, a bright star suddenly appeared in the constellation Cassiopeia. At first it was even brighter than Venus, but then it began to grow dim. After 18 months, it faded from view.

Modern astronomers recognize this event as a supernova explosion, the violent death of a certain type of star (see Chapter 12). In the sixteenth century, however, the prevailing opinion was quite different. Teachings dating back to Aristotle and Plato argued that the heavens were permanent and unalterable. Consequently, the "new star" of 1572 could not really be a star at all, because the heavens do not change; it must instead be some sort of bright object quite near Earth, perhaps not much farther away than the clouds overhead. A 25-year-old Danish astronomer named Tycho Brahe realized that straightforward observations might reveal the distance to this object.

It is everyone's common experience that when you walk from one place to another, nearby objects appear to change position against the background of more distant objects. Furthermore, the closer an object is, the more you have to change the angle at which you observe it as you move. This apparent change in position of the object with the changing position of the observer is called **parallax** (Figure 2-4).

Brahe reasoned as follows: If the new star is nearby, then its position should shift against the background stars over the course of a night because the Earth's rotation changes our viewpoint, as shown in Figure 2-5a. His careful observations failed to disclose any parallax, and so the new star had to be far away, farther from Earth than anyone had imagined (Figure 2-5b). Brahe summarized his findings in a small book, *De stella nova* ("On the New Star"), published in 1573.

Brahe's astronomical records were soon to play an important role in the development of a heliocentric cosmology. From 1576 to 1597, Brahe made comprehensive observations, measuring planetary positions with an accuracy of 1 arcmin, about as precise as is possible with the naked eye. Upon his death in 1601, most of these invaluable records fell into the hands of his gifted assistant, Johannes Kepler.

**Figure 2-4**   **Parallax** Nearby objects are viewed at different angles from different places. These objects also appear to be in a different place with respect to more distant objects when viewed at the same time by observers located at different positions. Both effects are called parallax, and they are used by astronomers, surveyors, and sailors to determine distances.

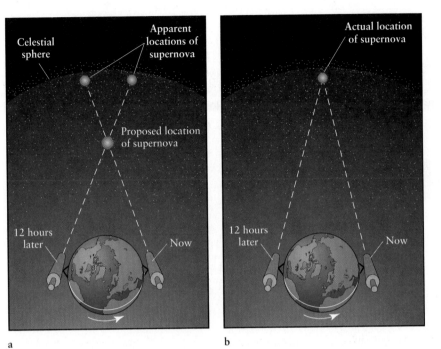

Celestial sphere

Apparent locations of supernova

Proposed location of supernova

12 hours later

Now

a

Actual location of supernova

12 hours later

Now

b

# KEPLER'S AND NEWTON'S LAWS

Until Johannes Kepler's time, astronomers had assumed that heavenly objects move in circles. For philosophical and aesthetic reasons, circles were considered the most perfect and most harmonious of all geometric shapes. However, assuming circular orbits failed to yield accurate predictions for the positions of the planets. For years, Kepler tried to find a shape for orbits that would fit Tycho Brahe's observations of the planets' positions against the background of distant stars. Finally, he began working with a geometric form called an **ellipse**.

## 2-3 Kepler's laws describe orbital shapes, changing speeds, and the lengths of planetary years

An ellipse can be drawn as shown in Figure 2-6a. Each thumbtack is at a **focus** (plural **foci**). The longest diameter across an ellipse, called the *major axis*, passes through both foci. Half of that distance is called the **semimajor axis**, whose length is usually designated by the letter *a*. In astronomy, the length of the semimajor axis is also the average distance between a planet and the Sun.

To Kepler's delight, the ellipse turned out to be the curve he had been searching for. Predictions of the locations of planets based on elliptical paths were in very close agreement with where the planets actually were. He published this discovery in 1609 in a book known today as *New Astronomy*. This important discovery is now considered the first of **Kepler's laws:**

*The orbit of a planet about the Sun is an ellipse with the Sun at one focus.*

Ellipses have two extremes. The roundest ellipse is a circle. The most elongated ellipse approaches being a straight line. The shape of a planet's orbit around the Sun is described by its **orbital eccentricity,** designated by the letter *e*, which ranges from 0 (circular orbit) to just under 1.0 (nearly a straight line). Figure 2-6b shows a sequence of ellipses and their associated eccentricities. Observations have since revealed that there is no object at the second focus for each elliptical planetary orbit.

Brahe's observations also showed Kepler that planets do not move at uniform speeds along their orbits. Rather, a planet moves most rapidly when it is nearest the Sun, a point on its orbit called **perihelion.** Conversely, a planet moves most slowly when it is farthest from the Sun, a point called **aphelion.**

After much trial and error, Kepler discovered a way to describe how fast a planet moves anywhere along its orbit. This discovery, also published in *New Astronomy*, is illustrated in Figure 2-7. Suppose that it takes 30 days for a planet to go from point *A* to point *B*. During that time, the line joining the Sun and the planet sweeps out a nearly triangular area (shaded in Figure 2-7). Kepler discovered that the line joining the Sun and the planet sweeps out the same area during any other 30-day interval. In other words, if the planet also takes 30 days to go

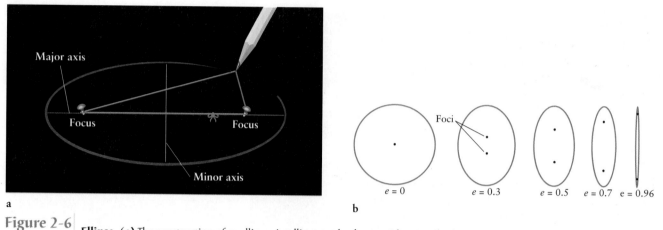

Major axis

Focus

Focus

Minor axis

a

Foci

$e = 0$     $e = 0.3$     $e = 0.5$    $e = 0.7$   $e = 0.96$

b

**Figure 2-6**    **Ellipse** **(a)** The construction of an ellipse: An ellipse can be drawn with a pencil, a loop of string, and two thumbtacks, as shown in this figure. If the string is kept taut, the pencil traces out an ellipse. The two thumbtacks are located at the two foci of the ellipse. **(b)** A series of ellipses with different eccentricities, *e*. Eccentricities range between 0 (circle) to just under 1.0 (virtually a straight line).

from point *C* to point *D*, then the two shaded segments in Figure 2-7 are equal in area. *Kepler's second law,* also called the **law of equal areas,** can be stated thus:

*A line joining a planet and the Sun sweeps out equal areas in equal intervals of time.*

Kepler's second law means that each planet's speed decreases as it moves from perihelion to aphelion. The speed then increases as the planet moves from aphelion toward perihelion.

Kepler was also able to relate a planet's year to its distance from the Sun. This discovery, published in 1619, stands out because of its impact on future developments in astronomy. Now called *Kepler's third law,* it predicts the planet's sidereal period if we know the length of the semimajor axis of the planet's orbit:

*The square of a planet's sidereal period around the Sun is directly proportional to the cube of the length of its orbit's semimajor axis.*

The relationship is easiest to use if we let *P* represent the sidereal period in Earth years and *a* represent the length of the semimajor axis measured in astronomical units (as we discussed in An Astronomer's Toolbox I-2). Now we can write Kepler's third law concisely as

$$P^2 = a^3$$

In other words, a planet closer to the Sun has a shorter year than does a planet farther from the Sun. Combining this with the second law reveals that planets closer to the Sun move more rapidly than those farther away. Using data from Tables 2-1 and 2-2, we can demonstrate Kepler's third law as shown in Table 2-3.

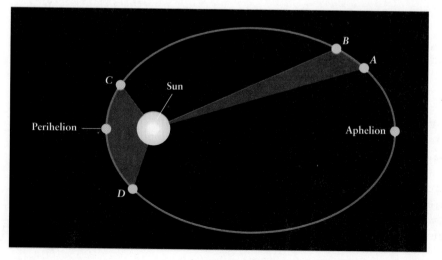

B

A

C

Sun

Perihelion

Aphelion

D

**Figure 2-7**    **Kepler's First and Second Laws**
According to Kepler's first law, every planet travels around the Sun along an elliptical orbit with the Sun at one focus. According to his second law, the line joining the planet and the Sun sweeps out equal areas in equal intervals of time.

Kepler's three laws apply not only to the planets orbiting the Sun but also apply whenever any object orbits another under the influence of their gravitational attraction. Thus, Kepler's laws apply to moons orbiting planets, artificial satellites orbiting the Earth, and even two stars revolving about each other. Throughout this book, we will see that Kepler's three laws have a wide range of practical applications.

## 2-4 Galileo's discoveries strongly supported a heliocentric cosmology

While Kepler was making rapid progress in central Europe, an Italian physicist was making equally dramatic observations in southern Europe. Galileo Galilei did not invent the telescope, but he was one of the first people to point the new device toward the sky and publish his

| | Sidereal period $P$ (yr) | Semimajor axis $a$ (AU) | $P^2$ | = | $a^3$ |
|---|---|---|---|---|---|
| Mercury | 0.24 | 0.390 | .06 | | 0.06 |
| Venus | 0.61 | 0.72 | 0.37 | | 0.37 |
| Earth | 1.00 | 1.00 | 1.00 | | 1.00 |
| Mars | 1.88 | 1.52 | 3.53 | | 3.51 |
| Jupiter | 11.86 | 5.20 | 140.7 | | 140.6 |
| Saturn | 29.46 | 9.54 | 867.9 | | 868.3 |
| Uranus | 84.01 | 19.19 | 7,058 | | 7,067 |
| Neptune | 164.79 | 30.06 | 27,160 | | 27,160 |
| Pluto | 248.54 | 39.53 | 61,770 | | 61,770 |

**Table 2-3**
**A Demonstration of Kepler's Third Law**

observations. He saw things that no one had ever imagined—mountains on the Moon and sunspots on the Sun. He also discovered that Venus exhibits phases.

After only a few months of observation, Galileo noticed that the apparent size of Venus as seen through his telescope was related to the planet's phase. Venus appears smallest at gibbous phase and largest at crescent phase. A geocentric cosmology could not explain why, but a heliocentric cosmology does. Galileo's observations therefore supported the conclusion that Venus orbits the Sun, not the Earth (Figure 2-8).

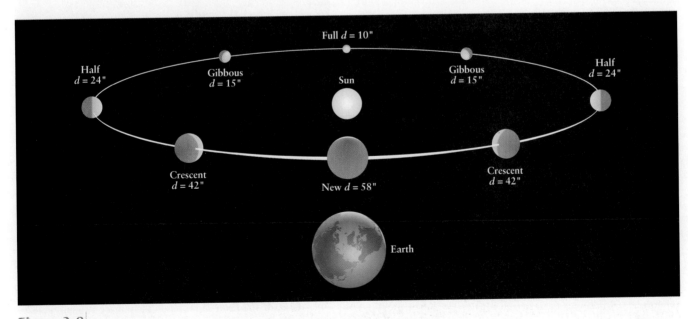

**Figure 2-8**  **The Changing Appearance of Venus**  This figure shows how the appearance (phase) of Venus changes as it moves along its orbit. The number below each view is the angular diameter ($d$) of the planet as seen from Earth, in arcseconds. Note that the phases correlate with the planet's angular size and its angular distance from the Sun, both as seen from Earth. These observations clearly support the idea that Venus orbits the Sun.

In 1610 Galileo also discovered four moons near Jupiter. In honor of their discoverer, these are today called the **Galilean moons** (or **satellites,** another term for moon). Galileo concluded that the moons are orbiting Jupiter because they move across from one side of the planet to the other. Confirming observations were made in 1620 (Figure 2-9). These observations all provided further proof that the Earth is not at the center of the universe. Like the Earth in orbit around the Sun, Jupiter's four moons obey Kepler's third law: The square of a moon's orbital period about Jupiter is directly proportional to the cube of its average distance from the planet.

Galileo's telescopic observations constituted the first fundamentally new astronomical data since humans began recording what they saw in the sky. In contradiction to then-prevailing opinions, these discoveries strongly supported a heliocentric view of the universe. Because Galileo's ideas could not be reconciled with certain passages in the Bible or with the writings of Aristotle and Plato, the Roman Catholic church condemned him, and he was forced to spend his latter years under house arrest "for vehement suspicion of heresy."

A major stumbling block prevented seventeenth-century thinkers from accepting Kepler's laws and Galileo's conclusions about the heliocentric cosmology. At that time, the relationships between matter, motion, and forces were not understood. People did not know about the gravitational force of the Sun, which keeps the planets in orbit. They did not know how planets, once they had started in orbit around the Sun, could keep moving. Once anything on Earth is put in motion, it quickly comes to rest. Why didn't the planets orbiting the Sun stop, too?

All those mysteries were soon explained by the brilliant, eccentric scientist Isaac Newton, who was born on Christmas Day in 1642, less than a year after Galileo died. In the decades that followed, Newton revolutionized science more profoundly than any person before him, and in doing so, he found physical and mathematical proof of the heliocentric cosmology.

b     R I V U X G

**Figure 2-9** **Jupiter and Its Largest Moons** (a) In 1610 Galileo discovered four "stars" that move back and forth across Jupiter. He concluded that they are four moons that orbit Jupiter much as our Moon orbits the Earth. This figure shows observations made by Jesuits in 1620. Compare the drawing labeled 13 with the twentieth-century photograph in (b). This photograph, taken by an amateur astronomer with a small telescope, shows the four Galilean satellites alongside an overexposed image of Jupiter. Each satellite would be bright enough to be seen with the unaided eye were it not overwhelmed by the glare of Jupiter.

a

## 2-5 Newton formulated three laws that describe fundamental properties of physical reality

Until the mid-seventeenth century, virtually all mathematical astronomy used the same approach. Astronomers from Ptolemy to Kepler worked *empirically*, that is, directly from data and observations. They adjusted their ideas and calculations until they finally came out with the right answers.

Isaac Newton introduced a new approach. He made just three assumptions, now called *Newton's laws of motion*, that he applied to all forces and bodies. He also found a formula for the force of **gravity**, the attraction between all objects due to their masses. Newton then showed that Kepler's three laws logically follow from these laws. Using his formula, Newton accurately described the observed orbits of the Moon, comets, and other objects in the solar system. His laws also apply to the motions of all bodies on the Earth.

*Newton's first law*, the **law of inertia**, states that

4 | *A body remains at rest or moves in a straight line at a constant speed unless acted upon by an unbalanced outside force.*

At first, this law might seem to conflict with your everyday experience. For example, if you shove a chair, it does not continue at a constant speed forever but rather comes to rest after sliding only a short distance. From Newton's viewpoint, however, an "unbalanced outside force" does indeed act on the moving chair, namely, friction between the chair's legs and the floor. Without friction, the chair would continue in a straight path at a constant speed. A **force** changes the motion of an object.

Newton's first law tells us that there must be an outside force acting on the planets. If there was no force acting on them, they would move away from the Sun along straight-line paths at constant speeds. In other words, they would leave their curved orbits. Because this does not happen, Newton concluded that some force confines the planets to their elliptical orbits. As we shall see, that force is gravity.

Newton's second assumption describes how a force changes the motion of an object. To appreciate the concepts of force and motion better, we must first understand the quantities that describe motion: speed, velocity, and acceleration.

Imagine an object in space. Push on the object and it begins to move. At any moment, you can describe the object's motion by specifying both its speed and direction. Speed and direction of motion together constitute the object's **velocity**. If you continue to push on the object, its speed will increase—it will accelerate.

**Acceleration** is the rate at which velocity changes with time. Because velocity involves both speed and direction, acceleration does not apply only to increases in speed. A slowing down, a speeding up, or a change in direction are all types of acceleration.

Suppose an object revolved about the Sun in a perfectly circular orbit. This body would have acceleration that involved only a change of direction. As this object moved along its orbit, its speed would remain constant, but its direction of motion would be continuously changing. Therefore, it would still be continuously accelerating.

Newton's second law says that the acceleration of an object is proportional to the force acting on it. In other words, the harder you push on an object, the greater the resulting acceleration. Newton's law also says that a greater mass pushed or pulled by a force accelerates more slowly than does an object of lesser mass pushed or pulled by the same force. That is why by pushing a child's wagon, you can accelerate it faster than you can accelerate a car by pushing on it. Newton's second law can be succinctly stated as an equation. If a force acts on an object, the object will experience an acceleration such that

$$\text{Force} = \text{mass} \times \text{acceleration}$$

5 | The **mass** of an object is a measure of the total amount of material in the object, which we measure in kilograms. For example, the mass of the Sun is $2 \times 10^{30}$ kg, the mass of a hydrogen atom is $1.7 \times 10^{-27}$ kg, and the mass of the author of this book is 83 kg. At rest, the Sun, a hydrogen atom, and I have these same masses regardless of where we happen to be in the universe.

It is important not to confuse the concept of mass with that of weight. **Weight** is the force with which an object is pulled down while on the ground (due to gravity's pull) or, equivalently, feels inside an accelerating rocket. Force is usually expressed in pounds or Newtons. For example, the force with which I am pressing down on the ground is 183 pounds.

But I weigh 183 pounds only on the Earth. I would weigh less on the Moon. Floating in space or orbiting in the Space Shuttle, I would be weightless, but my mass would always be the same. An astronaut in the Shuttle would still have to push me with a force to get me to move. Whenever we describe the properties of planets, stars, or galaxies, we speak of their masses, never of their weights.

Newton's final assumption, called *Newton's third law*, is the famous *law of action and reaction*:

*Whenever one body exerts a force on a second body, the second body exerts an equal and opposite force on the first body.*

For example, if I weigh 183 pounds, then I press down on the floor with a force of 183 pounds. Newton's third

## DISCOVER FOR YOURSELF — Astronomy's Foundation Builders

In the two centuries between 1500 and 1700, human understanding of the motion of celestial bodies and the nature of the gravitational force that keeps them in orbit surged forward as never before. Theories related to this subject were developed by brilliant thinkers, whose work established and verified the heliocentric model of the solar system and the role of gravity.

**Nicolaus Copernicus (1473–1543)** Copernicus was born in Torun, Poland, the youngest of four children.

He pursued his higher education in Italy, where he received a doctorate in canon law and studied medicine. Copernicus developed a heliocentric theory of the known universe, and published his work in 1543 under the title *De revolutionibus orbium coelestium*. Copernicus became ill and died just after receiving the first copy.

**Tycho Brahe (1546–1601) and Johannes Kepler (1571–1630)** Brahe, depicted within a portrait of Kepler, was born to nobility in the Danish city of Knudstrup, which is now part of Sweden. At age 20 he lost part of his nose in a duel and wore a metal replacement thereafter. In 1576 the Danish king

Frederick II built Brahe an astronomical observatory that Tycho named Uraniborg. He rejected Copernicus's heliocentric theory and the Ptolemaic geocentric system and devised a halfway theory called the "Tychonic system." According to Brahe's theory, the Earth is stationary, with the Sun and Moon revolving around it, while all the other planets revolve around the Sun. Brahe died in 1601.

Kepler was educated in Germany, where he spent three years studying mathematics, philosophy, and theology. In 1596, Kepler published a booklet in which he attempted to mathematically predict the planetary orbits. Although this idea was altogether wrong, its boldness and originality attracted the attention of Tycho Brahe, whose staff Kepler joined in 1600. Kepler deduced his three laws from Brahe's observations.

**Galileo Galilei (1564–1642)** Born in Pisa, Italy, Galileo studied medicine and philosophy at the University of Pisa. He abandoned medicine in favor of mathematics and returned to the university as a professor of mathe-

matics. There Galileo formulated his famous law of falling bodies: All objects fall with the same acceleration regardless of their weight. In 1609 he constructed a telescope and made a host of discoveries that contradicted the teachings of Aristotle and the Roman Catholic church. He summed up his life's work on motion, acceleration, and gravity in the book *Dialogues Concerning Two New Sciences*, published in 1637.

**Isaac Newton (1642–1727)** Although he delighted in constructing mechanical devices— sundials, model windmills, a water clock, and a mechanical carriage—Newton showed no

exceptional academic ability at Cambridge University, where he received a bachelor's degree in 1665. While pursuing experiments in optics, Newton constructed a refracting telescope and also discovered that white light is actually a mixture of all colors. His major work on forces and gravitation was the tome *Philosophiae naturalis principia mathematica*, which appeared in 1687. In 1704, Newton published his second great treatise, *Opticks*, in which he described his experiments and theories about light and color. Upon his death in 1727, Newton was buried in Westminster Abbey, the first scientist to be so honored.

---

law says that the floor is also pushing up against me with an equal force of 183 pounds. (If it were not, I would fall through the floor.) In the same way, Newton realized that because the Sun is exerting a force on each planet to keep it in orbit, each planet must also be exerting an equal and opposite force on the Sun. As each planet accelerates toward the Sun, the Sun in turn accelerates toward each planet.

The Sun is pulling the planets, so why don't they fall onto it? **Conservation of angular momentum,** a fundamental consequence of Newton's first law of motion, provides the answer. Angular momentum is a measure of how

## AN ASTRONOMER'S TOOLBOX 2-2

### Gravitational Force

From Newton's law of gravitation, if two objects having masses $m_1$ and $m_2$ are separated by a distance r, then the gravitational force F between them is

$$F = G(m_1 m_2 / r^2)$$

In this formula, G is the **universal constant of gravitation**, whose value has been determined from laboratory experiments:

$$G = 6.67 \times 10^{-11} \ m^3/(kg \cdot s^2)$$

The equation $F = G(m_1 m_2 / r^2)$ gives, for example, the force from the Sun on the Earth and, equivalently, from the Earth on the Sun. If $m_1$ is the mass of the Earth ($6.0 \times 10^{24}$ kg), $m_2$ is the mass of the Sun ($2.0 \times 10^{30}$ kg), and r is the distance from the Earth to the Sun ($1.5 \times 10^{11}$ m),

$$F = 3.6 \times 10^{22} \ kg \cdot m/s^2$$

This number can then be used in Newton's second law, F = ma, to find the acceleration of the Earth due to the Sun. This yields:

$$a_{Earth} = F/m_1 = 6.0 \times 10^{-3} \ m/s^2$$

Newton's third law says that the Earth exerts the same force on the Sun, so the Sun's acceleration due to the Earth's gravitational force is

$$a_{Sun} = F/m_1 = 1.8 \times 10^{-8} \ m/s^2$$

In other words, the Earth pulls on the Sun, causing the Sun to move toward it. Because of the Sun's greater mass, however, the amount that the Earth accelerates the Sun is more than 300,000 times less than the amount the Sun accelerates the Earth.

much energy is stored in an object due to its rotation and revolution. As the orbiting planets fall Sunward, their angular momentum provides them with motion perpendicular to that infall, meaning that planets continually fall toward the Sun, but they continually miss it. Because their angular momentum is conserved, planets neither spiral into the Sun nor away from it. Conservation of angular momentum states that an object's angular momentum remains constant unless acted on by an outside force.

Angular momentum depends on three things: how fast the body rotates or revolves, how much mass it has, and how spread out that mass is. The greater a body's angular motion or mass, or the more the mass is spread out, the greater its angular momentum. Consider, for example, twirling ice skaters. They rotate with a constant mass, practically free of outside forces. When spinning skaters wish to rotate more rapidly, they decrease the spread of their mass distribution by pulling their arms in closer to their bodies. According to the conservation of angular momentum, as the spread of mass decreases, either the amount of mass or the rotation rate must increase. Because the skaters don't change mass, they spin faster. In astronomy, we encounter many instances of the same law, as giant objects, like stars, contract.

We have now reconstructed the essential relationships between matter and motion. Scientific belief in the heliocentric cosmology still requires a force to hold the planets in orbit around the Sun and the moons in orbit around the planets. Newton identified that, too.

## 2-6 Newton's description of gravity accounts for Kepler's laws

Isaac Newton did not invent the idea of gravity. An educated seventeenth-century person would understand that some force pulls things down to the ground. It was Newton, however, who gave us a precise description of the action of gravity, or *gravitation*, as it is more properly called. Using his first law, Newton proved mathematically that the force acting on each of the planets is directed toward the Sun. This discovery led him to suspect that the force pulling a falling apple straight down to the ground is the same as the force on the planets that is always aimed straight at the Sun.

Newton succeeded in formulating a mathematical model describing the behavior of the gravitational force that keeps the planets in their orbits (presented in An Astronomer's Toolbox 2-2). In words, Newton's **universal law of gravitation** states:

*Two bodies attract each other with a force that is directly proportional to the product of their masses and inversely proportional to the square of the distance between them.*

In other words, gravitational force decreases with distance: Move twice as far away from a body and you feel only one-quarter of the force from it that you felt before.

Using his law of gravity, Newton found that he could mathematically explain Kepler's three laws. For example, whereas Kepler discovered by trial and error that the period of orbit, $P$, and average distance between two bodies, $a$, are related by $P^2 = a^3$, Newton demonstrated mathematically that this equation follows logically from his law of gravitation.

Newton also discovered that some objects have nonelliptical orbits around the Sun. His equations led him to conclude that the orbits of some objects are **parabolas** and **hyperbolas** (Figure 2-10). For example, comets hurtling toward the Sun from the depths of space often follow parabolic or hyperbolic orbits.

Newton's ideas turned out to be applicable in an incredibly wide range of situations. The orbits of the planets and their satellites could now be calculated with unprecedented precision. Using Newton's laws, mathematicians proved that the Earth's axis of rotation must precess because of the gravitational pull of the Moon and the Sun on the Earth's equatorial bulge (recall Figure 1-16).

In addition, Newton's laws and mathematical techniques were used to predict new phenomena. Newton's friend, Edmund Halley, was intrigued by historical records of a comet that was sighted about every 76 years. Using Newton's methods, Halley worked out the details of the comet's orbit and predicted its return in 1758. It was first sighted on Christmas night of that year, and to this day the comet bears Halley's name (Figure 2-11).

Perhaps the most dramatic confirmation of Newton's ideas was their role in the discovery of the eighth planet in our solar system. The seventh planet, Uranus, had been discovered accidentally by William Herschel in 1781 during a telescopic survey of the sky. Fifty years later, however, it was clear that Uranus was not following the orbit predicted by Newton's laws. Two mathematicians, John Couch Adams in England and Urbain-Jean-Joseph Lever-

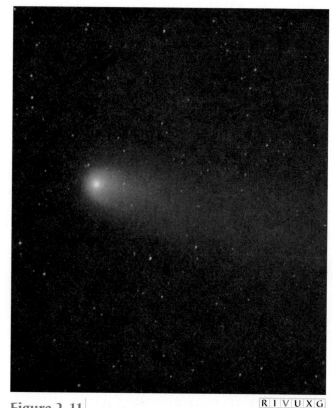

**Figure 2-11** R I V U X G

**Halley's Comet** Halley's Comet orbits the Sun with an average period of about 76 years. During the twentieth century, the comet passed near the Sun twice—once in 1910 and again, shown here, in 1986. The comet will pass close to the Sun again in 2061.

rier in France, independently calculated that the deviations of Uranus from its predicted orbit could be explained by the gravitational pull of a then unknown, more distant planet. Each man predicted that the planet would be found at a certain location in the constellation of Aquarius in September 1846. A telescopic search on September 23, 1846, revealed Neptune less than 1° from its calculated

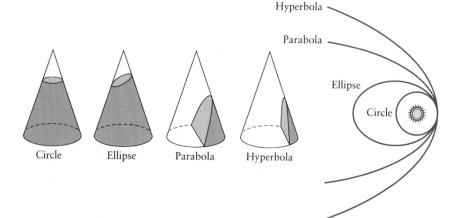

**Figure 2-10** **Conic Sections** A conic section is any one of a family of curves obtained by slicing a cone with a plane, as shown in this figure. The orbit of one body about another can be an ellipse, a parabola, or a hyperbola. Circular orbits are possible because a circle is just an ellipse for which both foci are at the same point.

Circle    Ellipse    Parabola    Hyperbola

Hyperbola

Parabola

Ellipse

Circle

R I V U X G

**Figure 2-12** **Uranus and Neptune** The discovery of Neptune was a major triumph for Newton's laws. In an effort to explain why Uranus (shown on the left, with three of its moons indicated by arrows) deviated from its predicted orbit, astronomers predicted the existence of Neptune (shown on the right, with one of its moons indicated by an arrow). Uranus and Neptune are nearly the same size; both have diameters about 4 times that of Earth.

position. Although sighted with a telescope, Neptune was really discovered with pencil and paper (Figure 2-12).

**Insight into science** **Quantify predictions**
Mathematics provides a language that enables science to make quantitative predictions that can be checked by anyone. For example, we have seen in this chapter how Kepler's third law and Newton's universal law of gravitation correctly predict the motion of objects under the influence of the Sun's gravitational attraction.

Over the years, Newton's ideas were successfully used to predict and explain motion here on Earth and throughout the universe. Even today, as we send astronauts into Earth orbit and send probes to the outer planets, Newton's equations are used to calculate the orbits and trajectories of the spacecraft.

It is a testament to Newton's genius that his three laws were precisely the basic ideas needed to understand the motions of the planets. Newton brought a new dimension of elegance and sophistication to our understanding of the workings of the universe.

## WHAT DID YOU THINK?

1  *What is the shape of the Earth's orbit around the Sun?* All planets have elliptical orbits around the Sun.

2  *Do the planets orbit the Sun at constant speeds?* No. The closer a planet is to the Sun in its orbit, the faster it is moving. It moves fastest at perihelion and slowest at aphelion.

3  *Do the planets all orbit the Sun at the same speed?* No. A planet's speed depends on its average distance from the Sun. The closest planet moves fastest, the most distant planet moves slowest.

4  *How much force does it take to keep an object moving in a straight line at a constant speed?* Unless an object is subject to an outside force, like friction, it takes no force at all to keep it moving in a straight line at a constant speed.

5  *How does an object's mass differ when measured on Earth and on the Moon?* Assuming the object doesn't shed or collect pieces, its mass remains constant whether on the Earth or on the Moon. Its weight, however, is less on the Moon.

## Key Words

## Key Ideas

✳ The ancient Greeks laid the groundwork for progress in science. Early Greek astronomers devised a geocentric cosmology, which placed the Earth at the center of the universe.

### Origins of a Sun-Centered Universe

✳ Copernicus's heliocentric (Sun-centered) theory simplified the general explanation of planetary motions compared to the geocentric theory.

✳ In a heliocentric cosmology, the Earth is but one of several planets that orbit the Sun.

✳ The sidereal period of a planet, which is measured with respect to the stars, is the planet's true orbital period. Its synodic period is measured with respect to the Sun as seen from the moving Earth (for example, from one opposition to the next).

### Kepler's and Newton's Laws

✳ Ellipses describe the paths of the planets around the Sun much more accurately than do circles. Kepler's three laws give important details about elliptical orbits.

✳ The invention of the telescope led Galileo to new discoveries, such as the phases of Venus and the moons of Jupiter, that supported a heliocentric view of the universe.

✳ Newton based his explanation of the universe on three assumptions now called Newton's laws of motion. These laws and his universal law of gravitation can be used to deduce Kepler's laws and to describe planetary motions with extreme accuracy.

✳ The mass of an object is a measure of the amount of matter in the object; its weight is a measure of the force with which the gravity of some other object pulls on the object's mass.

✳ In general, the path of one astronomical object about another, such as that of a comet about the Sun, is an ellipse, a parabola, or a hyperbola.

## Review Questions

1 How did Copernicus explain the retrograde motions of the planets?

2 Which planets can never be seen at opposition? Which planets can never be seen at inferior conjunction?

3 At what configuration (superior conjunction, greatest eastern elongation, etc.) would it be best to observe Mercury or Venus with an Earth-based telescope? At what configuration would it be best to observe Mars, Jupiter, or Saturn? Explain your answers.

4 What is the difference between the synodic and sidereal periods of a planet?

5 What are Kepler's three laws? Why are they important?

6 In what ways did the astronomical observations of Galileo support a heliocentric cosmology?

7 How did Newton's approach to understanding planetary motions differ from that of his predecessors?

8 What is the difference between mass and weight?

9 Why is the discovery of Neptune a major confirmation of Newton's universal law of gravitation?

10 Why does an astronaut have to exert a force on a weightless object in order to move it?

## Advanced Questions

The answers to all computational problems, which are preceded by an asterisk (*), appear at the end of the book.

11 Is it possible for an object in the solar system to have a synodic period of exactly one year? Explain your answer.

12 Explain qualitatively (in words) the systematic decrease in the synodic periods of the planets from Mars outward, as shown in Table 2-1.

*13 A line joining the Sun and an asteroid was found to sweep out 5.2 square astronomical units of space in 1994. How much area was swept out in 1995? In five years?

*14 A comet moves in a highly elongated orbit about the Sun with a period of 1000 years. What is the length of the semimajor axis of the comet's orbit? What is the farthest the comet can get from the Sun?

*15 The orbit of a spacecraft about the Sun has a perihelion distance of 0.5 AU and an aphelion distance of 3.5 AU. What is the spacecraft's orbital period?

*16 Look up orbital data for the four largest moons of Jupiter on the Internet, in the current issue of a reference from the U.S. Naval Observatory, such as the *Astronomical Almanac* or *Astronomical Phenomena,* or in such magazines as *Sky &*

*Telescope* and *Astronomy*. Demonstrate that these data obey Kepler's third law.

17  Make a diagram of Jupiter's phases as seen from Earth and as seen from Saturn.

18  In what direction (left or right, eastward or westward) across the celestial sphere do the planets normally appear to move as seen from Australia? In what direction is retrograde motion as seen from there?

## Discussion Questions

19  Which planet would you expect to exhibit the greatest variation in apparent brightness as seen from Earth? Explain your answer.

20  Use two thumbtacks, a loop of string, and a pencil to draw several ellipses. Describe how the shapes of the ellipses vary as you change the distance between the thumbtacks.

## What If ...

*21  The Earth was 2 AU from the Sun? What would the length of the year be? What about if the Earth was ½ AU from the Sun?

*22  The Earth was moved to a distance of 10 AU from the Sun? How much stronger or weaker would the Sun's gravitational pull be on the Earth?

## Observing Projects

23  It is quite probable that within a few weeks of your reading this chapter one of the planets will be in opposition or at greatest eastern elongation, making it readily visible in the evening sky. Using the Starry Night computer program, the Internet, or consulting a reference book, such as the current issue of the *Astronomical Almanac* or the pamphlet entitled *Astronomical Phenomena* (both published by the U.S. government), select a planet that is at or near such a configuration. At that configuration, would you predict the planet to be moving rapidly or slowly from night to night against the background stars? Verify your predictions by observing the planet once a week for a month, recording your observations on a star chart.

24  If Jupiter happens to be visible in the evening sky, observe the planet with a small telescope on five consecutive clear nights. Record the positions of the four Galilean satellites by making nightly drawings, just as the Jesuit priests did in 1620 (see Figure 2-9). From your drawings, can you tell which moon orbits closest to Jupiter and which orbits farthest? Is there a night when you could see only three of the moons? What do you suppose happened to the fourth moon on that night?

25  Using the Starry Night software, switch off the daylight and set your latitude to ±90° (North or South Pole). Center and lock on Jupiter and zoom in until the moons are plainly visible. Note the positions of the moons. Step forward in increments of one hour and draw the positions of the moons at each timestep. From your drawings, can you tell which moon orbits closest to Jupiter and which orbits farthest

away? Are there timesteps when you could see only three of the Galilean moons? What do you suppose happened to the fourth moon at those times?

26  If Venus happens to be visible in the evening sky, observe the planet with a small telescope once a week for a month. On each night, make a drawing of the phase that you see. From your drawings can you determine if the planet is nearer or farther from the Earth than the Sun is? Do your drawings show any changes in the shape of the phase from one week to the next? If so, can you deduce if Venus is coming toward us or moving away from us?

27  Using your Starry Night software, set the latitude to ±90° (North or South Pole), center and lock on Venus, turn off the daylight and the horizon, and zoom in until your can see the planet clearly. Draw the current phase of Venus. Adjust the timestep to 7 days and draw Venus to scale with the first drawing at each succeeding timestep. From your drawings, can you determine if the planet is nearer or farther from the Earth than the Sun is? Do your drawings show any changes in the planet's phase throughout the week? If so, can you deduce from this information whether Venus is coming toward us or moving away from us?

28  Perform project 27 using Mars instead of Venus. If you have done project 27, compare your results for the two planets. Why are the cycles of phases as seen from Earth different for the two planets?

**Mauna Kea Observatories**
Astronomers prefer to build ground-based observatories on isolated mountaintops far from city lights, and where the air is dry, stable, and cloud-free. Mauna Kea Observatories, on the island of Hawaii, is home for a dozen optical and infrared telescopes. The twin Keck Telescopes are visible in the center of the photograph. The island of Maui is in the distance.

# Light and Telescopes

## *In this chapter you will discover*

✳ the nature of visible light and other types of electromagnetic radiation

✳ how telescopes collect and focus light

✳ the limitation of telescopes

✳ why different types of telescopes are used for different types of research

✳ how astronomers are developing new generations of high-technology telescopes

✳ how astronomers use the entire spectrum of electromagnetic radiation to observe the stars and other astronomical objects and events

## WHAT DO YOU THINK?

1  What is light?

2  Which type of electromagnetic radiation is most dangerous to life?

3  What is the main purpose of a telescope?

4  Why do stars twinkle?

5  What type(s) of electromagnetic radiation can telescopes currently detect?

The telescope is the single most important tool of astronomy. Using a telescope, we see objects in space far more brightly, clearly, and at greater distances than we can with the naked eye. Telescopes have played a major role in revealing the universe since Galileo viewed the craters on the Moon four centuries ago.

Refracting telescopes, which use large lenses to collect incoming starlight, were popular with astronomers 100 years ago. Modern astronomers strongly prefer reflecting telescopes, which gather light with large curved mirrors. In either case, the main purpose of a telescope is to collect as much light as possible. Astronomers attach a variety of equipment to telescopes to record incoming starlight.

# THE NATURE OF LIGHT

Visible light is the form of electromagnetic energy to which our eyes are sensitive. But what exactly is visible light? How is it produced? What is it made of? How does it move through space? Scientists have struggled with these questions for the past four hundred years.

## 3-1  Early discoveries explained white light and revealed the speed of light

The first major breakthrough in understanding light came from a simple experiment performed by Isaac Newton in the late 1600s. He passed a beam of sunlight through a glass prism, which spread the light out into the colors of the rainbow, as shown in Figure 3-1a.

This rainbow, called a **spectrum** (*plural* **spectra**), suggested to Newton that white light is actually a mixture of all colors. In Newton's day, most people erroneously believed that the colors were somehow added to the light by the glass. Newton disproved this idea by focusing light through a second prism. Because only white light emerged from this second prism, he showed that the second prism had reassembled the colors of the rainbow to give back the original beam of sunlight (Figure 3-1b).

Light of any color travels incredibly quickly, far faster than sound. This is why we see lightning before we hear the accompanying thunderclap. But does light travel instantaneously from one place to another, or does it move with a measurable speed?

It is interesting to see how the discovery of the Galilean moons set the stage for another discovery, the finite speed of light. The first evidence for the finite speed of light came in 1675, when Ole Rømer, a Danish astronomer, carefully timed eclipses of Jupiter's moons (Figure 3-2). Rømer discovered that the moment at which a moon enters Jupiter's shadow depends on the distance between the Earth and Jupiter. When Jupiter is in opposition, that is, when Jupiter and the Earth are on the same side of the Sun, the Earth-Jupiter distance is relatively short compared to when Jupiter is near conjunction. At opposition, Rømer found that eclipses occur slightly earlier than expected, while they occur slightly later than predicted by Kepler's laws when Jupiter is near conjunction.

Rømer correctly concluded that light takes more time to travel longer distances across space. The greater the distance to Jupiter, the longer the image of an eclipse takes to reach our eyes. From his timing measurements, Rømer concluded that it takes 16½ minutes for visible light to traverse the diameter of the Earth's orbit (2 AU). Incidentally, Rømer's interpretation of the data makes sense only if both the Earth and Jupiter orbit the Sun. His experiment therefore supported the heliocentric cosmology but not the geocentric one.

Because no one then knew the length of the astronomical unit (Earth's average distance to the Sun), Rømer could not actually compute the speed of light. The first accurate laboratory measurements of the speed of visible light were performed in the mid-1800s. The constant speed of light in a vacuum, usually designated by the letter $c$, is 299,792.458 km/s, which we generally round to

$$c = 3.0 \times 10^5 \text{ km/s} = 1.86 \times 10^5 \text{ mi/s}$$

(Standard abbreviations for units of speed, such as km/s for kilometers per second and mi/s for miles per second will be used throughout the rest of this book.) Light traveling through air, water, glass, or any other substance always moves more slowly than it does in a vacuum.

The value $c$ is a fundamental property of the universe. It appears in equations that describe, among other things, atoms, gravity, electricity, and magnetism. Furthermore, according to Einstein's special theory of relativity (Chapter 13), nothing can travel faster than the speed of light in a vacuum.

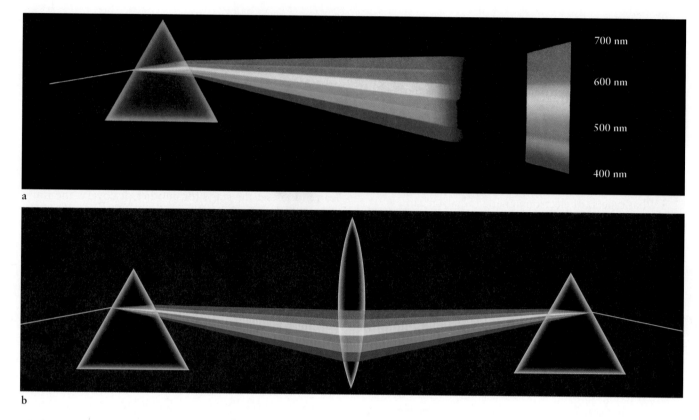

a

b

**Figure 3-1** **A Prism and a Spectrum** **(a)** When a beam of white light passes through a glass prism, the light is separated or refracted into a rainbow-colored band called a spectrum. The numbers on the right side of the spectrum indicate wavelengths in nanometers ($10^{-9}$ m). **(b)** Drawing of Newton's experiment showing that colors do not come from the glass of a prism. The second prism undoes the effects of the first prism, rather than adding more color, which it should if the prism creates the color.

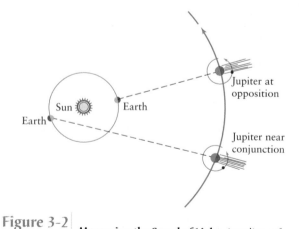

**Figure 3-2** **Measuring the Speed of Light** An eclipse of Jupiter's moon seen at opposition (Earth and Jupiter on the same side of the Sun) appears to occur earlier than an eclipse seen near conjunction (Earth and Jupiter on opposite sides of the Sun). The differences in apparent times of the eclipses are due to the extra time it takes light to travel the additional distance when the planets are near conjunction. The actual time of the eclipse is determined using Kepler's laws.

## 3-2 The complex nature of light became apparent only in the twentieth century

In the mid-1600s, the Dutch scientist Christiaan Huygens proposed that light travels in the form of waves. Isaac Newton, on the other hand, performed many experiments in optics that convinced him that light is composed of tiny particles of energy.

The English physicist Thomas Young demonstrated in 1801 that light has a wave nature. Young showed that the shadows of objects in light of a single color are not crisp and sharp. Instead, the boundary between illuminated and shaded areas is overlaid with patterns of closely spaced dark and light bands. These patterns are similar to the patterns produced by water waves passing through a barrier in the ocean (Figure 3-3). Young showed that a wavelike description of light could explain the results of his experiment but Newton's particle theory could not. Analyses of similar experiments soon produced overwhelming evidence for the wavelike behavior of light.

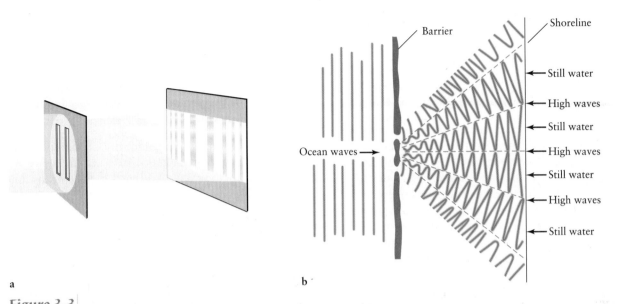

**Figure 3-3**  **Electromagnetic Radiation Travels as Waves** **(a)** Thomas Young's interference experiment shows that light of a single color passing through a barrier with two slits creates alternating light and dark patterns on a screen. **(b)** The intensity of light on the screen is analogous to the height of water waves that strike the shore after passing through a barrier with two openings. In certain places, ripples from both openings reinforce each other to produce extra high waves. At intermediate locations along the shoreline, ripples and troughs cancel each other, producing still water.

Further insight into the wave character of light came from calculations by the Scottish physicist James Clerk Maxwell in the 1860s. Maxwell succeeded in describing all the basic properties of electricity and magnetism in four equations. By combining these equations, Maxwell demonstrated that electrical and magnetic effects should travel through space together in the form of waves. Maxwell's suggestion that these waves are observed as light was soon confirmed by a variety of experiments. Because of its electric and magnetic properties, visible light is a form of *electromagnetic radiation* (Figure 3-4).

Newton showed that sunlight is actually composed of all the colors of the rainbow. Young and others showed that light travels as waves. What makes the colors of the rainbow distinct from each other? The answer is surprisingly simple: The only difference between different colors is the distance between two successive wave crests in the light wave. This distance is called the **wavelength** of the light, usually designated by the lowercase Greek letter $\lambda$ (lambda; see Figure 3-4). Maxwell's calculations predicted that light waves all travel at the same speed in empty space, regardless of wavelength.

The wavelengths of all colors are extremely small, less than a thousandth of a millimeter. To express these tiny distances conveniently, scientists use a unit of length called the *nanometer* (nm), where 1 nm = $10^{-9}$ m. Experiments demonstrated that visible light has wavelengths covering the range from about 400 nm for the shortest wavelength of violet light to about 700 nm for the longest wavelength of red light. Intermediate colors of the rainbow fall between these wavelengths (see Figure 3-5). The complete spectrum of colors from longest wavelength to shortest is red, orange, yellow, green, blue, and violet.

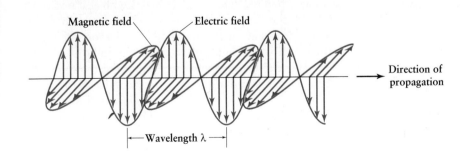

**Figure 3-4**  **Electromagnetic Radiation** All forms of electromagnetic radiation (radio waves, infrared radiation, visible light, ultraviolet radiation, X rays, and gamma rays) consist of oscillating electric and magnetic fields perpendicular to each other that move through empty space at a speed of $3 \times 10^5$ km/s. The distance between two successive crests is called the wavelength of the light.

## 3-3 Einstein showed that light sometimes behaves as particles and sometimes as waves

In 1905, just as scientists were becoming comfortable with the wave nature of light, Albert Einstein proposed that light sometimes behaves as particles! Einstein used this idea to explain the photoelectric effect: Shorter wavelengths of light can knock some electrons off the surfaces of metals while longer wavelengths of light cannot. Einstein knew that electrons are bound onto a metal's surface by electric forces and that it takes energy to overcome those forces. Because some colors (or, equivalently, wavelengths) can remove the electrons and others cannot, the electrons must receive different amounts of energy from different colors of light. But how? Einstein proposed that light travels as waves enclosed in discrete packets, called **photons**, and that photons with different wavelengths have different amounts of energy. *Specifically, the shorter the wavelength, the higher a photon's energy.*

$$\text{Photon energy} = \frac{\text{Planck's constant} \times \text{speed of light}}{\text{wavelength}}$$

where Planck's constant (named for the German physicist Max Planck) has the value $6.6 \times 10^{-34}$ kg·m²/s (see An Astronomer's Toolbox 3-1). Einstein's theory means that light can act both as waves and as particles.

All photons with the same wavelength are identical to each other, and, therefore, every photon of a given wavelength carries the same amount of energy as every other photon of that wavelength. The energy delivered by a photon is either enough to rip an electron off the surface of the metal or it is not. There is no middle ground, as there would be if light of one color came in packages with different numbers of waves so that photons of the same color could carry different amounts of energy. If this were the case, some photons of a certain color would have enough energy to free an electron, while other photons of the same color would not. Experimentally, this never happens. The concept of photons as just described has now been thoroughly proved, and astronomers incorporate it in their model of light.

## 3-4 Light is only one type of electromagnetic radiation

Visible light has a narrow range of wavelengths, from about 400 to 700 nm. But Maxwell's equations placed no length restrictions on the wavelengths of electromagnetic radiation. Researchers therefore began to look for forms of this radiation outside the range of wavelengths to which the cells of the human retina respond.

Around 1800 the British astronomer William Herschel discovered **infrared radiation** in an experiment with a prism. When he held a thermometer just beyond the red end of the visible spectrum, the thermometer registered a temperature increase, indicating that it was being heated by an invisible form of energy. Infrared radiation was identified as electromagnetic radiation with wavelengths slightly longer than red light. In experiments with electric sparks in 1888, the German physicist Heinrich Hertz first succeeded in producing electromagnetic radiation a few centimeters in wavelength, now known as **radio waves.** In 1895 Wilhelm Roentgen invented a machine that produces electromagnetic radiation with wavelengths shorter than 10 nm, now called **X rays.** Modern versions of Roentgen's machine are found in medical and dental offices and airport security checkpoints. Over the years, forms of radiation have been discovered with many other wavelengths.

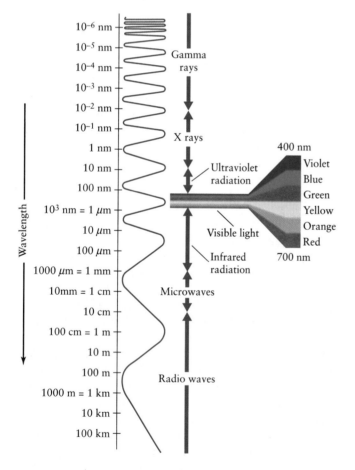

**Figure 3-5** | **The Electromagnetic Spectrum** The full array of all types of electromagnetic radiation is called the electromagnetic spectrum. It extends from the shortest-wavelength gamma rays to the longest-wavelength radio waves. Visible light forms only a tiny portion of the full electromagnetic spectrum. Note that 1 μm (one micrometer) is $10^{-6}$ meters.

## AN ASTRONOMER'S TOOLBOX 3-1

### Photon Energies

All photons with the same energy are identical: Their energy depends solely on the photon's wavelength. To find any photon's energy we use the equation

$$E = \frac{hc}{\lambda}$$

where Planck's constant $h$ is $6.6 \times 10^{-34}$ kg·m²/s, the speed of light $c$ is 300,000 km/s, and the photon's wavelength is $\lambda$.

*Example:* A photon of red light has a wavelength 700 nm. What is its energy? *Note:* All distances in the following equation must be converted to the same units, such as meters.

$$E_{red} = \frac{(6.6 \times 10^{-34}\ \text{kg} \cdot \text{m}^2/\text{s})(300,000\ \text{km/s})}{700\ \text{nm}}$$

$$E_{red} = 2.8 \times 10^{-19}\text{J}$$

(A joule, abbreviated J, is a unit of energy.) Each photon of red light with wavelength 700 nm has an energy of $2.8 \times 10^{-19}$ J.

*Compare!* In 1 second a 25-watt light bulb emits 25 J.

*Self-Test:* A photon has energy $4.9 \times 10^{-19}$ J. Calculate its wavelength. Referring to Figure 3-1, what is this photon's color?

---

We now know that visible light occupies only a tiny fraction of the full range of possible wavelengths, collectively called the **electromagnetic spectrum.** As shown in Figure 3-5, the electromagnetic spectrum stretches from the longest-wavelength radio waves, through infrared radiation, visible light, ultraviolet radiation, and X rays, to the shortest-wavelength photons, gamma rays. On the long-wavelength side of the visible spectrum, infrared radiation covers the range from about 700 nm to 1 mm. Astronomers interested in infrared radiation often express wavelength in *micrometers* or *microns* (abbreviated μm), where 1 μm = 1000 nm = $10^{-6}$ m. From roughly 1 mm to 10 cm is the range of microwaves, which are sometimes considered to be infrared radiation and sometimes radio waves.

At wavelengths shorter than those of visible light, **ultraviolet (UV) radiation** extends from about 400 nm down to 10 nm. Next are X rays, with wavelengths between about 10 and 0.01 nm, and beyond them are **gamma rays.** It should be noted that these boundaries are approximate and are primarily used as convenient divisions in the electromagnetic spectrum, which is actually continuous.

These various types of electromagnetic radiation share many basic properties. For example, they are all photons, they all travel at the speed of light, and they all sometimes behave as particles and sometimes as waves. But because of their different wavelengths (and therefore different energies), they interact very differently with matter. Your body tissues are nearly transparent to X rays but not to visible light; your eyes respond to visible light but not to infrared radiation; your radio detects radio waves but not ultraviolet radiation.

The Earth's atmosphere is relatively transparent to both visible light and radio waves, meaning that both pass through to reach ground-based telescopes sensitive to these forms of electromagnetic radiation. Astronomers say that the atmosphere has *windows* for these parts of the electromagnetic spectrum (Figure 3-6). Infrared radiation has a limited window through the atmosphere. We detect this radiation as heat.

Likewise, the longest-wavelength ultraviolet radiation, called UV-A, has a limited window. This radiation causes tanning and sunburns. Ozone ($O_3$) in the Earth's atmosphere screens out intermediate-wavelength ultraviolet radiation. This *ozone layer* is being depleted by human-made chemicals, such as chlorofluorocarbons (CFCs). As a result, more intermediate-wavelength ultraviolet, called UV-B, is reaching the Earth's surface, and these highly energetic photons damage living tissue, causing skin cancer and glaucoma, among other diseases. The Earth's atmosphere is completely opaque to the other types of electromagnetic radiation, meaning they do not reach the Earth's surface. (This is a good thing, because short-wavelength ultraviolet radiation (UV-C), X rays, and gamma rays are devastating to living tissue. Gamma rays are the deadliest.) Observations of these wavelengths must be performed high in the atmosphere or, ideally, from space.

Detecting electromagnetic radiation is the essence of observational astronomy. Until recently, photons were the only sources of detailed astronomical information that we had. However, in the past three decades, astronomers have built detectors of nonelectromagnetic energies and particles from space. Chapters 9 and 13 discuss two of these devices—neutrino detectors and gravity wave detectors.

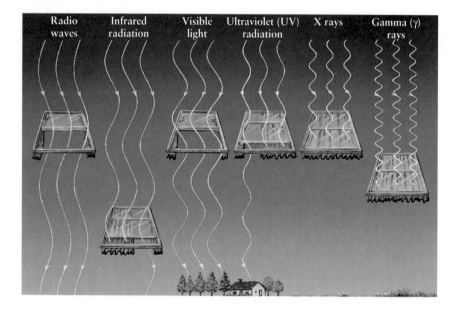

**Figure 3-6** **Windows Through the Atmosphere** The Earth's atmosphere allows different types of electromagnetic radiation to penetrate into it in varying amounts. Visible light and radio waves reach all the way to the Earth's surface. Some infrared and ultraviolet radiation also can reach the ground. The other types of radiation are absorbed or reflected by the gases in the air at different characteristic altitudes. While the atmosphere does not have actual "windows," astronomers use the term to characterize the passage of radiation through it.

# Optics and Telescopes

Since the time of Galileo, astronomers have been designing instruments to collect more light than the human eye can collect on its own. Collecting more light enables us to see things more brightly, in more detail, and at a greater distance. As we will see, there are just two basic types of telescopes—those that collect light through lenses and those that collect it from mirrors. Let's begin with the earliest telescopes, which use lenses.

## 3-5 A refracting telescope uses a lens to concentrate incoming light

Although light travels at about 300,000 km/s in a vacuum, it moves more slowly through a dense substance such as glass. As light enters the glass, it slows abruptly, much like a person walking from a boardwalk onto a sandy beach: Her pace suddenly slows as she steps from the smooth pavement onto the sand. And just as the person stepping back onto the boardwalk easily resumes her original pace, light exiting a piece of glass resumes its original speed.

As a result of changing speed, light also changes direction as it passes from one transparent medium into another—a phenomenon called **refraction.** You see refraction every day when looking through windows. Imagine a stream of photons from a star entering a window, as shown in Figure 3-7a. Astronomers call such photon flows *light rays*. As a light ray goes from the air into the glass, the light ray's direction changes so that it is more perpendicular to the surface of the glass than it was before entering. Once inside the glass, the light ray travels in a straight line. Upon emerging from the other side, the light

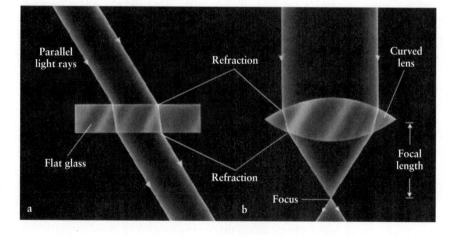

**Figure 3-7** **Refraction Through Flat and Curved Glass** **(a)** A light ray entering a flat piece of glass is bent or refracted to an angle closer to perpendicular to the surface than the angle at which it was originally traveling. As the ray leaves the piece of glass, it is bent away from the perpendicular. **(b)** When the glass is curved to form a parabolic convex lens, parallel light rays converge to a focus. The distance from the lens to the focus is the focal length of the lens.

ray bends once again, resuming its original direction and speed. The net effect is only a slight displacement. This is why what we see through a window is a good representation of what lies beyond the glass.

A telescope lens uses refraction to collect light. Unlike a window, lenses have curved surfaces (Figure 3-7b). These curves force the light rays to emerge from the lens in different directions than they had before entering the lens. Different rays striking the top surface of the lens at different places are refracted by different amounts: If the lens is shaped correctly, parallel light rays start converging once they enter the glass. As the light emerges from the glass, it converges further.

Light rays meet, or "come to a focus," at a point called the **focal point** of the lens. Its distance from the lens is the **focal length** (Figure 3-7b). Actually, only *parallel* light rays all meet at the focal point. This occurs for light from sources that are extremely far away, like the stars (Figure 3-8 shows why). If the object is close enough to be more than just a dot as seen through the telescope, then all the light from it does not converge at the focal point but rather focuses along a surface, the **focal plane** (Figure 3-9). The Moon and planets are examples of objects with images that extend over the focal plane.

A **refracting telescope,** or **refractor** (Figure 3-10) is an arrangement of two lenses used to gather light. The resulting image is a brighter, and often a bigger and clearer, view of the object. The lens at the top of the telescope, called the **objective lens,** has a large diameter and long focal length. Its purpose is to collect as much light as possible. The lens at the bottom of the telescope, the **eyepiece lens,** is smaller and has a short focal length. It restraightens the light rays after they have passed through the focal point or focal plane of the objective lens, making them parallel once again. Because the light is now more concentrated, it is brighter than before it entered the telescope.

**Figure 3-8** **Parallel Light Rays from Distant Objects** As light travels away from any object, the light rays, all moving in straight lines, separate. By the time light has traveled trillions of miles, only the light rays moving in parallel tracks are still near each other.

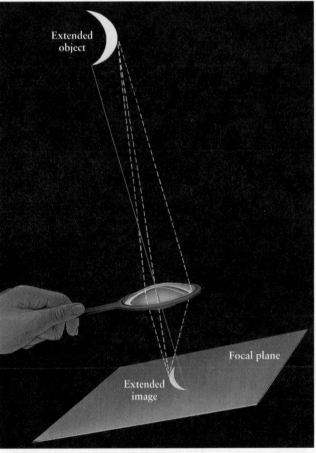

**Figure 3-9** **Extended Objects Create a Focal Plane** Light from objects larger than points in the sky does not all converge to the focal point of a lens. Rather, the object creates an image at the focal length in what is called the focal plane.

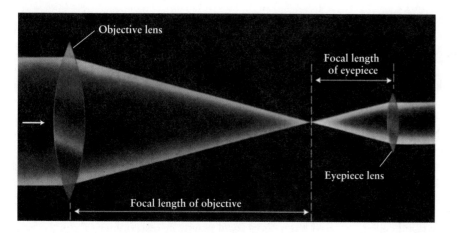

**Figure 3-10** **Essentials of a Refracting Telescope** A refracting telescope consists of a large, long-focal-length objective lens that collects and focuses light rays and a small, short-focal-length eyepiece lens that restraightens the light rays. The lenses work together to brighten, resolve, and magnify the image formed at the focal plane of the objective lens.

## 3-6 Telescopes brighten, resolve, and magnify

A telescope's most important function is to provide the astronomer with as many photons from the object as possible. The more photons, the more information the astronomer can extract—even if the object still appears as a pinpoint. For example, given enough photons, astronomers can measure the relative intensities of different wavelengths emitted by a star, which provides information about its temperature, chemical composition, age, and motion.

The observed brightness of any object depends on the total number of photons collected from it, which in turn depends on the area of the telescope's objective lens. A large objective lens intercepts and focuses more light than does a small objective lens. A large objective lens can therefore produce brighter images and detect fainter stars than a small objective lens can for an equal exposure time (Figure 3-11). No wonder telescopes are sometimes called "photon buckets."

> **Insight into science** **Costs and benefits** Science now relies heavily on technology to conduct experiments or make observations. The cost of cutting-edge astronomical observations may run to hundreds of millions of dollars. The return on such investments is a better understanding of how the universe works, how we can harness its capabilities, and our place in it.

The **light-gathering power** of a telescope is directly related to the area of the telescope's objective lens (or primary mirror; see Section 3-8). Recall that the area and diameter of a circle are related by

$$Area = \frac{\pi \times diameter \times diameter}{4}$$

where π (pi) is about 3.14.

Consequently, a lens with twice the diameter of another lens has 4 times the area of the smaller lens and therefore collects 4 times as much light as the smaller one. For example, a 36-cm-diameter lens has 4 times the area of an 18-cm-diameter lens. Therefore, the 36-cm telescope has 4 times the light-gathering power of a telescope half its size. The general rule is: *Double the diameter, quadruple the light-gathering power.*

The second most important function of any telescope is to reveal more details of the extended objects under study. A large telescope increases the sharpness of the image and the degree of detail that can be seen. **Angular resolution** measures the clarity of images (Figure 3-12). Poor angular resolution causes images of galaxies or planets to be fuzzy and blurred. A telescope with good angular resolution produces images that are sharp and crisp.

The angular resolution of a telescope is measured as the arc angle between two adjacent stars whose images can just barely be distinguished by that telescope. The smaller the angle, the sharper the image. Large, modern telescopes, like the Keck telescopes in Hawaii which we will discuss later in this chapter, have angular resolutions better than 0.1 arcsec. As a general rule: *Double the diameter, double the detail that can be seen.*

The final function of a telescope—often the least important one—is to make objects appear larger. This property is called **magnification**. The magnification of a refracting telescope is equal to the focal length of the objective lens divided by the focal length of the eyepiece lens:

$$Magnification = \frac{focal\ length\ of\ the\ objective}{focal\ length\ of\ the\ eyepiece}$$

For example, if the objective of a telescope has a focal length of 100 cm and the eyepiece has a focal length of 0.5 cm, then the magnifying power of the telescope is

$$Magnification = \frac{100\ cm}{0.5\ cm} = 200$$

This property is usually expressed as 200X.

**Figure 3-11** **Light-Gathering Power** Because a large lens intercepts more starlight than a small lens, a large lens produces a brighter image. The same principle applies to telescopes that collect light using a primary mirror rather than an objective lens. The two photographs of the Andromeda galaxy were taken through telescopes with different diameters and were exposed for equal lengths of time at equal magnification.

**Figure 3-12** **Resolution** The larger the diameter of a telescope's objective lens or primary mirror, the greater the detail the telescope can resolve. These two images of the Andromeda galaxy, taken through telescopes with different diameters, show this difference. Increasing the exposure time of the smaller diameter telescope (left), will only brighten the image, not improve the resolution.

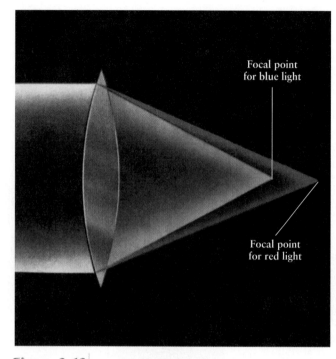

**Figure 3-13** **Chromatic Aberration Spreads Colors Out**
Light of different colors passing through a lens is refracted by different amounts. This effect is called chromatic aberration. As a result, different-colored objects have different focal lengths. Consequently, the images seen through uncorrected lenses are blurred.

**Figure 3-14** **An Achromatic Lens Corrects for Chromatic Aberration** A second lens (not the eyepiece) that refracts colors by different amounts than does the objective lens can bring all colors into focus at the same focal length.

There is a limit to the magnification of any telescope. Try to magnify beyond that limit and the image becomes distorted. As a rule, *double the objective lens's diameter, double the telescope's maximum magnification.*

## 3-7 Refracting telescopes have drawbacks

For convenience in manufacturing, objective lenses for refractors have spherical surfaces. However, a spherical lens distorts images, a problem called **spherical aberration.** This is only the first drawback of using objective lenses in large research telescopes.

The amount of refraction that light undergoes varies with color (recall Figure 3-1). Because different colors have different focal lengths, any lens distorts the colors of images (Figure 3-13), a problem called **chromatic aberration.** This problem is corrected with a *compound lens* composed of two different types of glass with different refractive properties. Because compound lenses in telescopes and cameras bring all the wavelengths into focus at the same focal length, they are called **achromatic lenses** (Figure 3-14).

A third problem, *sagging* of large lenses, became apparent in the nineteenth century. Despite their apparent rigidity, the shapes of glass structures, like large lenses,

distort under the pull of the Earth's gravity. This distortion occurs because the lens can only be supported around its edge, which is very thin. The thick center of a lens weighs it down and causes the glass to deform. As a result, the image is distorted. Moreover, as the telescope *tracks,* or follows, a star, the orientation of the telescope changes and, along with it, the amount of distortion. There is no way of knowing the exact distortion that will occur. As a result, gravity causes the images of stars to move during a single observation, causing them to appear blurry on film or other recording medium.

*Unwanted refractions* are a fourth problem. Lenses are ground from large, thick disks of glass that are formed by pouring molten glass into a mold. As the liquid glass cools, gas in it becomes trapped, creating air bubbles in the solidified disk. When the lens is then ground into shape, any air bubbles inside it create unwanted and unpredictable extra refractions that blur the images. Part of a lensmaker's expertise was to choose a volume of glass for grinding with as few air bubbles as possible.

A fifth problem is that glass is *opaque* to certain ranges of wavelengths. These wavelengths of light cannot pass through the glass. Even visible light is dimmed substantially in passing through the slab of glass at the front

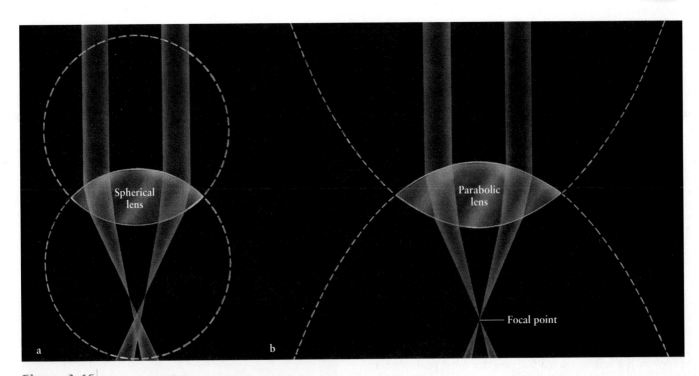

**Figure 3-15** **Spherical Aberration and Parabolic Lenses** **(a)** Different parts of a spherical lens refract light to different focal points. **(b)** The ideal shape for a lens is a parabola. A parabolic lens focuses all the light passing through it at the same focal length.

of a refractor, and ultraviolet radiation is largely absorbed by a glass lens.

We have already mentioned that spherical lenses produce smeared images (Figure 3-15a). Spherical aberration is overcome by focusing light through parabolic surfaces (Figure 3-15b), but grinding parabolic lenses is so difficult that for practical purposes, objective lenses always have spherical surfaces. Spherical aberration can also be minimized by making the objective lens so thin that its spherical surfaces nearly coincide with parabolic surfaces. A thin lens, however, has an extremely long focal length, which is why research refracting telescopes are many meters long (Figure 3-16). In fact, no major refracting telescopes were constructed in the twentieth century.

Nineteenth-century master opticians devoted their lives to overcoming the problems inherent in refracting telescopes, and several magnificent refractors were constructed in the late 1800s. The largest, completed in 1897 and located at the Yerkes Observatory near Chicago (see Figure 3-16), has an objective lens 102 cm (40 in.) in diameter with a focal length of 19 1/3 m (63.5 ft). The second largest refracting telescope, located at Lick Observatory near San Jose, California, has an objective lens of 91 cm (36 in.) in diameter.

**Figure 3-16** **A Large Refracting Telescope** This giant refracting telescope, built in the late 1800s, is housed at Yerkes Observatory near Chicago. The objective lens is 102 cm (40 in.) in diameter, and the telescope tube is 19 1/3 m (63½ ft) long.

## 3-8 Reflecting telescopes use mirrors to concentrate incoming starlight

Prompted by the problem of chromatic aberration, early in the eighteenth century Isaac Newton set about to replace the objective lens with a curved mirror to collect light. Called **reflecting telescopes** or **reflectors**, telescopes with mirrors easily overcome most of the problems inherent in refracting telescopes.

A mirror collects light using reflection rather than refraction. In Figure 3-17, the ray of light strikes a flat mirror, and we imagine a perpendicular line coming out of the mirror at that point. According to the principle of **reflection**, the angle between the arriving light ray and the perpendicular (dashed line) is always equal to the angle between the reflected light ray and the perpendicular. The rule is the same regardless of the light's wavelength or the mirror's shape.

Using this principle, Newton realized that a concave, parabolic mirror will cause parallel light rays to converge to a focal point, as shown in Figure 3-18b. The distance between the reflecting surface and the focal point, where the image of the distant object is formed, is the focal length of the mirror. The magnifying power of such a reflecting telescope is calculated in the same way as for a refractor: The focal length of the large or **primary mirror** is divided by the focal length of the eyepiece.

In order to view the image, Newton placed a small, flat mirror at a 45° angle in front of the focal point, as

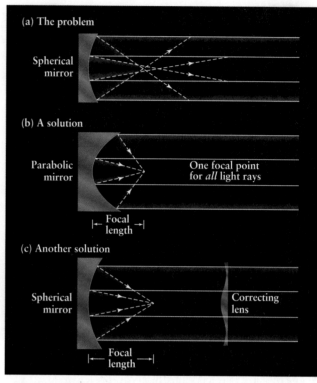

**Figure 3-18** **Spherical Aberration** **(a)** Different parts of a spherically concave mirror reflect light to slightly different focal points. This effect, spherical aberration, causes image blurring. This problem can be overcome by **(b)** using a parabolic mirror or **(c)** using a Schmidt corrector plate (lens) in front of the telescope.

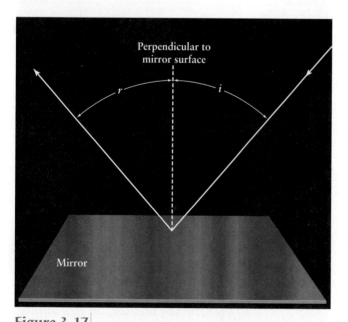

**Figure 3-17** **Reflection** The angle at which a beam of light strikes a mirror (the angle of incidence *i*) is always equal to the angle at which the beam is reflected from the mirror (the angle of reflection *r*).

sketched in Figure 3-19a. This **secondary mirror** reflects the light rays to one side of the telescope, and the astronomer views the image through an eyepiece lens. A telescope with this optical design is still called a **Newtonian reflector.**

Newtonian telescopes are very popular with amateur astronomers, because they are convenient to use while the observer is standing up. However, they are not used in research observatories because they are lopsided. If astronomers attach their often heavy and bulky research equipment onto its side, the telescope sags and distorts the image in unpredictable ways.

Three basic designs exist for the reflecting telescopes used in research. In the first design, the astronomer actually sits at the undeflected focal point, directly in front of the primary mirror, and controls the light-collecting equipment located there. This arrangement is called a **prime focus** (Figures 3-19b). All telescopes must be kept at the same temperature as the outdoors to prevent a temperature difference from expanding or contracting the telescope, thereby distorting the image. It is therefore often a chilling experience to ride throughout a mountaintop winter night in the open prime focus "observing cage."

In a second popular design, the **Cassegrain focus,** a hole is drilled directly through the center of the primary mirror. A convex secondary mirror placed between the primary mirror and its focal point reflects the light rays back through the hole (Figure 3-19c). This curved secondary mirror extends the telescope's focal length. Light-gathering equipment is bolted to the bottom of the telescope, and the light is brought into focus in it. This design has an advantage over Newtonian telescopes, in that the attached equipment is balanced and does not distort the telescope frame and, hence, the image.

The third design is handy for long and bulky optical equipment that cannot be mounted directly on the telescope or for observations that benefit from extremely long focal lengths. In this design (Figure 3-19d), a series of mirrors channels the light rays away from the telescope to a remote focal point called the **coudé focus** (named after a French word meaning "bent like an elbow").

Reflecting telescopes have numerous advantages over refracting ones. First, they avoid the problem of chromatic aberration, because, unlike household mirrors, all telescope mirrors have coated top surfaces, and light never enters the glass at all. Second, the mirrors, which can weigh tons, hardly warp at all, because they can be supported from underneath. Third, a mirror maker needs only to find a surface, rather than an entire volume, that is free of bubbles. Finally, because light does not enter the glass, the problem of a lens's opacity to different wavelengths never arises.

## 3-9  Reflecting telescopes also have limitations

The mirrored surfaces of research telescopes are polished so smoothly that the highest bumps are less than 1/500 the thickness of a human hair. Nonetheless, reflecting telescopes aren't perfect; there are several prices to pay for the advantages reflectors offer over refractors. Two of the most important are *blocked light* and *spherical aberration.* Let's consider each problem briefly.

You have probably noticed that the secondary mirror of a reflector blocks some incoming light, one unavoidable price that astronomers must pay. Typically, a secondary mirror prevents about 10% of the incoming light from reaching the primary mirror. This problem is addressed by constructing primary mirrors with sufficiently large surface areas to compensate for the loss of light. You might think as well that because light is missing from the center of the telescope due to blockage by the secondary, a corresponding central "hole" appears in the images. However, this problem does not occur, because light from all parts of each object enters all parts of the telescope (Figure 3-20).

To make a reflector, an optician traditionally grinds and polishes a large slab of glass into the appropriate concave shape. It is much easier to grind a spherical mirror than a more ideal parabolic one. However, light entering a spherical telescope mirror at different distances from the mirror's center comes into focus at different focal lengths,

**Figure 3-19** **Reflecting Telescopes** Four of the most popular optical designs for reflecting telescopes: **(a)** Newtonian focus (popular among amateur astronomers); and the three major designs used by researchers—**(b)** prime focus, **(c)** Cassegrain focus, and **(d)** coudé focus.

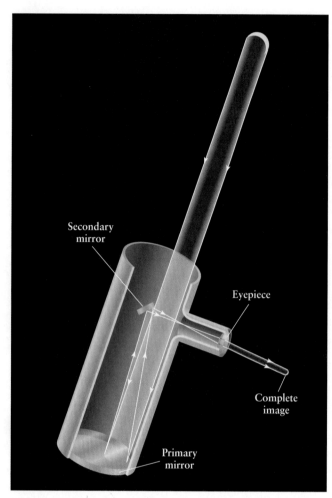

Secondary
mirror

Eyepiece

Complete
image

Primary
mirror

**Figure 3-20**   **The Secondary Mirror Does Not Create a Hole
in the Image**  Because the light rays from distant objects are parallel,
light from the entire object reflects off all parts of the mirror. Therefore,
every part of the object sends photons to the eyepiece. This figure shows
the reconstruction of the entire Moon from light passing through just
part of this telescope. The same drawing applies everywhere on the
primary mirror that is not blocked by the secondary mirror.

as occurs for a spherical lens, and images taken with all
such telescopes appear blurry (see Figure 3-18a).

Spherical aberration in reflecting telescopes with spherical primaries may be overcome by including a thin cor-
recting lens, called a **Schmidt corrector plate**, at the top of
the telescope (see Figure 3-18c). The light coming into the
telescope is refracted by the plate just enough to compensate for spherical aberration and to bring all the light into
focus at the same focal length. These correctors have the
added benefit of focusing light from a larger angle in the
sky than would be in focus without the plate. A Schmidt
corrector plate enables astronomers to map large areas of
the sky with relatively few photographs at moderately
high magnification. In other words, the Schmidt corrector
plate acts like a wide-angle lens on a camera. However,

the plate does not allow for as much magnification as a
telescope with a parabolic mirror.

While parabolic primary mirrors have been meticu-
lously ground over the past century, the advent of com-
puter-controlled grinding and rotating furnaces (Figure
3-21), in which the liquid glass actually spins into a parabolic shape, have now made it economical to cast parabolic mirrors with diameters of several meters.

## 3-10 Earth's atmosphere hinders astronomical research

Besides the problems and restrictions inherent in the optics
of telescopes, the Earth's atmosphere also introduces difficulties. The problems arise because the air is turbulent and
filled with impurities. You have probably seen turbulence
on a hot day while driving in a car, when the road ahead
appears to shimmer. Blobs of air, heated by the Earth,
move upward to create this effect. Light passing through
such a blob is refracted, because each hot air mass has a
different density than the cooler air around it. Hence,
because each blob behaves like a lens, images of objects
beyond it appear distorted.

The atmosphere over our heads is similarly moving
and changing density, and the starlight passing through it is
similarly refracted. Because the air density changes rapidly,
the resulting changes in refraction make the stars appear to
change brightness and position rapidly, an effect we see as
**twinkling**. When photographed through large telescopes,
twinkling smears out a star's image, causing it to look like
a disk rather than a pinpoint of light. Astronomers use the
expression "seeing" to describe how much twinkling is
occurring and, therefore, how smeared out are the images
from their telescopes.

The angular diameter of a star's smeared-out image,
called the **seeing disk**, is a realistic measure of the best
possible resolution. The size of the seeing disk varies
from one observatory site to another. At Mount Palomar
in California, the seeing disk is roughly one arcsec (1″).
The best conditions on Earth, with a seeing disk of 0.2″,
have been reported at the observatory on the 14,000-ft
summit of Mauna Kea, the tallest volcano on the island
of Hawaii (pictured in the photo that opens this chapter).

**Insight into science**   **Research requires patience**
Seeing—indeed, most observing in science—is rarely
ideal. Besides such natural phenomena, which are
beyond their control, scientists must also contend
with equipment failures, late deliveries of parts, and
design flaws. Patience is not a virtue in cutting-edge
science; it is a necessity. Next time you visit an
observatory, ask astronomers how the seeing has
been lately, but be prepared for a series of expletives!

a

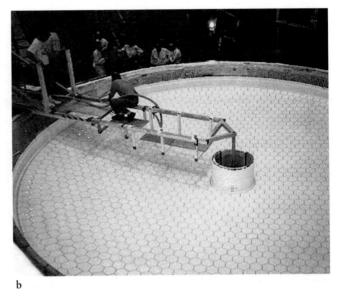

b

**Figure 3-21** **Rotating Furnace for Making Parabolic Telescope Mirrors** (a) To make each 8.4-meter primary mirror for the Large Binocular Telescope on Mount Graham, in Arizona, 40,000 pounds of glass are loaded into a rotating furnace and heated to 1450 K (2150°F). (b) After melting, spinning, and cooling, the mirror's parabolic surface is ready for final smoothing and coating with a highly reflective material.

Without the effects of the Earth's atmosphere, stars do not twinkle. As a result, photographs taken from telescopes in space reveal stars as much finer points, and more detail for extended objects (Figure 3-22). The Hubble Space Telescope, as we will see often in this book, now achieves magnificent resolution.

Light pollution from cities poses another problem for Earth-based telescopes (Figure 3-23). Keep in mind that the larger the primary mirror, the more light is gathered and therefore the more information astronomers can ob-

tain. The 5-m (200 in.) telescope at the Palomar Observatory between San Diego and Los Angeles, California, was the first truly great large telescope, providing astronomers with invaluable insights into the universe for decades. However, light pollution from the two cities now fills the night sky, seriously reducing the ability of that telescope to collect light from the stars. Not surprisingly, the best observing sites in the world are high on mountaintops—above smog, water vapor, and clouds—and far from city lights.

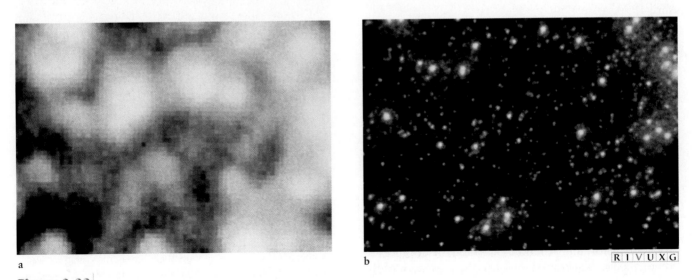

a

b

R I V U X G

**Figure 3-22** **Effects of Twinkling** The same star field photographed with (a) a ground-based telescope, which is subject to twinkling, and (b) the Hubble Space Telescope, which is free from the effects of twinkling.

**Figure 3-23**    **Light Pollution** These two images of Tucson, Arizona, were taken from the Kitt Peak National Observatory, which is 38 linear miles away. They show the dramatic growth in groundlight output between 1959 (top) and 1989 (bottom). Since 1972, light pollution, a problem for many observatories around the world, has been at least partially controlled by a series of local ordinances.

## 3-11  The Hubble Space Telescope provides stunning details about the universe

For decades, astronomers dreamed of a major observatory in space. Such a facility would eliminate the image distortion created by twinkling and by poor atmospheric transparency due to pollution, volcanic debris, and water vapor. It could operate 24 hours a day and over a wide range of wavelengths—from the infrared through the visible range and far out into the ultraviolet part of the spectrum. This was the mission of the Hubble Space Telescope (HST), which was carried aloft by the Space Shuttle in April 1990.

Soon after HST was placed in orbit, astronomers discovered that the telescope's 2.4-m primary mirror lacked the proper curvature, which caused its star images to be surrounded by a hazy glow. During a repair mission in December 1993, astronauts installed corrective optics that eliminated the problem (Figure 3-24). Now HST has a resolution of better than 0.1″, which is better than can be obtained by telescopes on the Earth's surface without the use of advanced technology (see section 3-12). Figure 3-25 compares HST pictures of part of a galaxy taken before and after the repair mission.

The observations taken by HST have staggered the imaginations of astronomers. It has dished up new observations related to the planets in our solar system, planetary systems forming around other stars, the distances to other galaxies, black holes, quasars, the formation of the earliest galaxies, and the age of the universe, among many other things. In the coming years, we can look forward to a wealth of crystal clear images of distant galaxies and interstellar clouds coming from HST and its successors, now under development.

**Figure 3-24**    **The Hubble Space Telescope (HST)** This photograph of HST hovering above the Space Shuttle's cargo bay was taken in 1993, at completion of the first servicing mission. During its 15-year lifetime, HST is studying the heavens at wavelengths from the infrared through the ultraviolet.

Figure 3-25  **A Comparison of HST Images**  These pictures of the nucleus of
the galaxy M100 compare HST images taken before and after the 1993 repair mission.
**(a)** Although atmospheric blurring is absent, HST's optical defect produced a "skirt"
around the nucleus. **(b)** After astronauts installed corrective optics, HST took this
sharp, clear picture of the galaxy.

## 3-12  Advanced technology is spawning a new generation of superb telescopes

The clarity of images taken by the Hubble Space Telescope may suggest that ground-based observational astronomy is a dying practice. However, two exciting ground-based techniques, called *active optics* and *adaptive optics*, promise to match the quality of Hubble—or better it!

**Active optics** finds the best orientation for the primary mirror in response to changes in temperature and the shape of the telescope mount. It adjusts the mirror every few seconds to help keep the telescope aimed at its target. With active optics, the New Technology Telescope in Chile and the Keck Telescopes in Hawaii routinely achieve resolutions as fine as 0.3" when the resolution is much worse for telescopes without active optics at the same sites.

**Adaptive optics** uses sensors to determine the amount of twinkling created by atmospheric turbulence. The stellar motion is neutralized by computer-activated, motorized supports that actually reshape either the primary mirror or a smaller mirror installed farther down the optical path of the telescope. Adaptive optics effectively eliminates atmospheric distortion and produces remarkably sharp images (Figure 3-26). Many large ground-based telescopes now

Figure 3-26  **Images from Earth and Space**  **(a)** Image of Saturn from an Earth-
based telescope without adaptive optics. **(b)** Image of Saturn from an Earth-based telescope
with adaptive optics. **(c)** Image of Saturn from the Hubble Space Telescope, which does not
incorporate adaptive optics technology.

use adaptive optics on at least some observations, resulting in images comparable to those from HST.

Until the 1980s, telescopes with primary mirrors of between 2 and 6 m were the largest and most powerful in the world. Now, new technologies in mirror building and computer control allow us to construct much larger telescopes. There are at least 38 reflectors around the world today with primary mirrors measuring 2 m or more in diameter. Among these are several with mirrors larger than 8 m in diameter, with more than half a dozen other very large telescopes under construction.

Because the cost of building very large mirrors is enormous, astronomers have recently devised less expensive ways to collect the same amount of light. One approach is to join several smaller mirrors together and aim them at the same focal point. The largest examples of this segmented-mirror technique are the 10-m (400 in.) Keck I and Keck II telescopes on the summit of Mauna Kea in Hawaii. Thirty-six hexagonal mirrors are mounted side by side in each telescope to collect the same amount of light as a single primary mirror 10 m in diameter (Figure 3-27). Another method combines images from different telescopes. For example, used together to observe the same object, the Keck Telescopes have the resolving power of a single 85-meter telescope.

## 3-13 Storing and analyzing light from space is key to understanding the cosmos

The invention of photography during the nineteenth century was a boon for astronomy. By taking a long exposure with a camera mounted at the focal point of a telescope, an astronomer could record extremely faint features that could not be seen just by looking through the telescope. The reason is that our eyes clear the images in them several times a second, whereas film adds up the intensity of all the photons affecting its emulsion. Brighter, higher-resolution images therefore reveal details in galaxies, star clusters, and planets that we could not otherwise see. Moreover, we can collect even more light by taking long exposures—hour-long exposures are quite routine.

Astronomers have long realized, however, that a photographic plate is an inefficient detector of light because it depends on a chemical reaction to produce an image. Typically, only 2% of the light striking a photographic plate triggers a reaction in the photosensitive material of

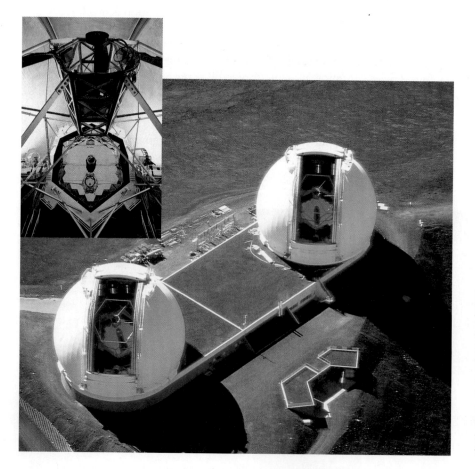

**Figure 3-27** | **The 10-m Keck Telescopes** Located on the (hopefully) dormant Mauna Kea volcano in Hawaii, these huge twin telescopes each consist of 36 hexagonal mirrors measuring 1.8 m (5.9 ft) across. Each Keck has the light-gathering, resolving, and magnifying ability of a single mirror 10 m in diameter. **Inset:** View down the Keck I telescope. The hexagonal apparatus near the top of the photograph shows the housing for the 1.4 m secondary mirror.

the plate. Thus, roughly 98% of the light falling onto a photographic plate is wasted.

Technology has changed all that. We have now replaced photographic film with highly efficient electronic light detectors called **charge-coupled devices (CCDs).** Each CCD is roughly the size of a large postage stamp (Figure 3-28). A CCD is divided into an array of small, light-sensitive squares called picture elements or, more commonly, **pixels.** For example, one of the latest CCDs has more than 16 million pixels arranged in 4096 rows by 4096 columns. This is about 1000 times more pixels per square centimeter than a typical computer screen.

When an image from a telescope is focused on the CCD, an electric charge builds up in each pixel in direct proportion to the intensity of the light falling on that pixel. When the exposure is finished, the charge on each pixel is read into a computer. The computer then transfers the image onto ordinary photographic film or a television monitor. CCDs commonly respond to 70% of the light falling on them; their resolution is better than that of film, and they respond more uniformly to light of different colors. Figure 3-29 shows one photograph and two CCD images of the same region of the sky, all taken

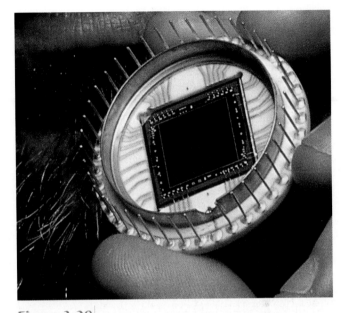

**Figure 3-28**   **A Charge-Coupled Device (CCD)** This tiny silicon square contains 16,777,216 light-sensitive pixels that store images in a one-piece CCD array. Electronic circuits transfer the data to a waiting computer.

a                                    b                                    c                RIVUXG

**Figure 3-29**   **Ordinary Photography Versus CCD Images** These three views of the same part of the sky, each taken with the same 4-m telescope, compare CCDs to photographic plates. **(a)** A negative print (black stars and white sky) of a photographic image. **(b)** A CCD image. Notice that many faint stars and galaxies virtually invisible in the ordinary photograph can be seen clearly in this CCD image. **(c)** This color view was produced by combining a series of CCD images taken through colored filters.

## DISCOVER FOR YOURSELF | Buying a Telescope

If you feel elated discovering treasures, you may want to consider viewing the night sky through powerful binoculars or a telescope. Both types of instruments enable you to find many beautiful, exciting objects in space. Binoculars are nice because you can quickly scan the heavens to see craters on the Moon, star clusters, the Great Nebula of Orion, and even the Andromeda galaxy. Telescopes open new vistas and even enable amateur astronomers to discover new comets, among other things, adding to our scientific database and immortalizing the discoverers' names.

Because the stars move continuously across the sky, it is essential to buy a telescope that can follow them reliably and easily. The tracking motion of the telescope is determined by the mount to which it is connected. The standard mount that enables you to track the stars by moving the telescope around one axis is called an *equatorial* mount. Tracking is more of a hit-or-miss proposition with alt-azimuth mounts than it is with an equatorial mount. Unless you are prepared to motorize and computerize the mount assembly, you should avoid alt-azimuth mounts. Whatever telescope you choose, you should be sure that the mount is light and compact enough so that you can conveniently transport the telescope.

Recall that objects appear brighter and are more highly magnified in larger-diameter telescopes than in smaller ones. You should buy the largest-diameter telescope you can afford. Active amateurs usually choose telescopes with diameters of at least 4 inches. However, keep in mind that the bigger the telescope, the heavier it is and the harder it is to transport.

You will also want to get a few different *eyepieces* (three is a good number to start with), so that you can look at large areas of the sky under low magnification and details of small areas under high magnification. As discussed in the text, the magnification of a telescope depends on the focal lengths of the objective lens and the eyepiece. The objective lens or primary mirror is fixed in the telescope. However, all telescopes come with removable eyepieces. Change the eyepiece with another of different focal length, and you change the telescope's magnification. Keep in mind, however, that increasing the magnification decreases the area of the sky that you see.

If you buy a telescope with the eyepiece positioned so that you have to crawl under it or climb a ladder to see through it, then you will find the experience less satisfying than if you can observe in a comfortable position. If the eyepiece is not easily accessible, you can buy a *diagonal mirror* (a right-angle mirror that goes between the telescope body and the eyepiece) to help correct the problem.

If you want to see especially large areas of the sky, you will want to purchase a telescope with a *Schmidt corrector plate*. The most common wide-field telescopes are of the Schmidt-Cassegrain design. These telescopes have a Schmidt corrector plate on top and a Cassegrain mirror on the bottom with their eyepieces located underneath.

It is also essential to have a *finder scope*, which is a small, low-magnification, very-wide-field telescope, with cross-hairs, attached directly onto your main telescope. If the finder and your main scope are well aligned, you can quickly zero in on an object.

If you plan to take photographs with a CCD camera or other instrument on your telescope or to show the cosmos to large groups of people, you will need a *tracking motor* to enable the telescope to follow the stars automatically. Tracking motors require their own power from batteries, an adaptor for a car cigarette lighter, or a 110-V outlet. The motor should also be able to run at high speed so you can slew (turn) the telescope rapidly from one object to the next. For photography, you will also need a *camera adapter*.

Viewing the Sun is very exciting, especially when you can observe sunspots. If you want to observe the Sun, it is essential that you buy a good *Sun filter*. **Never, ever look at the Sun directly, either through a telescope or with your naked eye. Doing so for even a second can lead to partial or total blindness!**

Finally, you will need a flashlight with a red gel coating over the lens. Because red light does not cause the pupils of your eyes to contract, they will remain dilated (wide open) while you use the red flashlight to inspect your equipment and your star charts.

This Schmidt-Cassegrain telescope, with its finder scope and motor-driven equatorial mount, is a favorite of dedicated amateur astronomers.

with the same telescope. Notice that many details visible in the CCD images are totally absent in the ordinary photograph.

# RADIO ASTRONOMY—AND BEYOND

Until the 1930s, all information that astronomers gathered from the universe was based on visible light. Scientists began to wonder if objects in the universe might also emit radio waves, infrared and ultraviolet radiation, and perhaps even X rays or gamma rays. Little did anyone realize back then the enormous range of objects and activities in the universe that emit one or more of these radiations without giving off any detectable visible light!

## 3-14 A radio telescope uses a large concave dish to reflect radio waves to a focal point

The first evidence of nonvisible radiation from outer space came from the work of a young radio engineer, Karl Jansky, working at Bell Laboratories. Using long antennas, Jansky was investigating the sources of radio static that affect short-wavelength radiotelephone communication. In 1932 he realized that a certain kind of radio noise is strongest when the constellation Sagittarius is high in the sky. Because the center of our Galaxy is located in the direction of Sagittarius, Jansky concluded that he was detecting radio waves from elsewhere in the Galaxy.

> **Insight into science  Think "outside the box"**
> Observations and experiments often require scientists to make connections between seemingly unrelated concepts. Jansky's proposal that some radio waves originate in space is an example of a scientist connecting apparently disparate scientific fields—astronomy and radio engineering.

**Radio telescopes** record radio signals from the sky. At first, astronomers were not enthusiastic about detecting radio noise from space, in part because of the poor angular resolution of early radio telescopes. The angular resolution of any telescope decreases as the wavelength increases. In other words, the longer the wavelength, the fuzzier the picture. Because radio radiation has very long wavelengths, astronomers back then thought that radio telescopes could produce only blurry, indistinct images.

The standard radio telescope today has a large, reflecting, concave dish that acts exactly like a mirror in an optical telescope (Figure 3-30). A small antenna tuned to the

**Figure 3-30**   **A Radio Telescope**  Similar to an optical telescope, the dish of this radio telescope serves to concentrate the radio waves. The dish of this radio telescope is 45.2 m (148 ft) in diameter. It is one of several large instruments at the National Radio Astronomy Observatory near Green Bank, West Virginia.

desired wavelength is located at the focal point. The incoming signal is relayed to amplifiers and recording instruments, which are typically located in a room at the base of the telescope's pier. Very large radio telescopes create sharper radio images because, as with optical telescopes, the bigger the dish, the better the angular resolution. For this reason, most modern radio telescopes have dishes more than 30 m in diameter. Nevertheless, even the largest radio dish in existence (305-m diameter in Arecibo, Puerto Rico) cannot come close to the resolution of the best optical telescopes. For example, a 6-m optical telescope has 2000 times better resolution than a 6-m radio telescope detecting radio waves of 1-mm wavelength.

A clever technique enables radio astronomers to produce high-resolution radio images. The idea behind **interferometry** is to combine the data received simultaneously by two or more telescopes. The telescopes can be kilometers or even continents apart. Incoming radio signals to each dish are made to "interfere," or blend together, so that the combined signal is sharp and clear. The result is impressive: The effective angular resolution is equivalent to that of one gigantic dish with a diameter equal to the distance between the telescopes.

Interferometry, exploited for the first time in the late 1940s, gave astronomers their first detailed views of "radio objects" in the sky. More recently, radio telescopes separated by thousands of kilometers have been linked together in **very-long-baseline interferometry (VLBI)**, producing images that are much sharper and crisper than even those from optical telescopes. The best angular resolution on Earth is obtained by combining radio data from telescopes on opposite sides of the Earth. In that case, features as small as 0.00001″ can be distinguished at radio wavelengths—10,000 times sharper than the best views from even the best optical telescopes. Radio telescopes are also being put into space and used in even longer-baseline interferometers.

One of the finest systems of radio telescopes began operating in 1980 in the desert near Socorro, New Mexico. Called the Very Large Array (VLA), it consists of 27 concave dishes, each 26 m (85 ft) in diameter. The 27 telescopes are positioned along the three arms of a gigantic Y spanning a distance of 36 km (22 mi). Working together, they can create radio images with 0.1″ resolution. Figure 3-31 shows only a portion of the VLA. This system produces radio views of the sky with resolutions comparable to that of the very best optical telescopes.

To make radio images more comprehensible, radio astronomers often use false colors or gray scales to display their radio views of astronomical objects. An example of the use of false colors is shown in Figure 3-32. The most intense radio emission is shown in red, the least intense in blue. Intermediate colors of the rainbow represent intermediate levels of radio intensity. Black indicates that there is no detectable radio radiation. Astronomers working at other nonvisible-wavelength ranges also frequently use false-color techniques to display views obtained from their instruments.

The Very Long Baseline Array (VLBA) consists of ten 25-m radio telescopes located across the United States from Hawaii to New Hampshire. With a maximum baseline of 8000 km, VLBA has a resolving power of .001″. It is making major contributions to astronomy, such as the

**Figure 3-31**   **The Very Large Array (VLA)** The 27 radio telescopes of the VLA system are arranged along the arms of a Y in central New Mexico. The telescopes can be moved so that the array can detect either wide areas of the sky (when they are close together, as in this photograph) or small areas with higher resolution (when they are farther apart).

discovery in 1994 of a black hole in a nearby galaxy (discussed in Chapter 13). The 1997 addition of a radio telescope in space triples the maximum baseline for radio telescope arrays.

a  RIVUXG   b  RIVUXG

**Figure 3-32**   **Optical and Radio Views of Saturn (a)** This picture was taken by a camera on board a spacecraft as it approached Saturn. The view was produced by sunlight reflected from the planet's cloudtops and rings. **(b)** This false-color picture, taken by the VLA, shows radio emission from Saturn at a wavelength of 2 cm.

Figure 3-33 | **The Infrared Space Observatory (ISO)** This satellite provided astronomers with more than 26,000 outstanding infrared images and other data about planets, comets, gas and dust around other stars and in interstellar space, galaxies, and the large-scale distribution of matter in the universe. Its work was completed in 1998.

## 3-15 Observations at other wavelengths are revealing previously invisible sights

As the success of radio astronomy mounted, astronomers started exploring the possibility of making observations at other nonvisible wavelengths, such as infrared. Because water vapor is the main absorber of infrared radiation from space, locating infrared observatories at sites of low humidity can overcome much of the atmosphere's hindrance. For example, the summit of Mauna Kea on Hawaii is exceptionally dry (most of the moisture in the air is below the height of Mauna Kea), and infrared observations are the primary function of NASA's 3-m telescope there.

The best way of avoiding water vapor is to place a telescope in orbit around the Earth. The 1983 Infrared Astronomical Satellite (IRAS) and the 1995 Infrared Space Observatory (ISO) (Figure 3-33) did much to reveal the full richness and variety of the infrared sky. For the first time, astronomers saw dark bands of dust in our galaxy, dust disks around nearby stars, and distant galaxies that emit most of their radiation at infrared wavelengths. These infrared satellites have located more than a quarter million infrared sources in the sky. Most of these features are invisible to optical telescopes.

The best observations of ultraviolet radiation are also made from space, because the Earth's atmosphere absorbs much of this radiation. During the early 1970s, both Apollo and Skylab astronauts used small telescopes above the Earth's atmosphere to give us some of our first views of the ultraviolet sky. Small rockets have also been used to place ultraviolet cameras briefly above the Earth's atmosphere. A typical ultraviolet view is shown in Figure 3-34, along with a corresponding infrared view from IRAS, a view in visible light, and a star chart.

Some of the finest ultraviolet astronomy has been accomplished by the International Ultraviolet Explorer (IUE), which was launched in 1978 and functioned until 1996. The satellite was built around a Cassegrain telescope with a 45-cm (18-in.) mirror and a total focal length of 6.74 m (22 ft). Observations covered the ultraviolet range from 116 to 320 nm. In 1992 the Extreme Ultraviolet Explorer (EUVE) was launched. It is sensitive to photons with wavelengths between 7 and 76 nm, at the short end of the ultraviolet spectrum. As with infrared observations, ultraviolet images reveal sights previously invisible and often unexpected.

Because neither X rays nor gamma rays penetrate the Earth's atmosphere, observations at these extremely short wavelengths must be made from space. Astronomers got their first look at the X-ray sky during brief rocket flights in the late 1940s. Several small satellites launched during the early 1970s viewed the entire X-ray and gamma-ray sky, revealing hundreds of previously

a    R I V U X G    b    R I V U X G

c    R I V U X G    d

**Figure 3-34** **Orion as Seen in Ultraviolet, Infrared, and Visible Wavelengths** **(a)** An ultraviolet view of the constellation Orion obtained during a brief rocket flight on December 5, 1975. The 100-s exposure captured wavelengths ranging between 125 and 200 nm. **(b)** A false-color view from the Infrared Astronomical Satellite uses color to display specific ranges of infrared wavelengths: red indicates long-wavelength radiation; green, intermediate-wavelength radiation; and blue, short-wavelength radiation. For comparison, **(c)** an ordinary optical photograph and **(d)** a star chart are included.

unknown short-wavelength sources, including several black hole candidates (see Chapter 13). X-ray telescopes have also been carried on Space Shuttle missions (see Figure I-3a). To date, thousands of X-ray sources have also been discovered all across the sky. Among these are stars, vast clouds of intergalactic gas, jets of gas emitted by galaxies, black holes, and quasars.

The electromagnetic radiation with the shortest wavelengths and the most energy are gamma rays. In 1991 the Compton Gamma Ray Observatory was carried aloft by the Space Shuttle. Named in honor of Arthur Holly Compton, an American physicist who made important discoveries about gamma rays, this orbiting observatory carries four instruments that are performing a variety of observations, giving us tantalizing views of the gamma-ray sky. However, its best resolution instrument is only accurate to about 5″.

5   We can now observe in virtually all parts of the electromagnetic spectrum. Before the twentieth century, astronomers had been like the blind person trying to describe the elephant. Our ancestors couldn't see the big picture of the universe. However, we are beginning to see it, and, as a result, our understanding of the cosmos is increasing dramatically. In this book we describe some of the most important observations that have been made and the theories that have arisen to explain them.

## WHAT DID YOU THINK?

1   *What is light?* Light, more properly "visible light," is one form of electromagnetic radiation. All electromagnetic radiations (radio waves, infrared radiation, visible light, ultraviolet radiation, X rays, and gamma rays) have both wave and particle properties.

2   *What type of electromagnetic radiation is most dangerous to life?* Gamma rays have the highest energies of all photons, and so they are the most dangerous to life. However, ultraviolet radiation from the Sun is the most common everyday form of dangerous electromagnetic radiation we encounter.

3   *What is the main purpose of a telescope?* A telescope is designed primarily to collect as much light as possible.

4   *Why do stars twinkle?* Rapid changes in the density of the Earth's atmosphere cause passing starlight to change direction, making the star appear to twinkle.

5   *What type(s) of electromagnetic radiation can telescopes currently detect?* Telescopes have been built that can observe the entire electromagnetic spectrum.

## Key Words

achromatic lens, 68
active optics, 75
adaptive optics, 75
angular resolution, 66
Cassegrain focus, 71
charge-coupled device
   (CCD), 77
chromatic aberration, 68
coudé focus, 71
electromagnetic spectrum, 63
eyepiece lens, 65

focal length, 65
focal plane, 65
focal point, 65
gamma ray, 63
infrared radiation, 62
interferometry, 79
light-gathering power, 66
magnification, 66
Newtonian reflector, 70
objective lens, 65
photon, 62

pixel, 77
primary mirror, 70
prime focus, 70
radio telescope, 79
radio wave, 62
reflecting telescope
   (reflector), 70
reflection, 70
refracting telescope
   (refractor), 65
refraction, 64

Schmidt corrector plate, 72
secondary mirror, 70
seeing disk, 72
spectrum (*plural* spectra), 59
spherical aberration, 68
twinkling, 72
ultraviolet (UV) radiation, 63
very-long-baseline
   interferometry (VLBI), 79
wavelength ($\lambda$), 61
X ray, 62

## Key Ideas

### The Nature of Light

* Photons, compact units of vibrating electric and magnetic fields, all carry energy through space at the same speed, "the speed of light" (300,000 km/s).

* Visible light, radio waves, infrared and ultraviolet radiation, X rays, and gamma rays are all forms of electromagnetic radiation.

* Visible light occupies only a small portion of the electromagnetic spectrum.

* The wavelength of visible light is associated with its colors. Wavelengths of visible light range from about 400 nm for violet light to 700 nm for red light.

* Infrared radiation and radio waves have wavelengths longer than those of visible light. Ultraviolet radiation, X rays, and gamma rays have wavelengths that are shorter.

### Optics and Telescopes

* A telescope's most important function is to gather as much light as possible. Its second function is to reveal the observed object in as much detail as possible. Often the least important function of a telescope is to magnify objects.

* Refracting telescopes, or refractors, produce images by bending light rays as they pass through glass lenses. Glass impurity, opacity to certain wavelengths, and structural difficulties make it inadvisable to build extremely large refractors.

* Reflecting telescopes, or reflectors, produce images by reflecting light rays from concave mirrors to a focal point or focal plane. Reflectors are not subject to many of the problems that limit the usefulness of refractors. Telescopes that employ advanced technologies, such as active or adaptive optics, produce extremely sharp images.

* Charge-coupled devices (CCDs) are often used at a telescope's focal point to record images.

* Earth-based telescopes are being built with active and adaptive optics. These advanced technologies yield resolving power comparable to the Hubble Space Telescope.

### Radio Astronomy—and Beyond

* Radio telescopes have large reflecting antennas (dishes) that are used to focus radio waves.

* Very sharp radio images are produced with arrays of radio telescopes linked together in a technique called interferometry.

* The Earth's atmosphere is transparent primarily to visible light and radio waves arriving from space, but it absorbs much of the electromagnetic radiation at other wavelengths.

* For observations at other wavelengths, astronomers depend upon telescopes carried above the atmosphere by rockets and satellites. Satellite-based observatories are giving us a wealth of new information about the universe and permit coordinated observation of the sky at all wavelengths.

## Review Questions

The answers to all computational problems, which are preceded by an asterisk (*), appear at the end of the book.

1 Describe refraction and reflection. Why do these processes enable astronomers to build telescopes?

2 Give everyday examples of refraction and reflection.

3 With the aid of a diagram, describe a refracting telescope.

*4 How much more light does a 3-m-diameter telescope collect than a 1-m-diameter telescope?

5 With the aid of a diagram, describe a reflecting telescope. Describe four different ways in which an astronomer can access the light collected by reflecting telescopes.

6 Explain some of the advantages of reflecting telescopes over refracting telescopes.

7 What are the three major functions of a telescope?

8 What is meant by the angular resolution of a telescope?

9  What limits the angular resolution of the 5-m telescope on Mount Palomar?

10  Why will very large telescopes of the future make use of multiple mirrors or ultrathin mirrors?

11  What is meant by adaptive optics? What problem does adaptive optics overcome?

12  Compare an optical reflecting telescope and a radio telescope. What do they have in common? How are they different?

13  Why can radio astronomers observe at any time during the day, whereas optical astronomers are mostly limited to observing at night?

14  Why must astronomers use satellites and Earth-orbiting observatories to study the heavens at X-ray or gamma-ray wavelengths?

15  Why did Rømer's observations of the eclipses of Jupiter's moons support the heliocentric, but not the geocentric, cosmogony?

## Advanced Questions

16  Advertisements for telescopes frequently extol magnifying ability. Is this a good criterion for evaluating telescopes? Explain your answer.

*17  The observing cage in which an astronomer sits at the prime focus of the 5-m telescope on Mount Palomar is about 1 m in diameter. What fraction of the incoming starlight is blocked by the cage? *Hint:* The area of a circle of diameter $d$ is $\pi d^2/4$, where $\pi = 3.14$).

*18  Compare the light-gathering power of the Palomar 5-m telescope to that of the fully dark-adapted human eye, which has a pupil diameter of about 5 mm.

19  Show by means of a diagram why the image formed by a simple refracting telescope is "upside down."

*20  Suppose your Newtonian reflector has a mirror with a diameter of 20 cm and a focal length of 2 m. What magnification do you get with eyepieces whose focal lengths are (a) 9 mm, (b) 20 mm, and (c) 55 mm?

21  Why do you think no major observatory has a Newtonian reflector as its primary instrument, whereas Newtonian reflectors are extremely popular among amateur astronomers?

22  Several telescope manufacturers build telescopes having a design called "Schmidt-Cassegrain." Consult the Internet or advertisements in such magazines as *Sky & Telescope* and *Astronomy* to see what these telescopes look like and what they cost. Why do you suppose they are very popular among amateur astronomers?

## Discussion Questions

23  Discuss the advantages and disadvantages of using a small telescope in Earth orbit versus a large telescope on a mountaintop.

24  If you were in charge of selecting a site for a new observatory, which factors would you consider important?

## What If ...

25  An observatory was established on the Moon? List the advantages and disadvantages for astronomy.

26  Telescopes were first invented today? What objects or areas of the sky would you recommend that astronomers explore first? Why?

## Observing Projects

27  Obtain a telescope along with several eyepieces of various focal lengths. If you can determine the telescope's focal length, calculate the magnifying powers of the eyepieces. Focus the telescope on some familiar object during the daytime, such as a distant lamppost or tree. **Do not focus on the Sun! Looking directly at the Sun can cause blindness.** Describe the image you see through the telescope. Is it upside down? How does the image move as you slowly and gently shift the telescope left and right, up and down? Examine the eyepieces, noting their focal lengths. By changing the eyepieces, examine the distant object under different magnifications. How does the field of view and the quality of the image change as you go from low power to high power?

28  On a clear night, view the Moon, a planet, and a star through a telescope using eyepieces of various focal lengths. (Use your Starry Night program or consult a source on the Internet or such magazines as *Sky & Telescope* or *Astronomy* to determine the phase of the Moon and the locations of the planets.) How do the images seem to change as you view with increasing magnification? Does it degrade at any point?

29  Many towns and cities have amateur astronomy clubs. Attend a "star party" hosted by your local club. People bring their telescopes to these gatherings and are delighted to show you their instruments and take you on a telescopic tour of the heavens. Such experiences can lead to a very enjoyable, lifelong hobby.

# Humans Had Infrared-Sensitive Eyes?

Our eyes are sensitive to less than a trillionth of 1% of the electromagnetic spectrum—what we call "visible light." But this minuscule resource provides an awesome amount of information about the universe. We interpret visible-light photons as the six colors of the rainbow—red, orange, yellow, green, blue, and violet. These colors combine to form all the others that make our visual world so rich. But the Sun actually emits photons of all wavelengths. So, what would we see if out eyes had evolved to sense another part of the spectrum?

**A Darker Vision** Gamma rays, X rays, and most ultraviolet radiation do not pass through the Earth's atmosphere, and the world would look dark, indeed, if our eyes were sensitive only to these wavelengths. Radio waves, in contrast, easily pass through our atmosphere. But in order to see the same detail from radio waves that we now see from visible-light photons, our eyes would require 10,000 times larger diameters, each about the size of a baseball field!

What about infrared radiation? While not all incoming infrared photons get through out atmosphere, short-wavelength ("near") infrared radiation passes easily through air. Depending on wavelength, the Sun emits between one-half and one–ten billionth as many infrared photons as optical photons. Fortunately, most of these are in the near infrared.

**Heat-Sensitive Vision** To see infrared photons, human eyes would need to be only 5 to 10 times larger. Some snakes have evolved infrared vision. Portable infrared "night vision" cameras and goggles are available to us humans. Because everything that emits heat emits infrared photons, infrared sight would be very useful. And not everything we see with infrared sight would be due just to reflected sunlight—hotter objects would be intrinsically brighter than cool ones. For example, seeing infrared would allow us to observe changes in a person's emotional state. Someone who is excited or angry often has more blood near the skin, and thus gives off more infrared (heat) than normal. Conversely, someone who is scared has less blood near the skin and thus emits less heat.

**Night Vision** The night sky would be a spectacular sight through infrared-sensitive eyes. Gas clouds in the Milky Way absorb visible light, thus preventing the light of distant stars from getting to the Earth. However, because most infrared radiation passes through these clouds, we would be able, unaided, to see distant stars that we cannot see today. On the other hand, the white glow of the Milky Way, which is caused by the scattering of starlight by interstellar clouds, would be dimmer, because the gas clouds do not scatter infrared light as much as they do visible light.

Our concept of stars would be different, too. Many stars, especially young, hot ones, are surrounded by co-coons of gas and dust that emit infrared radiation. This dust is heated by the nearby stars. Instead of appearing as pinpoints, many stars would appear to be surrounded by wild strokes of color, and we would have an Impressionist sky.

R I V U X G

The infrared (heat) from this kissing couple has been converted into visible light colors so that we can interpret the invisible radiation. The hottest regions are white, with successively cooler areas shown in yellow, orange, red, green, sky blue, dark blue, and violet.

R I V U X G

**Stellar Spectra**
Starlight passing through a prism or diffraction grating spreads into its component colors. From such spectra we can learn an incredible amount about stars, including their masses, surface temperatures, diameters, chemical compositions, and motions toward or away from us. This image, called a spectrogram, shows the spectra of stars in the Hyades star cluster. As you study astronomy, note the distinct differences between the intensities of the various colors emitted by different stars.

# The Origin and Nature of Light

## *In this chapter you will discover*

* ✳ the origins of electromagnetic radiation

* ✳ that stars with different surface temperatures emit different intensities of electromagnetic radiation

* ✳ that astronomers can determine the chemical compositions of stars and interstellar clouds by studying the wavelengths of electromagnetic radiation they absorb or emit

* ✳ how to tell whether an object in space is moving toward or away from Earth

1. How hot is a "red hot" object compared to objects glowing other colors?

2. What color is the Sun?

3. Can carbon dating be used to determine the age of space debris found on Earth?

Seeing objects in space is one thing, understanding their physical properties—temperature, chemical composition, mass, size, and motion—is something else altogether. Because almost all our knowledge about space comes from electromagnetic radiation, we now delve deeper into the nature of light. By understanding how light and matter interact, astronomers glean an amazing amount of information about the heavenly bodies.

# BLACKBODY RADIATION

We begin our own study of this interaction with a perfectly familiar property of matter—temperature.

## 4-1 Peak color shifts to shorter wavelengths as an object is heated

Imagine that you have taken hold of an iron rod in a completely darkened room. You cannot see the rod, but you can feel its warmth in your hands. The rod is emitting enough infrared radiation (heat) for you to feel, but too few visible photons for your eyes to see. Now imagine that there is also a propane torch in the room. You light it and stick the tip of the iron rod in the flame for several seconds, until it begins to glow (Figure 4-1a).

The rod's first visible color is red—it glows "red hot." Heated a little more, the rod appears orange and brighter (Figure 4-1b). Heated more, it appears yellow and brighter still. After even more heating, the rod appears white-hot and brighter yet (Figure 4-1c). If it didn't melt, further heating would make the rod appear blue and even brighter. This "thought experiment" shows how the light any object emits changes with its temperature:

1. As an object heats up, it gets brighter because it emits more electromagnetic radiation.

2. The dominant color or wavelength of the emitted radiation changes with temperature. A cool object emits most of its energy at long wavelengths, such as infrared or red. A hotter object emits most of its energy at shorter wavelengths, such as blue, violet, or even ultraviolet.

Although the iron rod appears to emit a single color depending on its temperature, it actually gives off all wavelengths of electromagnetic radiation. So do stars, rocks, animals, and most other things in nature. However, a single wavelength of emitted radiation, denoted $\lambda_{max}$, dominates. When the object is cool, like a rock or animal, $\lambda_{max}$ is a radio or infrared wavelength. When it is hot enough, $\lambda_{max}$ is in the range of visible light, giving a hot object its characteristic color. Sufficiently hot stars emit $\lambda_{max}$ in the ultraviolet part of the electromagnetic spectrum.

Iron bars and stars are good approximations to an important class of theoretical objects that astrophysicists call **blackbodies**. A blackbody absorbs all the electromagnetic radiation that strikes it. None of the incoming radiation is reflected or scattered off its surface. The incoming radiation heats up the blackbody, which then reemits the energy it has absorbed, but *not* with the same intensities at each wavelength as it received.

a                b                c

**Figure 4-1**  **Heating a Bar of Iron**
This sequence of drawings shows the changing appearance of a bar of iron as it is heated. As the temperature increases, the amount of energy radiated by the bar increases. The color of the bar also changes, because, as the temperature goes up, the dominant wavelength of light emitted by the bar decreases.

*Intensity* measures the energy at a given frequency emitted by each element of a surface per second. *The intensity of each wavelength emitted by a blackbody is determined solely by its temperature.* Profiles of the intensity of blackbody radiation at different wavelengths, called **blackbody curves,** are shown for three temperatures in Figure 4-2. (Temperatures throughout this book are expressed in Kelvins. If you are not familiar with the Kelvin temperature scale, please review Toolbox A-1, "Temperature Scales," at the back of the book.)

Unlike our iron rod, stars produce their own electromagnetic radiation rather than just absorbing and reradiating light from an outside source. Even so, stars behave like blackbodies, and the self-generated radiation they emit closely follows the idealized blackbody curves. We can, therefore, measure a real star's surface temperature from the amount of each color or wavelength it emits. To do that, we need to make our observations quantitative; that is, to express them mathematically.

## 4-2 The intensities of different emitted colors reveal a star's temperature

**Wien's law** is the mathematical relationship between the location of the color peak for each curve in Figure 4-2 and that blackbody's temperature. In 1893 the German physicist Wilhelm Wien found that

*The dominant wavelength of radiation emitted by a blackbody is inversely proportional to its temperature.*

In other words, the hotter any object becomes, the shorter its $\lambda_{max}$, and vice versa. Wien's law proves very useful in computing the surface temperature of a star, because the star's size and brightness are often unknown. All we need to know is the dominant wavelength of the star's electromagnetic radiation.

A blackbody curve describes in detail the electromagnetic radiation a star emits. In 1879 the Austrian physicist Josef Stefan observed that

*An object emits energy at a rate proportional to the fourth power of its temperature in Kelvins.*

In other words, if you double the temperature of an object (for example, from 500 to 1000 K), the energy emitted from each square meter of the object's surface each second increases by a factor of $2^4$, or 16 times. If you triple the temperature (for example, from 500 to 1500 K), the rate at which energy is emitted increases by a factor of $3^4$, or 81 times. Stefan's experimental results were put on firm theoretical ground by Ludwig Boltzmann in 1885. The intensity-temperature relationship for blackbodies is named the **Stefan-Boltzmann law** in their honor.

The Stefan-Boltzmann law and Wien's law are powerful tools for understanding the stars. If you would like to

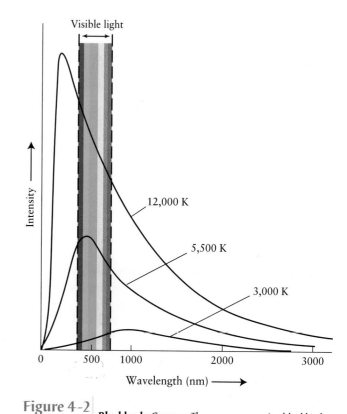

**Figure 4-2** **Blackbody Curves** Three representative blackbody curves are shown here. Each curve shows the intensity of radiation at every wavelength emitted by a blackbody at a particular temperature. Note that hotter objects emit more of every wavelength and that the peak of emitted radiation changes with temperature.

learn how to use the two radiation laws to make predictions, see An Astronomer's Toolbox 4-1. However, they describe only two basic properties of *blackbody radiation,* namely the relative rate at which it is emitted and the location of its peak. Because blackbodies give off all wavelengths of electromagnetic radiation, the blackbody curves, such as those already shown in Figure 4-2, give a more complete picture. These curves show the intensities of electromagnetic radiation emitted at different wavelengths by a blackbody with a given temperature. As we will see in the next section, the *differences* between an ideal blackbody curve and the curves seen from actual stars reveal such things as stellar chemistry and motions of stars toward or away from us, among other things.

Figure 4-3 shows how the intensity of sunlight varies with wavelength. The blackbody curve for a body with a temperature of 5800 K is also plotted in Figure 4-3. Note how the observed intensity curve for the Sun closely follows the ideal blackbody curve at most wavelengths. Because the observed intensity curves for most stars and the idealized blackbody curves are so closely correlated, the laws of blackbody radiation can be applied to starlight.

# AN ASTRONOMER'S TOOLBOX 4-1

## The Radiation Laws

Wien's law can be stated as a simple equation. If $\lambda_{max}$, the wavelength of maximum flux, is measured in meters, then

$$\lambda_{max} = \frac{2.9 \times 10^{-3}}{T}$$

where $T$ is the temperature of the blackbody measured in Kelvins.

*Example:* The Sun emits energy over a wide range of wavelengths, but the maximum intensity of sunlight is at a wavelength of about 500 nm, or $5 \times 10^{-7}$ m. From Wien's law, we can calculate the Sun's surface temperature as

$$T = \frac{2.9 \times 10^{-3}}{5 \times 10^{-7}} = 5800 \text{ K}$$

The total energy emitted from each square meter of an object's surface each second is called the **energy flux**, $F$. In this context, flux means "rate of flow." Using this concept, we can write the Stefan-Boltzmann law as

$$F = \sigma T^4$$

where $T$ is the temperature of the object in Kelvins and $\sigma$ (lowercase Greek sigma) is a number called the Stefan-Boltzmann constant:

$$\sigma = 5.67 \times 10^{-8} \text{J}/(\text{m}^2 \cdot \text{K}^4 \cdot \text{s})$$

where J are the energy units Joules, m is the unit meters, K is degrees Kelvin, and s is the unit seconds.

Because $F$ is the energy emitted per second from each square meter of an object, multiplying $F$ by the surface area, $4\pi r^2$, where r is the object's radius, yields the total energy emitted by the object each second. Denoted $L$, and called **luminosity**, we have

$$L = F \cdot 4\pi r^2 = \sigma T^4 4\pi r^2$$

*Compare!* The Stefan-Boltzmann law has several applications. Suppose you observe two stars of equal size, one with a surface temperature of 10,000 K and the other with the same surface temperature as the Sun (5800 K). You can use the Stefan-Boltzmann law to determine how much brighter the hotter star is:

$$\frac{F_{hotter}}{F_{colder}} = \frac{\sigma \times 10,000^4}{\sigma \times 5800^4} \approx 8.8$$

So, the hotter star emits 8.8 times as much energy from each square meter of its surface than does the cooler star. Because the stars are the same size, the hotter one is therefore 8.8 times brighter than the cooler one.

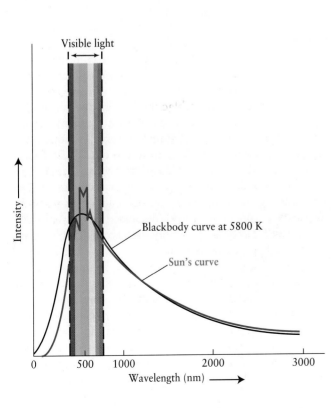

The peak of the intensity curve for the Sun is at a wavelength of about 500 nm, which is in the blue-green part of the visible spectrum.

By the end of the nineteenth century, physicists realized that they had reached an impasse in understanding electromagnetic radiation. While intensities and peak wavelengths were understood, all attempts to explain the characteristic shapes of blackbody curves had failed. A breakthrough finally came in 1900, when the German physicist Max Planck derived a mathematical formula for the blackbody curves. He assumed that electromagnetic

**Figure 4-3**    **The Sun as a Blackbody** This graph compares the intensity of sunlight (in red) over a wide range of wavelengths with the intensity of radiation from a blackbody at a temperature of 5800 K. The Earth's atmosphere scatters shorter wavelength photons more than it scatters longer wavelength ones. As seen from Earth, this lowers the intensity and shifts the blackbody peak to the right from where it actually occurs. To avoid these effects, the measurements of the Sun's intensity in this figure were made in space. The Sun mimics a blackbody remarkably well.

Table 4-1
Some Properties of Electromagnetic Radiation

| | Wavelength (nm) | Photon energy (eV)* | Blackbody temperature (K) |
|---|---|---|---|
| Radio | $>10^7$ | $<10^{-4}$ | $<0.03$ |
| Microwave† | $10^7$ to $4 \times 10^5$ | $10^{-4}$ to $3 \times 10^{-3}$ | 0.03 to 30 |
| Infrared | $4 \times 10^5$ to $7 \times 10^2$ | $3 \times 10^{-3}$ to 2 | 30 to 4100 |
| Visible | $7 \times 10^2$ to $4 \times 10^2$ | 2 to 3 | 4100 to 7300 |
| Ultraviolet | $4 \times 10^2$ to $10^1$ | 3 to $10^3$ | 7300 to $3 \times 10^6$ |
| X ray | $10^1$ to $10^{-2}$ | $10^3$ to $10^5$ | $3 \times 10^6$ to $3 \times 10^8$ |
| Gamma ray | $<10^{-2}$ | $>10^5$ | $>3 \times 10^8$ |

Note: > means greater than, < means less than.
*1 eV = 1.6 × 10⁻¹⁹ J.
†Microwaves, listed here separately, are often classified as radio waves or infrared radiation.

radiation is emitted in separate packets of energy. Light, a wave, behaves like a beam of particles. It was a remarkable result verified in 1905 by Albert Einstein, who called these particles of light photons.

We can now explain our thought experiment with the iron bar in terms of photons. Recall from Chapter 3 that the energy carried by a photon of light is inversely proportional to its wavelength. In other words, long-wavelength photons, such as radio waves, carry little energy.

Short-wavelength photons, like X rays and gamma rays, carry much more energy. The relationship between the energy of a photon and its wavelength is called **Planck's law** (you may want to refer back to An Astronomer's Toolbox 3-1.)

Together, Planck's law and Wien's law relate the temperature of an object to the energy of the photons it emits (Table 4-1). A cool object emits primarily long-wavelength photons that carry little energy, while a hot object gives off

## DISCOVER FOR YOURSELF  The Color of the Sun

Different people perceive the Sun to have different colors. To many it appears white, to others yellow. Still others, who notice it at sunset, believe it to be orange or even red. But we have seen that the Sun actually gives off all colors. Moreover, the peak in the Sun's spectrum falls between blue and green. Why doesn't the Sun appear turquoise? Several factors affect our perception of its color.

Before reaching our eyes, visible sunlight passes through the Earth's atmosphere. Certain wavelengths are absorbed and reemitted by the molecules in the air, a process called *scattering*. Violet light is scattered most strongly, followed in decreasing order by blue, green, yellow, orange, and red. That means that more violet, blue, and green photons are scattered by the Earth's atmosphere than are yellow photons. The intense scattering of violet, blue, and green has the effect of shifting the peak of the Sun's intensity entering our eyes from blue-green toward yellow. (The sky is blue because of the strong scattering of blue light—it isn't violet because the Sun emits many fewer violet photons than blue ones.)

The perception of a yellow Sun is further enhanced by our eyes themselves. Our eyes do not see all colors equally well. Rather, the light-sensitive cones in our eyes

each respond to one of three ranges of colors, which are centered on red, yellow, and blue wavelengths. None of the cones is especially sensitive to blue-green photons. By adding together the color intensities detected by the three types of cones, our brains *recreate* color. After combining all the light it can, the eye is most sensitive to the yellow-green part of the spectrum. We see blue and orange less well, and violet and red most poorly. Although the eye sees yellow and green light about equally well, the dominant color from the Sun is yellow because the air scatters green light, and our eyes are relatively insensitive to blue-green. Therefore, a casual glance at the Sun leaves the impression of a yellow object.

A longer look is extremely dangerous. **Don't try it!** Hypothetically, such a glance would leave the impression of a white Sun. The Sun's light is so intense that it would saturate the color-sensitive cones in our eyes; our brains interpret such saturation of the cones as white.

At sunrise and sunset we see an orange or red Sun. This occurs because close to the horizon the Sun's violet, blue, green, and even yellow photons are strongly scattered by the thick layer of atmosphere through which they travel, leaving the Sun looking redder and redder as it sets.

mostly short-wavelength photons that carry much more energy. In later chapters, we will find these ideas invaluable for understanding how stars of various temperatures interact with gas and dust in space.

But blackbody radiation doesn't tell the whole story of the colors emitted by objects. Even at room temperature the things around us display every color on an artist's palette. Why?

## DISCOVERING SPECTRA

In 1814 the German optician Joseph von Fraunhofer repeated Newton's classic experiment of shining a beam of sunlight through a prism (recall Figure 3-1). But Fraunhofer magnified the resulting rainbow-colored spectrum. He discovered that the solar spectrum contains hundreds of fine dark lines, which became known as **spectral lines.** Fraunhofer counted more than 600 such lines, and today Physicists have detected more than 30,000 of them. Hundreds of spectral lines are visible in the photograph of the Sun's spectrum shown in Figure 4-4.

By the mid-1800's, chemists discovered that they could produce spectral lines in the laboratory. Around 1857 the German chemist Robert Bunsen invented a special gas burner that produces a clean, colorless flame. Certain chemicals are easy to identify by the distinctive colors emitted when bits of the chemical are sprinkled into the flame of a Bunsen burner.

Bunsen's colleague, Gustav Kirchhoff, suggested that light from the colored flames could best be studied by passing it through a prism (Figure 4-5). The chemists

**Figure 4-5** **The Kirchhoff-Bunsen Experiment** In the mid-1850s, Kirchhoff and Bunsen discovered that when a chemical substance is heated and vaporized, the resulting spectrum exhibits a series of bright spectral lines. In addition, they found that each chemical element produces its own characteristic pattern of spectral lines.

promptly discovered that the spectrum from a flame consists of a pattern of thin, bright spectral lines against a dark background (not a blackbody spectrum!). They next found that *each chemical element produces its own characteristic pattern of spectral lines.* Thus was born in 1859 the technique of **spectral analysis,** the identification of chemical substances by their spectral lines.

**Figure 4-4** **The Sun's Spectrum** You can see numerous spectral lines in this photograph of the Sun's spectrum. The spectrum, which is one continous band, was cut into convenient segments to fit on this page.

## 4-3 Each chemical element produces its own unique set of spectral lines

A chemical **element** is a fundamental substance because it cannot be broken down into more basic units and still retain its properties. By the mid-1800s, chemists had already identified such familiar elements as hydrogen, oxygen, carbon, iron, gold, and silver. Spectral analysis promptly led to the discovery of additional elements, many of which are quite rare.

After Bunsen and Kirchhoff recorded the prominent spectral lines of all the known elements, they began to discover other spectral lines in mineral samples. In 1860, for example, they found a new line in the blue portion of the spectrum of mineral water. After chemically isolating the previously unknown element responsible for the line, they named it cesium (from the Latin *caesius,* meaning gray-blue). The next year a new spectral line in the red portion of the spectrum of a mineral sample led to the discovery of the element rubidium (from *rubidus,* for red).

During a solar eclipse in 1868, astronomers found a new spectral line in the light coming from the upper atmosphere of the Sun while the main body of the Sun was hidden by the Moon. This line was attributed to a new element, which was named helium (from the Greek *helios,* meaning Sun). Helium was not actually discovered on the Earth until 1895, when it was identified in gases obtained from a uranium mineral.

A list of the chemical elements is most conveniently displayed in the form of a **periodic table** (Figure 4-6). Each element has a unique atomic number (described later in this chapter), and the elements are arranged in the periodic table by their atomic numbers. With a few exceptions, this sequence also corresponds to increasing average mass of the atoms of the elements. Thus, hydrogen (symbol H) with atomic number 1 is the lightest element. Iron (Fe) has atomic number 26 and is a moderately heavy element.

All the elements in a single vertical column of the periodic table have similar chemical properties. For example, the elements in the far right column are all gases at room

# Periodic Table of the Elements

**Figure 4-6** | **Periodic Table of the Elements** The periodic table is a convenient listing of the elements arranged according to their atomic numbers and chemical properties. Due to naming disagreements, elements 110-113 do not yet have universally accepted names.

R I V U X G

Figure 4-7  **Iron in the Sun's Atmosphere** The upper spectrum is a portion of the Sun's spectrum from 420 to 430 nm. Numerous dark spectral lines are visible. The lower spectrum is a corresponding portion of the spectrum of vaporized iron. Several bright spectral lines can be seen against the black background. The fact that the iron lines coincide with some of the solar lines proves that there is some iron (albeit a very tiny amount) in the Sun's atmosphere.

temperature and normal air pressure, and they rarely react chemically with other elements.

In addition to the 92 naturally occurring elements, Figure 4-6 lists the artificially produced elements. All the human-made elements are heavier than uranium (U), and all are highly radioactive, meaning that they spontaneously decay into lighter elements shortly after being created in the laboratory.

> **Insight into science  Seek relationships** Finding how properties relate often provides invaluable insights into how new things work. The periodic table, for example, enables scientists to determine the properties of similar chemical elements. Keep an eye out for other such relationships throughout this book.

Because each chemical element produces its own unique pattern of spectral lines, scientists can determine the chemical composition of a remote astronomical object by identifying the spectral lines in its spectrum. For example, Figure 4-7 shows a portion of the Sun's spectrum along with the spectrum of an iron sample taken here on Earth. No other chemical has iron's particular pattern of spectral lines. It is iron's own distinctive "fingerprint." Because the spectral lines of iron also appear in the Sun's spectrum, we can safely conclude that the Sun's atmosphere contains some vaporized iron.

Bunsen and Kirchhoff collaborated in designing and constructing the first **spectroscope.** This device consists of a prism and several lenses that magnify the spectrum so that it can be closely examined. After photography was invented, scientists preferred to produce a permanent photographic record of spectra. A device for photographing a spectrum is called a **spectrograph,** and this instrument is among the astronomer's most important tools.

In its basic form, a spectrograph consists of a slit, two lenses, and a prism arranged to focus the spectrum of an astronomical object onto a small photographic plate, as shown in Figure 4-8. This optical device typically mounts

at the focal point of a telescope. The image of the object to be examined is focused on the slit. After the spectrum of a star or galaxy has been photographed, the exposed part of the photographic plate is covered, and light from a known source, such as a gas composed of helium, neon, and argon, is focused on the slit. This "comparison spectrum" is photographed next to the spectrum of the object, as in Figure 4-7. Because the wavelengths of the spectral lines in the comparison spectrum are already known from laboratory experiments, these lines can be used to identify and measure the wavelengths of the lines in the spectrum of the star or galaxy under study.

There are drawbacks to this type of spectrograph. A prism does not spread colors evenly: Blue and violet portions of the spectrum are spread out more than the red portion. In addition, because the blue and violet wavelengths must pass through more glass than the red wavelengths (see Figure 3-1), light is absorbed unevenly across the spectrum. Indeed, a glass prism is opaque to ultraviolet wavelengths.

A better device for breaking starlight into the colors of the rainbow is a **diffraction grating,** a piece of glass on

Figure 4-8  **A Prism Spectrograph** This optical device uses a prism to break up the light from an object into a spectrum. The lenses focus that spectrum on a photographic plate.

which thousands of closely spaced lines are cut. Some of the finest diffraction gratings have as many as 10,000 lines per centimeter. The spacing of the lines must be very regular. Light rays reflected from different parts of the diffraction grating interfere with each other to produce a spectrum. Observatories now record spectra with charge-coupled devices (CCDs; see Figures 3-28 and 3-29). Figure 4-9 shows the design of a modern diffraction grating spectrograph.

The spectral data is converted by computer to a graph that plots light intensity against wavelength. Dark lines in the rainbow-colored spectrum appear as depressions or valleys on the graph, while bright lines in the spectrum appear as peaks. For example, Figure 4-10 shows both a picture and a plot of a spectrum for hydrogen in which five dark spectral lines appear.

## 4-4 The brightnesses of spectral lines depend on conditions in the spectrum's source

During his pioneering experiments with spectra, Kirchhoff sometimes saw dark spectral lines, called *absorption lines,* among the colors of the rainbow. In other experi-

**Figure 4-10**  **Spectrum of Hydrogen Gas**  When a CCD is placed at the focus of a spectrograph, a rainbow-colored spectrum is recorded. The spectrum is converted by computer into a graph of intensity versus wavelength. Note that the dark spectral lines appear as dips in the intensity-versus-wavelength curve.

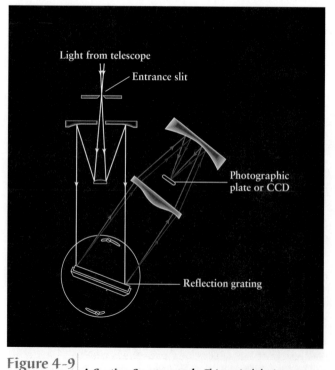

**Figure 4-9**  **A Grating Spectrograph**  This optical device uses a grating to break up the light from an object into a spectrum. An arrangement of lenses and mirrors focuses that spectrum onto a CCD.

ments, he saw bright spectral lines, called *emission lines,* against an otherwise dark background (Figure 4-5). By the early 1860s, Kirchhoff had discovered the conditions under which these different types of spectra are observed. His description is summarized today as **Kirchhoff's laws:**

**Law 1**  A hot object or a hot, dense gas produces a **continuous spectrum**—a complete rainbow of colors without any spectral lines. (This is a blackbody spectrum.)

**Law 2**  A hot, rarefied gas produces an **emission line spectrum**—a series of bright spectral lines against a dark background.

**Law 3**  A cool gas in front of a continuous source of light produces an **absorption line spectrum**—a series of dark spectral lines among the colors of the rainbow.

Figure 4-11 shows how absorption and emission lines are forme.d. Absorption lines are seen if the background is hotter than the gas. Emission lines are seen if the background is cooler. Note that the bright lines in the emission spectrum of a particular gas occur at exactly the same wavelengths as the dark lines in the absorption spectrum of that gas.

Consider, for example, the spectrum of the Sun. We know that the Sun's surface emits a continuous, black-

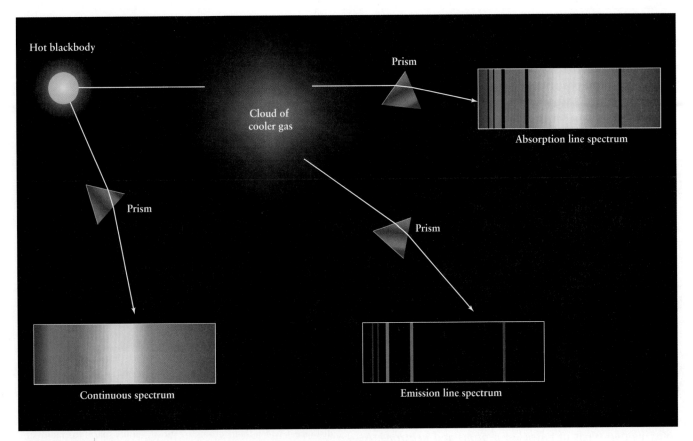

**Figure 4-11**    **Continuous, Absorption Line, and Emission Line Spectra** This schematic diagram summarizes how different types of spectra are produced. A hot, glowing object emits a continuous spectrum. If this source of light is viewed through a cool gas, dark absorption lines appear in the resulting spectrum. When the same gas is viewed against a cold, dark background, its spectrum consists of just bright emission lines.

body spectrum, but here on Earth many absorption lines are seen in it (see Figure 4-4). Kirchhoff's third law explains why. There must be a cooler gas between the surface of the Sun and the Earth. In fact, there are two: the Sun's atmosphere and the Earth's atmosphere. With these observations in hand, we turn now to exploring why the different spectra occur.

## ATOMS AND SPECTRA

An atom is the smallest particle of a chemical element that still has the properties of that element. At the time of Kirchhoff's discoveries, scientists knew that all matter is composed of atoms, but they did not know their structures. Scientists could see that atoms of a gas somehow extract light of specific wavelengths from white light that passes through the gas, leaving dark absorption lines, and they could even guess that the atoms then radiate light of precisely the same wavelengths—the bright emission lines (see Figure 4-7). But traditional theories of electromagnetism could not explain this phenomenon. The answer came early in the twentieth century with the development of quantum mechanics and nuclear physics.

## 4-5 An atom consists of a small, dense nucleus surrounded by electrons

The first important clue about the internal structure of atoms came in 1910 from Ernest Rutherford, a chemist and physicist from New Zealand. Rutherford and his colleagues at the University of Manchester in England were investigating the recently discovered phenomenon of radioactivity. Over time, a radioactive element naturally and spontaneously transforms into another element by emitting particles. Certain radioactive elements, such as uranium and radium, were known to emit such particles with considerable speed. It seemed plausible that a beam of these high-speed particles could penetrate a thin sheet of gold. Rutherford and his associates found that almost all the particles do pass through the gold sheet with little or no deflection. To the surprise of the experimenters, however, an occasional particle bounced right back. It must have struck something very dense indeed.

Rutherford was quick to realize the implications of his experiment. He correctly reasoned that most of the mass of an atom is concentrated in a compact lump that occupies only a small part of the atom's volume. Most of the radioactive particles pass freely through the nearly empty space that makes up most of the atom, but a few particles happen to strike the dense mass at the center and rebound.

Rutherford proposed a new model for the structure of an atom. According to this model, a massive, positively charged **nucleus** at the center of the atom is orbited by tiny, negatively charged *electrons* (Figure 4-12). Rutherford concluded that at least 99.98% of the mass of an atom is concentrated in the nucleus, whose diameter is only about one–ten thousandth the diameter of the atom.

Further research revealed that the nucleus of an atom contains two types of particles: *protons* and *neutrons*. A proton has almost the same mass as a neutron, and each proton or neutron is about 2000 times more massive than an electron. A proton has a positive electric charge and a neutron has no charge. Because like charges repel each other, protons do not naturally stay bound together; they try to move as far away from each other as possible. Neutrons help keep the protons bound together. Conversely, opposite charges attract, and it is this attraction that keeps electrons in orbit around the nucleus, held there by the positively charged protons.

The number of protons in an atom's nucleus determines what element that atom is. Each element is assigned an **atomic number** that equals the number of protons it contains. Of the naturally occurring elements, a hydrogen nucleus always has 1 proton, a helium nucleus always has 2, and so forth, up to uranium with 92 protons in its nucleus (see Figure 4-6).

In contrast, the number of neutrons in the nuclei of different atoms of the same element may vary. For example, oxygen, the eighth element in the periodic table, with an atomic number of 8, always has eight protons, but it may have eight, nine, or ten neutrons. These three slightly different kinds of oxygen are called **isotopes.** The isotope of oxygen with eight neutrons is by far the most abundant variety.

Many elements have isotopes that are radioactive. For example, carbon with six neutrons, $C^{12}$, is stable, while

**Figure 4-12** | **Rutherford's Model of the Atom** Electrons orbit the atom's nucleus, which contains most of the atom's mass. The nucleus contains two types of particles: protons and neutrons. Because the nucleus and the electrons are so small compared to the distance between them, atoms are mostly empty space. The rapid movements of the electrons make the nuclei appear to be surrounded by clouds.

carbon with eight neutrons, $C^{14}$, is unstable. $C^{14}$ decays into nitrogen with seven neutrons, $N^{14}$. To learn how radioactive decay is used to determine the ages of different objects, see An Astronomer's Toolbox 4-2.

Normally, the number of electrons orbiting an atom is equal to the number of protons in the nucleus, thus making the atom electrically neutral. Astronomers denote neutral atoms by writing the atomic symbol followed by the roman numeral I. For example, neutral hydrogen is written as H I and neutral iron is Fe I.

When an atom contains a different number of electrons than protons, the atom is called an **ion.** The process of creating an ion is called **ionization.** Ions are denoted by the atomic symbol followed by a roman numeral that is one greater than the number of missing electrons. Ionized hydrogen (missing its one electron) is denoted H II, while ionized iron with seven electrons missing is denoted Fe VIII. One way to create ions from neutral atoms is *photoionization,* in which photons of sufficiently high energy literally rip an electron completely out of orbit and into the space between atoms. Ionization occurs in stars, as seen by the fact that many of the spectral lines for stars correspond to ionized atoms.

# AN ASTRONOMER'S TOOLBOX 4-2

## Radioactivity and the Ages of Objects

The isotopes of many elements are radioactive, meaning that the elements spontaneously transform into other elements. Each radioactive isotope has a distinctive *half-life*, the time it takes half of the initial concentration of the isotope to transform into another element. After two half-lives, a radioactive isotope is reduced to $\frac{1}{2} \times \frac{1}{2}$ or $\frac{1}{4}$ of its initial concentration (see the accompanying figure). Among the most important radioactive elements for determining the age of objects in astronomy is the isotope of uranium with 146 neutrons, $U^{238}$. The half-life of $U^{238}$ decaying into lead is 4.5 billion years.

To determine the time since an object, such as a piece of space debris discovered on Earth, solidified, scientists estimate how much lead it had when it formed. Then they measure the amounts of uranium and lead it contains now. Subtracting the amount of original lead, they use the amount of uranium and lead with the graph to determine how long ago the object formed.

*Example:* Suppose a piece of space debris discovered on Earth was determined to have equal amounts of lead and uranium. How long ago did this debris form? Assuming that it originally had no lead, we see from the chart that a 1-to-1 mix of lead to uranium occurs 4.5 billion years after the object formed. This is one half-life of uranium.

[3] *Compare!* This process works with any radioactive isotope. However, some isotopes have such short half-lives that they are not useful in astronomy. For example, the well-known carbon dating used to determine the ages of ancient artifacts on Earth is of little use in astronomy because $C^{14}$ has a half-life of only 5730 years. Because so much of it has decayed away by then, carbon dating is useful for fewer than 100,000 years, usually a period over which little of astronomical importance occurs.

**The Transformation of Uranium into Lead**  This figure shows the rate that 1 kg of uranium decays into lead, as described in the text. It contains 1/8 kg of uranium after 13.5 billion years.

---

Atoms can also share electrons and, by doing so, remain bound together. Such groups are called **molecules.** They are the essential building blocks of all complex structures, including life.

## 4-6 Spectral lines occur when an electron jumps from one energy level to another

The challenge of reconciling Rutherford's atomic model with the observations of spectral analysis was undertaken by the young Danish physicist Niels Bohr, who joined Rutherford's group at Manchester in 1911. Bohr began by trying to understand the structure of hydrogen, the lightest of the elements. The simplest hydrogen atom consists of a single electron and a single proton. Hydrogen has a visible spectrum consisting of a pattern of lines that begins at 656.28 nm and ends at 364.56 nm. The longest-wavelength spectral line is called $H_\alpha$, the second $H_\beta$, the third $H_\gamma$, and so forth, ending with the shortest-wavelength line, $H_\infty$, at 364.56 nm. (The first few lines of the series are identified by Greek-letter subscripts; the remainder are identified by numerical subscripts.) The closer you get to 364.56 nm, the more spectral lines you see.

This spectral pattern had been described mathematically in 1885 by Johann Jakob Balmer, an German schoolteacher. By trial and error, Balmer discovered a formula for calculating the wavelengths of the hydrogen lines. Because of his discovery, the spectral lines of hydrogen at visible wavelengths are called *Balmer lines*, and the entire pattern from $H_\alpha$ to $H_\infty$ is called the *Balmer series*. The spectrum of the star shown in Figure 4-13 exhibits more than two dozen Balmer lines, from $H_{13}$ through $H_{40}$.

**Figure 4-13** **Balmer Lines in the Spectrum of a Star** This portion of the spectrum of a star called HD 193182 shows more than two dozen Balmer lines. HD are the initials of Henry Draper. Draper and succeeding astronomers cataloged this and many other stars. The series converges at 364.56 nm, just to the left of $H_{40}$. This star's spectrum also contains the first 12 Balmer lines ($H_\alpha$ through $H_{12}$), but they are not visible in this particular spectrogram.

Bohr's goal was to mathematically derive Balmer's formula from basic laws of physics. He began by assuming that the electron in a hydrogen atom moves around the nucleus only in certain specific orbits. As shown in Figure 4-14, it is customary to label these orbits $n = 1$, $n = 2$, $n = 3$, and so on. They are called the *Bohr orbits*.

Bohr argued that for an electron to jump, or to make a **transition,** from one orbit to another, the hydrogen atom must gain or lose a specific amount of energy. An electron jump from an inner orbit to an outer orbit requires energy; an electron jump from an outer orbit to an inner one releases energy. The energy gained or released by the atom when the electron changes orbits is the difference in energy between these two orbits.

According to Planck and Einstein, the packet of energy gained or released is a photon, whose energy is inversely proportional to its wavelength. Using these ideas, Bohr mathematically derived the formula that Balmer had dis-

covered by trial and error. Furthermore, Bohr's discovery elucidated the meaning of the Balmer series: All the Balmer lines are produced by electron transitions between the second Bohr orbit ($n = 2$) and higher orbits ($n = 3, 4, 5$, and so forth). As an example, Figure 4-15 shows the electron transition that gives rise to the $H_\alpha$ spectral line, which has a wavelength of 656.28 nm.

In addition to giving the wavelengths of the Balmer series, Bohr's formula correctly predicts the wavelengths of other series of spectral lines that occur at nonvisible wavelengths. For example, electron transitions between the lowest Bohr orbit ($n = 1$) and all higher orbits also produce spectral lines. These transitions create the *Lyman series,* which is entirely in the ultraviolet wavelengths. At infrared wavelengths is the *Paschen series,* which arises out of transitions to and from the third Bohr orbit ($n = 3$). Additional series exist at still longer wavelengths.

Bohr's ideas also help explain Kirchhoff's laws. Each spectral line corresponds to one specific transition between the orbits of the electrons of a particular element. An absorption line is created when an electron jumps from an inner orbit to an outer orbit, extracting the required photon from an outside source of energy, such as the continuous spectrum of a hot, glowing object. An emission line is produced when an electron transitions to a lower orbit and emits a photon.

Today's view of the atom owes much to the Bohr model, but it is enhanced in certain ways. The modern theory of atoms is called **quantum mechanics,** a branch of physics dealing with photons and subatomic particles that was first developed during the 1920s. As a result of this work, physicists have moved away from the concept that electrons are solid particles with planetlike orbits about the nucleus. Instead, electrons are now known to have both wave and particle properties; they are said to occupy certain allowed **energy levels** in the atom.

An extremely useful way of displaying the structure of an atom is an energy level diagram, such as that shown in Figure 4-16 for hydrogen. The lowest energy level, called the **ground state,** corresponds to the $n = 1$ Bohr

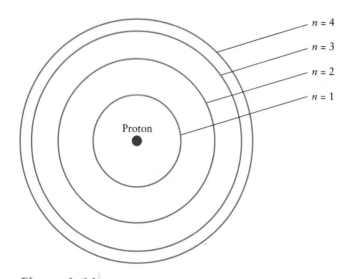

**Figure 4-14** **Bohr Model of the Hydrogen Atom** According to Bohr's model of the atom, an electron circles the nucleus only in allowed orbits $n = 1, 2, 3$, and so on. The first four Bohr orbits are shown here.

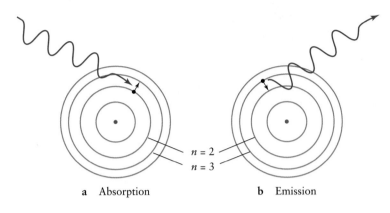

a　Absorption　　　　　　b　Emission

**Figure 4-15** **The Absorption and Emission of an H$_\alpha$ Photon** This schematic diagram, drawn according to the Bohr model of the atom, shows what happens when a hydrogen atom absorbs or emits an H$_\alpha$ photon, which has a wavelength of 656.28 nm. (**a**) The photon is absorbed by the atom as the electron jumps from orbit $n = 2$ up to orbit $n = 3$. (**b**) The photon is emitted by the atom as the electron falls from orbit $n = 3$ down to orbit $n = 2$.

orbit. An electron can jump from the ground state up to the $n = 2$ level only if the atom absorbs a photon of wavelength 121.6 nm. The energy of a photon is usually expressed in electron volts (abbreviated eV). The 121.6 nm photon has an energy of 10.19 eV, and so the energy level $n = 2$ is shown on the diagram as having an energy 10.19 eV above the energy of the ground state, which is usually assigned a value of 0 eV. Similarly, the $n = 3$ level is 12.07 eV above the ground state and so forth, up to the $n = \infty$ level at 13.6 eV. These states with more energy than the ground state are collectively called *excited states*.

If the atom absorbs a photon with an energy greater than 13.6 eV, an electron from the ground state will be knocked completely out of the atom. This is the process of photoionization mentioned above.

## 4-7 Spectral lines shift due to the relative motion between the source and the observer

Christian Doppler, a professor of mathematics in Prague, pointed out in 1842 that wavelength is affected by motion. As shown for the observer on the left in Figure 4-17, the wavelengths of electromagnetic radiation from an approaching source are compressed. The circles represent waves emitted from consecutive wave peaks as the source moves along. Because each successive wave is emitted from a position slightly closer to you, you see a shorter wavelength than you would if the source were stationary. All the spectral lines in the spectrum of an approaching source, regardless of its distance, are therefore shifted toward the short-wavelength (blue) end of the spectrum. This phenomenon is called a **blueshift.**

Conversely, electromagnetic waves from a receding source are stretched out. You see a longer wavelength than you would if the source were stationary. All the spectral lines in the spectrum of a receding source, regardless of its distance, are shifted toward the longer-wavelength (red) end of the spectrum, producing a **redshift.** In either case, the effect of relative motion on wavelength is called the **Doppler shift.** This effect varies directly with approaching or receding speed: An object approaching twice as fast as another has its colors blueshifted twice as much. An Astronomer's Toolbox 4-3 gives a numerical example.

The speed determined from the Doppler shift is called the **radial velocity,** because the motion is along our line of sight, or along the "radius" drawn from Earth to the star. Of course, the star may well have motion perpendicular to our line of sight, across the celestial sphere. This **proper motion** does not affect the perceived wavelength

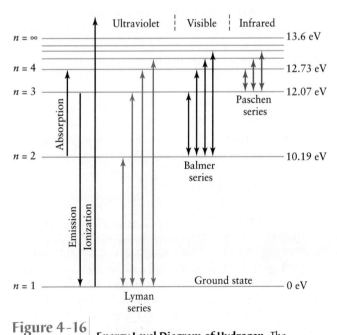

**Figure 4-16** **Energy Level Diagram of Hydrogen** The structure of the hydrogen atom is conveniently displayed in a diagram showing the energy levels above the ground state. A variety of electron jumps, or transitions, are shown, including those that produce some of the most prominent lines in the hydrogen spectrum.

# AN ASTRONOMER'S TOOLBOX 4-3

## The Doppler Shift

Suppose that $\lambda_0$ is the wavelength of a spectral line from a stationary source. This is the wavelength that a reference book would list or a laboratory experiment would yield. If the source is moving, this particular spectral line is shifted to a different wavelength $\lambda$. The size of the wavelength shift is usually written as $\Delta\lambda$ (delta lambda), where $\Delta\lambda = \lambda - \lambda_0$. This $\Delta\lambda$ is the difference between the wavelength that you actually observe in the spectrum of a star or galaxy and the wavelength listed in reference books.

Doppler proved that the wavelength shift is governed by the simple equation

$$\Delta\lambda/\lambda = v/c$$

where v is the speed of the source measured along the line of sight between the source and the observer. As usual, c is the speed of light ($3 \times 10^5$ km/s).

*Example:* The spectral lines of hydrogen appear in the spectrum of the bright star Vega as shown in the figure. The prominent hydrogen line $H_\alpha$ has a normal wavelength of 656.285 nm, but in Vega's spectrum the line is located at 656.255 nm. The wavelength shift is −0.030 nm, and so the star is approaching us with a speed of −14 km/s. The minus sign indicates that the star is moving toward us.

**The Spectrum of Vega**   Several Balmer lines are seen in this photograph of the spectrum of Vega, the brightest star in the constellation Lyra (the Harp). All the spectral lines are shifted very slightly toward the blue side of the spectrum, indicating that Vega is approaching us.

and cannot be determined by Doppler shift. Proper motion is determined by measuring a star's motion relative to background stars (Figure 4-18). The star with the greatest proper motion as seen from Earth is Barnard's star (Figure 4-19).

Proper motions are so small that they can be measured only for relatively nearby stars in our Galaxy. However, the radial velocity of virtually every object in space can be determined regardless of their distances from us. For example, Doppler shift measurements of the spectra of hot

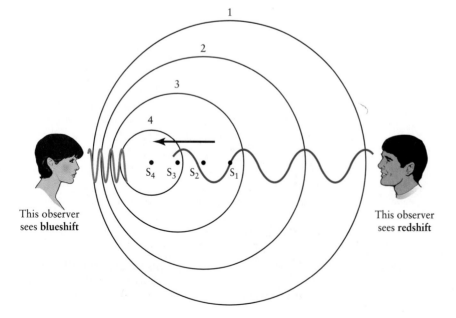

This observer sees **blueshift**

This observer sees **redshift**

**Figure 4-17** | **The Doppler Effect**
Wavelength is affected by motion between the light source and an observer. A source of light is moving toward the left. The four circles (numbered 1 through 4) indicate the location of light waves that were emitted by the moving source when it was at points $S_1$ through $S_4$, respectively. Note that the waves are compressed in front of the source but stretched out behind it. Consequently, wavelengths appear shortened (blueshifted) if the source is moving toward the observer and lengthened (redshifted) if the source is moving away from the observer. Motion perpendicular to an observer's line of sight does not affect wavelength.

gases on the Sun's surface reveal that they rise and fall. Doppler shift measurements of stars in double star systems give crucial data about the speeds of the stars orbiting around each other. Doppler measurements of the spectra of distant galaxies enable us to determine the rate at which the entire universe is expanding. In later chapters we will refer to the Doppler shift whenever we need to convert an observed wavelength shift into a speed.

> **Insight into science** **Remote science**
> Astronomical objects are so remote and their activities are often so complex that astrophysicists must use a tremendous amount of physics to interpret observations. For example, a single spectrum can contain information about stars, extrasolar planets, interstellar gas, the Earth's motion and atmosphere, and the performance of the observing telescope and the equipment attached to it. All of these factors must be understood theoretically and accounted for.

With the work of people like Planck, Einstein, Rutherford, Bohr, and Doppler, the interchange between astronomy and physics came full circle. Modern physics was born when Newton set out to understand the motions of the planets. Two and a half centuries later, physicists in their laboratories discovered the basic properties of electromagnetic radiation and the structures of atoms. As we will see in later chapters, the fruits of their labors have important implications for astronomy even today.

**August 24, 1893**

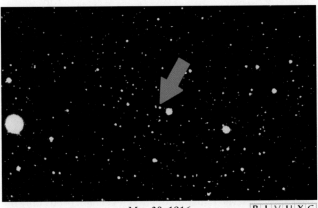

**May 30, 1916** R I V U X G

**Figure 4-19** **Barnard's Star** These two photographs, taken 22 years apart, show the proper motion of Barnard's star in the constellation of Ophiuchus. In addition to having the largest known proper motion (10.3″ per year), Barnard's star is one of the stars closest to Earth.

## WHAT DID YOU THINK?

1. *How hot is a "red hot" object compared to objects glowing other colors?* Of all objects that glow visibly from heat stored or generated inside them, those that glow red are the coolest. Those that glow blue are hottest.

2. *What color is the Sun?* The Sun emits all wavelengths of electromagnetic radiation. The colors it emits most intensely are in the blue-green part of the spectrum. Since the human eye is less sensitive to blue-green than to yellow, we see the Sun as yellow.

3. *Can carbon dating be used to determine the age of space debris found on Earth?* No. Carbon dating is only reliable for objects that formed within the past 100,000 years. Space debris found on Earth formed more than 4 billion years ago.

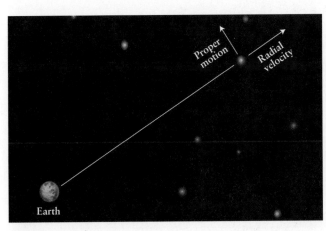

**Figure 4-18** **Radial and Proper Motions of a Star** The speed of a star toward or away from the Earth is its radial velocity. This motion creates a Doppler shift in the star's spectrum. The motion of the star across the sky, perpendicular to our line of sight, is called proper motion. Proper motion does not affect the star's spectrum.

## Key Words

| | | |
|---|---|---|
| absorption line spectrum, 94 | energy level, 98 | proper motion, 99 |
| atomic number, 96 | ground state, 98 | quantum mechanics, 98 |
| blackbody, 87 | ion, 96 | radial velocity, 99 |
| blackbody curve, 88 | ionization, 96 | redshift, 99 |
| blueshift, 99 | isotope, 96 | spectral analysis, 91 |
| continuous spectrum, 94 | Kirchhoff's laws, 94 | spectral lines, 91 |
| diffraction grating, 93 | luminosity, 89 | spectrograph, 93 |
| Doppler shift, 99 | molecule, 97 | spectroscope, 93 |
| element, 92 | nucleus (of an atom), 96 | Stefan-Boltzmann law, 88 |
| emission line spectrum, 94 | periodic table, 92 | transition, 98 |
| energy flux, 89 | Planck's law, 90 | Wien's law, 88 |

## Key Ideas

✳ By studying the wavelengths of electromagnetic radiation emitted and absorbed by an astronomical object, astronomers can learn the object's temperature, chemical composition and movement through space.

### Blackbody Radiation

✳ A blackbody is a hypothetical object that perfectly absorbs electromagnetic radiation at all wavelengths. Because a blackbody does not reflect any electromagnetic radiation from outside sources, the radiation that it does emit depends only on its temperature. Stars closely approximate blackbodies.

✳ Wien's law states that the dominant wavelength of radiation emitted by a blackbody is inversely proportional to its temperature. The intensities of radiation emitted at various wavelengths by a blackbody at a given temperature are shown by a blackbody curve.

✳ The Stefan-Boltzmann law relates the temperature of a blackbody to the rate at which it radiates energy.

### Discovering Spectra

✳ Spectroscopy—the study of electromagnetic spectra—provides important information about the chemical composition of remote astronomical objects.

✳ Kirchhoff's three laws of spectral analysis describe the conditions under which absorption lines, emission lines, and a continuous spectrum can be observed. Spectral lines serve as distinctive "fingerprints" for the chemical elements and chemical compounds comprising a light source.

### Atoms and Spectra

✳ An atom consists of a small, dense nucleus (composed of protons and neutrons) surrounded by electrons that can occupy only certain energy levels.

✳ The spectral lines of a particular element correspond to the various electron transitions between allowed energy levels of that element. When an electron shifts from one energy level to another, a photon of the appropriate energy (and hence a specific wavelength) is absorbed or emitted by the atom.

✳ The spectrum of hydrogen at visible wavelengths consists of the Balmer series, which arises from electron transitions between the second energy level of the hydrogen atom and higher levels.

✳ When an atom loses one or more electrons it is said to be charged, or ionized. One way for this to occur is for an electron to absorb an energetic photon and thereby fly free from its atom.

✳ The spectral lines of an approaching light source are shifted toward shorter wavelengths (a blueshift); the spectral lines of a receding light source are shifted toward longer wavelengths (a redshift).

✳ The equation describing the Doppler effect states that the size of a wavelength shift is proportional to the radial velocity between the light source and the observer.

## Review Questions

1 What is a blackbody? What does it mean to say that a star behaves almost like a blackbody? If stars behave like blackbodies, why are they not black?

2 What is Wien's law? How could you use it to determine the temperature of a star's surface?

3 What is the Stefan-Boltzmann law? Why are astronomers interested in it?

4 Using Wien's law and the Stefan-Boltzmann law, explain the changes in color and brightness that are observed as the temperature of a hot, glowing object increases.

5  Describe the experimental evidence that supports the Bohr model of the atom.

6  Explain how the spectrum of hydrogen is related to the structure of the hydrogen atom.

7  Why do different elements have different patterns of lines in their spectra?

8  What is the Doppler shift, and why is it important to astronomers?

9  Explain why the Doppler shift tells us only about the motion directly along the line of sight between a light source and an observer.

## Advanced Questions

The answers to all computational problems, which are preceded by an asterisk (*), appear at the end of the book.

*10  Approximately how many times around the world could a beam of light travel in 1 second?

*11  The bright star Regulus in the constellation of Leo (the Lion) has a surface temperature of 12,200 K. Approximately what is the dominant wavelength ($\lambda_{max}$) of the light it emits?

*12  The bright star Procyon in the constellation of Canis Minor (the Little Dog) emits the greatest intensity of radiation at a wavelength ($\lambda_{max}$) of 445 nm. Approximately what is the surface temperature of the star in Kelvins?

*13  The wavelength of $H_\beta$ in the spectrum of the star Megrez in the Big Dipper is 486.112 nm. Laboratory measurements demonstrate that the normal wavelength of this spectral line is 486.133 nm. Is the star coming toward us or moving away from us? At what speed?

*14  In the spectrum of the bright star Rigel, $H_\alpha$ has a wavelength of 656.331 nm. Is the star coming toward us or moving away from us? How fast?

*15  Imagine driving down a street toward a traffic light. How fast would you have to go so that the red light would appear green?

## Discussion Questions

16  Compare the technique of identifying chemicals by their spectral line patterns with that of identifying people by their fingerprints.

17  Suppose you look up at the night sky and observe some of the brightest stars with your naked eye. Is there any way of telling which stars are hot and which are cool? Explain.

18  How can we exclude the Earth's atmosphere as the source of the iron absorption lines in Figure 4-7?

## What If ...

*19  The Sun was twice its actual diameter, but still had the same surface temperature? At what wavelength would that new Sun emit its radiation most intensely? How many times brighter than the present Sun would the new Sun be? How might things on Earth be different under the bigger Sun?

20  All the nearby stars were observed to have blueshifted spectra? What conclusions could we draw? What other measurement would be useful to have for each star and why?

## Observing Projects

21  As soon as weather conditions permit, observe a rainbow. Confirm that the colors are in the same order as shown in this chapter's opening photo, page 86.

22  Turn on an electric stove or toaster oven. Watch as the color and brightness of the heating element changes. Confirm that red is the color of the coolest glowing objects, and that as an object heats up it gets brighter and its color shifts from red to orange to yellow.

23  Obtain a glass prism or a diffraction grating, available from science museum stores, catalogs, or your college physics department. Look through the prism or grating at various light sources, such as an ordinary incandescent lightbulb, a neon sign, and a mercury vapor street lamp. **DO NOT LOOK AT THE SUN! Looking directly at the Sun causes blindness.** Do you have any trouble seeing spectra? What do you have to do to see a spectrum? Describe the differences in the spectra of the various light sources you observed.

# II

# BUILD YOUR FOUNDATION

## The Solar System

*In the chapters ahead you will discover*

* how the solar system formed from a cloud of interstellar gas and dust

* why the environment of the early solar system was much more violent than it is today

* that the planets fall into groups with similar properties

* how astronomers characterize each planet's "personality" by its size, chemical composition, rotation rate, atmosphere, and features

* that the moons throughout the solar system are extremely varied

* that the solar system also contains debris—asteroids, meteoroids, and comets—left over from its formation

* why the Sun shines

RIVUXG

**The Planets of Our Solar System**
**This photomontage shows the nine planets of our solar system. At the top (from left to right) are Mercury, Venus, Earth, and Mars. At the bottom (from right to left) are Jupiter, Saturn, Uranus, Neptune, and Pluto. (The images are not reproduced to the same scale. See Figure II-9 for a scale drawing.)**

## WHAT DO YOU THINK?

1  How many stars are there in the solar system?

2  Was the solar system created as a direct result of the birth of the universe?

3  How long has the Earth existed?

4  Is Pluto always the farthest planet from the Sun?

5  What typical shapes do moons have?

For as long as people have looked up at the heavens, they have wondered about the Sun, the Moon, and the planets visible to the naked eye. Telescopes revealed many new features of these objects, as well as the existence of other planets and moons, along with asteroids, meteoroids, and comets, all orbiting the Sun. The Sun and all the bodies that orbit it make up our **solar system.**

1  The solar system has only one star, the Sun. The stars you see in the night sky, plus billions of stars too far away to see with the naked eye, are members of a larger system called the *Milky Way Galaxy.* The solar system is also part of the Milky Way, a topic we discuss in Chapter 14.

Did the solar system form all at once, or did it come together by serendipity? How were its varied building blocks of rock, metal, ice, and gas created? How has the solar system changed since its birth, and what does its history tell us about the planets we see today? Within the past few decades, telescopes and space probes, along with the principles of modern science, have finally provided answers to these age-old questions. Our new wealth of information is giving us a rapidly growing understanding of our nearest neighbors in space.

## II-1  The solar system formed from a cloud of cold gas and dust

Today we picture the solar system forming from a vast cloud of interstellar gas and dust, called the **solar nebula.** This cloud condensed 4.6 billion years ago under the influence of its own gravitational force to form the planets, the Sun, and other bodies. We can determine where the matter comprising the solar nebula came from by examining the chemical elements found in the bodies of the solar system and in the stars.

Hydrogen and helium are by far the most abundant elements in the universe. Together these elements account for 98% of the mass of all the material in existence—all the other elements combined account for only 2%. However, the Earth's mass contains less than 0.15% hydrogen and helium. Somehow the early solar system was enriched with the heavier elements we see today—oxygen, silicon, aluminum, iron, carbon, calcium, and others. Although these elements are rare in the universe as a whole, they are commonplace on Earth, and they are found to varying extents in all the other objects in the solar system. Our picture of the emerging solar nebula must explain these differences.

There is a good reason for the overwhelming abundance of hydrogen and helium throughout the universe. Most astronomers believe that the universe formed between 13 and 18 billion years ago with a violent event called the *Big Bang.* Only the lightest elements—hydrogen, helium, and a tiny amount of lithium—emerged when the cosmos was formed. The first stars, composed only of these three elements, condensed out of this primeval matter, probably within a few hundred million years after the Big Bang.

2  The difference in chemical composition between the early universe and the solar system shows that the solar system did not form as a *direct* result of the Big Bang. The solar system came billions of years later, and the heavier elements came to the solar nebula from stars that existed and exploded before the solar system formed.

Stars use various elements, such as hydrogen, helium, and carbon, to create energy. In this process, called *fusion,* lighter elements are transformed into heavier ones, such as hydrogen becoming helium or helium becoming carbon. We call fusion a *thermonuclear reaction* because the tremendous heat generated in the centers of stars is what enables nuclei of lighter elements there to combine, or fuse, into heavier elements. We explore details of the creation of this heavier matter in Chapters 11 and 12.

Near the ends of their lives, most stars cast matter out into space to form clouds of interstellar gas and dust. Often a star's outer layers are gradually expelled (Figure II-1), but some stars end their lives with spectacular detonations called *supernovae,* which blow the stars apart. Either way, the space between the stars becomes filled with gas containing mostly hydrogen and helium but gas that is now also enriched with heavy elements created in the stars. The solar nebula was a fragment of a larger cloud of such gas and dust.

Sometimes, the explosive force of a supernova compresses preexisting interstellar gas and dust from earlier star destruction. A new star and planetary system then condenses out of this enriched interstellar matter (Figure II-2). *We are literally made of star dust!*

R I V U X G

**Figure II-1** | **A Star Losing Mass** The old star HD 65750 in the southern constellation Carina (the Ship's Keel) is one of many stars rapidly shedding material from its outer layers. The *nebulosity* (cloudiness) around the star is caused by starlight scattered from dust grains that condensed from the material cast off by the star. The four lines (called *diffraction spikes*) and partial ring are created by structures within the telescope (the star is located where the diffraction spikes intersect).

## II-2 Gravity and heat shaped the young solar system

To find out more about the solar system's origins, we look for clues in its pieces. Especially valuable is the interplanetary debris we know as asteroids, meteoroids, and comets. Some of these bodies are believed to be essentially unchanged remnants from the formation of the solar system. Other clues to the origin of our solar system have recently been found in distant stars, where planetary systems like our own are developing.

Initially it was very cold inside the solar nebula—well below the freezing point of water. Ice and ice-coated dust grains composed of heavy elements were scattered abundantly across this vast volume, which had a diameter of at least 100 AU and a total mass about 2 to 3 times the mass of the Sun. Hydrogen and helium gas, accompanied by this ice and dust, fell toward the center of the solar nebula under the influence of the gravitational attraction of other gas in the nebula. As a result, the density and pressure at the center of the nebula began to increase, producing a concentration of matter called the **protosun**.

As the nebula continued to contract, atoms at its center collided with one another with increasing speed and frequency. Such collisions created heat, causing the temperature deep inside the solar nebula to soar. Therefore, the first heat in the solar system came from colliding gas, not from fusion, which began later.

The planets and other bodies orbiting the protosun formed because the solar nebula had a slight amount of rotation. Without this initial rotation, everything in the solar nebula would have fallen straight into the protosun, leaving no matter behind to form the planets. Mathematical studies show that the combined effects of gravity

R I V U X G

**Figure II-2** | **A Dusty Region of Star Formation** These young stars in the constellation Orion are still surrounded by much of the gas and dust from which they formed. The bluish, wispy nebulosity is caused by starlight reflecting off abundant interstellar dust grains. These grains are made of heavy elements (such as carbon, silicon, and iron) produced by earlier generations of stars. Astronomers hypothesize that the solar system formed in this way.

and rotation transform even an irregular cloud fragment (which the solar nebula probably was) into a rotating disk with a warm center and cold edges, as shown in Figure II-3. The disk shape explains why the orbits of the planets are all nearly in the same plane.

While we cannot see our solar system as it was back then, astronomers have found disks of gas and dust surrounding other young stars, like those shown in Figure II-4. These systems are believed to be undergoing the same initial stage of evolution that we are describing here for our solar system.

Temperatures around the young protosun soon began to climb. The rising temperatures vaporized all the com-

mon icy substances in the inner region of the solar nebula and pushed light gases like hydrogen and helium outward. In this inner region, where the Earth orbits today, mostly heavier elements remained. These heavy elements were eventually to compose the four inner planets: Mercury, Venus, Earth, and Mars.

## II-3 Innumerable collisions occurred in the early solar system

The formation of the inner planets was dominated by the impacts of small, rocky particles. Initially, neighboring

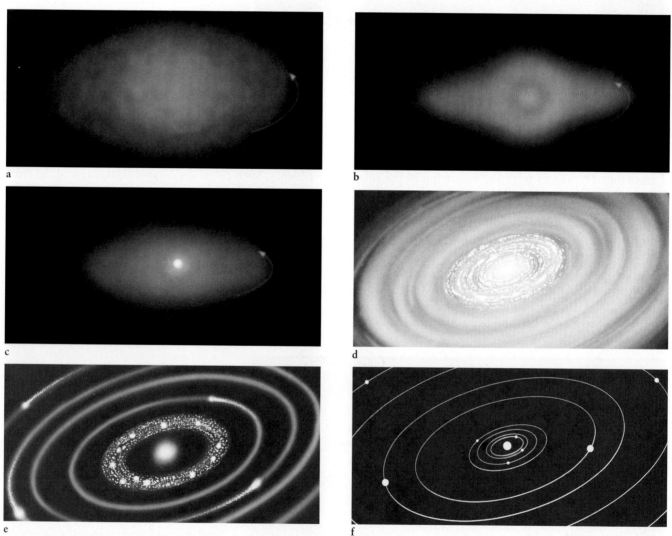

**Figure II-3** **The Birth of the Solar System** This sequence of drawings shows six stages in the formation of the solar system. **(a)** A slowly rotating cloud of interstellar gas and dust begins to contract because of its own gravity. **(b)** A central condensation, the protosun, forms as the cloud flattens and rotates faster. **(c)** A flattened disk of gas and dust surrounds the protosun, which has begun to shine. **(d)** The Sun's rising temperature removes the gas from the inner regions, leaving dust and larger debris revolving in place. **(e)** The planets have established dominance in their regions of the solar system. **(f)** The solar system as it appears today.

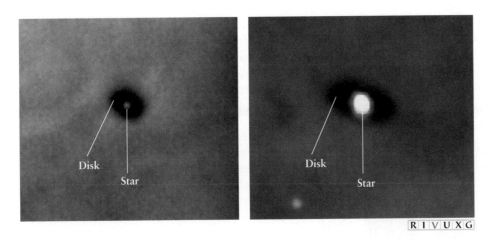

R I V U X G

**Figure II-4** **Young Circumstellar Disks of Matter** These Hubble Space Telescope images show circumstellar disks surrounding two stars 1500 light-years away from Earth in the Orion Nebula. These stars, both less than 10 million years old, are believed to be surrounded by protoplanetary gas and dust.

dust grains and pebbles in the solar nebula collided and stuck together. Then, over a period of a few million years, these accumulations of dust and pebbles coalesced into larger objects called **planetesimals,** with diameters of about 10 km.

During the next stage, the planetesimals collided. Because of their mutual gravitational attraction, they coalesced into still larger objects called **protoplanets.** This accumulation of material is called **accretion.** Computer simulations of the inner solar system based on Newton's laws (see Section 2-5) have shown that accretion continues for roughly 100 million years and should lead to the formation of fewer than half a dozen planets (Figure II-5). The agreement between predictions and our inner solar system is striking.

 **Insight into science   Computers aid analysis** Many equations are so complex that they require computer analysis to enable us to understand their implications. An example is the physics of the formation of the planets and the Sun.

While the inner regions of the solar system were heating up, temperatures in the outer regions of the solar nebula remained quite cool. Ice and ice-coated dust, along with hydrogen and helium gas, were able to survive in these cooler regions. The large outer planets—Jupiter, Saturn, Uranus, and Neptune—also probably formed from the

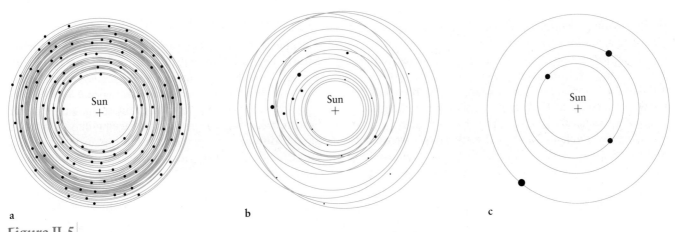

a                    b                    c

**Figure II-5**   **Accretion of the Inner Planets** These three drawings show the results of a computer simulation of the formation of the inner planets. (**a**) The simulation begins with 100 planetesimals. (**b**) After 30 million years, these planetesimals have coalesced into 22 proto-planets. (**c**) This final view is for an elapsed time of 441 million years, but the formation of the inner planets in this simulation is essentially complete after only 150 million years.

RIVUXG

**Figure II-6**   **Our Moon** This photograph, taken by astronauts in 1972, shows thousands of the craters produced by impacts of rocky debris left over from the formation of the solar system. Age-dating of lunar rocks brought back by the astronauts indicates that the Moon is about 4.5 billion years old. Most of the lunar craters were formed during the Moon's first 800 million years of existence, when the rate of bombardment was much greater than it is now.

accretion of planetesimals. For each, a rocky core even bigger than the Earth apparently served as a "seed," and for about a million years the core became coated with additional rock and gas. When the gas-rich envelope became as massive as the rocky core, the gas began to accumulate at a runaway pace. Thereafter, the envelope pulled in all the gas in its vicinity, creating a huge hydrogen-rich shell surrounding a core of rocky material.

Because of their gaseous shells, the outer planets today include abundant quantities of hydrogen, helium, methane, ammonia, and water. Many of the moons of the outer planets are also partially composed of these low-density substances.

During the millions of years that the planets were forming, so, too, was the Sun. During this time the temperature and pressure at the center of the contracting protosun continued to climb. Finally, the center of the protosun became hot enough to ignite thermonuclear reactions. Now, hydrogen fused into helium in its core, thereby releasing huge amounts of energy, and the Sun was born. Sunlike stars take approximately 100 million years to form from a nebula and settle down, which means that the Sun became a full-fledged

star at roughly the same time that the accretion of the inner planets was complete. Indeed, radiation from the Sun prevented new planetesimal formation, thereby limiting the sizes of the planets.

[3] According to radioactive dating of the oldest debris ever discovered from space, the solar system was essentially in the form we know today just under 4.6 billion years ago. But "essentially" is not "completely." While the planets were in place, only a few of the dozens of moons were in orbit, and innumerable pieces of rocky and icy debris were on collision courses with the planets. Indeed, evidence indicates that our Moon was created by such a collision (see Chapter 5).

Once formed, our Moon's surface was scarred by numerous impacts (Figure II-6). (Note that we capitalize the word *moon* only when we are referring to the Earth's Moon.) These scars, called **craters**, are also found on those planets and moons that do not have appreciable atmospheres or geological activity that would otherwise erase these features. Indeed, the Moon's virtually airless environment has preserved important information about the early history of the solar system.

Radioactive dating of Moon rocks brought back by the Apollo astronauts indicates that the rate of impacts declined dramatically about 3.8 billion years ago. Since that time, impact cratering has proceeded at a very low rate. Thus, most of the craters on the Moon and planets were formed during the first 800 million years of the solar system's history, as the young planets swept up rocky debris left over from the solar nebula. Eventually, biological evolution occurred on Earth, and possibly on Mars and on Jupiter's moon Europa.

## II-4   Comparisons among the nine planets show distinct similarities and significant differences

The nine planets are Mercury, Venus, Earth, Mars, Jupiter, Saturn, Uranus, Neptune, and Pluto. As shown in Figure II-7, the orbits of the four inner planets—Mercury, Venus, Earth, and Mars—are crowded close to the Sun. In contrast, the orbits of the four large outer planets—Jupiter, Saturn, Uranus, and Neptune—are widely spaced at greater distances from the Sun. Pluto is usually orbiting at the far fringes of the part of the solar system inhabited by planets.

[4] Kepler's laws (see Chapter 2) showed us that all the planets have elliptical orbits. Even so, most of their orbits are nearly circular. The exceptions are Mercury and Pluto, whose orbits are noticeably elliptical. In fact, Pluto's orbit sometimes takes it nearer to the Sun than its neighbor, Neptune. In 1979, Pluto passed inside Neptune's orbit.

**Figure II-7**  **The Solar System**  This scale drawing shows the distribution of planetary orbits around the Sun. The four inner planets are located close to the Sun; the five outer planets orbit at much greater distances from the Sun. The viewpoint of this figure makes the orbits appear much more oval-shaped than they really are. Seen from above the disk of the solar system, most of the orbits appear *nearly* circular, as shown for Neptune in Figure II-8. All orbits are counterclockwise because the view is from above the Earth's north pole.

Neptune was the most distant planet from the Sun until February 1999, when Pluto's elliptical orbit once again took it farther out than Neptune (Figure II-8). Pluto will be farther from the Sun than Neptune for about the next 230 years.

All the planetary orbits except that of Pluto lie in planes close to the ecliptic. Pluto's orbit has a conspicuous tilt (called an *orbital inclination*) compared with the orbits of the other planets (see Figure II-7). Table II-1 lists some orbital characteristics of the nine planets.

Size is the most obvious physical difference among the planets. The smallest of the four giant outer planets, Neptune, is nearly 4 times larger than the Earth, the largest of the four inner planets, in diameter. First place among the giant planets goes to Jupiter, whose diameter is about 11 times bigger than that of Earth. Pluto is the smallest of all the planets, and is even smaller than the seven largest moons, including our own. Figure II-9 shows the Sun and the planets drawn to the same scale. The diameters of the planets are given in Table II-2.

Mass is another characteristic that distinguishes the inner and outer planets. The four inner planets have low masses compared with the giant planets. Again, first

**Figure II-8**  **Pluto's Orbit Is Highly Elliptical**  Pluto has the most elliptical orbit of any planet. Indeed, it spends part of the time closer to the Sun than Neptune. It last moved inside Neptune's orbit in 1979 and moved back outside in 1999. Note that Pluto and Neptune will never collide because their orbits are synchronized by their gravitational interaction to keep them apart.

### Table II-1
### Orbital Characteristics of the Planets

|  | Average distance from Sun (AU) | Average distance from Sun (10⁶ km) | Orbital period (yr) |
|---|---|---|---|
| Mercury | 0.39 | 58 | 0.24 |
| Venus | 0.72 | 108 | 0.62 |
| Earth | 1.00 | 150 | 1.00 |
| Mars | 1.52 | 228 | 1.88 |
| Jupiter | 5.20 | 778 | 11.86 |
| Saturn | 9.54 | 1427 | 29.46 |
| Uranus | 19.19 | 2871 | 84.01 |
| Neptune | 30.06 | 4497 | 164.79 |
| Pluto | 39.53 | 5914 | 248.54 |

place goes to Jupiter, whose mass is 318 times greater than Earth's (see Table II-2).

Size and mass can be combined in a very useful way to provide information about the chemical composition of a planet (or any other object, for that matter). Matter composed of heavy elements, like iron or lead, has more particles compressed into the same volume than matter composed of light elements, like hydrogen, helium, or carbon. Therefore, objects composed primarily of heavier elements have a greater **average density** than objects composed primarily of lighter elements. Average density is given by the equation

$$\text{Average density} = \frac{\text{total mass}}{\text{total volume}}$$

The chemical composition (kinds of elements) of any object therefore determines how much space the object occupies per unit mass. For example, a kilogram of iron takes up less space (is denser) than a kilogram of water, even though both have the same mass. Average density is expressed in kilograms per cubic meter. To help us grasp this concept, we compare the average densities of the planets to something familiar, namely water, which has an average density of 1000 kg/m³.

The four inner planets have high average densities compared to water (see Table II-2). In particular, the average density of the Earth is 5520 kg/m³. Because the density of typical surface rock is only about 3000 kg/m³, the Earth must contain a large amount of material inside it that is denser than surface rock.

In sharp contrast, the outer planets have relatively low average densities. For example, Saturn has an average density less than that of water. Their low densities support the belief that the giant outer planets are composed primarily of such light elements as hydrogen and helium.

Pluto is again an oddity. Although it is smaller than any of the inner planets, its average density falls between those of the Earthlike and the giant outer planets. Pluto is probably composed of a mixture of rock and ice, because its average density is between that of rock and ice.

**Insight into science**  **Astronomical measurement**
The laws of physics allow us to infer things we cannot measure directly. For example, we derive information about planetary densities from their masses and volumes. Kepler's laws yield the mass of each planet from the periods of their moons' orbits, while the measured diameters of the planets yield their volumes. As shown in the equation above, dividing mass by volume yields the average density.

The planets fall roughly into four groups. The four inner planets form the first group, followed by Jupiter and Saturn as the second, Uranus and Neptune as the third, and Pluto as the fourth. A group of one? In fact, Pluto actually appears to be the largest of a group of myriad distant bodies in the solar system. Indeed, at least

### Table II-2
### Physical Characteristics of the Planets

|  | Diameter (km) | Diameter (Earth = 1) | Mass (kg) | Mass (Earth = 1) | Average density (kg/m³) |
|---|---|---|---|---|---|
| Mercury | 4,878 | 0.38 | $3.3 \times 10^{23}$ | 0.06 | 5430 |
| Venus | 12,100 | 0.95 | $4.9 \times 10^{24}$ | 0.81 | 5250 |
| Earth | 12,756 | 1.00 | $6.0 \times 10^{24}$ | 1.00 | 5520 |
| Mars | 6,786 | 0.53 | $6.4 \times 10^{23}$ | 0.11 | 3950 |
| Jupiter | 142,984 | 11.21 | $1.9 \times 10^{27}$ | 317.94 | 1330 |
| Saturn | 120,536 | 9.45 | $5.7 \times 10^{26}$ | 95.18 | 690 |
| Uranus | 51,118 | 4.01 | $8.7 \times 10^{25}$ | 14.53 | 1290 |
| Neptune | 49,528 | 3.88 | $1.0 \times 10^{26}$ | 17.14 | 1640 |
| Pluto | 2,300 | 0.18 | $1.3 \times 10^{22}$ | 0.002 | 2030 |

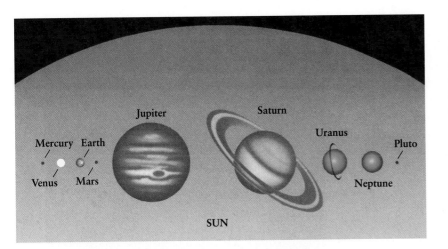

**Figure II-9** **The Sun and the Planets** This drawing shows the nine planets in front of the disk of the Sun, with all ten bodies drawn to the same scale. The four planets orbiting nearest the Sun (Mercury, Venus, Earth, and Mars) are small and made of rock and metal. The next four planets (Jupiter, Saturn, Uranus, and Neptune) are large and composed primarily of light elements, such as hydrogen and helium. Pluto has roughly equal amounts of rock and ice.

70 small bodies orbiting the Sun beyond Pluto have been identified, but few details about them are known.

From their average densities, we can draw conclusions about the composition of all the planets. The four inner ones, composed primarily of rock and metal, are called **terrestrial planets,** because they resemble the Earth (in Latin, *terra*). Mountains, craters, canyons, and volcanoes are common on their hard, rocky surfaces. Jupiter and Saturn are composed primarily of hydrogen and helium, Uranus and Neptune have extremely large quantities of water as well as much hydrogen and helium, and Pluto is probably a mix of rock and ice. Vast, swirling cloud formations dominate the appearance of Jupiter, Saturn, Uranus, and Neptune. The montage of photographs that opens this Foundation section shows the distinctive appearances of the different planets.

The different surfaces (or the upper cloud layers) of the planets reflect different amounts of light back out into space. The fraction of incoming light returning directly into space is measured by each body's **albedo.** An object that reflects no light has an albedo of 0.0; for example, powdered charcoal has an albedo of nearly 0.0. An object that reflects all the light incident upon it (a high-quality mirror comes close) has an albedo of 1.0. The albedo multiplied by 100 gives the percentage of light directly scattered off that body. The three inner planets with solid surfaces exposed to space (Mercury, Earth, and Mars) have albedos of .37 or less. The five planets surrounded by clouds have albedos of .47 or more. While Pluto's albedo of .5 is relatively high, we don't yet know much about its surface.

Every planet except Mercury and Venus has moons (sometimes called *natural satellites*). There are at least 67 moons in the solar system, and more are still being discovered. Unlike our Moon, most moons are irregularly shaped, more like potatoes than spheres. We will discover that there is as much variety among moons as there is among planets.

## II-5 Minor debris from the formation of the solar system still exists

In addition to the nine planets and their moons, many other objects orbit the Sun. Between the orbits of Mars and Jupiter are believed to be millions of such bodies, called **asteroids,** most of them smaller than a kilometer across. This region is called the **asteroid belt.** Asteroids are composed primarily of metal and rock. The largest asteroid, Ceres, has a diameter of about 900 km. The next largest, Pallas and Vesta, are each about 500 km in diameter. Still smaller ones are increasingly numerous. There are many thousands of kilometer-sized asteroids. A closeup picture of the asteroid Gaspra, taken by a spacecraft on its way to Jupiter, is shown in Figure II-10. In addition

R I V U X G

**Figure II-10** **An Asteroid** This picture of the asteroid Gaspra was taken in 1991 by the *Galileo* spacecraft on its way toward Jupiter. The asteroid measures $12 \times 20 \times 11$ km. Many, many similar chunks of rock orbit the Sun between the orbits of Mars and Jupiter.

# DISCOVER FOR YOURSELF | The Formation of the Solar System

### 4.7 billion years ago

✳ Solar nebula begins collapsing.
✳ Protosun starts growing from in-falling gas.

### 4.69 billion years ago

✳ Disk of matter forms around the young protosun.
✳ As regions of the disk clump and collide, planets begin to form.

### 4.6 billion years ago

✳ Formation of meteoroids, asteroids, comets, and planets essentially complete.
✳ Sun starts nuclear fusion and brightens.
✳ Sun's energy causes remaining gas and dust to dissipate into interstellar space.
✳ Meteoroids, asteroids, and comets continue to strike planets and the Sun.

### 3.8 billion years ago

✳ Most planetesimals have become parts of planets or the Sun.
✳ Frequent major impacts end.

to the asteroids between Mars and Jupiter, other asteroids have highly elliptical orbits that take them across the paths of some of the planets. Some asteroids have their own moons, such as asteroid Ida, which is orbited by heavily cratered Dactyl.

Pieces of rocky and metallic debris even smaller than asteroids are called **meteoroids.** Typical meteoroids are boulder-sized or smaller and apparently exist throughout the disk of the solar system, although most are in the asteroid belt. Most meteoroids are probably small fragments of asteroids that broke off when larger bodies collided.

Quite far from the Sun, well beyond the orbit of Pluto, are hundreds of billions of chunks of ice called **comets.** Some comets orbit in the same plane as the planets, but others are distributed in a sphere around the Sun. Many comets have highly elongated orbits that occasionally bring them close to the Sun. When this happens, the Sun's radiation vaporizes some of the comet's ices, producing long, flowing tails (Figure II-11).

Icy debris is also believed to have been discovered orbiting other stars. Studying this material is one method astronomers use in their quest to discover planets outside the solar system.

## II-6  Planets have been discovered outside our solar system

Astronomers are discovering planets around an ever-growing number of stars. These discoveries began with visual images showing disks of material orbiting a few stars, such as Beta Pictoris. The disk around Beta Pictoris is warped, a sign that planets may exist there as well. The planets would tug on different areas of the disk differently, thereby creating the warp (Figure II-12).

Planets outside the solar system are called *extrasolar planets.* Observing them in orbit around other stars is challenging, because even high-albedo planets reflect less than a millionth as much light as a star emits. Extrasolar planets are usually too dim and too close to their parent stars to be seen easily. Hence, scientists look for planets by indirect observation, primarily by detecting their gravitational tugs on the stars they orbit.

Some planets are discovered by measuring the radial velocity (defined in Chapter 4) of their stars, as measured by the Doppler shifts of the stars' spectra. This motion toward or away from Earth is created by the gravitational pull of the planet. The Doppler shift changes cyclically, and the length of time of one cycle is the period of the planet's orbit.

Other planets are discovered from variations in a star's proper motion, or motion among the background stars (see Chapter 4). A planet's attraction causes its star to deviate from motion in a straight line. This *astrometric* method of discovery looks for just such a wobble.

Finally, planets are located by changes in the brightness of their stars. As a planet passes between us and its star, it partially eclipses the star.

At least fourteen extrasolar planets have been discovered by these indirect methods—more than the number

R I V U X G

**Figure II-11**   **A Comet** The solid part of a typical comet is an irregularly shaped chunk of ice 10 km in diameter. When a comet passes near the Sun, solar radiation vaporizes some of the comet's ices, and the resulting gases and dust form one or two tails millions of kilometers long. This photograph shows Comet Kohoutek in 1974.

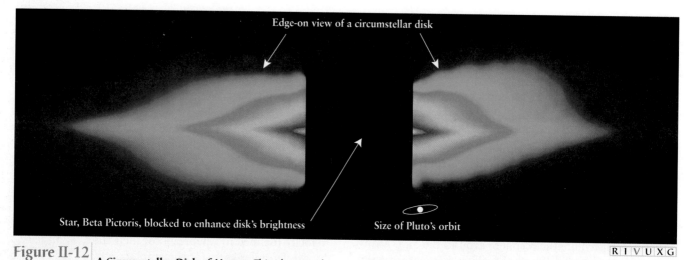

Edge-on view of a circumstellar disk

Star, Beta Pictoris, blocked to enhance disk's brightness

Size of Pluto's orbit

R I V U X G

**Figure II-12** | **A Circumstellar Disk of Matter** This photograph shows a disk of material orbiting the star called Beta Pictoris (blocked out in this image) 50 ly from Earth. A few hundred million years old, this disk is believed to be composed primarily of iceberglike bodies orbiting the star. Notice that the edge-on disk is warped. This is believed to be due to the gravitational tug of at least one planet orbiting Beta Pictoris inside the inner edge of the disk (in the central region that is blocked out).

of planets known in our solar system! They orbit a breathtaking variety of stars, including Earth-sized, burned-out hulks of stars much more massive than the Sun. Some known planets, however, do orbit stars much like our own.

The astrometric method also tells us the mass of a planet. The two lowest-mass planets known with confidence have 3 or 4 times the mass of the Earth, and one may have been observed with a mass only slightly greater than that of our Moon. The vast majority, however, range between a half and a few times the mass of Jupiter, and they orbit surprisingly close to their stars (Figure II-13). From our model of how our solar system formed, these planets should have formed much farther from their stars

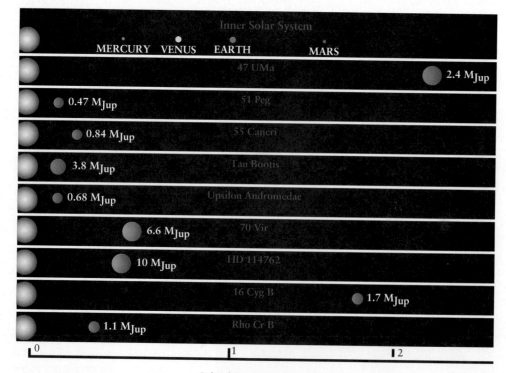

Inner Solar System

MERCURY   VENUS   EARTH        MARS

47 UMa                                              2.4 M$_{Jup}$

0.47 M$_{Jup}$   51 Peg

0.84 M$_{Jup}$   55 Cancri

3.8 M$_{Jup}$   Tau Bootis

0.68 M$_{Jup}$   Upsilon Andromedae

6.6 M$_{Jup}$   70 Vir

10 M$_{Jup}$   HD 114762

16 Cyg B                              1.7 M$_{Jup}$

1.1 M$_{Jup}$   Rho Cr B

0                            1                         2

Orbital Semimajor Axis (AU)

**Figure II-13** | **Planets Around Normal Stars** This figure shows the separation between extrasolar planets and nine Sunlike stars. The corresponding star names are given in the middle of each line. Our solar system, at the top, shows how close many of these large extrasolar planets are to their stars.

than they are at present. Astronomers are working to explain how planets can migrate inward.

While none of the planets discovered by these indirect methods has yet been observed directly, one extrasolar planet is believed to have been seen. But it is not orbiting a star! Rather, it is being ejected from its birthplace in orbit around a pair of stars (Figure II-14). As we discuss in Chapter 10, such star pairs are extremely common. Computer simulations reveal that most planets in such systems have unstable orbits. Many of these planets should eventually be ejected, as shown in Figure II-14.

All the *known* extrasolar planets are probably lifeless. Among them, the terrestrial type planets are orbiting only tiny stellar remnants. These stars exploded long ago with such titanic force that any life on their planets would have been annihilated. Meanwhile, the high-mass planets are, like Jupiter, probably composed primarily of hydrogen and helium. Life outside the solar system may still exist. We just have to find the right planets.

Of course, we are most likely to discover extraterrestrial life on extrasolar planets similar to Earth orbiting at about 1 AU from stars similar to the Sun. And the search is on, thanks to the power of adaptive optics and space telescopes. Planets are brighter at infrared wavelengths, compared to their stars, than they are at visible wavelengths. Will an infrared telescope therefore point us to the first traces of distant life?

We begin a detailed exploration of the solar system by examining the two bodies we know best—the Earth and its Moon. By understanding them, we will be better able to make some sense of the remarkably alien neighboring worlds we encounter thereafter. Indeed, we will discover that the properties of some planets, moons, and other objects in the solar system are so different from the Earth and its Moon as to make direct comparison almost meaningless.

## WHAT DID YOU THINK?

1   *How many stars are there in the solar system?* The solar system has only one star, the Sun.

2   *Was the solar system created as a direct result of the birth of the universe?* No. All matter was created by the Big Bang. However, much of the material that exists in our solar system was processed inside stars that evolved before the solar system existed. The solar system formed billions of years after the Big Bang occurred.

3   *How long has the Earth existed?* The Earth formed along with the rest of the solar system 4.6 billion years ago.

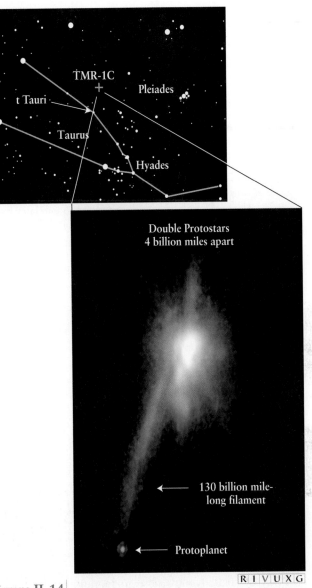

Double Protostars
4 billion miles apart

← 130 billion mile-long filament

← Protoplanet

R I V U X G

**Figure II-14** **A Candidate Extrasolar Planet** This Hubble Space Telescope image shows an apparent planet being flung from the pair of stars around which it was formed. Present calculations suggest that the planet's mass is 2 to 3 times the mass of Jupiter. The planet is presently $6 \times 10^9$ km from its birthplace, moving away at 10 km/s. This system is 450 ly from Earth.

4   *Is Pluto always the farthest planet from the Sun?* From 1979 to 1999, Neptune was the farthest planet from the Sun. This was because Pluto's orbit is highly eccentric, bringing that planet inside Neptune's orbit for about 20 years once every 250 years.

5   *What typical shapes do moons have?* While some moons are spherical, most look roughly like potatoes.

## Key Words

accretion, 109

albedo, 113

asteroid, 113

asteroid belt, 113

average density, 112

comet, 115

crater, 110

meteoroid, 115

planetesimal, 109

protoplanet, 109

protosun, 107

solar nebula, 106

solar system, 106

terrestrial planet, 113

## Key Ideas

✳   Hydrogen, helium, and traces of lithium, the three lightest elements, were formed shortly after the creation of the universe. The heavier elements were produced much later by stars and cast into space when the stars died. By mass, 98% of the matter in the universe is hydrogen and helium.

✳   The solar system formed 4.6 billion years ago from a swirling, disk-shaped cloud of gas, ice, and dust called the solar nebula.

✳   The four inner planets formed through the accretion of dust particles into planetesimals and then into larger protoplanets. The four large outer planets probably formed through the runaway accretion of gas onto rocky protoplanetary cores.

✳   The Sun formed at the center of the solar nebula. After about 100 million years, the temperature at the protosun's center was high enough to ignite thermonuclear reactions. For 800 million years after the Sun formed, impacts of asteroid-like objects on the young planets dominated the early history of the solar system.

✳   The four inner planets of the solar system share many characteristics and are distinctly different from the four giant outer planets and from Pluto.

✳   The four inner, terrestrial planets are relatively small, have high average densities, and are composed primarily of rock and metal.

✳   Jupiter and Saturn have large diameters and low densities and are composed primarily of hydrogen and helium. Uranus and Neptune have large quantities of water as well as much hydrogen and helium. Pluto, the smallest of the nine planets, is of intermediate density owing to its rock-ice composition.

✳   Asteroids are rocky and metallic debris in the solar system larger than about a kilometer in diameter. Meteoroids are smaller pieces of such debris. Comets are debris that contain both ice and rock.

✳   At least fourteen extrasolar planets have been discovered orbiting other stars and one that was ejected from its star system.

## Review Questions

1  Why was Pluto recently closer to the Sun than Neptune?

2  Why do astronomers believe that the solar nebula was rotating?

3  Describe three methods that exist for discovering extrasolar planets.

## Advanced Questions

4  What two properties of a planet must be known in order for its average density to be determined? How are these properties determined?

5  How can Neptune have more mass but a smaller diameter than Uranus? *Hint:* See Table II-2.

## What If ...

6 The Earth had formed around one of the very first generation stars created in the universe? What chemical elements would the Earth then be composed of? Could we humans exist on that version of Earth? Why or why not?

7 The Earth had a highly elliptical orbit, like Pluto? What would be different about the Earth and life on it?

## Observing Projects

8 Planetarium software can help you get an overview of our solar system. We will learn more about it in the next chapters. For now, however, open your Starry Night program and double click on the planet list to locate at least one of the inner and one of the outer planets. Are you surprised by their brightness relative to the stars? Which, if any, planets are visible in the night sky this month where you live?

# An Astronomer's Almanac

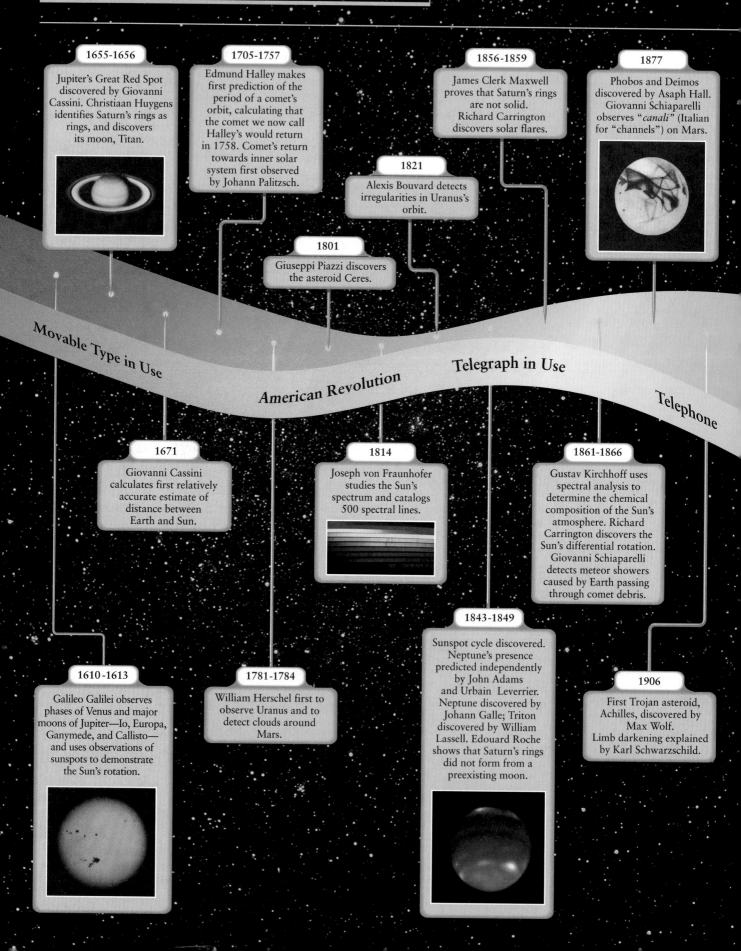

**1655-1656**
Jupiter's Great Red Spot discovered by Giovanni Cassini. Christiaan Huygens identifies Saturn's rings as rings, and discovers its moon, Titan.

**1705-1757**
Edmund Halley makes first prediction of the period of a comet's orbit, calculating that the comet we now call Halley's would return in 1758. Comet's return towards inner solar system first observed by Johann Palitzsch.

**1856-1859**
James Clerk Maxwell proves that Saturn's rings are not solid. Richard Carrington discovers solar flares.

**1877**
Phobos and Deimos discovered by Asaph Hall. Giovanni Schiaparelli observes *"canali"* (Italian for "channels") on Mars.

**1821**
Alexis Bouvard detects irregularities in Uranus's orbit.

**1801**
Giuseppi Piazzi discovers the asteroid Ceres.

Movable Type in Use

American Revolution

Telegraph in Use

Telephone

**1671**
Giovanni Cassini calculates first relatively accurate estimate of distance between Earth and Sun.

**1814**
Joseph von Fraunhofer studies the Sun's spectrum and catalogs 500 spectral lines.

**1861-1866**
Gustav Kirchhoff uses spectral analysis to determine the chemical composition of the Sun's atmosphere. Richard Carrington discovers the Sun's differential rotation. Giovanni Schiaparelli detects meteor showers caused by Earth passing through comet debris.

**1843-1849**
Sunspot cycle discovered. Neptune's presence predicted independently by John Adams and Urbain Leverrier. Neptune discovered by Johann Galle; Triton discovered by William Lassell. Edouard Roche shows that Saturn's rings did not form from a preexisting moon.

**1610-1613**
Galileo Galilei observes phases of Venus and major moons of Jupiter—Io, Europa, Ganymede, and Callisto—and uses observations of sunspots to demonstrate the Sun's rotation.

**1781-1784**
William Herschel first to observe Uranus and to detect clouds around Mars.

**1906**
First Trojan asteroid, Achilles, discovered by Max Wolf. Limb darkening explained by Karl Schwarzschild.

# THE SOLAR SYSTEM

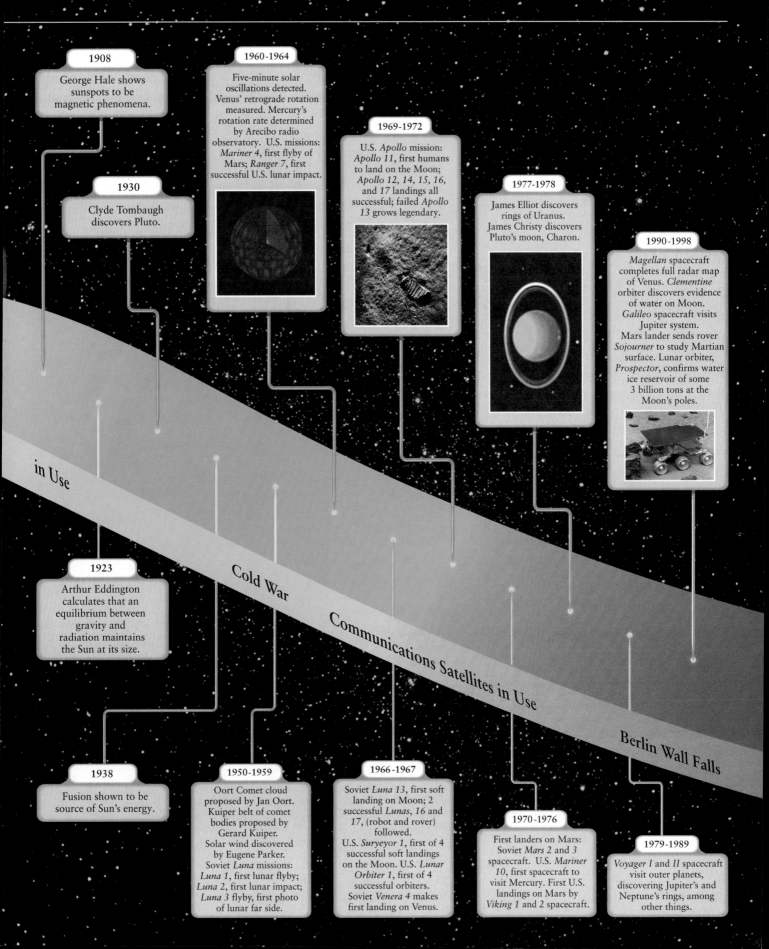

**1908**
George Hale shows sunspots to be magnetic phenomena.

**1930**
Clyde Tombaugh discovers Pluto.

**1960-1964**
Five-minute solar oscillations detected. Venus' retrograde rotation measured. Mercury's rotation rate determined by Arecibo radio observatory. U.S. missions: *Mariner 4*, first flyby of Mars; *Ranger 7*, first successful U.S. lunar impact.

**1969-1972**
U.S. *Apollo* mission: *Apollo 11*, first humans to land on the Moon; *Apollo 12, 14, 15, 16,* and *17* landings all successful; failed *Apollo 13* grows legendary.

**1977-1978**
James Elliot discovers rings of Uranus. James Christy discovers Pluto's moon, Charon.

**1990-1998**
*Magellan* spacecraft completes full radar map of Venus. *Clementine* orbiter discovers evidence of water on Moon. *Galileo* spacecraft visits Jupiter system. Mars lander sends rover *Sojourner* to study Martian surface. Lunar orbiter, *Prospector*, confirms water ice reservoir of some 3 billion tons at the Moon's poles.

in Use

Cold War

Communications Satellites in Use

Berlin Wall Falls

**1923**
Arthur Eddington calculates that an equilibrium between gravity and radiation maintains the Sun at its size.

**1938**
Fusion shown to be source of Sun's energy.

**1950-1959**
Oort Comet cloud proposed by Jan Oort. Kuiper belt of comet bodies proposed by Gerard Kuiper. Solar wind discovered by Eugene Parker. Soviet *Luna* missions: *Luna 1*, first lunar flyby; *Luna 2*, first lunar impact; *Luna 3* flyby, first photo of lunar far side.

**1966-1967**
Soviet *Luna 13*, first soft landing on Moon; 2 successful *Lunas*, 16 and 17, (robot and rover) followed. U.S. *Surveyor 1*, first of 4 successful soft landings on the Moon. U.S. *Lunar Orbiter 1*, first of 4 successful orbiters. Soviet *Venera 4* makes first landing on Venus.

**1970-1976**
First landers on Mars: Soviet *Mars 2* and *3* spacecraft. U.S. *Mariner 10*, first spacecraft to visit Mercury. First U.S. landings on Mars by *Viking 1* and 2 spacecraft.

**1979-1989**
*Voyager I* and *II* spacecraft visit outer planets, discovering Jupiter's and Neptune's rings, among other things.

# Earth and Moon

## *In this chapter you will discover*

✳   why the Earth provides such an ideal environment for life

✳   that the Earth's insides are churning, its surface is constantly in motion, and its atmosphere has evolved dramatically

✳   how Earth's magnetic field helps protect life from high-energy particles from space

✳   that the Moon's surface has been pounded by countless impacts from space debris

✳   how the Moon causes the Earth's tides

✳   that both the Earth and Moon have two major types of surface features

✳   that water has apparently been discovered at the Moon's poles

**The Earth and the Moon**   Earth is a dynamic planet whose surface is covered mostly with water and whose nitrogen-oxygen atmosphere supports life. In contrast, the Moon is a barren, desolate world. Because the Moon has virtually no atmosphere and no liquid water, lunar rocks have not been subjected to weathering by wind and water. They thus preserve information about the early history of the solar system. The Earth's diameter is about 4 times the Moon's.

RIVUXG

## WHAT DO YOU THINK?

1  Will the ozone layer, which is now being depleted, naturally replenish itself?

2  Who was the first person to walk on the Moon and on what Apollo space mission did this landing occur?

3  Do we see all parts of the Moon's surface at some time throughout the lunar cycle of phases?

4  Does the Moon rotate, and, if so, how fast?

5  What causes the ocean tides?

6  When does the spring tide occur?

Suppose you arrive in our solar system from a distant star system, looking for inhabitable worlds. Your spacecraft approaches the inner solar system on the opposite side of the Sun from the Earth. You encounter Mars, with its thin, chilled, unbreathable atmosphere and barren desert landscapes. Not very promising. The next planet you encounter, Venus, is enshrouded with corrosive clouds hiding a menacingly hot surface. Finally, you spy Earth, orbiting the Sun between these two relatively forbidding planets—one too cold, the other too hot.

Everchanging white clouds pirouette above the browns and blues of Earth's continents and oceans. Its close companion, the Moon, pockmarked by countless craters, causes those oceans to move up and down in a ceaseless rhythm. Is this Earth a world that your funding agency back home can use to justify more astronomical research?

## EARTH: A DYNAMIC, VITAL WORLD

A space traveler might first study the sunlight reflected from the Earth's shifting clouds and its surface. The Earth has an average albedo of 0.37, meaning it reflects about 37% of the sunlight it receives back into space. That reflection would tell an interstellar traveler that the Earth is indeed an appealing world to investigate, because it indicates large quantities of liquid water as well as dry land. Water, the nearly universal solvent in which life formed, covers about 71% of the Earth's surface.

Beneath the clouds, our visitor would find a geologically active world. Earthquakes shake many regions of it. Volcanoes pour huge quantities of molten rock from inside the Earth onto the surface, and gas from within it vents into the atmosphere. Some mountains are still rising, while others are wearing away. Water flow erodes topsoil and carves river valleys. Rain and snow help rid the atmosphere of dust particles. And life teems virtually everywhere, making the Earth unique in the solar system. Figure 5-1 details Earth's important physical and orbital

## ⊕ Earth's Vital Statistics

| | |
|---|---|
| Mass | $6.0 \times 10^{24}$ kg (1 M$_\oplus$) |
| Radius | 6378 km = 1 R$_\oplus$ |
| Average density | 5520 kg/m$^3$ |
| Orbital eccentricity | 0.017 |
| Inclination of equator to plane of orbit | 23.5° |
| Sidereal period of revolution (yr) | 365.26 d (1.00 yr) |
| Average distance from Sun | $1.496 \times 10^8$ km (1 AU) |
| Sidereal rotation period | 0.997 d |
| Solar rotation period (day) | 1.00 d |
| Albedo (average) | 0.37 |

R I V U X G

**Figure 5-1**    **A View of Earth's Surface**  An oasis in the forbidding void of space, the Earth is a world of unsurpassed beauty and variety. Everchanging cloud patterns drift through its skies. More than two-thirds of its surface is covered with oceans. This liquid water, in combination with a huge variety of chemicals from its lands, led to the formation and evolution of life over most of the planet's surface.

properties. The symbol ⊕ in Figure 5-1 and hereafter is astronomer's shorthand for "Earth."

Essential to the existence of complex life is an atmosphere containing oxygen molecules. The Earth's early atmosphere contained no oxygen, and we begin our survey of the Earth with a study of its atmosphere.

## 5-1 The Earth's atmosphere has evolved over billions of years

The Earth's atmosphere is unique among the planets. The air we breathe is predominantly a 4-to-1 mixture of nitrogen to oxygen, two gases that are found only in small amounts in the atmospheres of other planets. This is, however, the third atmosphere to envelop the Earth.

The first one, composed of trace remnants of hydrogen and helium left over from the formation of the solar system, didn't last very long. These gases are too light to stay near the Earth (consider what happens to helium balloons). Heated by sunlight, these gases quickly dissipated into space.

The gases of the second atmosphere came from inside the Earth through volcanoes and fissures. It was composed primarily of carbon dioxide and water along with some nitrogen. In fact, there was roughly 100 times as much gas in this second atmosphere as there is today. This atmosphere began changing when the oceans formed from water coming out of the planet as well as from impacts of water-rich comets.

The oceans absorbed about half of this atmosphere. (Water holds a lot of carbon dioxide, as you can see by the amount of carbon dioxide fizz in a can of soda.) As life evolved in the oceans, much of that dissolved carbon dioxide was transformed into the shells of many creatures, and then, as the animals died, their carbon-rich shells sank to the ocean bottoms. Shells piled up and compressed the shells underneath them into rock, such as limestone.

Early plant life in the oceans and then on the shores removed most of the remaining carbon dioxide in the air by converting it into oxygen and the nutrients that the plants needed to survive. The most efficient of such conversion mechanisms is photosynthesis, which today helps maintain the balance between carbon dioxide and oxygen in the air.

The first oxygen that returned to the air, a by-product of plant life activity, did not stay there for long. Oxygen is highly reactive, and early atmospheric oxygen thus combined quickly with many elements on the Earth's surface, notably iron, to form new compounds. (In Chapter 6, we will see similar iron compounds on Mars.)

Eventually, after all the minerals that could combine with oxygen had done so, the atmosphere began to fill

with this gas. With the carbon dioxide gone, the air became the nitrogen-oxygen mixture that we breathe today. The nitrogen was a remnant of the previous, carbon dioxide–rich atmosphere. While it was a tiny fraction of that earlier atmosphere, it now dominates the thinner air of today.

Because it has mass, the air is pulled down toward the Earth by gravity, thereby creating a pressure on the surface. We formally define pressure as a force acting over some area:

$$\text{Pressure} = \frac{\text{force}}{\text{area}}$$

The average atmospheric pressure at sea level is 14.7 pounds per square inch, or 1 atmosphere (atm). This pressure is due to the weight of the air pushing down. The atmospheric pressure decreases smoothly with increasing altitude, falling by roughly half with every 5.5 km of altitude. Seventy-five percent of the mass of the atmosphere lies within 11 km (approximately 7 mi, or 36,000 ft) of the Earth's surface.

All Earth's weather—clouds, rain, sleet, and snow—occurs in this lowest region of the atmosphere, called the **troposphere.** Commercial jets generally fly at the top of the troposphere to minimize buffeting and jostling from the denser air below. If the Earth were the size of a typical classroom globe, the troposphere would only be as thick as a sheet of paper wrapped around its surface.

Starting at the Earth's surface, atmospheric temperature initially decreases upward, reaching a minimum of 218 K (−67°F) at the top of the troposphere. Above this level is the region called the **stratosphere,** which extends from 11 to 50 km (approximately 7 to 31 mi) above the Earth's surface. This is the realm of the **ozone layer.**

**Insight into science   Question your assumptions** The importance of word meanings in science cannot be overstressed. The word "layer" in the term "ozone layer" suggests a narrow region, perhaps a few meters thick. Not so. The ozone layer extends in altitude over some 40 kilometers of the atmosphere!

Ozone molecules (three oxygen atoms, $O_3$, bound together) are created in the stratosphere when ultraviolet radiation from the Sun occasionally splits $O_2$ up there into two separate oxygen atoms. Each resulting oxygen atom then combines with different $O_2$ molecules to create ozone. Once created, ozone efficiently absorbs solar ultraviolet rays, thereby heating the air in this layer while preventing most of this lethal radiation from reaching the

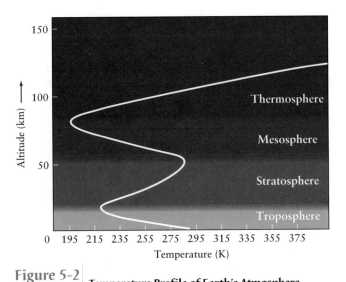

**Figure 5-2** **Temperature Profile of Earth's Atmosphere** The atmospheric temperature changes with altitude because of the way sunlight interacts with various gases at different heights.

the depletion rate succeed, the ozone layer will naturally replenish itself, and the ozone holes will probably stop occurring.

Note in Figure 5-2 that above the stratosphere lies the **mesosphere.** Atmospheric temperature again declines as one moves up through the mesosphere, reaching a minimum of about 200 K (–103°F) at an altitude of about 80 km (50 mi). This minimum marks the bottom of the **thermosphere,** above which the Sun's ultraviolet light ionizes atoms, producing charged particles that reflect radio waves. You can tune in distant AM radio stations because their transmissions bounce off this region and return to Earth. On the other hand, FM radio stations use much higher frequencies that pass through this layer without bouncing back. That is why you must be quite near an FM station in order to receive it.

## 5-2 Plate tectonics produce major changes to the Earth's surface

Landmasses protrude through Earth's oceans. These are regions of the Earth's outermost layer, or **crust,** and are composed of relatively low-density rock that literally floats on denser material below it. By studying the various kinds of rocks at a particular location, geologists can deduce the history of that site. For example, whether an area was once covered by an ancient sea or flooded by lava from volcanoes is readily apparent from the kinds of rocks that are present.

Beneath the Earth's oceans, the crust has distinct segments, indicating that the continents are separate bodies. We all believe, or at least hope, that the land under our feet will remain secure and unchanged. But the segmented appearance of the Earth's crust reveals that the planet is constantly changing. Earthquakes and volcanoes hint at activity hidden below the Earth's crust.

Anyone who carefully examines the shapes of Earth's continents (Figure 5-3) might conclude that landmasses move. Eastern South America, for example, looks like it would fit snugly against western Africa. Between 1912 and 1915, the German scientist Alfred Wegener published his observations on the remarkable fit between landmasses on either side of the Atlantic Ocean, an idea proposed by Isaac Newton two centuries earlier. Wegener was inspired to advocate the theory of *continental drift*—that the continents on either side of the Atlantic Ocean have drifted apart.

Wegener argued that a single, gigantic supercontinent called Pangaea (meaning "all lands") began to break up and drift apart some 200 million years ago. Initially, most geologists ridiculed Wegener's ideas. Although it was generally accepted that the continents do "float" on denser

Earth's surface. The temperature therefore increases upward through the stratosphere to about 285 K (50°F) at its top (Figure 5-2).

Not much ozone actually exists in the ozone layer. If all of it was compressed to the density of the air we breathe, it would be a layer only a few millimeters (about one-eighth of an inch) thick. It is enough, however, to protect us from the Sun's ultraviolet radiation. Because the Earth's early atmosphere lacked ozone, ultraviolet radiation penetrated to the planet's surface, where it provided much of the energy necessary for life to evolve in the oceans. After life developed, however, the formation of the ozone layer by the same ultraviolet radiation was essential in lessening the amount of ultraviolet radiation that reached the surface, thereby allowing life-forms to leave the oceans and survive on land.

Today, the ozone layer is still vitally essential, because the energy it absorbs would otherwise cause skin cancer and various eye diseases. In sufficiently high doses, ultraviolet radiation even damages the human immune system. The ozone layer is being depleted today by human-made chemicals, such as chlorofluorocarbons (CFCs). At present, ultraviolet levels in the mid-latitudes of the northern hemisphere are decreasing by about 4% per decade. Much more rapid and dramatic drops in the ozone layer over the Earth's poles, especially over the south pole, have been observed for more than two decades. These seasonal *ozone holes* typically extend over 24 million square kilometers, or 4.6% of the Earth's surface area.

Fortunately, as explained above, sunlight creates ozone. If current international efforts aimed at slowing

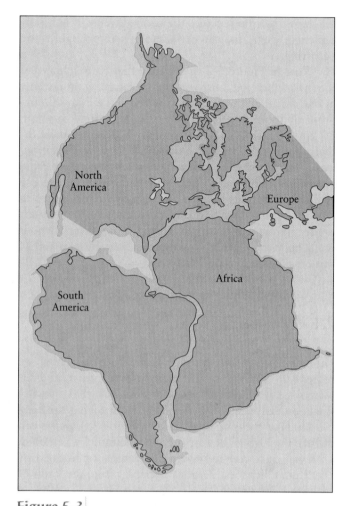

**Figure 5-3** **Comparing the Continents** Africa, Europe, Greenland, North America, and South America fit together as though they were once joined. This fit is especially convincing if the edges of the continental shelves (shown in orange) are matched rather than matching today's shorelines. Alfred Wegener used this fit to support his theory of continental drift.

**Insight into science  From proof to acceptance** Scientific theories may take years or decades of testing and replication before they are accepted. Newton speculated about continental drift in the late seventeenth century. Wegener had proposed plate tectonics by 1915. His theory was not verified until the 1950s and was not accepted until the late 1960s.

Seafloor spreading provided just the physical mechanism that had been missing from the theory of continental drift. It established Wegener's theory of crustal motion, which was developed into the modern theory of **plate tectonics.**

The Earth's surface is made up of about a dozen major tectonic plates that move relative to each other. In recent years, geologists have uncovered evidence that points to a whole succession of supercontinents that once broke apart and then reassembled. Pangaea is only the most recent supercontinent in this cycle, which repeats on average every 500 million years (Figure 5-5).

The boundaries between plates are the sites of some of the most impressive geological activity on our planet. Earthquakes tend to occur at the boundaries of the Earth's crustal plates, where the plates are colliding, separating, plunging under one another, or sliding against each other. The boundaries between tectonic plates (Figure 5-6) stand out clearly when the epicenters of earthquakes are plotted

rock beneath them, few geologists could accept that entire continents move across the Earth's face at speeds as great as several centimeters per year. Wegener and other "continental drifters" could not explain what forces could be shoving the massive continents around or how less-dense continental rock had gotten through the dense rock of the ocean floor.

Then, in the mid-1950s, long, underwater mountain ranges were discovered. The Mid-Atlantic Ridge, for example, stretches all the way from Iceland to Antarctica (Figure 5-4). Careful examination revealed that molten rock from the Earth's interior is being forced upward there, pushing the ocean floor apart on either side of the ridge. This **seafloor spreading** is in fact pushing the Americas away from Europe and Africa at a speed of roughly 3 cm per year.

**Figure 5-4** **The Mid-Atlantic Ridge** This artist's rendition of the floor of the North Atlantic Ocean shows an unusual mountain range in the middle of the ocean floor. Called the Mid-Atlantic Ridge, these mountains are created by lava seeping up from the Earth's interior along a rift that extends from Iceland to Antarctica.

a  200 million years ago

b  180 million years ago

c  Today

**Figure 5-5** | **The Supercontinent Pangaea  (a)**
The continents of today are pieces of what was once a bigger, united body called Pangaea. **(b)** Geologists now argue that Pangaea must have first split into two smaller supercontinents, which they call Laurasia and Gondwanaland. **(c)** These bodies later separated into the continents of today. Gondwanaland split into Africa, South America, Australia, and Antarctica, while Laurasia divided to become North America and Eurasia.

**Figure 5-6** | **The Earth's Major Tectonic Plates**  The Earth's surface is divided into a dozen or so rigid plates that move relative to each other. The boundaries of the plates are the scenes of violent seismic and geologic activity, such as earthquakes, volcanoes, rising mountain ranges, and sinking seafloors. The arrows indicate whether plates are moving apart (↔), together (→←), or sliding past one another (⇌). **Inset left:** The Separation of Two Plates. The plates that carry Egypt and Saudi Arabia are moving apart, leaving the trench that contains the Red Sea. This view, taken by astronauts in 1966, shows the northern Red Sea on the right. Egypt is at the bottom, Saudi Arabia is at the upper right, and the Sinai Peninsula dominates the center. **Inset right:** The Collision of Two Plates. The plates that carry India and China are colliding. As a result, the Himalayas are being thrust upward. In this photograph, taken by astronauts in 1968, India is on the left and Tibet is on the right. Mount Everest is one of the snow-covered peaks near the center.

on a map. The San Andreas Fault, running along the West Coast of the United States, is an example of a fault where the North American and Pacific plates are rubbing against each other.

Great mountain ranges, such as those along the western coast of South America, are thrust up by ongoing collisions between plates. Where old crust is pushed back down into the Earth, we find deep trenches, such as the ones off the coasts of Japan and Chile. Figure 5-6 shows two well-known geographic features that resulted from separating and colliding tectonic plates, respectively.

But what powers all this activity? Some enormous energy source has caused shifting among landmasses for billions of years. The answer to that question leads us to a scientific model of the Earth's interior.

## 5-3 Earth's interior consists of a rocky mantle and an iron-rich core

Geologists have calculated that the Earth was entirely molten soon after its formation, about 4.6 billion years ago. The violent impact of space debris, along with energy released by the breakup of radioactive elements, heated and melted the young Earth. The same process of atoms breaking apart, called *nuclear fission*, creates heat used to generate electricity in nuclear power plants.

Iron and the other dense elements sank toward the center of the young, molten Earth, just like a rock sinks in a pond. At the same time, less dense materials were forced upward toward the surface. The process, called **planetary differentiation**, produced a layered structure within the Earth: a very dense central **core** surrounded by a **mantle** of less dense minerals, which in turn was surrounded by a thin crust of relatively light minerals (Figure 5-7a).

Differentiation explains why most of the rocks you find on the ground are composed of lower-density elements, like silicon and aluminum. The iron, gold, lead, and other denser elements found on the Earth today actually had to return to the surface via volcanoes and other lava flows. Despite this return of heavier elements to the surface, the average density of crustal rocks is 3000 kg/m$^3$, considerably less than the average density of the Earth as a whole (5520 kg/m$^3$).

The violent impacts that occurred as the Earth formed generated tremendous heat inside our planet. The temperature in the Earth rises steadily from about 290 K on the surface to nearly 5000 K at the center. Radioactivity in the Earth's interior helps replenish heat lost through the planet's surface. In addition to temperature, pressure also increases with increasing depth below the Earth's surface. The deeper you go, the more mass presses down on you. These factors shape the Earth's inner layers.

You might think that temperatures of 5000 K (9000°F) would melt virtually any substance. The mantle, in particular, is composed largely of minerals rich in iron and magnesium, both of which have melting points of only slightly more than 1250 K at the Earth's surface. However, the melting point of anything also depends on the pressure to which it is subjected: The higher the pressure, the higher the melting point. The high pressures within the Earth (1.4 million atm at the bottom of the mantle) preserve a solid mantle down to a depth of about 2900 km.

Geologists have deduced basic properties of the Earth's interior by studying the response of the planet to earthquakes. Earthquakes produce a variety of **seismic waves**, vibrations that travel through the Earth either as ripples like ocean waves or by compressing matter like sound waves. Geologists use sensitive **seismographs** to detect and record these vibratory motions. The varying den-

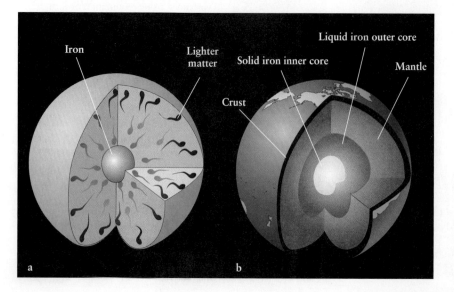

**Figure 5-7** | **Cutaway Model of the Earth**
The early Earth was probably a homogeneous mixture of elements with no continents or oceans. **(a)** While molten, iron sank to the center and light material floated upward to form a crust. **(b)** As a result, Earth has a dense iron core and a crust of light rock, with a mantle of intermediate density between them.

Rift    Ocean floor    Mountain range    Continent

Convection currents

**Figure 5-8** | **The Mechanism of Plate Tectonics** Convection currents in the Earth's interior are responsible for pushing around rigid plates on its crust. New crust forms in oceanic rifts, where lava oozes upward between separating plates. Mountain ranges and deep oceanic trenches are formed where plates collide. Note that not all tectonic plates move together or apart.

sity and composition of the Earth's interior affects the direction of seismic waves traveling through the Earth. By studying the deflection of these waves, geologists have been able to determine properties of the Earth's interior.

By 1906, the year of the great San Francisco earthquake, analysis of seismic recordings led to the discovery that the Earth has a molten iron core with a radius of about 3500 km (2150 mi). For comparison, the overall radius of our planet is 6378 km (3963 mi). More careful measurements in the 1930s revealed that the molten core actually surrounds a solid core with a radius of about 1250 km (775 mi). The cores are composed of roughly 80% iron and 20% lighter elements, such as silicon. The interior of our planet therefore has a curious structure: a liquid region sandwiched between a solid inner core and a solid mantle (see Figure 5-7b).

A sophisticated computer model of the Earth's interior predicts that the solid inner core is rotating faster than the rest of the planet. Seismic evidence supports the model that the core goes around about 2 degrees more per year than the surface. Refined models and observations like these are finally revealing the secrets of the Earth's interior.

The tremendous heat and molten rock inside the Earth provide the mechanism for plate tectonics. The energy to drive the plate motion is the heat inside the planet. That heat, from the formation of the planet and from the decay of radioactive elements, affects the Earth's surface today. This heat transport is accomplished by **convection,** a process in which lighter, hotter material rises upward, while denser, cooler material sinks downward. You witness convection every time you see simmering soup. Heated from below, blobs of hot soup move upward. Once this soup reaches the surface, it gives off the heat it carries, cools, becomes denser, and sinks back down.

Heated from below, parts of the Earth's upper mantle are hot enough to have an oozing, plastic flow (like stretch-

ing Silly Putty). As sketched in Figure 5-8, molten rock, or *magma,* seeps upward along cracks or rifts in the ocean floor, which are found where plates are separating. Cool crustal rock sinks back down into the Earth where plates collide. The continents ride on top of the plates as they are pushed around by the convection currents beneath them.

Supercontinents like Pangaea apparently sow the seeds of their own destruction by blocking the flow of heat from the Earth's interior. As soon as a supercontinent forms, temperatures beneath it rise. As heat accumulates, the supercontinent domes upward and cracks. Overheated molten rock wells up to fill the resulting rifts, which continue to widen as pieces of the fragmenting supercontinent are pushed apart.

The Earth's molten, iron-rich interior affects the Earth's exterior in another important way, one that has fewer obvious signs than tectonic plate motion. It is the creation of a magnetic field that extends through the Earth's surface and out into space. As we will see next, that magnetic field results from the convection of molten metal inside the Earth combined with the Earth's rotation.

## 5-4 The Earth's magnetic field shields us from the solar wind

If you have ever used a compass, you have seen the effect of the Earth's magnetic field. This planetary field is quite similar to the field that surrounds a bar magnet (Figure 5-9a). Magnetic fields are created by electrical charges in motion. Thus, the motion of your compass is caused by electrical charges moving inside the Earth. (The wires in your house also carry moving charges called electric currents that, in turn, create magnetic fields of their own.)

Many geologists believe that convection of molten iron in the Earth's outer core combined with our planet's

a

R I V U X G

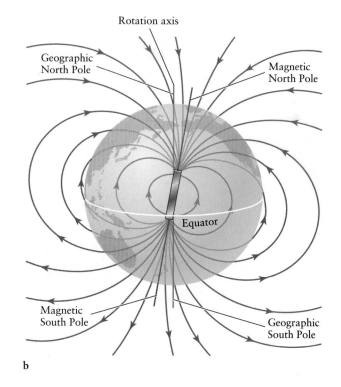

b

## Figure 5-9

**The Earth's Magnetic Field (a)** The magnetic field of a bar magnet is revealed by the alignment of iron filings on paper. **(b)** Generated in the Earth's molten, metallic core, the Earth's magnetic field extends far into space. Note that the field is not aligned with the Earth's rotation axis. We will see similar misalignments, often much more exaggerated, in other planets' magnetic fields.

rotation creates electric currents, which in turn create the Earth's magnetic field. The details of this so-called **dynamo theory** of Earth's magnetic field are still being developed.

Magnetic fields virtually always form complete loops (see Figure 5-9b). The magnetic fields near the Earth's south rotation pole loop out tens of thousands of kilometers into space and then return near the Earth's north rotation pole. The places where the magnetic fields pierce the Earth's surface most intensely are called the north and south magnetic poles. These poles move daily. Evidence in solidified lava also reveals that the Earth's magnetic field actually reverses or flips on an irregular schedule ranging from tens of thousands to hundreds of thousands of years. The last reversal was about 600,000 years ago. Today, the magnetic and rotation poles of the Earth are about 11.3° apart; other planets have much larger angles between their magnetic and rotation poles. Figure 5-10 is a scale drawing of the magnetic fields around the Earth, which comprise the Earth's **magnetosphere**.

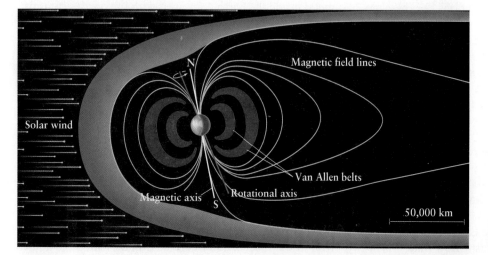

## Figure 5-10

**Earth's Magnetosphere** A slice through the Earth's magnetic field, which surrounds the entire planet, carves out a cavity in space that excludes charged particles ejected from the Sun, called the solar wind. The Earth is located in this cavity. Most of the particles of the solar wind are deflected around the Earth by the fields in a turbulent region colored blue in this drawing. Because of the strength of the Earth's magnetic field, our planet can trap charged particles in two huge, doughnut-shaped rings called the Van Allen belts (in red).

Our planet's magnetic field protects the Earth's surface from bombardment by energetic particles from space. Most of these particles originate in the Sun (see Chapter 9) and are called the **solar wind.** The solar wind is an erratic flow of charged particles away from the Sun's upper atmosphere. Near the Earth, the particles in the solar wind move at speeds of approximately 400 km/s, or about a million miles per hour.

Magnetic fields can change the direction of moving charged particles. The solar wind particles heading toward the Earth are deflected by our planet's magnetic field. The field is strong enough to trap these charged particles in two huge, doughnut-shaped rings called the **Van Allen radiation belts,** which surround the Earth (see Figure 5-10). These belts, discovered in 1958 during the flight of America's first successful Earth-orbiting satellite, were named after the physicist James Van Allen, who insisted that the satellite carry a Geiger counter to detect charged particles.

> **Insight into science**  Pace of discovery
> Discoveries such as the Van Allen belts remind us of how much remains to be learned, even about our own planet. Scientists continue to make such fundamental discoveries for many reasons. The technology necessary to make certain discoveries, for example, may just have been developed. Or our understanding of the laws of physics may be incomplete. Sometimes the equations are so complex that their implications are not yet clear. Progress in science requires unquenchable curiosity and unfailing openmindedness.

Sometimes the Van Allen belts overload with particles. The particles then leak through the magnetic fields at their weakest points and cascade down into the Earth's upper atmosphere, usually in a ring-shaped pattern. As these high-speed, charged particles collide with gases in the upper atmosphere, the gases fluoresce (give off light) like the gases in a neon sign. The result is a beautiful, shimmering display called the **northern lights (aurora borealis)** or the **southern lights (aurora australis),** depending on the hemisphere in which the phenomenon is observed (Figure 5-11).

Occasionally a violent event on the Sun's surface called a **coronal mass ejection** sends a burst of protons and electrons straight through the Van Allen belts and into the atmosphere. The resulting auroral display can be exceptionally bright (see Figure 5-11) and can often be seen over a wide range of latitudes and longitudes. Such events also disturb radio transmissions and can damage communications satellites and electrical transmission lines.

a  R I V U X G

b  R I V U X G

c  R I V U X G

**Figure 5-11**  **The Northern Lights (Aurora Borealis)** A deluge of charged particles from the Sun can overload the Van Allen belts and cascade toward the Earth, producing aurorae that can be seen over a wide range of latitudes. Aurorae typically occur 100 to 400 km above the Earth's surface. **(a)** View of aurora from the deep space satellite *Dynamics Explorer 1.* **(b)** View of aurora from the Space Shuttle. **(c)** View of aurora from Earth.

# THE MOON AND TIDES

Earth's only natural satellite provides one of the most dramatic sights in the nighttime sky. The Moon is so large and so nearby that some of its surface features are readily visible to the naked eye. Without a telescope, you can easily see dark gray and light gray areas that cover vast expanses of the Moon (Figure 5-12). Although people have long known the Moon's effects on ocean tides, its surface began to be understood only with the observations of Galileo.

In Galileo's day, most people believed that the Moon had a smooth face. Galileo's telescope revealed mountains towering above its barren surface. Was the Moon's geological history as active as that of the Earth? What are the Moon's surface features telling us about its evolution? Our model of the Moon starts with a view of its surface.

## 5-5  The Moon's surface is covered with craters, plains, and mountains

Perhaps the most familiar and characteristic features on the Moon are its craters (Figure 5-13). Through an Earth-based telescope some 30,000 craters are visible, with diameters ranging from 1 km to more than 100 km. Following a tradition established in the seventeenth century, the most prominent craters are named after scientists, mathematicians, and philosophers, such as Kepler, Copernicus, Pythagoras, Plato, and Aristotle. Close-up photographs from lunar orbit reveal millions of craters too small to be seen with Earth-based telescopes. Indeed, extreme close-up photos of the Moon's surface reveal countless microscopic craters (Figure 5-14).

Earth, by comparison, has only about 200 known impact craters. Many craters on Earth have been drawn inside our planet by plate tectonic motions and thereby obliterated. Many craters of a few kilometers in diameter and smaller have been worn away by weathering effects of wind and water. The Earth's atmosphere vaporized innumerable space rocks that would otherwise have created small craters here. Moreover, those that did hit had been slowed down by the atmosphere so that they produced smaller craters than they would have otherwise. We will discuss Earth's craters further in Chapter 8.

Virtually all lunar craters, both large and small, are the result of bombardment by meteoritic material. Nearly all these craters are circular, indicating that they were not merely gouged out by high-speed rocks. Instead, the violent collisions with the Moon's surface vaporized the rapidly moving debris, and the resulting explosions of hot gas and debris produced the round-rimmed craters we observe today.

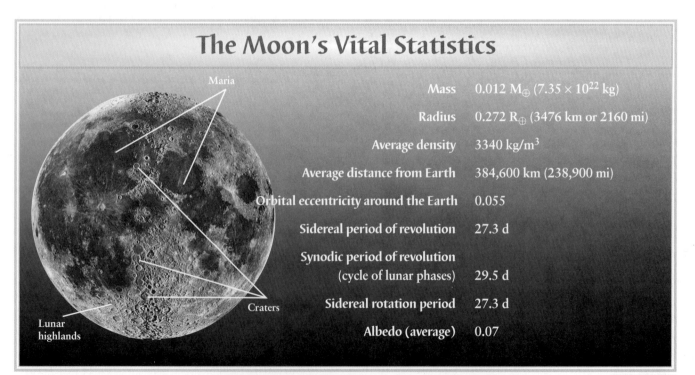

## The Moon's Vital Statistics

| | |
|---|---|
| Mass | 0.012 $M_\oplus$ ($7.35 \times 10^{22}$ kg) |
| Radius | 0.272 $R_\oplus$ (3476 km or 2160 mi) |
| Average density | 3340 kg/m$^3$ |
| Average distance from Earth | 384,600 km (238,900 mi) |
| Orbital eccentricity around the Earth | 0.055 |
| Sidereal period of revolution | 27.3 d |
| Synodic period of revolution (cycle of lunar phases) | 29.5 d |
| Sidereal rotation period | 27.3 d |
| Albedo (average) | 0.07 |

Labels: Maria, Craters, Lunar highlands

R I V U X G

**Figure 5-12**  **The Moon**  Our Moon is one of seven large satellites in the solar system. The Moon's diameter of 3476 km (2160 mi) is slightly less than the distance from New York to San Francisco. This photograph is a composite of first quarter and last quarter views, in which long shadows enhance the surface features.

The largest of the maria is Mare Imbrium (Sea of Showers). It is roughly circular and measures 1100 km (700 mi) in diameter. Although the maria seem quite smooth in telescopic views from the Earth, close-up photographs from lunar orbit reveal that they contain small craters and occasional cracks called **rilles** (Figure 5-15). Rilles are the collapsed roofs of underground lava rivers. When the flowing lava solidified, it shrank and the roofs fell in on the space created.

As we will discuss further in Section 5-8, the same side of the Moon always faces the Earth. One of the surprises stemming from the earliest lunar exploration is that only one small mare graces the Moon's far side (Figure 5-16). Except for that lone mare, craters cover the entire side of the Moon that faces away from the Earth.

Detailed measurements by astronauts in lunar orbit demonstrated that the maria on the Moon's Earth-facing side are 2 to 5 km below the average lunar elevation. In contrast, the cratered terrain on the lunar far side is typically 4 to 5 km above the average lunar elevation. The flat, low-lying, dark gray maria cover only 17% of the lunar surface. The remaining 83% is composed of light gray, heavily cratered, mountainous regions called **highlands** (Figure 5-17).

Lunar mountain ranges that rim the vast maria basins suggest ancient impacts far more violent than those that

**Figure 5-13** **Details of a Lunar Crater** This photograph, taken from lunar orbit by astronauts in 1969, shows a typical view of the Moon's heavily cratered far side, the hemisphere we never see from Earth. The large crater near the middle of the picture is approximately 80 km (50 mi) in diameter. Note the crater's central peak and the numerous tiny craters that pockmark the surrounding lunar surface.

Meteoritic impact causes material from the crater site to be ejected onto the surrounding surface. This pulverized rock is called an **ejecta blanket.** You can see some of the more recent ejecta blankets, which are lighter colored than the older ones (see Figure 5-12). The blanket darkens with time as its surface roughens from impacts by the solar wind. The lightness of its ejecta is one clue astronomers use to determine how long ago a crater formed.

Craters larger than about 20 km also form central peaks (see Figure 5-13). These occur because the impact compresses the crater floor so much that afterward the crater rebounds and pushes the peak upward. As the peak goes up, the crater walls collapse and form terraces.

Besides the craters, the most obvious characteristic of the Moon visible from Earth is that its surface is various shades of gray. Most prominent are the large, dark gray plains called **maria** (pronounced "mar-ee-uh"). The singular form of this term, **mare** (pronounced "mar-ay"), means "sea" in Latin and was introduced in the seventeenth century when observers using early telescopes thought these features were large bodies of water. We know now that no water exists in liquid form on our satellite. Nevertheless, we retain these poetic, fanciful names, from Mare Tranquillitatis (Sea of Tranquillity) and Mare Nubium (Sea of Clouds) to Mare Nectaris (Sea of Nectar) and Mare Serenitatis (Sea of Serenity).

**Figure 5-14** **A Microscopic Lunar Crater** This photograph made with a microscope shows tiny microcraters less than 1 mm across on a piece of moon rock.

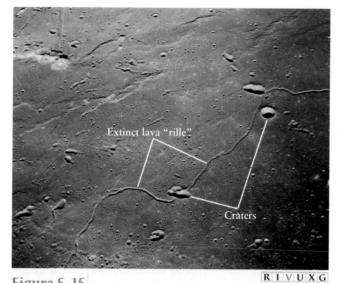

**Figure 5-15** R I V U X G **Details of a Lunar Mare** Close-up views of the lunar surface reveal numerous tiny craters and rilles on the maria. Astronauts in lunar orbit took this photograph of Mare Tranquillitatis (Sea of Tranquillity) in 1969 while searching for potential landing sites for the first manned landing.

**Figure 5-17** R I V U X G **Mare Imbrium and the Surrounding Highlands** Mare Imbrium, the largest of the 14 dark plains that dominate the Earth-facing side of the Moon, is ringed by lighter-colored highlands strewn with craters and towering mountains. The highlands were created by asteroid impacts pushing land together.

produced typical craters. One such mountain range can be seen around the edges of Mare Imbrium (see Figure 5-17). Although they take their names from famous terrestrial ranges, such as the Alps, the Apennines, and the Carpathians, the lunar mountains were not formed in the same way as mountains on Earth. Analyses of Moon rocks and observations from lunar orbit imply that violent impacts of huge asteroids thrust up the lunar highlands.

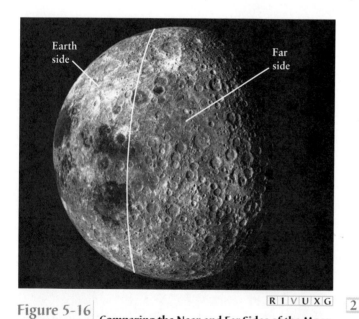

**Figure 5-16** R I V U X G **Comparing the Near and Far Sides of the Moon** On the left side of this photograph are three maria visible from the Earth. In this view, taken by the crew of *Apollo 16*, the far side of the Moon is more uniformly and more heavily cratered than the side facing the Earth. The curve indicates the separation between the near and far sides of the Moon.

## 5-6 Visits to the Moon yielded invaluable information about its history

Reaching the Moon became a national obsession for the United States in the early 1960s. A series of successful space missions by the former Soviet Union spurred President John F. Kennedy's goal of landing a person on the Moon by the end of that decade. The Apollo missions focused our technologies and, in the end, the drama of space travel and the compelling views from space of the oasis that is our Earth helped bring all of humanity a little closer together. Amid all the excitement, a prodigious amount of scientific research about the Moon was accomplished during the Apollo program.

The first of six manned lunar landings, *Apollo 11*, set down on Mare Tranquillitatis on July 20, 1969. Astronaut Neil Armstrong was the first human to set foot on the Moon. In the same year, *Apollo 12* also set down in a maria. Human voyages into space encountered a setback when *Apollo 13* experienced a nearly fatal explosion en route to the Moon. Fortunately, a titanic effort by the

Figure 5-18    **An Apollo Astronaut on the Moon** *Apollo 17* astronaut Harrison Schmitt enters the Taurus-Littow Valley on the Moon. Here an enormous boulder had once slid down a mountain, fracturing on the way. This final Apollo mission landed in the most rugged terrain of any Apollo flight.

Figure 5-19    **The Regolith** The Moon's surface is covered by a layer of powdered rock that was created by billions of years of bombardment by space debris. Called the regolith, this surface material sticks together like wet sand, as illustrated by this *Apollo 11* astronaut's bootprint.

astronauts and ground personnel brought it safely home. *Apollo 14* through *Apollo 17* took on progressively more challenging lunar terrain. The period of human visitation to our Moon lasted fewer than four years, culminating in 1972, when *Apollo 17* landed amid rugged mountains just east of Mare Serenitatis (Figure 5-18).

Astronauts discovered that the lunar surface is covered with a layer of fine powder and rock fragments. This layer, ranging in thickness from 1 to 20 m, is called the **regolith** (from the Greek, meaning "blanket of stone"; Figure 5-19). It formed during 4.5 billion years of relentless meteoritic bombardment. We don't refer to this layer as soil, because the term *soil* suggests the presence of decayed biological matter, which is not found on the Moon's surface. The rough, powdered regolith absorbs most of the light incident on it. Therefore, the Moon's albedo is only 0.07, meaning that it reflects only 7% of the light striking it.

A major goal of the Apollo program was to study the geology of the Moon and help determine its history. The astronauts brought back a total of 382 kg (842 lb) of lunar rocks, which have provided important information about the early history of the Moon. Lunar rock from the maria is solidified lava, which tells us that the Moon's surface was once molten. Indeed, these Moon rocks are composed mostly of the same minerals that are found in volcanic rocks on Hawaii and Iceland. The rock of these low-lying lunar plains is called **mare basalt** (Figure 5-20). About 17% of the Moon's surface rock is basalt.

In contrast to the dark mare basalt, the lunar highlands are covered with a light-colored rock called **anorthosite** (Figure 5-21). On Earth, anorthositic rock is found

Figure 5-20    **Mare Basalt** This 1.53-kg (3.38-lb) specimen of mare basalt was brought back by *Apollo 15* astronauts in 1971. Small holes that cover about a third of its surface suggest that gas must have been dissolved in the lava from which this rock solidified. When the lava reached the air less lunar surface, bubbles formed as the pressure dropped and the gas expanded. Some of the bubbles were frozen in place as the rock cooled.

**Figure 5-21** **Anorthosite** The lunar highlands are covered with this ancient type of rock, which is believed to be the material of the original lunar crust. This sample's dimensions are 18 × 16 × 7 cm. While this sample is medium grey, other anorthosites retrieved from the Moon have been white, while others are darker grey than this one. It was brought back by Apollo 16 astronauts.

comet and asteroid impacts. Current estimates are that at least 3 billion tons of water are encased there.

The presence of water ice greatly simplifies the prospect of human settlement of the Moon, because its ice can provide us not only with liquid water but also with oxygen and rocket fuel. (The Space Shuttle's primary rocket engine is fueled by liquid oxygen and liquid hydrogen. Separating the Moon's water into liquid oxygen and liquid hydrogen would create the same fuel.) An existing water reservoir makes the prospect of establishing a lunar colony much less expensive and more feasible than lugging all that water from Earth or dragging a water-rich comet to the Moon.

Astronauts found no traces of life on the Moon. Life as we know it requires liquid water, of which the Moon has none. The Moon has an atmosphere of sodium gas so thin as to be considered an excellent vacuum compared to the density of the air we breathe. An atmosphere is held around a planet or moon by that body's gravitational force. Our Moon has so little gravitational attraction (about one-sixth that of the Earth) that it is unable to retain its tenuous atmosphere, which therefore continuously leaks into space. The atmosphere is renewed by sunlight striking the Moon, releasing more gas from its rocks. The Earth, in contrast, has enough mass to retain such gases as nitrogen, oxygen, carbon dioxide, and argon.

only in very old mountain ranges, such as the Adirondacks in the eastern United States. Compared to the mare basalts, which have more of the heavier elements, like iron, manganese, and titanium, anorthosite is rich in calcium and aluminum. Anorthosite is therefore less dense than basalt. Many highland rocks brought back to Earth are **impact breccias** (Figure 5-22), which are composites of different rock fused together as a result of meteorite impacts.

By carefully measuring trace amounts of radioactive elements, geologists determined that lunar rocks on the maria are between 3.1 and 3.8 billion years old, while typical anorthositic specimens from the highlands are between 4.0 and 4.3 billion years old. All these ancient specimens probably represent samples of the Moon's original crust (Figure 5-23).

In 1994 the lunar-orbiting *Clementine* spacecraft made an incredible discovery: the first evidence of frozen water on the Moon. Located in craters near the Moon's south pole, this ice is not in danger of being evaporated by sunlight, because, like sunlight at Earth's poles, the Sun deposits very little energy at the lunar poles. In 1998 the *Lunar Prospector* spacecraft confirmed *Clementine*'s discovery and also found evidence for ice at the Moon's north pole. The ice probably was deposited on the Moon during

**Figure 5-22** **Impact Breccias** These rocks are created from shattered debris fused together under high temperature and pressure. Such conditions prevailed immediately following impacts of space debris on the Moon's surface.

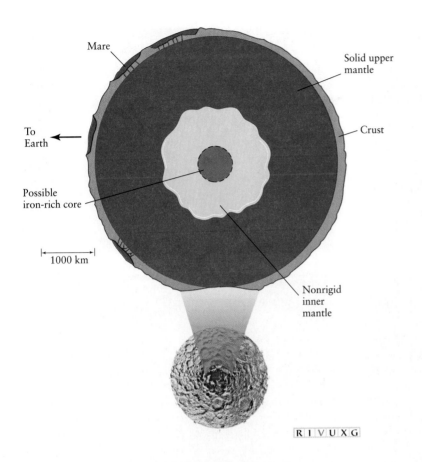

Mare
Solid upper mantle
To Earth
Crust
Possible iron-rich core
1000 km
Nonrigid inner mantle

R I V U X G

**Figure 5-23** **The Moon's Interior**
Based on seismic experiments left on the Moon by Apollo astronauts, we know that the Moon has a crust and mantle, and probably has a core. The lunar crust has an average thickness of about 60 km on the Earth-facing side but about 100 km on the far side. The crust and solid upper mantle extend to about 800 km, where the nonrigid inner mantle begins. If the Moon has a core, its radius is 400 km or less. Although the main features of the Moon's interior are analogous to those of the Earth, the proportions are quite different. **Inset:** Note the exposed mantle in the south polar region, taken by the *Clementine* spacecraft. The crust was apparently stripped away by an impact.

## 5-7 The Moon probably formed from debris cast into space when a huge planetesimal struck the young Earth

Before the Apollo program, scientists debated three different theories about the origin of the Moon. The **fission theory** holds that the Moon was pulled out from a rapidly rotating proto-Earth. This theory does not explain why the Moon has virtually no water incorporated in its rocks, as revealed by the bone-dry samples brought back by Apollo astronauts. The **capture theory** posits that the Moon was formed elsewhere in the solar system and then drawn into orbit about the Earth by gravitational forces. However, it is physically very difficult for a planet to capture a large moon. Furthermore, because bodies formed in different places have different overall chemical compositions, this theory does not explain the similar geochemistries of the Moon and the Earth's surface. The **cocreation theory** proposes that the Earth and the Moon were formed near each other at the same time. However, this theory fails to explain why, compared to the Earth, the Moon has less of the denser elements, such as iron, and why certain types of rocks found on the Moon are not found on Earth.

A fourth theory, first proposed in the 1980s, applies the established fact of age-old collisions in the solar system

to the formation of the Moon. Called the **collision-ejection theory**, it proposes that the newly formed Earth was struck by a Mars-sized object that literally splashed some of the Earth's surface layers into orbit around the young planet. Evidence from Moon rocks places this event at around 4.5 billion years ago. A computer simulation of this cataclysm is shown in Figure 5-24. This orbiting matter became a short-lived disk that eventually condensed into the Moon, just like planetesimals condensed to form the Earth.

The collision-ejection theory is consistent with many of the known facts about the Moon. For example, rock vaporized by the impact would have been depleted of volatile (easily evaporated) elements and water, leaving the parched Moon rocks we now know. Furthermore, most of the debris from the collision would lie near the plane of the ecliptic, as long as the orbit of the impacting asteroid had also been in that plane. Indeed, the impact of an object large enough to create the Moon could also have tipped the Earth's axis of rotation and so inaugurated the seasons.

The surface of the newborn Moon probably stayed molten for millennia, due to heat released during the impact of rock fragments falling onto the young satellite and the decay of short-lived radioactive isotopes. As the Moon gradually cooled, low-density lava floating on the

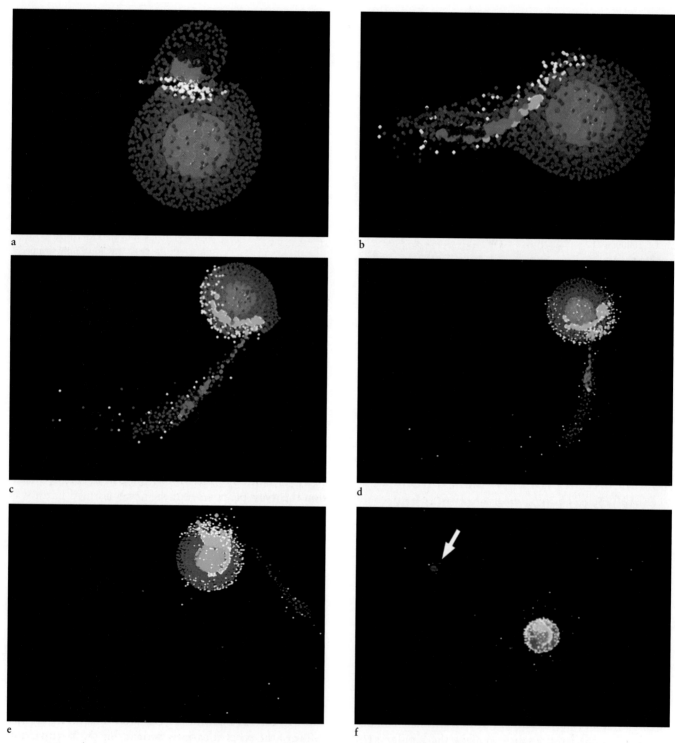

**Figure 5-24** **The Moon's Creation** This supercomputer simulation models the creation of the Moon from material ejected by the impact of a large asteroid with the young Earth. To follow the ejected material as it moves away from the Earth, successive views pan out to show increasingly larger volumes of space. Blue and green indicate iron from the cores of the Earth and the asteroid; red, orange, and yellow indicate rocky mantle material. In this simulation, the impact ejects both mantle and core material, but most of the dense iron falls back onto the Earth. The surviving ejected rocky matter coalesces to form the Moon (indicated by arrow) during its first orbit of the Earth.

Moon's surface began to solidify into the anorthositic crust that exists today in the highlands.

By correlating the ages of Moon rocks with the density of craters at the sites where they were collected, geologists have found that the rate of impacts on the Moon changed over the ages. The ancient, heavily cratered lunar highlands are evidence of intense bombardment that dominated the Moon's early history. As shown in Figure 5-25, this barrage extended from 4.5 billion years ago, when the Moon's surface first solidified, until about 3.8 billion years ago. This period of impacts was punctuated by a severe pounding 4 billion years ago, part of a solar-system-wide reign of terror.

The frequency of impacts gradually tapered off as meteoroids and planetesimals were swept up by the newly formed planets. Recorded among the final scars at the end of this crater-making era are the impacts of more than a dozen objects, each measuring at least 100 km across. As these huge rocks pounded the young Moon, they may have blasted gaping craters in what would later become the maria. Meanwhile, heat from the decay of long-lived radioactive elements like uranium and thorium began to remelt the inside of the Moon. Then, from 3.8 to 3.1 billion years ago, great floods of molten rock gushed up from the lunar interior through the weak spots in the crust created by the large impacts. Lava filled the impact basins and created the maria we see today.

Relatively little has happened on the Moon since those ancient times. A few fresh large craters have been formed, but the astronauts visited a world that has remained largely unchanged for more than 3 billion years. At that time, the biological history of our own planet was just beginning, as the first organisms began populating the Earth's oceans.

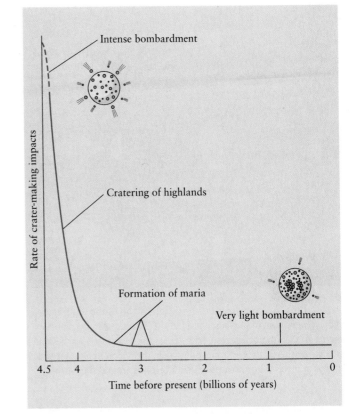

**Figure 5-25**  **The Rate of Crater Formation**  The first 700 million years of the Moon's history were dominated by frequent crater-making impacts. Near the end of this intense bombardment, several large impacts gouged out the mare basins. For the past 3.8 billion years, the impact rate has been relatively low.

## 5-8  Gravitational forces produce the tides and keep the same side of the Moon always facing the Earth

If you observe the Moon from Earth over many nights, you always see the same maria and highlands. Even as the Moon goes through its phases, the same surface features face the Earth (see Figure 1-18). Only from spacecraft have we seen the "far side."

The far side of the Moon is often, and incorrectly, called the "dark side" of the Moon. Whenever we view less than a full Moon, some sunlight is falling on its far side. Indeed, throughout each cycle of lunar phases all parts of the Moon get equal amounts of sunlight.

Although the same side of the Moon always faces the Earth, the Moon does rotate. In fact, it has **synchronous rotation:** It rotates on its axis at the same rate that it

orbits the Earth (Figure 5-26). To see why, hold your arm out with your palm facing you. Your palm represents the side of the Moon facing the Earth. Standing still, swing your arm horizontally always keeping the palm facing you. The only way you can do this is if your hand turns (rotates) at exactly the same rate that it swings (revolves). To understand the origin of the Moon's synchronous rotation, we must investigate the gravitational forces acting between the Earth and the Moon.

Figure 5-27a shows the gravitational force from the Moon at several locations on the Earth. Recall from Chapter 2 that gravitational force decreases with distance. Therefore, the side of the Earth closest to the Moon gets a greater gravitational pull than the Earth's core does. Similarly, less attraction from the Moon occurs on the Earth's far hemisphere than at the Earth's core. These differences in gravitational pull at different places create **tidal forces.**

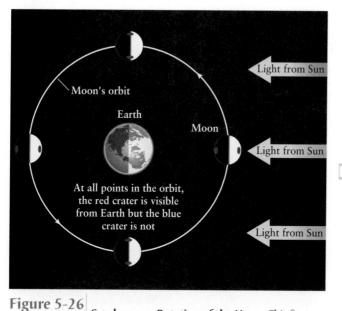

**Figure 5-26** **Synchronous Rotation of the Moon** This figure shows the motion of the Moon around the Earth as seen from above the Earth's north polar region. In order for the Moon to keep the same side facing the Earth as it orbits our planet, the Moon must rotate on its axis at precisely the same rate that it revolves around the Earth.

5 As sketched in Figure 5-27b, tidal forces deform the oceans, causing them to rise at some places around the equator and to settle elsewhere. There are two areas of high tide on the Earth at any given time, one on the side closest to the Moon, the other on the opposite side of the Earth. The high tide on the far side of the Earth from the Moon occurs because the water there is "left behind" as the Moon pulls the solid center of the Earth toward it. Low tides occur when the Moon is near the horizon. Because there are high tides on opposite sides of the Earth, there are two cycles of high and low tides on Earth each day. These cycles do not span exactly 24 hours, because during that time the Moon is moving around the Earth, changing the location of the high tide.

6 The Sun also raises tides in the oceans, but only half as high as the Moon. Although the Sun is much more massive than the Moon, the Sun is nearly 400 times farther away and gravitational force rapidly drops off with distance. When the Sun, Moon, and Earth are aligned, the Sun and Moon pull in the same direction, and the tides are highest. These **spring tides** (so-called not because they only occur in the spring—they occur in all seasons—but, from the German, *springen*, meaning to "spring up") occur at every new and full Moon. At first quarter and third quarter, the Sun and Moon form a right angle as seen from the Earth. The gravitational forces they exert compete with each other, and so the tidal distortion is the least pronounced. The smaller tidal shifts on these days are called **neap tides** (Figure 5-28).

When the Moon was young, Earth's gravity created huge tides of molten rock up to 60 ft high there. The Earth's tidal force acted to keep the Moon's high tide directly between the Moon's center and the center of the Earth. As a result, the Moon's rotation rate became the same as its revolution rate. When it solidified, the Moon was locked into this synchronous rotation.

a                                                                         b

**Figure 5-27** **Tidal Forces** The Moon induces tidal forces on the Earth because the strength of the Moon's gravity varies over the Earth's surface. **(a)** Orange arrows indicate the strength and direction of the Moon's gravitational pull at selected points on the Earth. **(b)** Ignoring the Earth's rotation and continents until Figure 5-28, this figure shows how two high tides are created on the Earth by the Moon's gravitational pull. Orange arrows indicate the strength and direction of the tidal forces acting on the Earth. At any location, the tidal force equals the Moon's gravitational pull at that point minus the gravitational pull of the Moon at the center of the Earth. The Sun has a weaker, but otherwise identical, effect.

Friction between the Earth
and its oceans makes the
Earth rotate more slowly

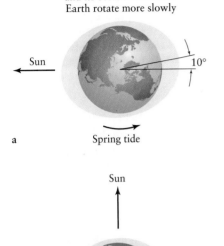

Sun ←

10°

**a**    Spring tide

Forward pull on the moon
makes it spiral outward

Moon

Sun ↑

10°

Moon ↑

**b**    Neap tide

**Figure 5-28**  **Tides on the Earth**  The gravitational forces of the Moon and Sun deform the oceans. The high tide closest to the Moon leads it around the Earth by about 10°. The greatest deformation (spring tides) occurs when the Sun, Earth, and Moon are aligned with the Sun and Moon either on the same or opposite sides of the Earth. The least deformation (neap tides) occurs when the Sun, Earth, and Moon form a right angle.

## 5-9  The Moon is moving away from the Earth

A century ago, Sir George Darwin (son of evolutionist Charles Darwin) proposed that the Moon is moving away from the Earth. To understand why, note in Figure 5-28 that *the high tide nearest the Moon is actually 10° ahead of the Moon in its orbit around the Earth.* This offset occurs because the Earth is rotating faster than the tide can move along an ocean. As the tides move westward to try to stay under the Moon, they encounter the eastward-moving landmasses, disrupting the water's motion. Gravitational pull from the high ocean tide nearest the Moon acts like a lever that pulls the Moon ahead in its orbit, giving it energy and thereby forcing it to spiral outward.

To test this concept, the *Apollo 12* astronauts placed on the Moon a set of reflectors, similar to the orange and red ones found on cars, to help measure the Moon-Earth distance. Pulses of laser light were fired at the Moon from the Earth (Figure 5-29), and the time it took the light to reach the Moon, bounce off the reflector array, and return to the Earth was carefully recorded. Knowing the speed of light and the time it takes for the light pulse to make the circuit, astronomers can calculate the distance to the Moon with an error of less than a meter. From such measurements over time, astronomers have established that the Moon is spiraling away from the Earth at a rate of 3.8 cm per year.

This means, of course, that the Moon used to be closer to the Earth. Although the initial distance of the Moon when it coalesced is not known, its present rate of recession tells us that it was at least half its present distance; more accurate calculations suggest one-tenth. At this distance, the tides on the Earth were a thousand times higher than they are today. (What effects would those tides have had on the geology of the young Earth?)

Where does the Moon get the energy necessary to spiral away from the Earth? The answer, of course, is from the Earth itself. Because the oceans move westward, while the planet rotates eastward, the oceans actually push on the continents opposite to the Earth's direction of rotation. The result is that *the tides are slowing the Earth's rotation.* The rotational energy lost by the Earth is gained by the Moon. Geologists calculate that when the Earth first formed, the day was only between 5 and 6 hours long!

Will the Earth's rotation ever slow down enough so that it will be in synchronous rotation with respect to the Moon? Yes, but billions of years before that happens, the Sun is going to self-destruct, so it doesn't really matter!

The Moon still inspires many questions and hides many secrets. The six American manned missions and three Soviet soft, robotic lunar landings have barely scratched the Moon's surface, bringing back samples from only nine locations. We still know very little about the Moon's far side and its poles. Is the Moon's interior molten? Does the Moon really have an iron-rich core? How old are the youngest lunar rocks? Did lava flows occur over western Oceanus Procellarum only two billion years ago, as crater densities there suggest? Is the Moon really geologically dead, or does it just appear dead because our examination of the lunar surface has been so cursory? Such questions can be answered only by returning to the Moon.

## WHAT DID YOU THINK?

1 *Will the ozone layer, which is now being depleted, naturally replenish itself?* Yes. Ozone is created continuously from normal oxygen molecules by their interaction with the Sun's ultraviolet radiation.

2 *Who was the first person to walk on the Moon and on what Apollo space mission did this landing occur?* Neil Armstrong was the first person to set foot on the Moon. He and Buzz Aldrin flew on the *Apollo 11* spacecraft piloted by Michael Collins. Armstrong and Aldrin set down the *Eagle* lander on the Moon on July 20, 1969.

3 *Do we see all parts of the Moon's surface at some time throughout the lunar cycle of phases?* No. Because the Moon's rotation around Earth is synchronous, we always see the same side. The far side of the Moon has been seen only from spacecraft that pass or orbit it.

4 *Does the Moon rotate, and, if so, how fast?* The Moon rotates at the same rate that it revolves around the Earth. If the Moon didn't rotate, then, as it revolved, we would see its entire surface from the Earth, which we don't.

5 *What causes the ocean tides?* The tides are created by gravitational forces, primarily from the Moon and, to a lesser extent, from the Sun.

6 *When does the spring tide occur?* Spring tides occur twice monthly, during each full and new Moon.

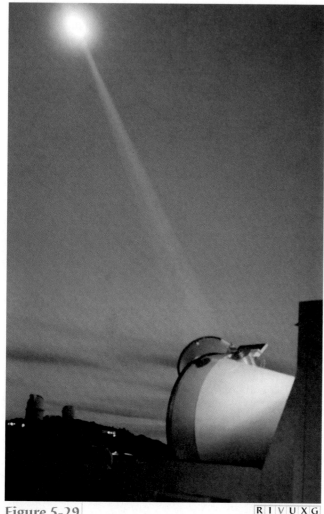

**Figure 5-29** | **Lunar Ranging** A pulse of laser light is fired through this McDonald Observatory telescope at the Moon; the light is then reflected back by the corner reflectors placed on the Moon by Apollo astronauts. From the time it takes the light to reach the Moon and reflect back to Earth, astronomers can determine the distance to the Moon to within 1 m. The long exposure required to show the laser beam in this photograph causes the overexposed, glowing Moon.

R I V U X G

## Key Words

## ⚕ Observing Projects

41   Use a telescope to observe the Moon. Compare the texture of the lunar surface you see on the maria with that of the lunar highlands. How does the visibility of details vary with distance from the terminator (the boundary between day and night on the Moon)?

42   If you live near the ocean, observe the tides to see how the times of high and low tides are correlated with the position of the Moon in the sky.

43   Using Starry Night set for today, center and lock on the Moon, turn off daylight and the horizon, and zoom until the Moon is about 3 inches across in the center of your screen. What phase is it in? Make a rough sketch of the lunar features you can see. Set the time increment to be 1 day per timestep. Step several days ahead, watching the phase change. Compare the features you see to those you have sketched. Besides seeing more or fewer of them, how have these features changed? Step through a complete cycle of phases. Describe whether and, if so, how the features change throughout this cycle.

# The Moon Didn't Exist?

The Moon—full and bright or crescent and mysterious—is a fixture in our sky and in our cultures. Throughout history, people have weaved myths about the Moon and its effects on everything from childbirth to stock market activity. Countless romances have begun under a full Moon, and entire nations have dedicated themselves to reaching it.

What if the Moon didn't exist? Would life have even developed here? If so, how would it be different? Would a self-aware species like ourselves have evolved?

**Delayed Origins** It would have been much more difficult for life to start evolving on a Moonless Earth. In the first place, the minerals from which life developed in our Earth's young oceans were swept down there in large part by gigantic tides, a thousand times higher than the tides of today, created by the Moon when it was some 10 times closer to Earth than at present. The newly formed Earth, spinning more than 4 times faster than at present, caused these tides to move miles inland and back out to sea every 2 hours or so.

A Moonless Earth would still have tides, created by the Sun. At most, however, these tides would be only one-third as high as Earth's *present* tides. Minerals would wash into the oceans incredibly slowly from the flow of rivers. As a result, it is likely that it would have taken much longer, perhaps hundreds of millions of years longer, for enough minerals to be dissolved for life to firmly establish itself.

**Harsh Conditions** Animal life's transition from oceans to land would also be much harder because of continuous winds, between 50 and 150 miles per hour, created by the planet's rapid rotation. The resulting waves (wind causes waves) would be enormous and perpetual. Fish with legs near shores would almost certainly be pounded to a pulp rather than being able to sedately walk onto land and then back into the water, as apparently happened on Earth.

But we see on Earth how life develops in the most incredible forms and places. It seems plausible, then, that sea life would find a way to make the transfer to dry land and continue to evolve there.

Allowing then for diverse terrestrial life on the Moonless Earth, what would be different about that life compared to life on our Earth? For one thing, creatures evolving there would have to withstand the perpetual pummeling from winds and the debris they carry. Turtle-like shells are one solution.

Obviously, there would be no eclipses nor moonlight—all clear nights would be equally dark and star-filled. Therefore, Earthlike nocturnal animals would be less successful at hunting, foraging, and traveling. Instead, these animals might evolve more enhanced senses to compensate for their inability to see visible light well at night.

**Rush Hour** Clearly, naked apes do not seem a likely bet on a Moonless Earth, nor do birds battling the ever-present winds. But given enough time, complex, even self-aware life likely would evolve. After all, we evolved because of the challenges faced by our ancestors, and Earth without a Moon would clearly provide challenges of its own.

Finally, the physiology of life on Earth evolved based on a 24-hour day. This is most evident in our biological clocks, or circadian rhythms, which are the internal mechanisms that regulate sleeping and waking, eating, and other cyclic activities. Faced with a 6-hour day, these circadian rhythms would be hopelessly out of sync with the natural world. In order to function on such a world, all its creatures would have to evolve biological clocks based on 6 hours. Considering all we have to do now, imagine what life would be like with only 6 hours in each day!

# 6

**Canyon Fog**
**Delicate fog created by the scant water vapor in Mars's atmosphere fills canyons of Labyrinthus Noctis at sunrise and sunset.**

# The Other Terrestrial Planets

## *In this chapter you will discover*

* Mercury, a Sun-scorched planet with a heavily cratered, Moonlike surface and an Earthlike, iron core

* Venus, perpetually shrouded in thick, poisonous clouds and mostly covered by gently rolling hills

* Mars, the inspiration for scientific and popular speculation about extraterrestrial life

## WHAT DO YOU THINK?

1    Is the temperature on Mercury, the closest planet to the Sun, higher than the temperature on Earth?

2    What planet is most similar to Earth?

3    What is the composition of the clouds surrounding Venus?

4    Does Mars have liquid water on its surface today?

5    Is life known to exist on Mars today?

Far from the Sun are planets so alien that comparisons to Earth are virtually meaningless. Closer to home, however, a space traveler from Earth might feel like an ordinary tourist. The craters on Mercury call to mind the Moon. The clouds of Venus look almost comforting, which belies their true chemical composition. The image that opens this chapter shows a feature on Mars that brings to mind the Grand Canyon. How can a riverbed exist on such an arid world?

As we begin our tour of the inner solar system, it will help to keep asking, while they look like things back home, are they really the same? See the table "The Inner Planets: A Comparison."

# MERCURY

When it is visible, Mercury is among the brightest objects in the sky. It is best seen at its *greatest elongation*, when it is farthest in angle from the Sun. When it rises before the Sun, Mercury heralds the ascending Sun in the brightening eastern sky as a "morning star." When Mercury rises after the Sun, it is seen only for a short time after sunset, hovering low over the western horizon as an "evening star."

## 6-1    Photographs from *Mariner 10* reveal Mercury's lunarlike surface

Telescopic photographs of Mercury taken from Earth reveal almost nothing about the planet's surface. It was only in 1974, when *Mariner 10* coasted past the planet, that astronomers got their first glimpse of Mercury's inhospitable but familiar-looking surface. Figure 6-1 lists Mercury's essential properties. *Mariner 10* photographed only 40% of Mercury's surface, and we still do not know what the other 60% looks like.

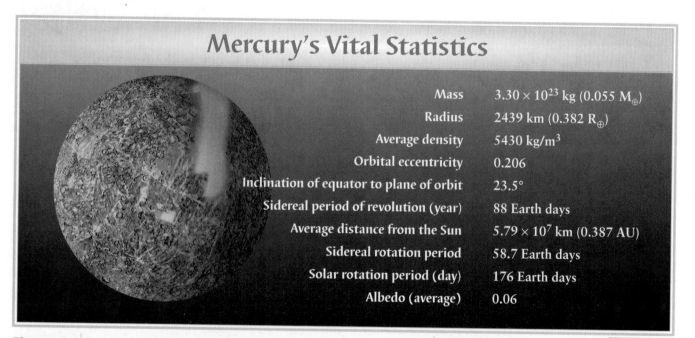

## Mercury's Vital Statistics

| | |
|---|---|
| Mass | $3.30 \times 10^{23}$ kg (0.055 $M_\oplus$) |
| Radius | 2439 km (0.382 $R_\oplus$) |
| Average density | 5430 kg/m$^3$ |
| Orbital eccentricity | 0.206 |
| Inclination of equator to plane of orbit | 23.5° |
| Sidereal period of revolution (year) | 88 Earth days |
| Average distance from the Sun | $5.79 \times 10^7$ km (0.387 AU) |
| Sidereal rotation period | 58.7 Earth days |
| Solar rotation period (day) | 176 Earth days |
| Albedo (average) | 0.06 |

**Figure 6-1**    *Mariner 10* **Mosaic of Mercury**  This image of Mercury is a composite of dozens of photos taken by the *Mariner 10* spacecraft. The blank region was not imaged by the spacecraft, nor was the opposite side of the planet. Although most of these images were recorded at distances of about 200,000 km (125,000 mi) from Mercury, their resolution is still vastly superior to that of the best Earth-based images.

R I V U X G

# The Inner Planets: A Comparison

| | Interior | Surface | Temperature | Atmosphere | Magnetic Field |
|---|---|---|---|---|---|
| Mercury | Not known whether liquid or solid | Heavy cratering, scarps | 700 K by day, 100 K by night | H, He, K, P, O; transient and tenuous | 0.1 times Earth's field |
| Venus | At least partially molten | Light cratering, mostly volcanic plains, gently rolling hills, some volcanos | 750 K | Mostly $CO_2$, N; 98 times denser than Earth's | None detected |
| Earth | Solid inner core, molten outer core, mantle | Very little cratering, continents and land at ocean floors, weathering, volcanos, global tectonic plates | 200–315 K | Mostly $N_2$, $O_2$ | Strong global field |
| Mars | Probably solid core | Moderate cratering, weathering, dormant volcanos, huge canyons | 160–280 K | Mostly $CO_2$, $N_2$; 0.006 times as dense as Earth's | Weak, local fields |

*For detailed numerical comparisons between planets, see Tables A-1 and A-2.*

Earth

Mercury

Mars

Venus

| Distance Line for the Inner Solar System | |
|---|---|
| From center of Sun to its surface | .0047 AU |
| From center of Sun to Mercury | .39 AU |
| From center of Sun to Venus | .72 AU |
| From center of Sun to Earth | 1.0 AU |
| From center of Earth to center of Moon | .000026 AU |
| From center of Sun to Mars | 1.5 AU |

0.0047 AU

Moon    0.000026 AU

0.39 AU          0.33 AU          0.28 AU          0.5 AU

Sun          Mercury          Venus          Earth          Mars

Astronomers conclude that most of the craters on both Mercury and the Moon were produced by impacts in the 800 million years after the solar system formed. As we saw in Chapter 5, the strongest evidence for this belief comes from analyzing and dating Moon rocks. Debris remaining after the planets were formed pounded these young worlds, gouging out most of the craters we see today. Like the Moon, Mercury has an exceptionally low albedo of 0.06 (6% reflection). (Ask yourself, why then is it so bright?)

Although first impressions of Mercury evoke a lunar landscape, closer scrutiny reveals decidedly nonlunar characteristics. Lunar craters (see Figure 6-2) are densely packed, with one overlapping the next. In sharp contrast, Mercury's surface has broad plains separating many craters and no sign of the large, localized, nearly crater-less maria seen on the Moon. Figure 6-3 shows a typical close-up view of Mercury.

As we learned in Chapter 5, the lunar maria were produced by extensive lava flows that occurred between 3.1 and 3.8 billion years ago. Ancient lava flows also probably formed the Mercurian plains. As large meteoroids punctured the planet's thin, newly formed crust, lava welled up from the molten interior to flood low-lying areas. The number of craters that pit Mercury's

plains suggest that the plains formed near the end of the era of heavy bombardment, just over 3.8 billion years ago. Mercury's plains are therefore older than most of the lunar maria, leaving more time for cratering to eradicate marialike features.

The most impressive feature discovered by *Mariner 10* was a huge circular region called the Caloris Basin (Figure 6-4), which measures 1300 km (810 mi) in diameter and lies along the *terminator* (the border between day and night) in the figure. It is surrounded by a 2-km-high ring of mountains, beyond which are relatively smooth plains. Like the lunar maria, the Caloris Basin was probably gouged out by the impact of a large meteoroid that penetrated the planet's crust. Because relatively few craters pockmark the lava flows that filled the basin, the Caloris impact must have occurred toward the end of the crater-making period.

The Caloris impact was a tumultuous event that shook the entire planet. Indeed, the collision even seems to have affected the side of Mercury directly opposite the Caloris Basin: That area has a jumbled, hilly region covering nearly half a million square kilometers, about twice the size of Wyoming. Figure 6-5 is a wide-angle view of this area. The hills, which appear as tiny wrinkles that cover most of the photograph, are about 5 to 10 km wide

**Figure 6-2** **Mercury and Our Moon** Mercury (left) and our Moon are shown here to the same scale. Mercury's diameter is 4878 km and the Moon's is 3476 km. For comparison, the distance from New York to Los Angeles is 3944 km (2451 mi). Mercury's surface is more uniformly cratered than that of the Moon. Daytime temperatures at the equator on Mercury reach 700 K (800°F), hot enough to melt lead or tin.

R I V U X G

**Figure 6-3** **Mercury's Craters and Plains** This view of Mercury's northern hemisphere was taken by *Mariner 10* as it sped past the planet in 1974. Numerous craters and broad plains cover an area 480 km (300 mi) wide.

**Figure 6-5** **Unusual, Hilly Terrain** What look like tiny, fine-grained wrinkles on this picture are actually closely spaced hills, part of a jumbled terrain that covers nearly 500,000 km² on the opposite side from the Caloris Basin. The large, smooth-floored crater, Petrarch, near the center of this photograph has a diameter of 170 km (106 mi). This impact crater was produced more recently than the Caloris Basin.

**Figure 6-4** **The Caloris Basin** *Mariner 10* sent back this view of a huge impact basin on Mercury's equator. Only about half of the Caloris Basin appears because it happened to lie on the terminator when the spacecraft sped past the planet. Although the center of the impact basin is hidden in the shadows (just beyond the left side of the picture), several semicircular rings of mountains reveal its extent.

and between 100 and 1800 m high. Geologists believe that energy from the Caloris impact traveled through the planet and became focused, like light through a lens, as it passed through Mercury. As this concentrated energy reached the far surface of the planet, jumbled hills were pushed up.

*Mariner 10* also revealed gently rolling plains and numerous, long cliffs, called **scarps**, meandering across Mercury's surface (Figure 6-6). The scarps are believed to have developed as the planet cooled. Almost everything that cools contracts. Therefore, as Mercury's mantle and molten iron core cooled and contracted, its surface moved inward. Because it is solid, Mercury's crust wrinkled as it contracted, forming the scarps. These features and the lack of recent volcanic activity suggest that the planet's interior is solid to a significant depth. Otherwise, lava would have leaked out as the scarps formed.

## 6-2 Mercury's temperature range is the most extreme in the solar system

By sending radio waves to Mercury and examining the reflected signal, astronomers recently made an exciting

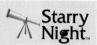
To learn more about the inner planets, start up your Starry Night program and put on the equatorial (Eq) and ecliptic (Ec) grids and the planet list.

**Mercury** In the planet list, double click on Mercury. Your view will change until Mercury is in the sky. Mercury should lock in the center of your screen. Where is the Sun? If Mercury is west of the Sun, set the time to 6 A.M. and run time forward or backward until Mercury is above the eastern horizon and the Sun is below the horizon. This shows Mercury as the "morning star." Now, run time forward at 3 minutes per timestep and you will see the white dot of Mercury fade. Conversely, if Mercury is east of the Sun, set the time to 6 P.M. and run the time forward or backward until the Sun has set but Mercury is still up. Mercury is now the "evening star." In either case, note how close Mercury always is to the horizon when it is visible in the night sky.

Making sure that you are "locked" on Mercury, hide the daylight and the horizon. Start time running at a rate of 1 sidereal day per timestep. Describe the motion of the Sun relative to Mercury. You can learn more about Mercury as seen from Earth by trying Observing Project 38 at the end of this chapter.

**Venus** Repeat the same procedures for Venus as we have described for Mercury. What similarities and differences do you see?

**Mars** Making sure that the horizon and daylight are on, set the time to midnight and lock on Mars. If it is below the horizon, run the program forward at a rate of 7 days per timestep until Mars is well above the horizon at night. Turn off the horizon and daylight and determine how far (in angle) the Sun is away from Mars. You can make this measurement by "dragging" the celestial sphere around and measuring angles from the grid. What is Mars's phase? Does the Sun ever get this far in angle from Mercury or Venus, as seen from Earth?

Relock on Mars and run the program at a rate of 1 sidereal day or more until the Sun is very close to Mars. Zoom in on the planet. What is Mars's phase now? You can learn more about Mars's unusual motion relative to Earth's orbit by trying Observing Project 39 at the end of this chapter.

**Figure 6-6** **Scarps on Mercury** A long, meandering cliff called Santa Maria Rupes runs from north to south across this *Mariner 10* image of a region near Mercury's equator. This cliff, called a scarp by geologists, is more than 1 km high and runs for several hundred kilometers. Note how the crater in the center of the image was distorted vertically when the scarp formed.

discovery—water ice near Mercury's poles. As on our Moon, sunlight is too weak at the poles to evaporate all the ice there at once.

Mercury's mass is quite low, only 5.5% that of the Earth. As on our Moon, the force of gravity on Mercury is too weak to hold a permanent atmosphere, but trace amounts of five different gases have been detected around Mercury. Scientists believe that hydrogen and helium gas there come from the Sun, while sodium and potassium gas escape from rocks inside the planet (a process called *outgassing*, which also occurs on the Earth). Oxygen observed in Mercury's atmosphere may come from polar ice that is slowly evaporating. All these gases drift into space and are continually being replenished.

Because of Mercury's slow rotation and minimal atmosphere, the differences in temperature between day and night are far more noticeable there than on the Earth. At noon on Mercury, the surface temperature is 700 K (800°F). At the terminator, where day meets night, the temperature is about 425 K (305°F), and on the night side, the temperature falls as low as 100 K (−280°F)! This range of temperature, 600 K (1080°F), on Mercury contrasts tremendously with the typical change in temperature of 11 K (20°F) between day and night on Earth.

## 6-3   Mercury has an iron core and a magnetic field like Earth

Mercury's average density of 5430 kg/m$^3$ is quite similar to Earth's (5520 kg/m$^3$). As we saw in Chapter 5, typical rocks on Earth's surface have a density of only about 3000 kg/m$^3$ because they are composed primarily of lightweight elements. The high average densities of both Mercury and our planet are caused by their iron cores.

Because Mercury is less dense than the Earth, you might conclude that Mercury has a lower percentage of iron, a heavy element, than our planet. In fact, it is the Earth's greater mass pressing inward that makes it denser. This inward force compresses the iron core, making it denser than it would otherwise be.

Mercury is actually the most iron-rich planet in the solar system. Figure 6-7 shows a scale drawing of its interior, where an iron core fills 42% of the planet's volume. Surrounding the core is a 600-km-thick rocky mantle. For comparison, the Earth's iron core occupies only 17% of

the planet's volume. Whether Mercury's core is at all molten is not yet known.

Events early in Mercury's history must somehow account for its high iron content. We know that the inner regions of the primordial solar nebula were incredibly hot. Perhaps only iron-rich minerals were able to withstand the solar heat there and these subsequently formed iron-rich Mercury. According to another theory, an especially intense outflow of particles from the young Sun stripped Mercury of its low- density mantle shortly after the Sun formed. A third possibility is that, during the final stages of planet formation, Mercury was struck by a large planetesimal. Computer simulations show that this cataclysmic collision would have ejected much of the lighter mantle (Figure 6-8).

As we saw in Chapter 5, electric currents flowing in a planet's liquid iron core create a planetwide magnetic field.

$t = 0$                          $t = 3$ min

$t = 6$ min                     $t = 30$ min

**Figure 6-8**   **The Stripping of Mercury's Mantle**   To account for Mercury's high iron content, one theory proposes that a collision with a massive asteroid stripped Mercury of most of its rocky mantle. These four images show a computer simulation of a head-on collision between proto-Mercury and a body one-sixth its mass. Both worlds are shattered by the impact, which vaporizes much of their rocky mantles. Mercury eventually reforms from the iron-rich debris left behind.

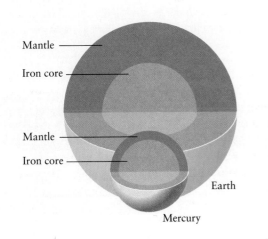

Mantle

Iron core

Mantle

Iron core

Earth

Mercury

**Figure 6-7**   **The Interiors of Mercury and Earth**   Mercury has the highest percentage of iron of any planet in the solar system. Its iron core occupies an exceptionally large fraction of its interior.

The Earth creates its electric current by rotating once a day. Mercury rotates 59 times more slowly (see Table 6-1). A solar day on Mercury (noontime to noontime) is actually 176 Earth days—twice as long as a year there! This is hardly fast enough to generate a magnetic field, and yet *Mariner 10* did find a weak magnetic field on Mercury. The origin of its present magnetic field remains a mystery.

# VENUS

Venus and Earth have almost the same mass, the same diameter, and the same average density. Indeed, if Venus was located at the same distance from the Sun as is the Earth, then it, too, might well have evolved life. However, Venus is 30% closer to the Sun than the Earth, and this one difference between the two planets leads to a host of others, making Venus inhospitable to life.

## 6-4  The surface of Venus is completely hidden beneath a permanent cloud cover

At nearly twice the distance from the Sun as Mercury, Venus is often easy to view without interference from the Sun's glare. At its greatest elongation, Venus is seen high above the western horizon after sunset, where, like Mercury, it is often called the "evening star." High in the eastern sky before sunrise, it is called the "morning star."

Venus is so easy to identify because it is often one of the brightest objects in the night sky. Only the Sun and the Moon outshine Venus at its greatest brilliance. Venus is often mistaken for a UFO, because when it appears low on the horizon, its bright light is strongly refracted by the Earth's atmosphere, making it appear to rapidly change color and position.

Unlike Mercury, Venus is intrinsically bright (Figure 6-9), because it is completely surrounded by light-colored, highly reflective clouds. Earth-based telescopic views cannot penetrate this thick, unbroken layer of clouds. Until 1962 we did not even know how fast Venus rotates. In the 1960s, however, both the United States and the former Soviet Union began sending probes to Venus. The Americans sent fragile, lightweight spacecraft into orbit near Venus (see Figure 6-9). The Soviets, who had more powerful rockets, sent massive vehicles directly into the Venusian atmosphere.

Building spacecraft that could survive the descent proved to be more frustrating than anyone had expected. Finally, in 1970 a Soviet probe managed to transmit data for a few seconds directly from the Venusian surface. Soviet missions during the 1970s measured a surface temperature of 750 K (900°F) and a pressure of 98 atm. This is the same force you would feel if you were swimming 0.93 km (3000 ft) underwater.

In contrast to Earth's nitrogen- and oxygen-rich atmosphere, 96% of Venus's thick atmosphere is carbon dioxide; the remaining 4% is mostly nitrogen. This atmosphere is remarkably similar to the early carbon dioxide–

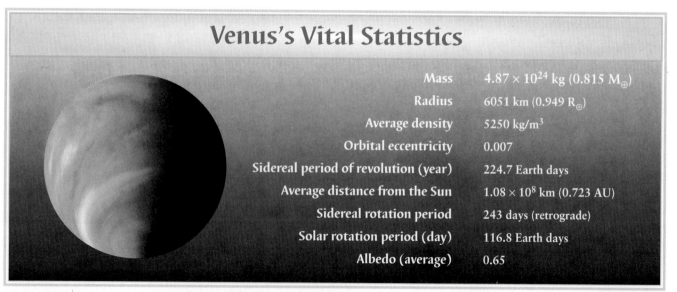

## Venus's Vital Statistics

| | |
|---|---|
| Mass | $4.87 \times 10^{24}$ kg ($0.815\ M_{\oplus}$) |
| Radius | 6051 km ($0.949\ R_{\oplus}$) |
| Average density | 5250 kg/m³ |
| Orbital eccentricity | 0.007 |
| Sidereal period of revolution (year) | 224.7 Earth days |
| Average distance from the Sun | $1.08 \times 10^{8}$ km (0.723 AU) |
| Sidereal rotation period | 243 days (retrograde) |
| Solar rotation period (day) | 116.8 Earth days |
| Albedo (average) | 0.65 |

R I V U X G

**Figure 6-9**    **Venus** Venus's thick cloud cover efficiently traps heat from the Sun, resulting in a surface temperature even hotter than that on Mercury. Unlike Earth's clouds, which are made of water droplets, Venus's clouds are very dry and contain droplets of concentrated sulfuric acid. This ultraviolet image was taken by the *Pioneer Venus Orbiter* in 1979.

rich atmosphere of the Earth. These gases were vented from inside Venus through volcanoes and other openings. Unlike the Earth, however, Venus has no liquid oceans to dissolve the carbon dioxide and so it remains in the atmosphere there today.

Soviet spacecraft also discovered that Venus's clouds are confined to a 20-km-thick layer located 50 to 70 km above the planet's surface. Below the clouds is a 20-km-thick layer of haze. Beneath this haze the Venusian atmosphere is remarkably clear all the way down to the surface. Standing on the surface of Venus, you would experience a perpetually cloudy day.

Unlike the clouds on Earth, which appear white from above, the cloudtops of Venus appear yellowish or yellow-orange to the human eye. These colors are typical of sulfur and its compounds. Indeed, spacecraft found substantial amounts of sulfur dust in Venus's upper atmosphere and sulfur dioxide and hydrogen sulfide at lower elevations. The clouds are composed of droplets of concentrated sulfuric acid! Because of the tremendous atmospheric pressure on Venus, the droplets do not fall as rain; instead, they remain suspended in thick mist.

Compelling evidence suggests that these sulfurous compounds come from active volcanoes. All of the major volcanic chemicals spewed by Earth's volcanoes have been detected in Venus's atmosphere. Because many of these substances are very short-lived, they must be constantly replenished by new eruptions. In fact, the abundance of sulfur compounds on Venus does vary, suggesting varied volcanic activity. Finally, spacecraft have also picked up radio bursts thought to be strokes of lightning, as seen in the plumes of erupting volcanoes on Earth.

The results of the early probing of Venus are summarized in Figures 6-10 and 6-11. Both pressure and temperature decrease smoothly with increasing altitude. From the changes in temperature and pressure above a planet's surface, we can understand the structure of its atmosphere. As shown in Figure 6-11, Venus's surface temperature is very high, higher even than that of Mercury.

## 6-5 The greenhouse effect heats Venus's surface

At first, no one could believe reports that the surface temperature on Venus was hotter than the surface temperature on Mercury, which, after all, is closer to the Sun. After some initial skepticism, however, astronomers quickly found a straightforward explanation—the **greenhouse effect.**

Perhaps you have had the experience of parking your car in the sunshine on a warm summer day. You roll up the windows, lock the doors, and go on an errand. After a few hours, you return to discover that the interior of

**Figure 6-10** **Pressure in the Venusian Atmosphere** The pressure at the Venusian surface is a crushing 98 atm (1440 lb/in.²). Above the surface, atmospheric pressure decreases smoothly with increasing altitude.

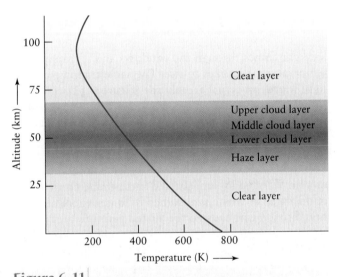

**Figure 6-11** **Temperature in the Venusian Atmosphere** The temperature of Venus's atmosphere increases smoothly from a minimum of about 173 K (−150°F) at an altitude of 100 km to a maximum of nearly 750 K (900°F) on the ground.

your automobile has become stiflingly hot, typically 20 K (36°F) warmer than the outside air temperature.

What happened to make your car so warm? First, sunlight entered your car through the windows. This radiation was absorbed by the dashboard and the upholstery, raising their temperatures. Because they become warm, your dashboard and upholstery emit infrared radiation, which you detect as heat and to which your car windows are opaque. This energy is therefore trapped inside your car and absorbed by the air and interior surfaces. As more

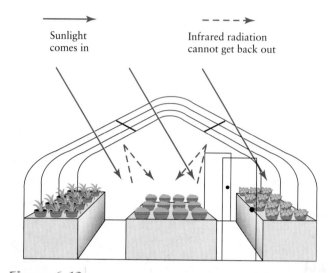

Sunlight comes in          Infrared radiation cannot get back out

**Figure 6-12** **The Greenhouse Effect** Incoming sunlight easily penetrates the windows of a greenhouse and is absorbed by objects inside. These objects reradiate energy at infrared wavelengths, to which the glass windows are opaque. The trapped infrared radiation causes the temperature inside the greenhouse to rise.

sunlight comes through the windows and is trapped, the temperature continues to rise. The same trapping of sunlight warms an actual greenhouse (Figure 6-12).

Carbon dioxide is responsible for a similar warming of Venus's atmosphere. Like your car windows, carbon dioxide is transparent to visible light but opaque to infrared radiation. On Earth, atmospheric water and carbon dioxide both contribute to the greenhouse effect in our air. Warmed by sunlight, the Earth emits infrared radiation, some of which is absorbed by water vapor and carbon dioxide gas, causing the air temperature to rise. The

small amounts of these gases in the Earth's atmosphere produce a comparatively gentle greenhouse effect. The heating due to the greenhouse effect is, however, increasing here as our atmosphere acquires more carbon dioxide from burning fossil fuels and other sources. The air retains this gas due to large-scale clear cutting of forests, which decreases the number of plants that can convert this gas into other carbon compounds and oxygen.

Although most of the visible sunlight striking the Venusian cloudtops is reflected back into space, enough light reaches the Venusian surface to heat it. The warmed surface in turn emits infrared radiation, which cannot escape through Venus's carbon dioxide–rich atmosphere. This trapped radiation produces the high temperatures found on Venus.

Without the greenhouse effect, the surface of Venus would have a noontime temperature of 465 K. Because of it, that temperature is actually a sweltering 750 K, which is hotter than the hottest spot on Mercury! Furthermore, the thick atmosphere keeps the night side of Venus at nearly the same temperature, unlike the night side of Mercury, where the temperature drops precipitously. This high temperature prevents liquid water from existing on Venus; its surface is bone dry.

## 6-6 Venus is covered with gently rolling hills, two "continents," and numerous volcanoes

Soviet spacecraft that landed on Venus provided us with intriguing close-up images of the planet's arid surface. Figure 6-13 is a view of Venus's regolith taken in 1981. Russian scientists believe that this region was covered

a

b

R I V U X G

**Figure 6-13** **The Venusian Surface** (a) This color photograph, taken by a Soviet spacecraft, shows rocks that appear orange because the light was filtered through the thick, sulfur-rich clouds. (b) By comparing the apparent color of the spacecraft to the color it was known to be, computers can correct for the sulfurous light. The actual color of the rocks is gray. In this view, the rocky plates covering the ground may be fractured segments of a thin layer of lava.

R I V U X G

**Figure 6-14**  **A Venusian Landscape**  Most of Venus is covered with plains and gently rolling hills produced by numerous lava flows. The crater in the foreground is 48 km (30 mi) in diameter. The volcano near the horizon is 3 km (1.9 mi) high. Note the cracks in the Venusian surface toward the right side of the image.

R I V U X G

**Figure 6-15**  **A Map of Venus**  These false-color radar maps of Venus and Earth show the large-scale surface features of the two planets. The equators extend across the middle of each map. Color indicates elevation—white for highest, blue for lowest. The elevated region in the north of Venus is Ishtar Terra, dominated by Maxwell Montes, the planet's highest mountain. The large, scorpion-shaped feature extending along the equator is Aphrodite Terra, a continentlike highland that contains several spectacular volcanoes.

with a thin layer of lava that fractured upon cooling to create the rounded, interlocking shapes seen in the photograph. Indeed, measurements by several spacecraft indicate that Venusian rock is quite similar to lava rocks called basalt, which are common on Earth and the Moon.

By far the best images of the Venusian surface came from the highly successful *Magellan* spacecraft that arrived at Venus in 1990. In orbit about the planet, *Magellan* sent radar signals through the clouds surrounding Venus. It mapped Venus using a radar altimeter that bounced microwaves off the ground directly below the spacecraft. By measuring the time delay of the radar echo, scientists determined the heights and depths of Venus's hills and valleys. As a result, astronomers have been able to construct a three-dimensional map of the planet. The resolution of this map is about 100 m. You could see a football stadium on Venus if there were any! (None were detected.)

Venus is remarkably flat compared to the Earth. More than 80% of Venus's surface is covered with volcanic plains and gently rolling hills created by numerous lava flows. Figure 6-14 is a computer-generated view showing a typical Venusian landscape.

Radar images revealed two large highlands, or "continents," rising well above the generally level surface of the planet (Figure 6-15). The continent in the northern hemisphere is Ishtar Terra, named after the Babylonian goddess of love and approximately the same size as Australia. Ishtar Terra is dominated by a high plateau ringed by towering mountains. The highest mountain is Maxwell

Montes, whose summit rises to an altitude of 11 km above the average surface. For comparison, Mount Everest on Earth rises 9 km above sea level.

The largest Venusian continent, Aphrodite Terra (named after the Greek goddess called Venus by the Romans) is a vast belt of highlands just south of the equator. Aphrodite is 16,000 km (10,000 mi) in length and 2000 km (1200 mi) wide, giving it an area about one-half that of Africa. The global view of the surface of Venus in

Figure 6-16 shows that most of Aphrodite Terra is covered by vast networks of faults and fractures.

Fewer than a thousand impact craters have been observed on Venus (Figure 6-17), compared to the hundreds of thousands seen on the Moon and Mercury. Today, of course, Venus's thick atmosphere heats, and thereby vaporizes, much of the infalling debris that would otherwise create craters. Because its atmosphere is thicker than Earth's, Venus clears such space junk falling toward it more efficiently than Earth does. However, there should have been a period shortly after Venus formed and before its atmosphere developed (from gases escaping from the planet's interior), when impacts were common.

The low number and incredibly random distribution of craters on Venus leads to the belief that the planet's surface is periodically erased and replaced. Because geologists have found no large-scale tectonic plate motion on Venus to refresh the surface, the current theory explaining the lack of extensive cratering is that the entire planet's surface melts! This would occur if the crust is very thick compared to the Earth's crust. A crust 300 km thick would insulate the interior so much that, heated by radioactive elements in it, the mantle could become hot enough to episodically melt the crust. It has been proposed that every 700 million years or so, the entire surface of Venus liquifies until the pent-up heat escapes, a new solid crust forms, and the process begins anew.

The sulfur content of the air and the traces of active volcanoes on the surface are strong clues that, like the Earth, Venus has a molten interior. Because the average density of Venus is similar to that of Earth, its core is pre-

R I V U X G

**Figure 6-16** **A "Global" View of Venus** A computer was used to map numerous *Magellan* images onto a simulated globe. Color is used to enhance small-scale structures. Extensive lava flows and lava plains cover about 80% of Venus's relatively flat surface.

dominantly iron. Currents in the molten iron should generate a magnetic field. However, none of the spacecraft sent there has detected one. This is plausible only if Venus rotates exceptionally slowly. In fact, the planet takes

**Figure 6-17** **Craters on Venus** Impact craters on Venus tend to occur in clusters, which suggests that they are formed from large, single pieces of infalling debris that are broken up in the atmosphere. Shown here is the triple-impact Crater Stein.

R I V U X G

## Mars's Vital Statistics

| | |
|---|---|
| Mass | $6.42 \times 10^{23}$ kg (0.107 $M_{\oplus}$) |
| Radius | 3393 km (0.532 $R_{\oplus}$) |
| Average density | 3950 kg/m$^3$ |
| Orbital eccentricity | 0.093 |
| Sidereal period of revolution (year) | 687 Earth days (1.88 Earth years) |
| Average distance from the Sun | $2.28 \times 10^8$ km (1.52 AU) |
| Sidereal rotation period | 24 h 37 min 22 s |
| Solar rotation period (day) | 24 h 39 min |
| Albedo (average) | 0.16 |

**Figure 6-18**   R I V U X G

**Mars Viewed from the Earth** This high-quality, Earth-based photograph of Mars was taken in 1997, when the Earth-Mars distance was 100 million km (62 million miles). This photograph portrays what you might typically see through a moderate-sized telescope under excellent observing conditions. Notice the polar ice caps.

116.8 Earth days to get from one sunrise to the next (that is, if you could see the Sun from Venus's surface).

Unlike the Earth, Venus has **retrograde rotation.** This means that the direction of Venus's orbit around the Sun (counterclockwise as seen from space far above the Earth's north pole) is *opposite* the direction of its rotation (clockwise as seen from the same vantage point). In other words, sunrise on Venus occurs in the west. Venus's rotation axis is tilted more than 177%, compared to the Earth's 23½% tilt. Because Venus's axis is within 3% of being perpendicular to the plane of its orbit around the Sun, the planet has no seasons. Although we do not know the cause of Venus's retrograde rotation, one likely explanation is that a major impact altered the rotation early in the planet's existence.

### 6-7 *Magellan's* last gasp revealed irregularities in Venus's upper atmosphere

*Magellan* finished mapping the surface of Venus in 1994. Instead of keeping it in orbit, NASA astronomers decided to give it one last mission. They fired braking rockets on the spacecraft, sending it spiraling down toward the planet's surface. By doing so, it was hoped that *Magellan* would transmit useful data about Venus's dense atmosphere before the spacecraft vaporized. On October 12,

1994, *Magellan* stopped transmitting, but before it did so, it sent back tantalizing information.

*Magellan* indicated that the density of Venus's atmosphere is lower at some altitudes than was expected and higher at other altitudes. This information will be useful in helping astronomers plan future missions to the cloud-shrouded planet.

## MARS

Mars is the only planet whose surface features can be seen through Earth-based telescopes. Its distinctive rust-colored hue makes it stand out in the night sky (Figure 6-18). When Mars is near opposition, even telescopes for home use reveal its seasonal changes. Dark markings on the Martian surface can be seen to vary, and prominent polar caps shrink noticeably during the spring and summer months (Figure 6-19).

### 6-8 Earth-based observations originally suggested that Mars might harbor life

The Dutch physicist Christiaan Huygens made the first reliable observations of Mars in 1659. Using a telescope of his own design, Huygens identified a prominent, dark

R I V U X G

**Figure 6-19** **Changing Seasons on Mars** During the Martian winter, the temperature drops so low that carbon dioxide freezes out of the Martian atmosphere. A thin coating of carbon dioxide frost covers a broad region around Mars's south pole. During summer in the southern hemisphere, the range of this south polar carbon dioxide cap decreases dramatically.

surface feature that reemerged about every 24 hours, suggesting a rate of rotation similar to that of Earth. Huygens's observations soon led to speculation about life on Mars because the planet seemed so similar to Earth.

In 1877, Giovanni Virginio Schiaparelli, an Italian astronomer, reported seeing 40 lines crisscrossing the Martian surface. He called these dark features *canali*, an Italian term meaning "channels." It was soon mistranslated into English as *canals*, implying the existence on Mars of intelligent creatures capable of substantial engineering feats. This speculation led Percival Lowell, who came from a wealthy Boston family, to finance a major new observatory near Flagstaff, Arizona. By the end of the nineteenth century, Lowell had allegedly observed 160 Martian "canals."

It soon became fashionable to speculate that the Martian canals formed an enormous, planetwide irrigation network to transport water from melting polar caps to vegetation near the equator. (The seasonal changes on Mars's dark surface markings can be mistaken for vegetation.) In view of the planet's reddish, desertlike appearance, Mars was thought to be a dying planet whose inhabitants must go to great lengths to irrigate their farm-

lands. No doubt the Martians would readily abandon their arid ancestral homeland and invade the Earth for its abundant resources. Hundreds of science fiction stories and dozens of monster movies owe their existence to the *canali* of Schiaparelli.

> **Insight into science** **Perception and reality** Our perceptual tendency to see patterns, even where they do not exist, makes it easy for us to leap to conclusions. Scientific images and photographs, however, require objective analysis.

On October 30, 1938, Orson Welles presented a radio program announcing the invasion of Earth by Martians. His broadcast, live from New York City, was inspired by the 1898 H. G. Wells book *War of the Worlds*. (You might ask your grandparents if they heard the show.) So realistic was the performance that it caused panic throughout the New York–New Jersey area. How close to the mark was H. G. Wells when it comes to life on Mars?

## 6-9 Early space probes to Mars found no canals but did find some controversial features

Spacecraft journeying to Mars since the 1970s have sent back pictures that clearly show numerous flat-bottomed craters (Figure 6-20), volcanoes, and canyons, but not one canal of a size consistent with those allegedly seen from Earth. Schiaparelli's *canali* were optical illusions, and H. G. Wells was completely off the mark. We have observed no cities nor roads nor other signs of technological civilization on the red planet. Intelligent Martian life has never existed.

> **Insight into science** **Fact and fiction** Science fiction may foreshadow real scientific discoveries. (Can you think of some examples?) But science also reins in the fantasies of the science fiction writer. A century ago, millions of people, including top scientists, believed that technologically advanced life existed on Mars.

Astronomers *have* photographed several surface features that at first glance could have been crafted by intelligent life-forms. In 1976, a hundred years after Schiaparelli's alleged discovery of *canali*, the *Viking Orbiter 1* spacecraft photographed a feature that appeared to be a humanlike face (Figure 6-21a). However, when the *Mars Orbiter* photographed the surface in greater detail and

**Figure 6-20** **Martian Craters** Numerous flat-bottomed craters are seen in this mosaic of images taken by a spacecraft orbiting Mars in 1980. The large, heavily cratered, circular area in the upper half of this view is known as Arabia. Carbon dioxide snow covers the floors of craters near the Martian south pole, toward the lower right in this picture.

from a different angle in 1998 (Figure 6-21b), it found no facial features. If anything, the same spot looked more like a giant shoeprint.

Scientists universally believe that the face and other alleged patterns were created by shadows on wind-blown hills. Other features, in the shape of pyramids, are consistent with the winds eroding softer rock around harder rock pushed upward from inside Mars billions of years ago. The same thing happens on Earth.

> **Insight into science　Keep it simple** Was the Martian "face" made by intelligent life or by wind-sculpted hills? When two alternative interpretations present themselves, scientists apply the principle of Occam's razor and always choose the one requiring the fewest unproved assumptions.

## 6-10　Probes to Mars found craters, volcanoes, and canyons

The partially eroded, flat bottoms of the Martian craters probably result from dust storms that frequently rage across the planet's surface. Over the past two centuries, astronomers have seen faint surface markings disappear under a reddish-orange haze as thin Martian winds have

a

b

**Figure 6-21** **In the Eye of the Beholder** These two images of the same site on Mars, taken 22 years apart, show how the apparent face in **(a)** changed to a more "natural-looking" feature in **(b)**. This transformation was due to weathering of the site, improved camera technology, and the change in angle at which the photograph was taken.

**Figure 6-22** **Canyons and Volcanoes on Mars** This high-altitude photograph shows a variety of the features on Mars, including broad, towering volcanoes (left) on the highland called Tharsis Bulge, impact craters (upper right), and vast, windswept plains. An enormous canyon system called Valles Marineris crosses horizontally just below the center of the image.

Mars appears to have a core about 3400 km in diameter. The planet, however, does not appear to have any plate tectonic activity, and one hot spot thus can keep pumping lava upward through the same vent for millions of years. One result is Olympus Mons—a single giant volcano rather than a long chain of smaller ones. The volcano's summit has collapsed to form a volcanic crater, called a **caldera,** large enough to contain the state of Rhode Island (see Figure 6-23). Calculations indicate that Mars's interior should be solid, so we do not expect to see any further volcanic activity there.

For reasons we do not yet understand, most of the volcanoes on Mars are in the northern hemisphere, but most of the impact craters are in the southern hemisphere. Between the two hemispheres, *Mariner 9* discovered a vast canyon running roughly parallel to the Martian equator (see Figures 6-22 and 6-24). In honor of the *Mariner* spacecraft that revealed so much of the Martian surface, this enormous chasm has been designated Valles Marineris.

Valles Marineris stretches over 4000 km, beginning with heavily fractured terrain in the west and ending with ancient cratered terrain in the east. If this canyon were located on Earth, it would stretch from New York to Los Angeles. Unlike the Grand Canyon, which it superficially resembles, Valles Marineris was not formed by water erosion. Some geologists suspect Valles Marineris is an ancient

stirred up finely powdered dust from the Martian regolith. Some storms actually obscure the entire planet. Over the ages, deposits of dust have filled in the crater bottoms, but 3 billion years of sporadic storms have not wiped out the craters; the Martian atmosphere is too thin to carry enough of this extremely fine-grained powder to eradicate the craters completely.

In 1971 the *Mariner 9* spacecraft went into orbit around Mars and began sending back detailed, close-up pictures that showed enormous volcanoes and vast canyons (Figure 6-22). The largest Martian volcano, Olympus Mons (Figure 6-23), covers an area as big as the state of Missouri and rises 24 km (15 mi) above the surrounding plains—nearly 3 times the height of Mount Everest. The highest volcano on Earth, Mauna Loa in the Hawaiian Islands, has a summit only 8 km above the ocean floor.

On Earth, the Hawaiian Islands are only the most recent additions to a long chain of volcanoes. They resulted from **hot-spot volcanism,** a process by which molten rock rises to the surface from a fixed hot region far below. The Pacific tectonic plate is slowly moving northwest at a rate of several centimeters per year. As a result, new volcanoes are created above the hot spot, while older ones move off and become extinct, eventually disappearing beneath the ocean.

**Figure 6-23** **The Olympus Caldera** This view of the summit of Olympus Mons is based on a mosaic of six pictures taken by one of the *Viking* orbiters. The caldera consists of overlapping, volcanic craters and measures about 70 km across. The volcano is wreathed in mid-morning clouds brought upslope by cool air currents. The cloudtops are about 8 km below the volcano's peak.

**Figure 6-24** **A Segment of Valles Marineris** This mosaic of *Viking* orbiter photographs shows a segment of Valles Marineris. The canyon is about 100 km (60 mi) wide in this region. The canyon floor has two major levels. The northern (upper) canyon floor is 8 km (5 mi) beneath the surrounding plateau, whereas the southern canyon floor is only 5 km (3 mi) below the plateau.

R I V U X G

fracture in the Martian crust somewhat similar to those caused by plate tectonics on Earth. The Martian canyons are up to 6 km (4 mi) deep and 190 km (120 mi) wide.

## 6-11 Surface features indicate that water once flowed on Mars

Despite disproving the theory that Mars has broad canals, Mars-orbiting spacecraft did reveal many features

that look like dried-up riverbeds (Figure 6-25) and lakes. Some of these riverbeds include intricate branched patterns and delicate channels meandering among flat-bottomed craters. Rivers on Earth invariably follow similarly winding courses (Figure 6-26).

The Martian riverbeds were totally unexpected because liquid water cannot exist today on Mars. Water is liquid over a limited range of temperature and pressure. At low temperatures, water becomes ice; at high temperatures, it becomes steam. The atmospheric pressure also

R I V U X G

**Figure 6-25** **Ancient River Channel on Mars?** Winding canyons on Mars, such as the one in this *Viking Orbiter I* image (on the right) appear to be at least partly due to sustained water flow. This belief is supported by the terraces seen on the canyon walls in the high resolution *Mars Orbiter* image (**inset**). Long periods of water flow require that the planet's atmosphere was once thicker and its climate more Earthlike. These conditions were necessary to enable liquid water to flow across its surface.

**Figure 6-26**    **River Flow on Earth** This image of a river in China shows the same snakelike curve as the river channels on Mars.

affects the state of water. If the pressure is very low, molecules easily escape from the liquid's surface, causing the water to vaporize. Because the pressure of Mars's atmosphere is only 0.6% that of the Earth's atmosphere, any liquid water on Mars today would furiously boil and rapidly evaporate into the thin Martian air.

Evidence that water once flowed on Mars comes from so-called *SNC meteorites* found on Earth (Figure 6-27). These space rocks are believed to have once been pieces of Mars because their chemistries are consistent with those of rocks studied on Mars's surface and because they contain trace gases in amounts found only in the current Martian atmosphere. The meteorites were ejected into space during especially powerful impacts on that planet's surface. They also contain water-soaked clay, which is not expected to be found on any object in the solar system besides Mars.

Where did the water on Mars come from and go to? Some of it was (and still is) frozen in the Martian polar caps, although their seasonal variations today are created by the freezing and evaporating of dry ice (carbon dioxide). The temperature at Mars's poles, typically 160 K (−170°F), keeps the water there permanently frozen, like the permafrost found in northern Asia and on Antarctica. Other frozen water probably lies elsewhere under the Martian surface. In the distant past, heat from meteoritic impacts or from volcanic activity occasionally melted this layer of permafrost on Mars. The ground then collapsed, and millions of tons of rock pushed the water to the surface. These flash floods could account for the riverbeds we see there today.

a

b

**Figure 6-27**    **A Piece of Mars on Earth** Mars's thin atmosphere does little to protect it from impacts. Some of the debris ejected from impact craters there apparently traveled to Earth. **(a)** This SNC meteorite was recovered in Antarctica. It shows strong evidence of having been exposed to liquid water on Mars, perhaps for hundreds of years. **(b)** These may be the fossil remains of primitive bacterial life on Mars.

## 6-12 Martian air is thin and often filled with dust

As on Venus, some 95% of Mars's thin atmosphere is composed of carbon dioxide. The remaining 5% consists of nitrogen, argon, and some traces of oxygen. The planet's gravitational force is just strong enough to hold all its gases but too weak to prevent most water vapor from evaporating into space. Consequently, the concentration of water vapor in Mars's atmosphere is 30 times lower than the concentration above the Earth. If all the water vapor could somehow be squeezed out of the Martian atmosphere, it would not fill even one of the five Great Lakes of North America.

Mars experiences Earthlike seasons because of a striking coincidence, first noted in the late 1700s by the German-born English astronomer William Herschel. Just as the Earth's equatorial plane is tilted 23½° from the plane of its orbit, Mars's equator makes an angle of about 25° with its orbit. However, the Martian seasons last nearly twice as long as Earth's, because Mars takes nearly two Earth years to orbit the Sun.

Unlike our blue sky, Mars's atmosphere is often pastel red, sometimes turning shades of pink and russet (Figure 6-28). All these colors are due to fine dust, blown from the planet's desertlike surface during powerful windstorms. The dust is iron oxide, familiar here on Earth as rust. Mars's sky changes color because the amount of dust in the air varies with the season. During the winter, carbon dioxide ice adheres to the dust particles and drags them to the ground. This helps clear and lighten the air. In the summer months, the carbon dioxide is not frozen, and the dust blown by surface winds remains aloft longer.

The sky color also changes over periods of many years. In 1995, for example, the amount of dust was observed to have dropped dramatically compared to that observed in the 1970s. The reason for such long-term changes is still under investigation.

Mars has an intriguing magnetic field. Unlike the Earth, Mars has no global dipole field. Instead, it has at least nine regions where local magnetic fields pierce its crust, as if created by giant bar magnets just below the surface. These local fields are 75 times weaker than the Earth's global field and appear to be remnants of an earlier, global Martian magnetic field that disappeared as the planet's interior cooled and solidified.

## 6-13 The *Viking* landers sent back close-up views of a barren Martian surface

The discovery in the 1960s that water once existed on the surface of Mars rekindled speculation about Martian life. Although it was clear that Mars has neither civilizations

Figure 6-28 | R I V U X G | **Mars's Rust-Colored Sky** Taken by the Mars *Pathfinder*, this stunning photograph of another world shows the *Sojourner* rover snuggled against a rock named Moe on the Ares Vallis. At the top of the image, the pink color of the Martian sky is evident. One of the two "Twin Peaks" can be seen in the background, about 1 km away.

nor fields of plants, microbial life-forms still seemed possible. Searching for Martian microbes was one of the main objectives of the ambitious and highly successful *Viking* missions.

The two *Viking* spacecraft were launched during the summer of 1975. Each spacecraft consisted of two modules—an orbiter and a lander. Almost a year later, both *Viking* landers set down on rocky plains north of the Martian equator.

The landers confirmed the long-held suspicion that the red color of the planet is due to large quantities of iron in its soil. Despite the high iron content of its crust, Mars has a lower average density ($3950 \text{ kg/m}^3$) than that of other terrestrial planets (more than $5000 \text{ kg/m}^3$ for Mercury, Venus, and Earth). Mars must therefore contain a lower percentage of iron than these other planets.

Each *Viking* lander was able to dig into the Martian regolith and retrieve rock samples for analysis

(Figure 6-29). Bits of the regolith were observed to cling to a magnet mounted on the scoop, indicating that the regolith contains iron (consistent with the assertion above that Mars is rich in iron oxides). Indeed, further analysis showed the rocks at both sites to be rich in iron, silicon, and sulfur. The Martian regolith can best be described as an iron-rich clay.

The *Viking* landers each carried a compact biological laboratory designed to test for microorganisms in the Martian soil. Three biological experiments were conducted, each based on the idea that living things alter their environment: They eat, they breathe, and they give off waste products. In each experiment, a sample of the Martian regolith was placed in a closed container, with or without a nutrient substance. The container was then examined for any changes in its contents.

The first data returned by the *Viking* biological experiments caused great excitement. In almost every case, rapid and extensive changes were detected inside the sealed containers. However, further analysis showed that these changes were due solely to nonbiological chemical processes. Apparently, the Martian regolith is rich in chemicals that effervesce (fizz) when moistened. A large amount of oxygen is apparently tied up in the regolith in the form of unstable chemicals called *peroxides* and *superoxides,* which break down in the presence of water to release oxygen gas.

The chemical reactivity of the Martian regolith probably comes from ultraviolet radiation that beats down on the planet's surface. Ultraviolet photons easily break apart molecules of carbon dioxide ($CO_2$) and water vapor ($H_2O$) by knocking off oxygen atoms, which then become loosely attached to chemicals in the regolith. Ultraviolet photons also produce ozone ($O_3$) and hydrogen peroxide ($H_2O_2$), which become incorporated in the regolith. In all these cases, the loosely attached oxygen atoms make the regolith extremely reactive.

Here on Earth, hydrogen peroxide is commonly used as an antiseptic. When you pour this liquid on a wound, it fizzes and froths as the loosely attached oxygen atoms chemically combine with organic material and thereby destroy germs. The *Viking* landers may have failed to detect any organic compounds on Mars, because the superoxides and peroxides in the Martian regolith make it literally antiseptic today.

Okay, if there is no sign of life on Mars now, what about signs of ancient life there from the days when water flowed? Curiously, the best evidence *may* lie on Earth. At least 12 meteorites believed to have originated on Mars have been identified (Section 6-11). When cut open, several of them have shown microscopic features that could be fossils of Martian bacterial life and their excretions (see Figure 6-27b). These features, no larger than 500 nm (1/100th the diameter of a human hair), are 30 times smaller than bacteria found on Earth. Only more detailed chemical analysis can convince biologists that SNC meteorites indeed contain fossils of alien life. Past claims of fossils from Mars have already been disproved, and it remains to be seen if the present ones hold up.

We have entered an exciting new era of Martian exploration. The 1997 visit by *Pathfinder* and its little rover, *Sojourner* (see Figure 6-28), to the Ares Vallis has revealed evidence of several floods at the mouth of a dried flood plain. The evidence includes layers of sediment, clumps of rock and sand stuck together like similar groupings created by water on Earth, rounded rocks worn down as they were dragged by the water, and rocks aligned by water flow. The rover performed some 20 chemical analyses of rocks in the landing area. Results indicate chemical compositions consistent with the SNC meteorites, supporting the belief that these latter bodies came from Mars. The soil is a rusty color, laden with sulfur. The source of this sulfur is a mystery, as is the large amount of silicon-rich silica discovered on rocks such as Barnacle Bill.

R I V U X G

**Figure 6-29** **Digging in the Martian Soil**
*Viking's* mechanical arm with its small scoop protrudes from the right side of this view of the *Viking 1* landing site. Several small trenches dug by the scoop in the Martian regolith appear near the left side of the picture.

Outlet                    Shoreline

100 km    Shoreline              Inflow
                                 channels          R I V U X G

**Figure 6-30**  **Evidence of Water on Mars**  A dry Martian lake,
photographed by the *Mars Global Surveyor* with a resolution of 1.5 m.
An excellent example of how geology and astronomy overlap, the
features of this dry lake are consistent with those found on lakebeds
on Earth.

The *Mars Global Surveyor*'s images have greatly
strengthened the belief that Mars once had liquid water,
as seen in Figure 6-30. This orbiting spacecraft has taken
visible light images of Mars's surface, seeing objects as
small as 1.5 m across. Its laser has mapped the planet to

a resolution of 330 m along the surface and 1 m in
height. *Surveyor* has also mapped heat emitted from
Mars and mineral deposits on its surface, along with the
planet's strange magnetic fields, discussed earlier. This
remotely sensed information is invaluable in helping
space scientists decide where to explore Mars's surface in
upcoming missions.

## 6-14  Mars's two moons look more like potatoes than spheres

Two tiny moons orbit close to Mars's surface. Phobos
(meaning "fear") and Deimos (meaning "panic") are so
small that they were not discovered until 1877. Potato-
shaped Phobos is the inner and larger of the two (Figure
6-31a). It rises in the west and gallops across the sky in
only five and a half hours, as seen from Mars's equator. It
is heavily cratered. Football-shaped Deimos is less cratered than Phobos (Figure 6-31b). As seen from Mars,
Deimos rises in the east and takes about three Earth days
to creep from one horizon to the other.

Phobos and Deimos were not formed like our Moon,
by splashing off Mars. Rather, they are captured plane-
simals. Both moons are in synchronous rotation as they
orbit the red planet.

**Insight into science**  **Imagine the moon**
The definitions we use for words based on our
everyday experience often fail us in astronomy.
For example, the word "moon" usually
creates an image of a spherical body, like our
Moon. In reality, most of the moons in the
solar system are unsymmetrical, like Phobos
and Deimos.

**Figure 6-31**  **Phobos and Deimos**
*Viking* orbiters provided these images of Mars's
moons. **(a)** Phobos, the larger of Mars's two
moons, is potato-shaped and measures
approximately 28 × 23 × 20 km. **(b)** Deimos is
less cratered than Phobos and measures
roughly 16 × 12 × 10 km.

R I V U X G

## WHAT DID YOU THINK?

1  *Is the temperature on Mercury, the closest planet to the Sun, higher than the temperature on Earth?* The temperature on the daytime side of Mercury is much higher than on Earth, but the temperature on the nighttime side of Mercury is much lower than on Earth, because Mercury rotates so slowly and has little atmosphere to retain heat.

2  *What planet is most similar to Earth?* Venus is most similar to Earth in size, chemistry, and distance from the Sun. Mars, more commonly believed to be Earth's "sister planet," is more similar to Earth (than Venus) only in its length of day and seasons.

3  *What is the composition of the clouds surrounding Venus?* The clouds are composed primarily of sulfuric acid.

4  *Does Mars have liquid water on its surface today?* No, but there are strong indications that it had liquid water in the distant past.

5  *Is life known to exist on Mars today?* No current life has yet been discovered on Mars.

## Key Words

caldera, 162          greenhouse effect, 155          hot-spot volcanism, 162          retrograde rotation, 159          scarp, 151

## Key Ideas

✳ All four inner planets are composed primarily of rock and metal, and thus they are classified as terrestrial.

### Mercury

✳ Even at its greatest orbital elongations, Mercury can be seen from Earth only briefly after sunset or before sunrise.

✳ The *Mariner 10* spacecraft passed near Mercury in the mid-1970s, providing pictures of its surface. The Mercurian surface is pocked with craters like the Moon's, but extensive, smooth plains lie between these craters. Long cliffs meander across the surface of Mercury. These scarps probably formed as the planet cooled, solidified, and shrank.

✳ The long-ago impact of a large object formed the huge Caloris Basin on Mercury and shoved up jumbled hills on the opposite side of the planet.

✳ Mercury has an iron core much like that of the Earth.

### Venus

✳ Venus is similar to the Earth in size, mass, and average density, but it is covered by unbroken, highly reflective clouds that conceal its other features from Earth-based observers.

✳ While most of Venus's atmosphere is carbon dioxide, its dense clouds contain droplets of concentrated sulfuric acid mixed with yellowish sulfur dust. Active volcanoes on Venus may be a constant source of this sulfurous veil.

✳ Venus's exceptionally high temperature is caused by the greenhouse effect, as the dense carbon dioxide atmosphere traps and retains heat emitted by the planet. The surface pressure on Venus is 98 atm, and the surface temperature is 750 K. Both temperature and pressure decrease as altitude increases.

✳ The surface of Venus is surprisingly flat, mostly covered with gently rolling hills. There are two major "continents" and several large volcanoes. The surface of Venus shows evidence of local tectonic activity but not the large-scale motions that play a major role in continually reshaping the Earth's surface.

### Mars

✳ Earth-based observers found that the Martian solar day is nearly the same as that of the Earth, that Mars has polar caps that expand and shrink with the seasons, and that the Martian surface undergoes seasonal color changes.

✳ A century ago observers reported networks of linear features that many perceived as canals. These observations led to speculation about self-aware life on Mars.

✳ The Martian surface has numerous flat-bottomed craters, several huge volcanoes, a vast equatorial canyon, and dried-up riverbeds—but no canals formed by intelligent life. Flash-flood features and dry riverbeds on the Martian surface indicate that water once flowed there.

✳ Liquid water would quickly boil away in Mars's thin present-day atmosphere, but the planet's polar caps contain some frozen water, and a layer of permafrost may exist beneath the regolith.

✳ The Martian atmosphere is composed mostly of carbon dioxide. The surface pressure is less than 0.01 atm.

✳ Chemical reactions in the regolith together with ultraviolet radiation from the Sun apparently act to sterilize the Martian surface.

✳ Mars has no global magnetic fields, but local fields pierce its surface in at least nine places.

✳ Features that may be fossil remains of bacteria have been found in several meteorites that are believed to have come from Mars.

✳ Mars has two potato-shaped moons, the captured planetesimals Phobos and Deimos. Both are in synchronous rotation with Mars.

## Review Questions

1  Why is Mercury so difficult to observe? When is the best time to see the planet? *Tip:* The Discover for Yourself box, "The Inner Solar System," on page 152 can help.

2  Compare the surfaces of Mercury and our Moon. How are they similar? How are they different?

3  Compare the interiors of Mercury and Earth. How are they similar? How are they different?

4  What kind of large-scale feature is found on Mercury? Why are the examples of this feature probably much older than tectonic features on the Earth?

5  Briefly describe at least one theory explaining why Mercury has a large iron core.

6  Astronomers often refer to Venus as the Earth's twin. What physical properties do the two planets have in common? In what ways are the two planets dissimilar?

7  Why is it hotter on Venus than on Mercury?

8  What is the greenhouse effect? What role does it play in the atmospheres of Venus and the Earth?

9  What evidence exists for active volcanoes on Venus?

10  Describe the Venusian surface. What kinds of geological features would you see if you could travel around the planet?

11  Why do astronomers believe that Venus's surface was not molded by the kind of tectonic activity that shaped the Earth's surface?

12  Why is Mars red?

13  When is the best time to observe Mars from Earth? *Tip:* The Discover for Yourself box, "The Inner Solar System," on page 152 can help.

14  Compare the cratered regions of Mercury, the Moon, and Mars. Assuming that the craters on all three worlds originally had equally sharp rims, what can you conclude about the environmental histories of these worlds?

15  How would you tell which craters on Mars were formed by meteoritic impacts and which by volcanic activity?

16  Compare the volcanoes of Venus, Earth, and Mars. Do you think hot-spot volcanism is or was active on all three worlds?

17  What geologic features indicate plate tectonic activity once occurred on Mars? The lack of what features indicate that the effects of plate tectonics there have been limited?

18  What is the current knowledge concerning life on Mars? Do you think that today Mars is as barren and sterile as the Moon? Why or why not?

## Advanced Questions

19  What evidence do we have that the surface features on Mercury were not formed during recent geologic history?

20  Venus takes 440 days to move from greatest western elongation to greatest eastern elongation, but it needs only 144 days to go from greatest eastern elongation to greatest western elongation. With the aid of a diagram, explain why.

21  As seen from Earth, the brightness of Venus changes as it moves along its orbit. Describe the main factors that determine Venus's variations in brightness as seen from Earth. *Tip:* See the discussion of Venus in Chapter 2.

22  How might Venus's cloud cover change if all of Venus's volcanic activity suddenly stopped? How might these changes affect the overall Venusian environment?

23  Compare Venus's continents with those on the Earth. What do they have in common? How are they different?

24  Explain why Mars has the longest synodic period of all the planets, although its sidereal period is only 687 days.

25  With carbon dioxide accounting for about 95% of the atmospheres of both Mars and Venus, why do you suppose there is little greenhouse effect on Mars today?

26  Could the polar regions of Mars reasonably be expected to harbor life-forms, even though the Martian regolith is sterile at the *Viking* lander sites?

# Discussion Questions

27 If you were planning a return mission to Mercury, what features and observations would be of particular interest to you?

28 If you were designing a space vehicle to land on Venus, what special features would you think necessary? In what ways would this mission and landing craft differ from a spacecraft designed for a similar mission to Mercury?

29 Suppose someone told you that the *Viking* mission failed to detect life on Mars simply because the tests were designed to detect terrestrial life-forms, not Martian life-forms. How would you respond?

30 Compare the scientific opportunities for long-term exploration offered by the Moon and Mars. What difficulties would there be in establishing a permanent base or colony on each of these two worlds?

31 Imagine you are an astronaut living at a base on Mars. Describe your day's activities, what you see, the weather, the spacesuit you are wearing, and so on.

# What If ...

32 Mars rotated once every 20 days rather than once every 1.026 days? What would be different?

33 The carbon dioxide content of the Earth's atmosphere increased? What would happen to the Earth? This is not a completely hypothetical question, because carbon dioxide levels *are* increasing today.

# Observing Projects

34 Refer to the accompanying table to determine the dates of the next two or three greatest elongations of Mercury. Consult such magazines as *Sky & Telescope* and *Astronomy* or use your Starry Night planetarium software to see if any of these greatest elongations is going to be especially favorable for viewing the planet. If so, make plans to be one of those rare individuals who has actually seen the innermost planet of the solar system. Set aside several evenings (or mornings) around the date of the favorable elongation to reduce the chances of being "clouded out." Select an observing site that has a clear, unobstructed view of the horizon where the Sun sets (or rises). Make arrangements to have a telescope at your disposal. Search for the planet on the dates you have selected and make a drawing of its appearance through your telescope.

### Greatest Elongations of Mercury

| Eastern | Western |
| --- | --- |
| 1999 October 24 | 1999 December 3 |
| 2000 February 15 | 2000 March 28 |
| 2000 June | 2000 July 27 |
| 2000 October 6 | 2000 November 15 |
| 2001 January 28 | 2001 March 11 |
| 2001 May 22 | 2001 July 9 |
| 2001 September 18 | 2001 October 29 |
| 2002 January 11 | 2002 February 21 |
| 2002 May 4 | 2002 June 21 |
| 2002 September 1 | 2002 October 13 |
| 2002 December 26 | 2003 February 4 |

35 Refer to the table below to see if Venus is currently near a greatest elongation. If so, view the planet through a telescope. Make a sketch of the planet's appearance. From your sketch, can you determine whether Venus is closer to us or farther from us than the Sun?

### Greatest Elongations of Venus

| Eastern | Western |
| --- | --- |
| 1999 June 11 | 1999 October 30 |
| 2001 January 17 | 2001 June 8 |
| 2002 August 22 | 2003 January 11 |

36 Observe Venus through a telescope once a week for a month and make a sketch of the planet's appearance on each occasion. From your sketches, can you determine whether Venus is approaching us or moving away from us?

37 Consult such magazines as *Sky & Telescope* and *Astronomy* or use your Starry Night planetarium software to determine Mars's location among the constellations. If Mars is suitably placed for observation, arrange to view the planet through a telescope. Draw a picture of what you see. What magnifying power seems to give you the best image? Can you distinguish any surface features? Can you see a polar cap or dark markings? If not, can you offer an explanation for Mars's bland appearance?

38 Using your Starry Night software, try the activities in this chapter's Discover for Yourself box, "The Inner Solar System," to follow Mercury's orbit as it nears the Sun. With the timestep

still set at 1 sidereal day, describe the motion of Mercury relative to the equatorial coordinate grid (that is, the celestial sphere). What can you say about the maximum and minimum angles on the celestial sphere between Mercury and the Sun? Periodically, stop the motion, zoom in on Mercury, and determine its phase. Can you find a pattern of phases?

39 The Discover for Yourself box on page 152 will introduce you to Mars's motion. With Starry Night, notice that the equatorial coordinate system usually flows smoothly past Mars. Run the program with a timestep of 2 sidereal days until the coordinate grid behind Mars begins to slow down. Unlock Mars by moving it to the left quarter of the screen. Then run the program again. Describe how Mars now moves relative to the coordinate system. Stop the program, run it backward through this entire period, and stop it. Set the timestep at 1 sidereal day and rerun it forward in time, this time using the single-step feature. Make a day-by-day drawing of Mars's motion during this time. What is the name of this motion?

# The Outer Planets

## *In this chapter you will discover*

✳ Jupiter, an active, vibrant, multicolored world more massive than all the other planets combined

✳ Jupiter's diverse system of moons

✳ Saturn, with its spectacular system of thin, flat rings and numerous moons

✳ Uranus and Neptune: similar to each other but quite different from Jupiter and Saturn

✳ tiny Pluto and its moon, Charon, more like a bound pair of planets than a planet-moon system

**Saturn Through the Hubble Space Telescope**
Glowing aurorae crown the poles of Saturn. This is just one example of the features of the outer planets we are now able to view with current, powerful telescopes and from spacecraft traveling into the outer realm of the planets.

R I V U X G

## WHAT DO YOU THINK?

1  Is Jupiter a "failed star" or almost a star?

2  What is Jupiter's Great Red Spot?

3  Does Jupiter have continents and oceans?

4  Is Saturn the only planet with rings?

5  Are the rings of Saturn solid ribbons?

It's easy to imagine a trip to Mars, exploring its vast canyons and icy polar regions. Even Venus and Mercury are best understood by carefully comparing them to the Earth and the Moon. We find fewer similarities between the Earth and the outer planets. The aurora, shown on the photograph of Saturn opening this chapter, is similar in shape and origin to the aurorae seen on Earth (see Figure 5-11a). However, Jupiter, Saturn, Uranus, and Neptune are so much larger, are rotating so much faster, and have such different chemical compositions from our world that we will seem indeed to have left the Earth far behind. Pluto is so small that the tidal force of its moon, Charon, has set the planet into synchronous rotation, just as the Earth has permanently turned one face of our Moon toward us.

Nothing on Earth suggests the swirling red and brown clouds of Jupiter, the everchanging ring system of Saturn, the golden surface of the moon Io, or the blue-green clouds of Uranus and Neptune. (See the table "The Outer Planets: A Comparison.") We begin our exploration of the outer planets with Jupiter, an alien world of unsurpassed splendor.

## JUPITER

Even viewed through a small telescope on Earth, Jupiter's appearance sets it apart from the terrestrial worlds (Figure 7-1). As seen from the *Voyager 1* and *Voyager 2* spacecraft, which flew by Jupiter within months of each other in 1979, and from the Hubble Space Telescope today, Jupiter's multicolored bands create a world of breathtaking beauty. Figure 7-2 lists Jupiter's essential properties.

## 7-1 Jupiter's rotation helps create colorful, global weather patterns

More than 1300 Earths could be packed into the volume of Jupiter, the largest planet in the solar system. Using the orbital periods of its moons and Kepler's laws, astronomers have determined that Jupiter is 318 times more massive than Earth. This means that Jupiter has more than twice as much mass as all the other planets combined. Nevertheless, it would have to be 75 times more massive still before it could generate its own energy like the Sun and therefore be classified as a star.

What we see of Jupiter are the tops of clouds that permanently cover it (Figure 7-2). Because it rotates about once every ten hours—the fastest of any planet—Jupiter's clouds are in perpetual motion and are confined to narrow bands. In contrast, winds on slower-rotating Earth wander over vast ranges of latitude (compare Figure 5-1).

Jupiter's cloud bands provide a backdrop for ovals and turbulent swirls, which are storms similar in structure to hurricanes on Earth (Figure 7-3). On Jupiter, these storms last from hours to centuries. In 1998, two 50-year-old storms on Jupiter were observed to merge into a single storm as large across as the Earth's diameter.

R I V U X G

**Figure 7-1**    **Jupiter as Seen from Earth**  Many features of Jupiter's clouds are visible from Earth, including multicolored parallel regions and several light and dark spots. The Great Red Spot appears in the lower, central region of the Jovian atmosphere.

# The Outer Planets: A Comparison

| | Interior | Surface | Rings | Atmosphere | Magnetic Field* |
|---|---|---|---|---|---|
| **Jupiter** | Terrestrial core, liquid metallic shell, liquid hydrogen mantle | No solid surface, atmosphere gradually thickens to liquid state, belt and zone structure, hurricanelike features | Yes | Primarily H, He | 19,000 × Earth's total field, 14 × stronger than Earth's surface field |
| **Saturn** | Similar to Jupiter, with bigger terrestrial core and less metallic hydrogen | No solid surface, less distinct belt and zone structure than Jupiter | Yes | Primarily H, He | 570 × Earth's total field, $2/3$ × Earth's surface field |
| **Uranus** | Terrestrial core, liquid water shell, liquid hydrogen and helium mantle | No solid surface, weak belt and zone system, hurricanelike features, color from methane absorption of red, orange, yellow | Yes | Primarily H, He, some CH4 | 50 × Earth's total field, 0.7 × Earth's surface field |
| **Neptune** | Similar to Uranus | Like Uranus | Yes | Primarily H, He, some CH4 | 35 × Earth's total field, 0.4 × Earth's surface field |
| **Pluto** | Unknown | Apparently rock and ice | No | Unknown | Unknown |

For detailed numerical comparisons between planets, see Appendix Tables A-1 and A-2.

*To see the orientations of these magnetic fields relative to the rotation axes of the planets, see Figure 7-31.

# Jupiter's Vital Statistics

| | |
|---|---|
| Mass | $1.90 \times 10^{27}$ kg (318 $M_\oplus$) |
| Equatorial radius | 71,490 km (11.2 $R_\oplus$) |
| Average density | 1330 kg/m$^3$ |
| Orbital eccentricity | 0.048 |
| Sidereal period of revolution (year) | 11.86 Earth years |
| Average distance from Sun | $7.78 \times 10^8$ km (5.20 AU) |
| Equatorial rotation period | 9 h 50 min 28 s |
| Internal sidereal rotation period | 9 h 55 min 30 s |
| Albedo (average) | 0.51 |

R I V U X G

**Figure 7-2** **Jupiter as Seen from a Spacecraft** This view was sent back from *Voyager 1* in 1979. Features as small as 600 km across can be seen in the turbulent cloudtops of this giant planet. The complex cloud motions surrounding the Great Red Spot are clearly visible.

Computers show us how the cloud features on Jupiter would look if the planet's atmosphere were unwrapped like a piece of paper (Figure 7-4). You can also see ripples, plumes, and light-colored wisps in these images.

One rust-colored oval, a hurricanelike storm called the **Great Red Spot** (Figure 7-5), is so large that it can be seen through a small telescope (at present it is about 25,000 km long by 12,000 km wide). Two Earths could easily fit side by side inside the storm. Heat welling upward from inside Jupiter has maintained this storm for more than three centuries. How fortunate we are that Earth's atmosphere sustains storms for only a few weeks.

Giovanni Cassini, a European astronomer, first observed the Great Red Spot in 1665. Twenty-five years later, he noticed that the speeds of Jupiter's clouds vary with latitude, an effect called **differential rotation**. Near the poles, the rotation period of Jupiter's atmosphere, 9 h 55 min 30 s, is 5 minutes longer than at the equator.

Spectra from Earth-based telescopes and from the *Galileo* probe sent into Jupiter's upper atmosphere in 1995 give more detail about Jupiter's atmosphere. More than 86% of its mass is hydrogen and 13% is helium. The remainder consists of traces of simple compounds such as methane ($CH_4$), ammonia ($NH_3$), and water vapor ($H_2O$). The descent of the probe from the *Galileo*

R I V U X G

**Figure 7-3** **Close-up of Jupiter's Atmosphere** The dynamic winds and complex chemical composition of Jupiter's atmosphere create this beautiful and complex banded pattern.

a *Voyager 1* view

b *Voyager 2* view

R I V U X G

**Figure 7-4** **Jupiter's Northern and Southern Hemispheres** Computer processing shows the entire Jovian atmosphere from (**a**) *Voyager 1* and (**b**) *Voyager 2*. A computer was used to "unwrap" the planet's atmosphere. Notice in both views the dark ovals in the northern hemisphere and the white ovals in the southern hemisphere. The banded structure is absent near the poles. Notice, too, that the Great Red Spot moved westward while the white ovals moved eastward during the four months between the two *Voyager* flybys.

R I V U X G

**Figure 7-5** **The Great Red Spot** Turbulence surrounding the Great Red Spot is clearly seen in this view. When this picture was taken in 1979, the Great Red Spot measured about 20,000 km long and about 10,000 km wide. For comparison, the Earth's diameter is 12,756 km. The prominent white oval south of the Great Red Spot has been observed since 1938.

spacecraft into Jupiter's atmosphere revealed wind speeds of up to 600 km/h, higher than expected air density and temperature, and lower than expected concentrations of water, helium, neon, carbon, oxygen, and sulfur. This probably occurred because the probe descended into a particularly arid region of the atmosphere, akin to the air over a desert on Earth.

Observations and calculations over the past several decades indicate that Jupiter has three major cloud layers (Figure 7-6). The *Galileo* probe failed to detect them. Astronomers hypothesize that *Galileo* descended through a particularly cloud-free region.

The uppermost Jovian cloud layer is composed of crystals of frozen ammonia. These crystals and the frozen water in Jupiter's clouds are white, so what chemicals create the subtle tones of brown, red, and orange? Some scientists think that sulfur compounds, which can assume many different colors depending on their temperature, play an important role. Others think that phosphorus might be involved, especially in the Great Red Spot.

Astronomers first determined Jupiter's overall chemical composition from its average density—only 1330 kg/m$^3$ — which implies that Jupiter is composed primarily of lightweight elements such as hydrogen and helium surrounding a relatively small core of metal and rock. (Re-

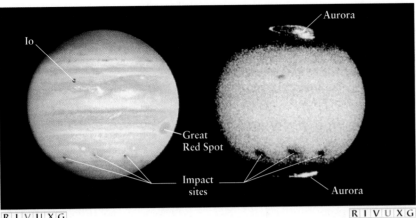

R I V U X G

Figure 7-9 **Three Impact Sites of Comet Shoemaker-Levy 9 on Jupiter**   Shown here are visible (left) and ultraviolet (right) images of Jupiter taken by the Hubble Space Telescope after three pieces of Comet Shoemaker-Levy 9 struck the planet. The astronomers had expected white remnants (the color of condensing ammonia or water vapor); the darkness of the impact sites may have come from carbon compounds in the comet debris. Note the aurorae in the ultraviolet image. Aurorae and lightning are common on Jupiter, due in part to the planet's strong magnetic fields and dynamic cloud motions.

R I V U X G

# JUPITER'S MOONS AND RINGS

Jupiter hosts at least 16 moons, a typical number for the four large planets. Galileo was the first person to observe the four largest moons, in 1610, seen through his meager telescope as pinpoints of light. He called them the "Medicean stars" to attract the attention of a wealthy Florentine patron of the arts and sciences. To Galileo, they provided evidence supporting the then-controversial Copernican cosmology; at that time, Western theologians asserted that all cosmic bodies orbited the Earth. The fact that the Medicean stars orbited Jupiter raised grave concerns in some circles.

To the modern astronomer, these moons are four extraordinary worlds, different both from the rocky terrestrial planets and from hydrogen-rich Jupiter. Now called collectively the **Galilean moons** or **Galilean satellites,** they are named after the mythical lovers and companions of the Greek god Zeus (called Jupiter by the early Romans). From the closest moon outward, they are Io, Europa, Ganymede, and Callisto.

These four worlds were photographed extensively by the *Voyager 1* and *Voyager 2* flybys and by the *Galileo* spacecraft (Figure 7-10). The two inner Galilean satellites, Io and Europa, are approximately the same size as our Moon. The two outer satellites, Ganymede and Callisto, are comparable in size to Mercury. Figure 7-10 presents comparative information about these six bodies.

## 7-4 Io's surface is sculpted by volcanic activity

Sulfury Io is among the most exotic moons in our solar system (Figure 7-11). With a density of 3570 kg/m³, it is neither terrestrial nor "Jovian" in chemical composition. It zooms through its orbit of Jupiter once every 1.8 days.

Like our Moon, Io is in synchronous rotation with its planet. Like the Earth, Io has a sizable iron core—it extends halfway out from the moon's center.

Images of Io reveal giant erupting volcanoes, such as the two labeled in Figure 7-11. These volcanoes are named after gods and goddesses associated with fire in Greek, Norse, Hawaiian, and other mythologies. Io also has numerous black "dots" on its surface, which apparently are dormant volcanic vents. Old lava flows radiate from many of these locations, which are typically 10 to 50 km in diameter and cover 5% of Io's surface.

**Insight into science** What's in a name? Following up on the preconceptions that common words create (see the Insight "Imagine the moon" in Chapter 6), objects with familiar names often have different characteristics than we expect. We tend to envision moons as inert, airless, lifeless, dry places similar to our Moon. As we will see throughout this chapter, starting with Io, some moons have active volcanoes, atmospheres, and probably vast, underground, liquid water oceans.

The existence of active volcanoes on Io had been predicted from analysis of the gravitational forces to which that moon is subjected. As it rapidly orbits Jupiter, Io repeatedly passes between Jupiter and one or another of the other Galilean satellites. These moons pull Io farther from Jupiter, changing the tidal forces acting on it from the planet. As the distance between Io and Jupiter varies, the resulting tidal stresses alternately squeeze and flex the moon. In turn, this constant tidal stressing heats Io's interior through friction, possibly generating as much energy inside Io as the detonation of 2400 tons of TNT every second. Molten rock eventually makes its way to the

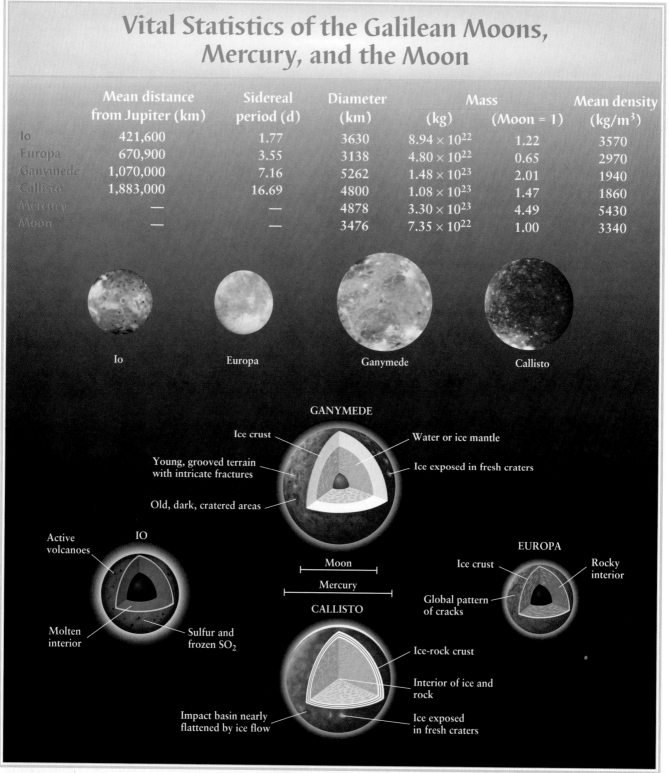

# Vital Statistics of the Galilean Moons, Mercury, and the Moon

| | Mean distance from Jupiter (km) | Sidereal period (d) | Diameter (km) | Mass (kg) | Mass (Moon = 1) | Mean density (kg/m³) |
|---|---|---|---|---|---|---|
| Io | 421,600 | 1.77 | 3630 | $8.94 \times 10^{22}$ | 1.22 | 3570 |
| Europa | 670,900 | 3.55 | 3138 | $4.80 \times 10^{22}$ | 0.65 | 2970 |
| Ganymede | 1,070,000 | 7.16 | 5262 | $1.48 \times 10^{23}$ | 2.01 | 1940 |
| Callisto | 1,883,000 | 16.69 | 4800 | $1.08 \times 10^{23}$ | 1.47 | 1860 |
| Mercury | — | — | 4878 | $3.30 \times 10^{23}$ | 4.49 | 5430 |
| Moon | — | — | 3476 | $7.35 \times 10^{22}$ | 1.00 | 3340 |

Io          Europa          Ganymede          Callisto

**GANYMEDE**
Ice crust
Water or ice mantle
Young, grooved terrain with intricate fractures
Ice exposed in fresh craters
Old, dark, cratered areas

Moon
Mercury

**IO**
Active volcanoes
Molten interior
Sulfur and frozen $SO_2$

**EUROPA**
Ice crust
Rocky interior
Global pattern of cracks

**CALLISTO**
Ice-rock crust
Interior of ice and rock
Impact basin nearly flattened by ice flow
Ice exposed in fresh craters

**Figure 7-10** **The Galilean Satellites** The four Galilean satellites are shown here to the same scale. Io and Europa have diameters and densities comparable to our Moon and are composed primarily of rocky material. Ganymede and Callisto are roughly as big as Mercury, but their low average densities indicate that each contains a thick layer of water and ice. These cross-sectional diagrams of the interiors of the Galilean moons show the probable internal structures of the four Galilean moons based on their average densities and on information from the *Galileo* mission.

R I V U X G

**Figure 7-11** **Io** This close-up view was taken by the *Galileo* spacecraft in 1997. Scientists believe that the range of colors results from surface deposits of sulfur ejected from Io's numerous volcanoes. The plume on the top, from the active Pillan Patera volcano, rises more than 190 km above the moon's surface. The plume from the volcano Prometheus rises up 100 km, casting a visible shadow. Prometheus has been active in every image taken of Io since the *Voyager* flybys of 1979.

moon's surface. Volcanoes and vents on Io eject roughly one trillion tons of matter each year.

Satellite instruments have identified sulfur and sulfur dioxide in the material erupting from Io's volcanoes. Sulfur is normally bright yellow. If heated and suddenly cooled, however, it forms molecules that assume a range of colors, from orange and red to black, which accounts for Io's tremendous range of colors (see Figure 7-11). Sulfur dioxide ($SO_2$) is an acrid gas commonly discharged from volcanic vents here on Earth and, apparently, on Venus. When eruptions on Io release this gas into the cold vacuum of space, this gas crystallizes into white flakes, which fall onto the surface and account for the moon's whitish deposits.

The *Galileo* spacecraft detected an atmosphere around Io. Composed of oxygen, sulfur, and sulfur dioxide, it is only one-billionth as dense as the air we breathe. Gases apparently ejected from Io's volcanoes have also been observed extending out into space and forming a doughnut-shaped region around Jupiter.

## 7-5 Europa may harbor liquid water below its surface

Images of Europa's ice and rock surface from *Voyager 2* and the *Galileo* spacecraft suggest that Jupiter's second-closest Galilean moon is likely to harbor liquid water (Figure 7-12). Europa orbits Jupiter every 3½ days and, like Io, is in synchronous rotation. The changing gravitational tug of Io creates stress inside Europa similar to the magma-generating distortion that Io undergoes. The

stress may create enough heat inside Europa to keep the water just a few kilometers below the moon's ice and rock surface in a liquid state.

Strong evidence for liquid water inside Europa comes from the ice floes seen in Figure 7-12. The surface features are similar to those seen in the Arctic region on Earth. The movement of the surface features creating the floes, along with swirls, strips, and ridges, appear to be driven by circulating water underneath the moon's surface, creating tectonic plate motion and tidal flexing. Also indicative of water is the moon's reddish color. The coloring may be due to salt deposits left after liquid water rose to the surface and evaporated.

*Galileo* spacecraft images suggest that some of Europa's features have moved within the past few million years,

R I V U X G

**Figure 7-12** **Europa** Imaged by the *Galileo* spacecraft, Europa's ice surface is covered by numerous streaks and cracks that give the satellite a fractured appearance. The streaks are typically 20 to 40 km wide. The upper inset shows the thin, disturbed icy crust of Europa, colored by dust from the impact crater Pwyll. The grooves and ridges in this image are typically 100 m across. The lower inset shows details of Pwyll Crater, 50 km in diameter.

and perhaps much more recently. Indeed, the *Galileo* spacecraft provides evidence for volcanism on Europa, strengthening the belief that this moon still has a liquid water layer. Replenishment of the surface by tectonic plate motion would explain why only a few small impact craters, such as crater Pwyll (see Figure 7-12), have survived.

Europa's average density of 2970 kg/m³ is slightly less than Io's. Half of its mass may be water. It also has a metallic core of much higher density. In 1995 astronomers discovered an extremely thin atmosphere containing mole-cular oxygen surrounding Europa. The density of this gas is about $10^{-11}$ times lower than the air we breathe. The oxygen may come from water molecules broken up on the moon's surface by ultraviolet radiation from the Sun.

## 7-6 Ganymede is larger than Mercury

Ganymede is the largest satellite in the solar system (Figure 7-13). Its diameter is larger than Mercury's, although its density of 1940 kg/m³ is much less than that of Mercury. Ganymede orbits Jupiter in synchronous rotation once every 7.2 days. Like its neighbor Europa, Ganymede has an iron-rich core, a rocky mantle, a weak magnetic field, a thin atmosphere, and a covering of dirty ice. It also features aurorae.

Like our Moon, Ganymede has two very different kinds of terrain. Dark, polygon-shaped regions are its oldest surface features, as judged by their numerous craters. Light-colored, heavily grooved terrain is found between the dark, angular islands. These lighter regions are much less cratered and therefore younger. Ganymede's grooved terrain consists of parallel mountain ridges up to 1 km high and spaced 10 to 15 km apart. These features suggest that the process of plate tectonics may have dominated Ganymede's early history. But unlike Europa, where tectonic activity still occurs today, tectonics on Ganymede bogged down three billion years ago as the satellite's crust froze solid.

One mechanism that may have created Ganymede's large-scale features is a bizarre property of water. Unlike most liquids, which shrink upon solidifying, water expands when it freezes. Seeping up through cracks in Ganymede's original crust, water thus forced apart fragments of that crust. This process could have produced jagged, dark islands of old crust separated by bands of younger, light-colored, heavily grooved ice.

## 7-7 Callisto bears the scars of a huge asteroid impact

Callisto is Jupiter's outermost Galilean moon. It orbits Jupiter in 16.7 days, and, like the other Galilean moons, Callisto's rotation is synchronous. Callisto is 91% as big and 96% as dense as Ganymede. It has a thin atmosphere of hydrogen and carbon dioxide. However, unlike the other Galilean moons, Callisto appears to be homogeneous inside, without an iron-rich core.

**Figure 7-13**    R I V U X G

**Ganymede** This side of Ganymede is dominated by a huge, dark, circular region called Galileo Regio, which is the largest remnant of Ganymede's ancient crust. Darker areas of the moon are older; lighter areas are younger, tectonically deformed regions. The white areas in and around some craters indicate the presence of water ice. **Inset:** The half-crater visible at the left of the inset is 19 km in diameter. Note in this *Galileo* spacecraft image the deep furrows in the moon's icy crust that probably resulted from crustal movement due to impacts and tectonic plate motion.

**Figure 7-14**    R I V U X G

**Callisto** The outermost Galilean satellite is almost exactly the same size as Mercury. Numerous craters pockmark Callisto's icy surface. Note the series of faint, concentric rings that cover the left third of the image. These rings outline a huge impact basin called Valhalla, which dominates the Jupiter-facing hemisphere of this frozen, geologically inactive world.

While numerous large impact craters are scattered over Callisto's dark, ancient, icy crust, it has very few craters smaller than 100 m across (Figure 7-14). Astronomers speculate that the smaller craters have disintegrated. Unlike Ganymede and Europa, Callisto has no younger, grooved terrain. The absence of grooved terrain suggests that tectonic activity never began there: The satellite simply froze too rapidly. It is bitterly cold on Callisto: *Voyager* instruments measured a noontime temperature of 155 K (−180°F), and the nighttime temperature plunges to 80 K (−315°F).

Callisto carries the cold, hard evidence of what happens when one astronomical body strikes another. *Voyager 1* photographed the huge impact basin, Valhalla, on Callisto (see Figure 7-14). An asteroid-sized object produced Valhalla Basin, which is located on Callisto's Jupiter-facing hemisphere. Like throwing a rock into a calm lake, ripples ran out from the impact site along Callisto's surface, cracking the surface and freezing into place for eternity. The largest remnant rings surrounding the impact crater have diameters of 3000 km. The probable interiors of the Galilean moons are shown in Figure 7-10.

## 7-8 Other debris orbits Jupiter as smaller moons and ringlets

Besides the four Galilean moons, Jupiter has at least 12 other moons and a set of tenuous ringlets. The non-Galilean moons are all irregular in shape and smaller

**Figure 7-15**    **Irregularly Shaped Amalthea**    Amalthea is actually more typical of Jupiter's moons than are the Galilean satellites. One of the four innermost satellites, Amalthea measures only 247 km along its longest dimension. The bright spot is inside a south polar crater named Gaea. Amalthea, discovered in 1892 by the American astronomer E. E. Barnard at the Lick Observatory, is named for the mythological goat that raised the infant Zeus.

than 250 km in diameter (Figure 7-15). Four of these moons are inside Io's orbit; the other eight are outside Callisto's orbit. The outer ones appear to be captured asteroids, while the inner ones are probably smaller pieces broken off a larger body.

As astronomers predicted, the cameras aboard *Voyager 1* discovered three ringlets around Jupiter, the brightest of which is seen in Figure 7-16, a *Voyager 2* image.

**Figure 7-16**    **Jupiter's Ring**    A portion of Jupiter's faint ring system, photographed by *Voyager 2*. The ring, which is closer to Jupiter than any of the planet's satellites, is probably composed of tiny rock fragments. The brightest portion of the ring is about 6000 km wide. The outer edge of the ring is sharply defined, but the inner edge is somewhat fuzzy. A tenuous sheet of material extends from the ring's inner edge all the way down to the planet's cloudtops.

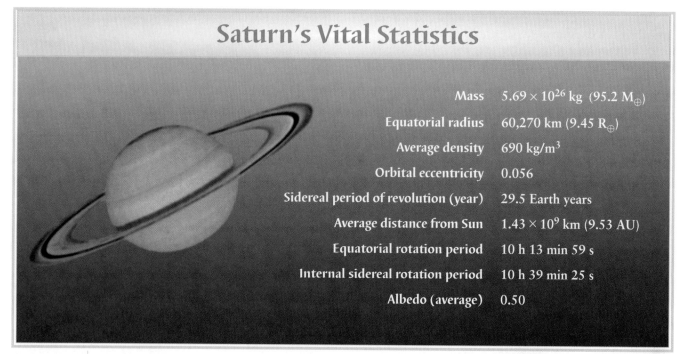

## Saturn's Vital Statistics

| | |
|---|---|
| Mass | $5.69 \times 10^{26}$ kg  (95.2 $M_\oplus$) |
| Equatorial radius | 60,270 km (9.45 $R_\oplus$) |
| Average density | 690 kg/m³ |
| Orbital eccentricity | 0.056 |
| Sidereal period of revolution (year) | 29.5 Earth years |
| Average distance from Sun | $1.43 \times 10^9$ km (9.53 AU) |
| Equatorial rotation period | 10 h 13 min 59 s |
| Internal sidereal rotation period | 10 h 39 min 25 s |
| Albedo (average) | 0.50 |

**Figure 7-17**    R I V U X G    **Saturn** *Voyager 2* sent back this image when the spacecraft was 34 million kilometers from Saturn. Note that you can see the planet through the gaps in the rings. This creates the illusion that the gaps are empty.

They are the darkest, simplest set of rings in the solar system. They consist of very fine dust particles that should be continuously kicked out of orbit by radiation from Jupiter and the Sun. Therefore, these rings are being replenished, probably by material from Io and the other moons. As we will see, Jupiter is only the first of four planets with rings.

## SATURN

Saturn, with its ethereal rings, presents the most spectacular image of all the planets (Figure 7-17). Giant Saturn has 95 times as much mass as the Earth, making it second in mass and size to Jupiter. Like Jupiter, Saturn has a thick, active atmosphere composed predominantly of hydrogen. It also has a strong magnetic field. Ultraviolet images from the Hubble Space Telescope reveal aurorae around Saturn like those seen around Jupiter (see the chapter-opening image and Figure 7-9). Saturn's vital statistics are listed in Figure 7-17.

### 7-9 Saturn's surface and interior are similar to those of Jupiter

Partly obscured by the thick, hazy atmosphere above them, Saturn's clouds lack the colorful contrast visible on

Jupiter. Nevertheless, photographs do show faint stripes in Saturn's atmosphere similar to Jupiter's belts and zones (Figure 7-18). Changing features there show that Saturn's

**Figure 7-18**    R I V U X G    **Belts and Zones on Saturn** *Voyager 1* took this view of Saturn's cloudtops at a distance of 1.8 million kilometers. Note that there is little swirling structure and substantially less contrast between belts and zones on Saturn than there is on Jupiter.

**Figure 7-19** **A New Storm on Saturn** An arrowhead-shaped storm was discovered on Saturn in July 1994. Located near Saturn's equator, it stretches 12,700 km across, making its diameter roughly the same as that of the Earth.

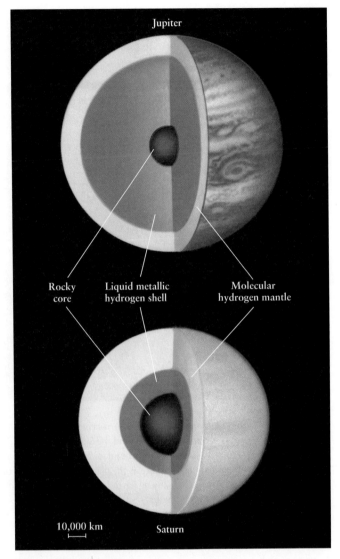

**Figure 7-20** **Cutaway of the Interiors of Jupiter and Saturn** The interiors of both Jupiter and Saturn are believed to have three regions: a terrestrial core, a metallic hydrogen shell, and a normal liquid hydrogen mantle. Their atmospheres are thin layers above the normal hydrogen, which boils upward, creating the belts and zones.

atmosphere, too, has differential rotation—ranging from 10 hours and 14 minutes at the equator to 10 hours and 40 minutes at high latitudes. Although Saturn lacks a long-lived spot like Jupiter's Great Red Spot, it does have storms, including a major new one discovered by the Hubble Space Telescope in November 1994 (Figure 7-19).

Astronomers infer that Saturn's interior structure resembles Jupiter's. A layer of molecular hydrogen just below the clouds surrounds a mantle of liquid metallic hydrogen and a solid, terrestrial core (Figure 7-20).

Because of its smaller mass, Saturn's interior is less compressed than Jupiter's. Saturn's rocky core is larger, and the pressure there is insufficient to convert as much hydrogen into a liquid metal. Saturn's rocky core is about 32,000 km in diameter, while its layer of liquid metallic hydrogen is 12,000 km thick (see Figure 7-20). To alchemists, Saturn was associated with the extremely dense element lead. This is wonderfully ironic in that at 690 kg/m$^3$, Saturn is the least dense body in the entire solar system.

## 7-10 Saturn's spectacular rings are composed of fragments of ice and ice-coated rock

Even when viewed from Earth through a small telescope (Figure 7-21), Saturn's magnificent rings are among the most spectacular objects in the nighttime sky. They are

tilted 27° from Saturn's plane of orbit, and so sometimes they are nearly edge-on to us, making them virtually impossible to see. Even when the rings are at their maximum tilt from our perspective, Saturn is so far away that our best Earth-mounted telescopes can reveal only their largest features. But even these have turned out to hold surprises. In 1675, Giovanni Cassini discovered one remarkable feature—a dark division in the rings. This 5000-km-wide gap, called the **Cassini division**, separates the dimmer **A ring** from the brighter **B ring**, which lies closer to the planet. By the mid-1800s, astronomers using improved telescopes detected a faint **C ring** just inside the B ring and barely visible in Figure 7-21. The Cassini division exists

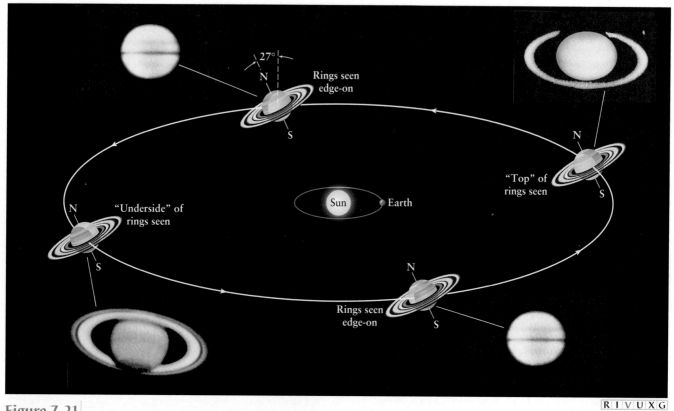

**Figure 7-21** **Saturn from the Earth** Saturn's rings are aligned with its equator, which is tilted 27° from the plane of Saturn's orbit around the Sun. Therefore, Earth-based observers see the rings at various angles as Saturn moves around its orbit. The plane of Saturn's rings and equator keep the same orientation in space as the planet goes around its orbit, just as the Earth does as it orbits the Sun. The accompanying Earth-based photographs show how the rings seem to disappear entirely about every 15 years.

R I V U X G

because the gravitational force from Saturn's moon Mimas combines with the gravitational force from the planet to keep the region clear of debris. Whenever matter drifts into the Cassini division, these two gravitational forces pull it out again, an effect called a **resonance.**

A second gap exists in the outer portion of the A ring (again, barely visible in Figure 7-21), named the **Encke division,** after the German astronomer Johann Franz Encke, who allegedly saw it in 1838. (Many astronomers have argued that Encke's report was erroneous, because his telescope was inadequate to resolve such a narrow gap.) The first undisputed observation of the 270-km-wide division was made by the American astronomer James Keeler in the late 1880s, with the newly constructed 36-in. refractor at the Lick Observatory in California. Unlike the Cassini division, the Encke division is kept clear because a small moon, Pan, orbits within it (Figure 7-22a).

Pictures from the *Voyager* spacecraft show that Saturn's rings are razor thin—less than 2 km thick according to recent estimates. This is quite amazing when you con-

sider that the total ring system has a width (from inner edge to outer edge) of more than 89,000 km.

Because Saturn's rings are also very bright, the particles that form them must be highly reflective. Astronomers had long suspected that the rings consist of ice and ice-coated rocks. Data from Earth-based observatories and the *Voyager* spacecraft have confirmed this suspicion. Because the temperature of the rings ranges from 93 K (−290°F) in the sunshine to less than 73 K (−330°F) in Saturn's shadow, frozen water is in no danger of melting or evaporating from the rings.

In order to determine the size of the particles in Saturn's rings, *Voyager* scientists measured the brightness of different rings from many angles as the spacecraft flew past the planet. They also measured changes in radio signals received from the spacecraft as it passed behind the rings. The largest particles in Saturn's rings are roughly 10 m across, although snowball-sized particles about 10 cm in diameter are more abundant than larger pieces. High-resolution images from *Voyager* revealed that the ring structure seen from Earth actually consists of hun-

a

b

**Figure 7-22** **Numerous Thin Ringlets Comprising Saturn's Rings**  **(a)** This *Voyager 2* image shows Saturn's tiny moon Pan orbiting within the Encke division. Pan exerts gravitational force on the ring particles to either side that help maintain the Encke division's 270-km width. **(b)** Details of Saturn's rings are visible in this photograph sent back by *Voyager 2*. It shows that the rings are actually composed of thousands of closely spaced ringlets. The colors are exaggerated by computer processing to show the different ringlets more clearly.

dreds upon hundreds of closely spaced thin bands, or **ringlets,** of particles (Figure 7-22b). Furthermore, myriad dust-sized particles fill the Cassini and Encke divisions.

The *Voyager* cameras also sent back the first high-quality pictures of the **F ring,** a thin set of ringlets just beyond the outer edge of the A ring. Two tiny satellites following orbits on either side of the F ring serve to keep the ring intact (Figure 7-23). The outer of the two satellites orbits Saturn at a slower speed than do the ice particles in the ring. As the ring particles pass near it, they receive a tiny, backward gravitational tug, which slows them down, causing

them to fall into orbits a bit closer to Saturn. Meanwhile, the inner satellite orbits the planet faster than the F ring particles. Its gravitational force pulls them forward and nudges the particles into a higher orbit. The combined effect of these two satellites is to focus the icy particles into a well-defined, narrow band about 100 km wide.

Because of their confining influence, these two moons, Prometheus and Pandora, are called **shepherd satellites.** Among the most curious features of the F ring is that the ringlets are sometimes braided and sometimes separate (Figure 7-24).

**Figure 7-23** **The F Ring and Its Two Shepherds**  Two tiny satellites, Prometheus and Pandora, each measuring about 50 km across, orbit Saturn on either side of the F ring. The gravitational effects of these two shepherd satellites confine the particles in the F ring to a band about 100 km wide.

**Figure 7-24** **Braided F Ring**  This photograph from *Voyager 1* shows several strands, each measuring roughly 10 km across, that comprise the F ring. The total width of the F ring is about 100 km.

Saturn's shell of liquid metallic hydrogen produces a planetwide magnetic field that apparently affects its rings. Saturn's slower rotation and much smaller volume of liquid metallic hydrogen produce a magnetic field only about 3% as strong as Jupiter's (but still 570 times stronger than Earth's). Data from spacecraft show that Saturn's magnetosphere contains radiation belts similar to those of Earth. Furthermore, dark **spokes** move around Saturn's rings (Figure 7-25); these are believed to be created by the magnetic field, which lifts charged particles out of the plane in which the rings orbit. Spreading the particles out decreases the light scattered from them and therefore makes the rings appear darker.

## 7-11 Titan has a thick, opaque atmosphere rich in nitrogen, methane, and other hydrocarbons

Saturn has more moons than any other planet, and only 7 of its 22 known moons are spherical. The rest are oblong, suggesting that they are captured asteroids. Saturn's largest moon, Titan, is second in size to Ganymede among the moons of the solar system, and the only moon to have a dense atmosphere. About ten times more gas lies above each square meter of Titan's surface than lies above each square meter of the Earth.

**Figure 7-26**    R I V U X G    **Titan**  This view of Titan was taken by *Voyager 2*. Very few features are visible in the thick, unbroken haze that surrounds this large satellite. The main haze layer is located nearly 300 km above Titan's surface.

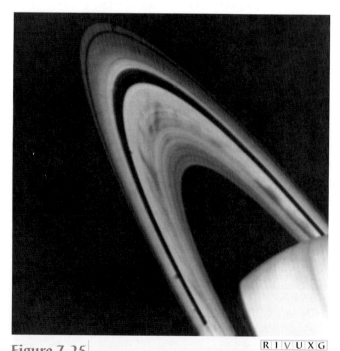

**Figure 7-25**    R I V U X G    **Spokes in Saturn's Rings**  Believed to be caused by Saturn's magnetic field temporarily lifting particles out of the ring plane, these dark regions move around the rings like the spokes on a rotating wheel.

Christiaan Huygens discovered Titan in 1655, the same year he proposed that Saturn has rings. By the early 1900s, several scientists had begun to suspect that Titan might have an atmosphere, because it is cool enough and massive enough to retain heavy gases.

Because of its atmosphere, Titan was a primary target for the *Voyager* missions. To everyone's surprise, the *Voyager*s spent hour after precious hour sending back featureless images (Figure 7-26). Unexpectedly, Titan's thick cloud cover completely blocked any view of its surface. The same dense haze allows little sunlight to penetrate—that moon's surface must be a dark, gloomy place.

*Voyager* data indicated that roughly 90% of Titan's atmosphere is nitrogen. Most of this nitrogen probably formed from the breakdown of ammonia ($NH_3$) by the Sun's ultraviolet radiation into hydrogen and nitrogen atoms. Because Titan's gravity is too weak to retain hydrogen, this gas has escaped into space, leaving behind ample nitrogen. The unbreathable atmosphere is about four times as dense as Earth's.

The second most abundant gas on Titan is methane, a major component of natural gas. Sunlight interacting with methane induces chemical reactions that produce a variety of other carbon-hydrogen compounds, or **hydrocarbons.** For example, spacecraft have detected small amounts of ethane ($C_2H_6$), acetylene ($C_2H_2$), ethylene ($C_2H_4$), and propane ($C_3H_8$) in Titan's atmosphere.

Ethane, the most abundant of these compounds, condenses into droplets as it is produced and falls to Titan's surface to form a liquid. Enough ethane may exist to create rivers, lakes, and even oceans on Titan. Nitrogen combines with these hydrocarbons to produce other compounds. Although one of these compounds, hydrogen cyanide (HCN), is a poison, some of the others are the building blocks of life's organic molecules.

Some molecules can join together in long, repeating molecular chains to form substances called **polymers,** of which plastics are the best known example. Many of the hydrocarbons and carbon-nitrogen compounds in Titan's atmosphere can form such polymers. Scientists hypothesize that droplets of lighter polymers remain suspended in Titan's atmosphere to form a mist, while heavier polymer particles settle down onto Titan's surface. As a result, any land that protrudes above Titan's ethane oceans is probably covered with a thick layer of sticky, tarlike goo. The Hubble Space Telescope observed further evidence for different surface features on Titan in 1994 (Figure 7-27).

We have little reason to suspect that life exists on Titan, however; its surface temperature of 95 K (−288°F) is prohibitively cold. Nevertheless, a more detailed study of the chemistry of Titan may shed light on the origins of life on Earth.

**Figure 7-27**    R I V U X G    **Surface Features on Titan**  The Hubble Space Telescope, using an infrared camera, was able to peer through most of Titan's haze (see Figure 7-26). The nature of the surface features is not yet known.

# URANUS

Uranus and its largest moons are so far from the Sun (19.2 AU) that from Earth they appear to be a small cluster of stars. No wonder they seem fixed in the heavens: A Uranian year equals 84 Earth years. Since its discovery in 1781, Uranus has only orbited the Sun just over 2½ times.

## 7-12  Uranus sports a hazy atmosphere and clouds

Until early 1996, observations indicated that Uranus, the fourth most massive planet in our solar system, has few notable features. The 1986 visit of *Voyager 2* to Uranus revealed a remarkably featureless world. It took the Hubble Space Telescope's infrared camera to find what *Voyager*'s visible light camera could not: Uranus has a system of belts and zones. Its hydrogen atmosphere has traces of methane along with a high-altitude haze under which is clear air and everchanging clouds that dwarf the typical cumulus clouds we see on Earth. They are towering and huge, each typically as large as Europe (Figure 7-28). Along with the rest of the atmosphere, the clouds go around the planet once every 16½ hours.

Uranus contains 14½ times as much mass as the Earth and is 4 times bigger in diameter (see Figure 7-28). Its outer layers are composed predominantly of gaseous hydrogen and helium. The temperature in the upper atmosphere of the planet is so low (about 73 K, or −330°F) that the methane and water there condense to form clouds of ice crystals. Because methane freezes at a lower temperature than water, methane forms higher clouds over Uranus. Methane efficiently absorbs red light, giving Uranus its blue-green color.

Earth-based observations show that Uranus rotates once every 17 hours 14 minutes on an axis of rotation that lies very nearly in the plane of the planet's orbit. With its axis of rotation tilted over by 98°, Uranus is one of only three planets with retrograde rotation (Venus and Pluto are the other two). From *Voyager* photographs, planetary scientists have concluded that each of the five largest Uranian moons has probably had at least one shattering impact. A catastrophic collision with an Earth-sized object may also have knocked Uranus on its side, as we see it today.

As Uranus orbits the Sun, its north and south poles alternately point almost directly toward or directly away from the Sun, producing exaggerated seasons (Figure 7-29). In the summertime, near Uranus's north pole, the Sun is almost directly overhead for many Earth years, at which time southern latitudes are subjected to a continuous, frigid winter night. Forty-two Earth years later, the situation is reversed. For more on this, see the essay "What If … Earth's Axis Tilted on the Ecliptic?" on page 40.

**Uranus's Vital Statistics**    **Neptune's Vital Statistics**

| | Uranus | | Neptune |
|---|---|---|---|
| Mass | $8.68 \times 10^{25}$ kg (14.5 $M_\oplus$) | Mass | $1.02 \times 10^{26}$ kg (17.1 $M_\oplus$) |
| Equatorial radius | 25,559 km (4.01 $R_\oplus$) | Equatorial radius | 24,764 km (3.88 $R_\oplus$) |
| Average density | 1290 kg/m³ | Average density | 1640 kg/m³ |
| Orbital eccentricity | 0.047 | Orbital eccentricity | 0.009 |
| Sidereal period of revolution (year) | 84.0 Earth years | Sidereal period of revolution (year) | 164.8 Earth years |
| Average distance from Sun | $2.87 \times 10^9$ km (19.2 AU) | Average distance from Sun | $4.50 \times 10^9$ km (30.1 AU) |
| Equatorial rotation period | 16 h 30 min (retrograde) | Equatorial rotation period | 19 h 6 min |
| Interior sidereal rotation period | 17 h 14 min (retrograde) | Interior sidereal rotation period | 16 h 7 min |
| Albedo (average) | 0.66 | Albedo (average) | 0.62 |

R I V U X G

**Figure 7-28** **Uranus, Earth, and Neptune** These images of Uranus, the Earth, and Neptune are to the same scale. Uranus and Neptune are quite similar in mass, size, and chemical composition. Both planets are surrounded by thin, dark rings, quite unlike Saturn's, which are broad and bright. The clouds on the right of Uranus are each the size of Europe.

From its mass and density (1290 kg/m³), astronomers conclude that Uranus's interior has three layers. The outer 30% of the planet is liquid hydrogen and helium, the next 40% inward is highly compressed liquid water (with some methane and ammonia), and the inner 30% is a rocky core (Figure 7-30). Indirect evidence for the water layer comes from the apparent deficiency of ammonia on Uranus. This gas dissolves easily in water, which would explain the scarcity of ammonia in the planet's atmosphere.

*Voyager 2* passed through the magnetosphere of Uranus, revealing that the planet's magnetic field is 50 times stronger than that of the Earth. That strength is reasonable, considering the planet's mass and rotation rate, but everything else about the magnetic field is extraordinary. It is remarkably tilted—59% from its axis of rotation—and does not even pass through the center of the planet (Figure 7-31).

Because of the large angle between the magnetic field of Uranus and its rotation axis, the magnetosphere of Uranus wobbles considerably as the planet rotates. Such a rapidly changing magnetic field will help us in explaining pulsars, a type of star we will study in Chapter 12.

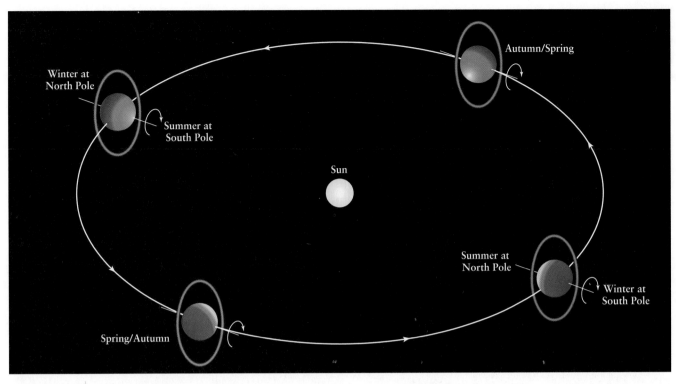

**Figure 7-29** **Exaggerated Seasons on Uranus** Uranus's axis of rotation is tilted so steeply that it lies nearly in the plane of its orbit. Seasonal changes on Uranus are thus greatly exaggerated. For example, during midsummer at Uranus's south pole, the Sun appears nearly overhead for many Earth years, while the planet's northern regions are subjected to a long, continuous winter night. Half an orbit later, the seasons are reversed.

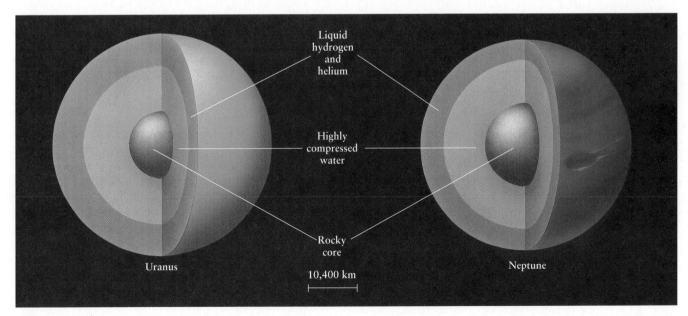

**Figure 7-30** **Cutaways of the Interiors of Uranus and Neptune** The interiors of both Uranus and Neptune are believed to have three regions: a terrestrial core surrounded by a liquid water mantle, which is surrounded in turn by liquid hydrogen and helium. Their atmospheres are thin layers at the top of their hydrogen and helium layers.

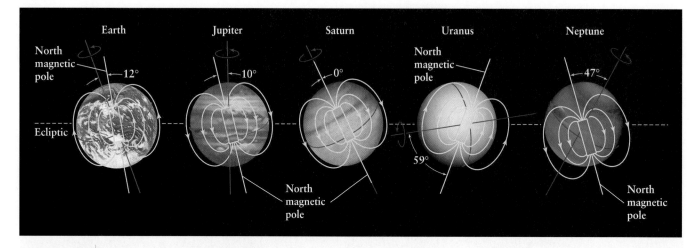

**Figure 7-31** **The Magnetic Fields of Five Planets** This drawing shows how the magnetic fields of Earth, Jupiter, Saturn, Uranus, and Neptune are tilted relative to their rotation axes. Note that the magnetic fields of Uranus and Neptune are offset from the centers of the planets and steeply inclined to their rotation axes. Jupiter, Saturn, and Neptune have north magnetic poles on the hemisphere where Earth has its south magnetic pole.

## 7-13 A system of rings and satellites revolves around Uranus

Nine of Uranus's thin, dark rings were discovered accidentally in 1977 when Uranus passed in front of a star. The star's light was momentarily blocked by each ring, thereby revealing their existence to astronomers (Figure 7-32).

Blocking the light of a more distant object, such as the star here, by something between it and us, such as Uranus's rings, is called an **occultation**. A picture taken while *Voyager* was in Uranus's shadow revealed more thin rings (Figure 7-33).

Most of Uranus's moons, like its rings, orbit in the plane of the planet's equator. Five of these satellites, rang-

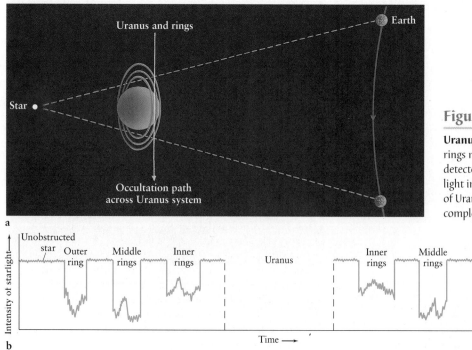

**Figure 7-32** **Discovery of the Rings of Uranus** (a) Light from a star is reduced as the rings move in front of it. (b) With sensitive light detectors, astronomers can detect the variation in light intensity. Such dimming led to the discovery of Uranus's rings. Of course, the star vanishes completely when Uranus occults it.

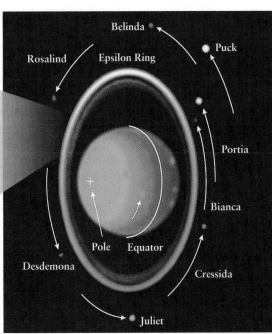

**Figure 7-33** **The Rings and Moons of Uranus** This image of Uranus, its rings, and eight of its moons was taken by the Hubble Space Telescope. **Inset:** Close-up of part of the ring system taken by *Voyager 2* when the spacecraft was in Uranus's shadow looks back toward the Sun. Numerous fine dust particles between the main rings gleam in the sunlight. Uranus's rings are much darker than Saturn's, and this long exposure revealed many very thin rings and dust lanes. The short streaks are star images blurred because of the spacecraft's motion during the exposure.

ing in diameter from 480 to nearly 1600 km, were known before the *Voyager* mission. However, *Voyager*'s cameras discovered ten additional satellites, each fewer than 50 km across. Several of these tiny, irregularly shaped moons are shepherd satellites whose gravitational pull confines the particles within the thin rings that circle Uranus.

The smallest of Uranus's five main satellites, Miranda, is the most fascinating and bizarre of its 17 known moons. Unusual wrinkled and banded features cover Miranda's surface (Figure 7-34). Its highly varied terrain suggests that it was once seriously disturbed. Perhaps a shattering impact temporarily broke it into several pieces that then recoalesced, or perhaps severe tidal heating, as we saw on Io, moved large pieces of its surface.

Miranda's core originally consisted of dense rock, while its outer layers were mostly ice. If a powerful impact did occur, blocks of debris broken off from Miranda drifted back together through mutual gravitational attraction. Recolliding with that moon, they formed a chaotic mix of rock and ice. In this scenario, the landscape we see today on Miranda is the result of huge, dense rocks trying to settle toward the satellite's center, forcing blocks of less dense ice upward toward the surface.

# NEPTUNE

Neptune is physically similar to Uranus (review Figures 7-28 and 7-30). Neptune has 17.1 times the Earth's mass, 3.88 times Earth's diameter, and a density of 1640 kg/m³. Unlike Uranus, however, cloud features can readily be

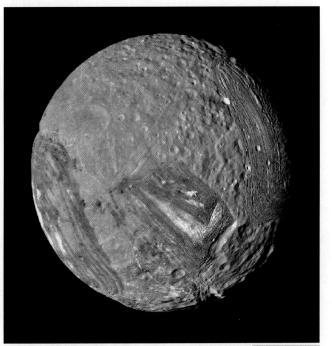

**Figure 7-34** **Miranda** The patchwork appearance of Miranda in this mosaic of *Voyager 2* images suggests that this satellite consists of huge chunks of rock and ice that came back together after an ancient, shattering impact by an asteroid or a neighboring Uranian moon. The curious banded features that cover much of Miranda are parallel valleys and ridges that may have formed as dense, rocky material sank toward the satellite's core. At the very bottom of the image—where a "bite" seems to have been taken out of the satellite—is a range of enormous cliffs that jut upward as high as 20 km, twice the height of Mount Everest.

discerned on Neptune. Its whitish, cirruslike clouds consist of methane ice crystals. The methane absorbs red light, leaving the planet's belts and zones with a banded, bluish appearance (Figure 7-35). Like Jupiter, the atmosphere of Neptune also experiences differential rotation. The winds on Neptune blow as fast as 2000 km/h—among the fastest in the solar system.

## 7-14 Neptune was discovered because it had to be there

Neptune's discovery is storied because it illustrates a scientific prediction leading to an expected discovery. In 1781 the British astronomer William Herschel discovered Uranus. Its position was carefully plotted, and by the 1840s, it was clear that even considering the gravitational effects of all the known bodies in the solar system, Uranus was not following the path predicted by Newton's and Kepler's laws. Either the theories behind these laws were wrong, or

**Figure 7-35** **Neptune's Banded Structure.** Several Hubble Space Telescope images at different wavelengths were combined to create this enhanced-color view of Neptune. The dark blue and light blue areas are the belts and zones, respectively. The dark belt running across the middle of the image lies just south of Neptune's equator. White areas are high-altitude clouds, presumably of methane ice. The very highest clouds are shown in yellow-red, as seen at the very top of the image. The green belt near the south pole is a region where the atmosphere absorbs blue light, perhaps indicating some differences in chemical composition.

there had to be another, yet-to-be-discovered body in the solar system pulling on Uranus.

Independent, nearly simultaneous calculations by an English mathematician, John Adams, and a French astronomer, Urbain Leverrier, predicted the same location for the alleged planet. That planet, Neptune, was located in 1846 by the German astronomer Johann Galle, within a degree or two of where it had to be to have the observed influence on Uranus.

In August 1989, nearly 150 years after its discovery, *Voyager 2* arrived at Neptune to cap one of NASA's most ambitious and successful space missions. Scientists were overjoyed at the detailed, close-up pictures and wealth of data about Neptune sent back to Earth by the spacecraft.

> **Insight into science** **Process and progress** Scientific theories make testable predictions (see Chapter 1). Based on the details of Uranus's orbit around the Sun, Newton's law of gravitation predicted that Uranus's orbit was being affected by the gravitational attraction of another planet. The law also predicted where that planet was located, leading to the discovery of Neptune.

At the time *Voyager 2* passed it, a giant storm raged in Neptune's atmosphere. Called the **Great Dark Spot**, it was about half as large as Jupiter's Great Red Spot. The Great Dark Spot (Figure 7-36) was located at about the same latitude on Neptune and occupied a similar proportion of Neptune's surface as the Great Red Spot does on Jupiter. Although these similarities suggested that similar mechanisms created the spots, the Hubble Space Telescope in 1994 showed that the Great Dark Spot had disappeared. Then, in April 1995, another storm developed in the opposite hemisphere.

Neptune's interior is believed to be very similar in composition and structure to that of Uranus: a rocky core surrounded by ammonia- and methane-laden water (see Figure 7-30). Also, as with Uranus, Neptune's magnetic axis, the line connecting its north and south magnetic poles, is tilted sharply from its rotation axis. In this case, the tilt is 47%. Again like Uranus, Neptune's magnetic axis does not pass through the center of the planet (see Figure 7-31).

## 7-15 Neptune has rings and has captured most of its moons

Like Uranus, Neptune is surrounded by a system of thin, dark rings (Figure 7-37). It is so cold at these distances

**Figure 7-36** **Neptune** This view from *Voyager 2* looks down on the southern hemisphere of Neptune. The Great Dark Spot, whose diameter at the time was about the same size as the Earth's diameter, is near the center of this picture. It has since vanished. Note the white, wispy methane clouds.

from the Sun that both planets' ring particles retain methane ice. Scientists speculate that eons of radiation damage have converted this methane ice into darkish carbon compounds, thus accounting for the low reflectivity of the rings.

Neptune has eight known moons. Seven have irregular shapes and highly elliptical orbits, which suggest that Neptune captured them. Triton, discovered in 1846, is spherical and was quickly observed to have a nearly circular, retrograde orbit around Neptune. It is difficult to imagine how a satellite and planet could form together but rotate in opposite directions. Indeed, only a few of the small outer satellites of Jupiter and Saturn have retrograde orbits, and these bodies are probably captured asteroids. Some scientists have therefore suggested that Triton may have been captured three or four billion years ago by Neptune's gravity.

After being captured, Triton most likely began in a highly elliptical orbit. However, with each revolution, the changing distance to Neptune would have stretched and flexed Triton. This activity would have made its orbit more circular. It would also have provided enough energy to melt much of the satellite's interior and obliterate Triton's original surface features, including craters. Triton's south polar region is shown in Figure 7-38. Note that very few craters are visible.

Triton does exhibit some surface features seen on other icy worlds, such as long cracks resembling those on

**Figure 7-37** **Neptune's Rings** Two main rings are easily seen in this view alongside overexposed edges of Neptune. Careful examination also reveals a faint inner ring. A fainter still sheet of particles, whose outer edge is located between the two main rings, extends inward toward the planet.

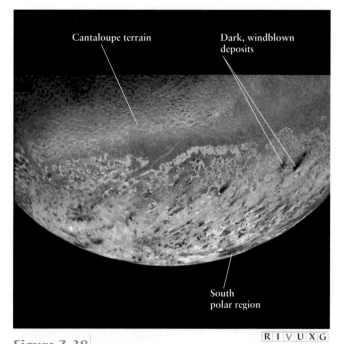

Cantaloupe terrain

Dark, windblown
deposits

South
polar region

**Figure 7-38**    R I V U X G

**Triton's South Polar Cap** Approximately a dozen high-resolution *Voyager 2* images were combined to produce this view of Triton's southern hemisphere. The pinkish polar cap is probably made of nitrogen frost. A notable scarcity of craters suggests that Triton's surface was either melted or flooded by icy lava after the era of bombardment that characterized the early history of the solar system.

Europa and Ganymede. Other features unique to Triton are quite puzzling. For example, near the top of Figure 7-38, you can see a wrinkled terrain that resembles the skin of a cantaloupe. Triton also has a few frozen lakes like the one shown in Figure 7-39. Some scientists have speculated that these lakelike features are the calderas of extinct ice volcanoes. A mixture of methane, ammonia, and water, which can have a melting point far below that of pure water, could have formed a kind of cold lava on Triton.

It is unlikely that any lava is flowing on Triton today, however, because the satellite is so very cold. *Voyager* instruments measured a surface temperature of 36 K (−395°F), making Triton the coldest world that our probes have ever visited. Nevertheless, *Voyager* cameras did glimpse two towering plumes of gas extending up to 8 km above the satellite's surface. Perhaps these plumes are nitrogen gas escaping through vents or fissures warmed by the feeble summer Sun.

In the same way that our Moon raises tides on Earth, Triton raises tides on Neptune. Whereas the tides on Earth cause our Moon to spiral outward, the tides on Neptune cause Triton (in its retrograde orbit) to spiral inward. Within the next quarter of a billion years, Triton will move close enough to Neptune for the planet to create tides on its moon's solid surface high enough to pull

Triton apart. Pieces of Triton will then literally float into space until the entire moon is demolished! By destroying Triton, Neptune will create a new ring system that will be much more substantial than its present one.

## PLUTO AND BEYOND

In 1930 the astronomer Clyde W. Tombaugh was searching for a giant planet that was allegedly affecting Neptune's orbit. Instead, Tombaugh discovered tiny Pluto (Figure 7-40), a planet far too small to have a noticeable gravitational effect on Neptune. Figure 7-41 lists Pluto's essential data.

Tombaugh recognized Pluto as a planet because it moved among the background stars from night to night, but he had no idea how strange its orbit is compared with those of the other planets. As discussed in Foundation II (see Figure II-8), Pluto's orbit is so elliptical that it is sometimes closer to the Sun than Neptune, as it was from 1979 to 1999. It is now farther away from the Sun than Neptune and will continue to be so for about the next 230 years.

Pluto's orbit is tilted with respect to the plane of the ecliptic more than any other planet. Refined observations show that Neptune's orbit is not being affected by another giant planet. No such planet has ever been found.

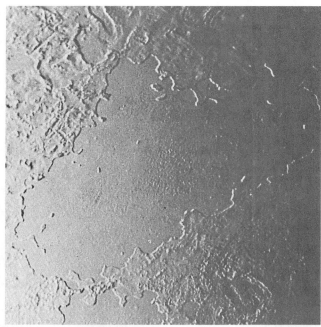

**Figure 7-39**    R I V U X G

**A Frozen Lake on Triton?** Some scientists believe that this lakelike feature is the caldera of an ice volcano. The flooded basin is about 200 km wide and 400 km long, an area about the size of West Virginia.

R I V U X G

**Figure 7-40** **Discovery of Pluto** Pluto was discovered in 1930 by searching for a dim, starlike object that slowly moves against the background stars. These two photographs were taken one day apart.

## 7-16 Pluto and its moon Charon are about the same size

Pluto was little understood for half a century until astronomers noticed that its image sometimes appears oblong (Figure 7-42). This observation led to the discovery of Pluto's only known moon, Charon, in 1978. From 1985 through 1990, the orbit of Charon was oriented so that Earth-based observers could watch it eclipse Pluto. Astronomers used observations of these eclipses to determine that Pluto is only twice as broad as its satellite: Its diameter is 2380 km and Charon's is 1190 km.

The Pluto-Charon system is more like a pair of planets in close orbit around each other than a typical planet-moon system. The average distance between Charon and Pluto is less than 1/20 the distance between the Earth and our Moon. Furthermore, Pluto always keeps the same side facing Charon. This is the only case in which a moon and a planet both have synchronous rotation with respect to each other. As seen from the satellite-facing side of Pluto,

# Pluto's Vital Statistics

| | |
|---|---|
| Mass | $1.30 \times 10^{22}$ kg ($0.002\ M_{\oplus}$) |
| Radius | 1190 km ($0.19\ R_{\oplus}$) |
| Average density | 1800 kg/m$^3$ |
| Orbital eccentricity | 0.249 |
| Sidereal period of revolution (year) | 248.6 Earth years |
| Average distance from Sun | $5.91 \times 10^9$ km (39.5 AU) |
| Sidereal rotation period | 6.39 Earth days (retrograde) |
| Albedo (average) | 0.50 |

R I V U X G

**Figure 7-41** **Pluto and Its Vital Statistics** This Hubble Space Telescope image of Pluto shows little detail but indicates that the major features of Pluto's surface each cover large amounts of its surface area.

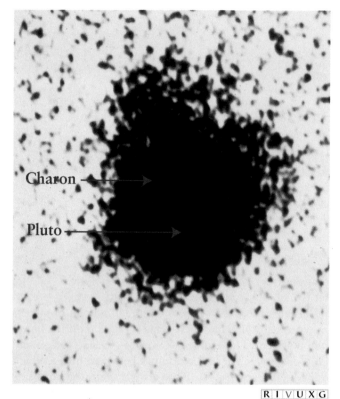

<br/>

Figure 7-42 | **Discovery of Charon**  Long ignored as just a defect in the photographic emulsion, the bump on the upper left side of this image of Pluto led astronomer James Christy to discover the moon Charon.

R I V U X G

Figure 7-43 | **Pluto and Charon**  This picture of Pluto and its moon, Charon, is a composite of Hubble Space Telescope images. It shows the greatest detail we have of Pluto's surface. Pluto and Charon are separated here by only 19,700 km. **Inset:** A map covering 85% of Pluto's surface. The contrast between dark and light strongly suggests regions covered with ice and regions covered with rocky material.

Charon neither rises nor sets, but instead hovers in the sky, perpetually suspended above the horizon.

The exceptional similarities between Pluto and Charon suggest that this binary system may have formed when Pluto collided with a body of similar size. Perhaps chunks of matter were stripped from this second body, leaving behind a mass, now called Charon, that was vulnerable to capture by gravity. Alternatively, perhaps gravity captured Charon into orbit during a close encounter between the two worlds.

Both of these are unlikely scenarios. For either to be feasible, many Plutolike objects must have existed in the outer regions of the young solar system. One astronomer estimates that there must have been at least a thousand Plutolike bodies in order for a collision or close encounter between two of them to have occurred at least once since the solar system formed 4.6 billion years ago.

The best pictures of Pluto and Charon were taken by the Hubble Space Telescope (Figure 7-43). Like Neptune's moon Triton, these worlds are probably composed of nearly equal amounts of rock and ice. The surface of Pluto shows more large-scale features than any object in the solar system other than the Earth. Examination of Hubble Space Telescope images has identified some 12 distinct regions.

## 7-17  Our solar system extends beyond Pluto

Ever since Pluto was discovered, astronomers have searched for other objects in our solar system. Neptune's captured moons support the idea of other bodies at the outskirts of the solar system, and astronomers have begun discovering them. These distant masses of ice and rock, which we will discuss further in Chapter 8, lie in a region now called the *Kuiper belt*, which is believed to extend out 500 AU from the Sun.

🚀 **Insight into science   Paradigm shift** Usually it takes overwhelming evidence contradicting established beliefs before scientists begin looking at old ideas in completely new ways. Such changes in fundamental beliefs are sometimes called *paradigm shifts.* The acceptance of tectonic plate motion is one such change. Until the past few years, astronomers took it for granted that Pluto is a planet. Perhaps another paradigm shift in the near future will reclassify tiny Pluto as the largest object in the Kuiper belt.

## WHAT DID YOU THINK?

1  *Is Jupiter a "failed star" or almost a star?* Jupiter has 75 times too little mass to shine as a star.

2  *What is Jupiter's Great Red Spot?* The Great Red Spot is a long-lived, oval cloud circulation similar to a hurricane on Earth.

3  *Does Jupiter have continents and oceans?* No. Jupiter is surrounded by a thick atmosphere primarily of hydrogen and helium that becomes liquid as one moves inward. The only solid matter in Jupiter is its core.

4  *Is Saturn the only planet with rings?* No. Four planets (Jupiter, Saturn, Uranus, and Neptune) have rings.

5  *Are the rings of Saturn solid ribbons?* Saturn's rings are all composed of thin, closely spaced ringlets consisting of particles of ice and ice-coated rocks. If they were solid ribbons, Saturn's gravitational tidal force would tear them apart.

## Key Words

| | | |
|---|---|---|
| A ring, 185 | F ring, 187 | polymer, 189 |
| B ring, 185 | Galilean moon (satellite), 179 | resonance, 186 |
| belt, 177 | Great Dark Spot, 194 | ringlet, 187 |
| C ring, 185 | Great Red Spot, 175 | shepherd satellite, 187 |
| Cassini division, 185 | hydrocarbon, 188 | spoke, 188 |
| differential rotation, 175 | liquid metallic hydrogen, 177 | zone, 177 |
| Encke division, 186 | occultation, 192 | |

## Key Ideas

### Jupiter and Saturn

✳ Jupiter is by far the largest and most massive planet in the solar system.

✳ Jupiter and Saturn probably have rocky cores surrounded by a thick layer of liquid metallic hydrogen and an outer layer of ordinary liquid hydrogen. Both planets have an overall chemical composition very similar to that of the Sun.

✳ The visible features of Jupiter exist in the outermost 100 km of its atmosphere. Saturn has similar features, but they are much fainter. Three cloud layers exist in the upper atmospheres of both Jupiter and Saturn. Because Saturn's cloud layers extend to a greater altitude, the colors of the Saturnian atmosphere appear muted.

✳ The colored ovals visible in the Jovian atmosphere represent gigantic storms, some of which (such as the Great Red Spot) are stable and persist for years or even centuries.

✳ Jupiter and Saturn have strong magnetic fields created by electric currents in the metallic hydrogen layer.

✳ Four large satellites orbit Jupiter. The two inner Galilean moons, Io and Europa, are roughly the same size as our Moon. The two outer moons, Ganymede and Callisto, are approximately the size of Mercury.

✳ Io is covered with a colorful layer of sulfur compounds deposited by frequent explosive eruptions from volcanic vents. Europa is covered with a smooth layer of frozen water crisscrossed by an intricate pattern of long cracks.

✳ The heavily cratered surface of Ganymede is composed of frozen water with large polygons of dark, ancient crust separated by regions of heavily grooved, lighter-colored, younger terrain. Callisto has a heavily cratered ancient crust of frozen water.

✳ Saturn is circled by a system of thin, broad rings lying in the plane of the planet's equator. Each major ring is composed of a great many narrow ringlets consisting of numerous fragments of ice and ice-coated rock.

## Uranus and Neptune

✳ Uranus and Neptune are quite similar in appearance, mass, size, and chemical composition. Each has a rocky core probably surrounded by a dense, watery mantle; the axes of their magnetic fields are steeply inclined to the axis of rotation; and both planets are surrounded by systems of thin, dark rings.

✳ Uranus is unique in that its axis of rotation lies nearly in the plane of its orbit, producing greatly exaggerated seasons on the planet.

✳ Uranus has five moderate-sized satellites, the most bizarre of which is Miranda.

✳ The largest satellite of Neptune, Triton, is an icy world with a tenuous nitrogen atmosphere. Triton moves in a retrograde orbit that suggests it was captured into orbit by Neptune's gravity, and its orbit is spiraling down toward Neptune.

## Pluto and Beyond

✳ Pluto, the smallest planet in the solar system, and its satellite, Charon, are icy worlds that may well resemble Triton.

✳ Other objects orbit the Sun beyond the orbit of Pluto. Dozens of these Kuiper belt objects have been observed.

## Review Questions

1  Describe the appearance of Jupiter's atmosphere. Which features are long-lived and which are fleeting?

2  What causes the belts and zones in Jupiter's atmosphere?

3  What is liquid metallic hydrogen? Which planets appear to contain this substance? What produces this form of hydrogen?

4  Compare and contrast the surface features of the four Galilean satellites, discussing their geologic activity and their evolution.

5  What energy source powers Io's volcanoes?

6  Why are numerous impact craters found on Ganymede and Callisto but not on Io or Europa?

7  Describe the structure of Saturn's rings. What are they made of?

8  Why do features in Saturn's atmosphere appear to be much fainter and "washed out" compared with features in Jupiter's atmosphere?

9  Explain how shepherd satellites operate. Is "shepherd satellite" an appropriate term for these objects? Explain.

10 Describe Titan's atmosphere. What effect has sunlight presumably had on Titan's atmosphere?

11 Describe the seasons on Uranus. Why are the Uranian seasons different from those on any other planet?

12 Briefly describe the evidence supporting the idea that Uranus was struck by a large planetlike object several billion years ago.

13 Why are Uranus and Neptune distinctly bluer than Jupiter and Saturn?

14 Compare the ring systems of Saturn and Uranus. Why were Uranus's rings unnoticed until the 1970s?

15 How do the orientations of Uranus's and Neptune's magnetic axes differ from those of the other planets?

16 Suppose you were standing on Pluto. Describe the motions of Charon relative to the horizon. Under what circumstances would you never see Charon?

17 Describe the circumstantial evidence supporting the idea that Pluto is one of a thousand similar icy worlds that once occupied the outer regions of the solar system.

18 Explain why Triton will never collide with Neptune, even though Triton is spiraling toward that planet.

19 Discovery of Pluto's moon, Charon, could have been made decades before it actually was. What caused the delay?

## Advanced Questions

20 Consult the Internet or such magazines as *Sky & Telescope* and *Science News* to determine what space missions are now under way. What data and pictures have they sent back that updates information presented in this chapter?

21 Long before the *Voyager* flybys, Earth-based astronomers reported that Io appeared brighter than usual for a few hours after emerging from Jupiter's shadow. From what we know about the material ejected from Io's volcanoes, explain this brief brightening of Io.

22 Compare and contrast Valhalla on Callisto with the Caloris Basin on Mercury.

23 Can we infer from naked-eye observations that Saturn is the most distant of the planets visible without a telescope? Explain.

24 As seen by Earth-based observers, the intervals between successive edge-on presentations of Saturn's rings alternate between 13 years 9 months and 15 years 9 months. Why do you think these two intervals are not equal?

25 Compare and contrast the internal structures of Jupiter and Saturn with the internal structures of Uranus and Neptune. Can you propose an explanation for the differences between these two pairs of planets?

26 Neptune has the third largest mass of all the planets, but Uranus has the third largest diameter. Reconcile these two facts.

## Discussion Questions

27 Suppose that you were planning a mission to Jupiter employing an airplanelike vehicle that would spend many days, even months, flying through the Jovian clouds. What observations, measurements, and analyses should this aircraft make? What dangers might it encounter, and what design problems would you have to overcome?

28 Speculate on the possibility that Europa, Ganymede, or Callisto might harbor some sort of marine life.

29 Suppose you were planning separate missions to each of Jupiter's Galilean moons. What questions would you want these missions to answer, and what kinds of data would you want your spacecraft to send back? Given the different environments on the four satellites, how would the designs of the four spacecraft differ?

30 NASA and the Jet Propulsion Laboratory have tentative plans to place spacecraft in orbit about Uranus and Neptune early in the twenty-first century. What kinds of data should be collected and what questions would you like to see answered by these missions?

31 Would you expect the surfaces of Pluto and Charon to be heavily cratered? Explain.

## What If ...

32 Jupiter, at its present location, were a star? What would Earth be like? *Hint:* Recall that to be a star, Jupiter would have to have 75 times more mass than it has today.

33 Jupiter had formed at one-third its present distance of 5.2 AU from the Sun? What would Earth be like?

34 Io were struck by another object of similar size? *Hint:* You can create a variety of different scenarios by imagining the impacting body striking from different directions and with different speeds and at different angles.

## Observing Projects

35 Consult such magazines as *Sky & Telescope* and *Astronomy* or your Starry Night software to determine whether Jupiter is currently visible in the night sky. If so, make arrangements to view the planet through a telescope. What magnifying power seems to give you the best view? Draw a picture of what you see. Can you see any belts and zones? How many? Can you see the Great Red Spot?

36 Observe Jupiter through a pair of binoculars. Can you see all four Galilean moons? Make a drawing of what you observe. To identify the moons, do Project 37.

37 Use the zoom feature of your Starry Night software to enlarge each of the Galilean moons as they appear tonight.

Turn off the timestep feature of the software by pressing the square on the lower left (timestep) window. Keeping each moon centered on the computer screen, you can make it big enough to see surface features. In this way, identify the moons by comparing their images on the screen and in this book.

38 Observe Jupiter through a telescope on three or four consecutive nights. Make a drawing each night showing the positions of the Galilean satellites relative to Jupiter. Record the time and date of each observation. To identify the moons, do Project 37 or consult the section called "Satellites of Jupiter" in the *Astronomical Almanac* for the current year to see if you can identify the satellites by name.

39 Consult such magazines as *Sky & Telescope* and *Astronomy* or your Starry Night software to determine whether Saturn is currently visible in the night sky. If so, view Saturn through a small telescope. Make a sketch of what you see. Estimate the angle at which the rings are tilted to your line of sight. Can you see the Cassini division? Can you see any belts or zones in Saturn's clouds? Do you observe a faint, starlike object near Saturn that might be Titan? What observations could you perform to test whether the starlike object is a Saturnian satellite? If Saturn is not visible, perform the observations described in this problem using your Starry Night software.

40 Determine whether Uranus or Neptune is currently visible in the night sky. If so, make arrangements to view them through a telescope. To help you find these planets, use your Starry Night software or the star chart published each January in *Sky & Telescope* showing the paths of Uranus and Neptune against the background stars. For more detailed images than you can get through a telescope, locate these planets with your Starry Night software and zoom to high resolution. Also, locate Triton with the software.

# We Lived on a Metal-Poor Earth?

Earth provides a wealth of building blocks necessary for the development and evolution of complex life-forms. More than 80 elements on or near Earth's surface combine in countless ways essential for its diversity of flora and fauna. Especially important for life on Earth are metals such as iron. The human body typically contains more than 3 grams of iron, mostly in the form of hemoglobin that helps transport oxygen through the bloodstream. A slight iron imbalance leads to anemia or toxicity. And without abundant metals, most of our technologies would never have developed.

But what if the solar system formed from an interstellar gas cloud containing fewer metals and high-mass elements, that is, a solar system with less iron, nickel, copper, and other metals crucial for life as we know it?

**Metal-Rich versus Metal-Poor** Earth formed with almost $6 \times 10^{21}$ metric tons of star matter, almost one-third of it iron. If the solar system formed from interstellar gas containing relatively few heavy elements, Earth would contain a much lower fraction of such elements as uranium, lead, iron, and nickel. The Earth would be comprised of a correspondingly higher fraction of lower-mass elements, such as carbon, nitrogen, oxygen, silicon, and aluminum. That version of Earth—let's call it Lithia—would be profoundly different from our planet.

**The Environment of Lithia** Without heavy elements like iron, Lithia's density, gravity, and magnetic fields are much lower than Earth's. Let's assume that the density of Lithia is similar to that of the Moon. (We'll assume that the Moon doesn't change in its characteristics or its distance from Lithia.)

If Lithia has the same radius as Earth, the force of gravity on Lithia's surface is only 60% that on Earth. That is, you would weigh 40% less on Lithia. Mobile life-forms that evolve on the metal-poor planet require much less muscle strength to get about and less bone density to resist the planet's gravity.

With few radioactive elements to provide heat, the core of Lithia cooled off quickly compared with the core of Earth. As a result, Lithia has much less heat flowing upward through the mantle and crust, so there most likely is no plate tectonic activity. Lithia also has less volcanic activity. But without plate motion to spread the lava flows, volcanoes will grow to higher elevations than on Earth. The lack of crustal motions also means that pockets of high-density material will not rise from below, forming the metal ores mined on Earth.

The Moon's orbit and the tides it induces are different, too. With Lithia having 60% of Earth's mass, the Moon at its present distance orbits once every 36 days instead of once every 29.5 days. Thus, the cycle of lunar phases is longer, but the cycle of lunar-induced tides is slightly shorter. Surprisingly, the tides have roughly the same height, because the tidal effects depend only on the Moon's gravity and not Earth's, and the distance between Earth and Lithia and their satellite is the same.

**Life on Lithia** The low abundance of iron and the resulting lack of a planetary magnetic field might preclude the development of complex life-forms on Lithia. And yet, the incredible diversity of life we find on Earth suggests that evolution on Lithia might well occur. If advanced life-forms evolve on Lithia before the Sun uses up its nuclear fuel and becomes a red giant, the chemical differences between Lithia and Earth dictate that they will differ profoundly from the complex life we find on Earth.

Without the heat of radioactive elements to keep Lithia's core molten for billions of years and with fewer metals, Lithia will have a much weaker magnetic field—if any at all—than Earth. Inhabitants can expect aurorae to grace the night sky continuously. On the other hand, high-energy particles and the radiation they create in the atmosphere will bombard any life-forms on the surface continuously and will also adversely affect the ozone layer. A larger amount of the Sun's harmful ultraviolet radiation will reach the surface. Without protection, life as we know it could not survive.

**A Barren Landscape** With a thin atmosphere and an abundance of lighter elements such as silicon (silicon dioxide is sand), Lithia might resemble a barren desert here on Earth.

R I V U X G

**8**

# Vagabonds of the Solar System

*In this chapter you will discover*

✳ asteroids and meteoroids, pieces of interplanetary rock and metal

✳ comets, debris that contains large amounts of ice

✳ space debris that falls through the Earth's atmosphere

✳ that an impact from space 65 million years ago probably caused the extinction of the dinosaurs—and most other life on Earth

✳ that a wayward asteroid could again threaten life on Earth

**Comet Hyakutake**
A comet is almost always named after the person who first sees it. The Japanese amateur astronomer Yuji Hyakutake first observed the comet named for him using binoculars on the morning of January 30, 1996. This photograph was taken two months later, in April 1996, when the comet extended more than 30° across the sky—more than 60 times the diameter of the full Moon as seen from Earth. Comet Hyakutake passed within 0.1 AU (15 million kilometers or 9 million miles) of Earth.

## WHAT DO YOU THINK?

1  Are the asteroids a planet that was somehow destroyed?

2  How far apart are the asteroids on average?

3  Why do comets have tails?

4  In which direction does a comet tail point?

5  What is a shooting star?

The formation of the solar system was not a tidy affair. Planetesimals collided by the billions to form the planets we see today and their moons. For nearly a billion years, pieces of smaller debris struck often, scarring and shaping their surfaces. This era of frequent collisions ended some 3.8 billion years ago, but innumerable debris still orbits the Sun, and dramatic impacts continue even today. Large, rocky bodies occasionally pass startlingly close to the Earth—closer even than the Moon. Myriad small ones penetrate the atmosphere daily. Comets weave glorious trails across the sky. Icy bodies such as these may once have provided Earth with water and other material essential to the evolution of life.

These leftovers, the vagabonds of the solar system, are the asteroids, meteoroids, and comets. We have already seen pieces of Comet Shoemaker-Levy 9 plunge into Jupiter. You may have spotted Comet Hale-Bopp during the summer of 1997. What do the characteristics of these interplanetary travelers reveal about the material in the solar system and about how life began here?

# ASTEROIDS

We know that the solar system formed from a rotating disk of gas and dust. The matter that had too much angular momentum to fall onto the protosun coalesced at varying distances into planetesimals. Many of these chunks of rock and metal eventually collided, forming the planets and larger moons. Others were captured whole by various planets as small, irregularly shaped moons, like Phobos and Deimos around Mars. However, numerous planetesimals still orbit the Sun today in splendid isolation. These are the **asteroids,** sometimes called *minor planets.*

## 8-1 Most asteroids orbit the Sun between Mars and Jupiter

On New Year's Day 1801, the Sicilian astronomer Giuseppi Piazzi was carefully mapping faint stars in the constellation of Taurus. He noticed a dim, previously uncharted "star" that shifted its position slightly over the next several nights. Uranus, Neptune, and Pluto were all discovered in the same way. Was Piazzi's object another planet? The answer is no. His lucky sighting was too small to qualify as a full-fledged planet. Rather, he was the first to discover an asteroid.

Later that year the orbit of this object was determined to lie between Mars and Jupiter. At Piazzi's request, the object was named Ceres (pronounced see-reez), after the patron goddess of Sicily. Ceres is spherical like the planets (Figure 8-1), but its diameter is a scant 940 km, only one-quarter the diameter of our Moon.

Earth

Moon

Ceres

R I V U X G

**Figure 8-1** **Comparison of Ceres with the Moon and the Earth** Ceres, the Moon, and the Earth are shown here to scale. Ceres, shown in this infrared photo (Earth and Moon appear in visible light) is the largest asteroid and is so small that it is not considered a planet. Because it does not orbit a body other than the Sun, it is also not classified as a moon.

In 1802 the German astronomer Heinrich Olbers discovered another faint, starlike object that moved against the background stars. He called it Pallas, after the Greek goddess of wisdom. Like Ceres, Pallas orbits the Sun in a low eccentricity (nearly circular) orbit between the orbits of Mars and Jupiter. Pallas is even dimmer and smaller than Ceres, with a diameter of only 600 km.

Only two more of these minor planets—Juno and Vesta—were found until the mid-1800s, when telescopes improved. Astronomers then began to stumble across many more asteroids between the orbits of Mars and Jupiter orbiting the Sun at distances of between 2 and 3½ AU. This region of the solar system is now called the **asteroid belt** (Figure 8-2). Asteroids whose orbits lie entirely within this region are called **belt asteroids.**

The next real breakthrough came in 1891, when the German astronomer Max Wolf applied photographic techniques to the search for asteroids. A total of 300 asteroids had been found up to that time, each painstakingly discovered by scrutinizing the skies for faint, uncharted "stars" whose positions shifted slowly from one night to the next. With the advent of astrophotography, however, the floodgates were opened. Astronomers could simply aim a camera-equipped telescope at the stars and take long exposures. If an asteroid happened to be in the field

of view, it left a distinctive trail on the photographic plate (Figure 8-3). Using this technique, Wolf alone discovered 228 asteroids.

By February 1999, official numbers had been assigned to 9999 asteroids, with more than 90,000 other observations awaiting confirmation as new asteroids. Astronomers estimate that more than a billion asteroids exist in the solar system.

**Insight into science**  **Confirming observations** New findings must be confirmed or replicated by other competent scientists before discoveries and observations are accepted by the scientific community. Therefore, asteroid observations need to be repeated and repeated in order to determine exact orbits and to eliminate the possibility that a sighting is of a known asteroid or other object.

Ceres alone accounts for about 30% of the mass of all the known asteroids combined. Only three asteroids—Ceres, Pallas, and Vesta—have diameters greater than 300 km. Thirty other asteroids have diameters between 200 and 300 km, and 200 more are bigger than

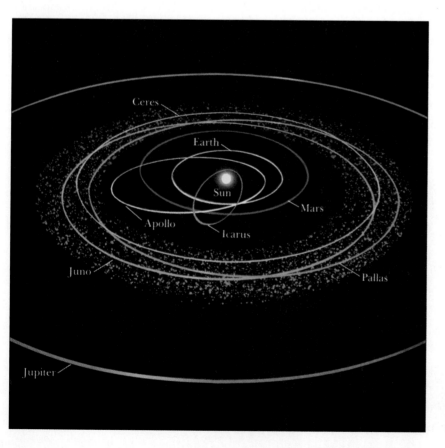

**Figure 8-2** **Asteroid Orbits** Most asteroids orbit the Sun in a 1½-AU-wide belt between the orbits of Mars and Jupiter. The orbits of belt asteroids Ceres, Pallas, and Juno are indicated. Some asteroids, such as Apollo and Icarus, have highly eccentric paths that cross Earth's orbit. Others, called the Trojan asteroids, follow the same orbit as Jupiter.

**Figure 8-3** **Discovering Asteroids** Two asteroids observed in a single photograph. Asteroids are often discovered via such long-exposure photographs.

periods, like the one in Figure 8-4. Note the gaps at simple fractions ( ⅓, ⅖, 3/7, and ½) of Jupiter's orbital period.

To understand the Kirkwood gaps, imagine an asteroid circling the Sun once every 5.93 years, exactly half of Jupiter's orbital period. On every second trip around the Sun, the asteroid lines up between Jupiter and the Sun. In other words, the asteroid always comes nearest to Jupiter at the very same point in its orbit. There, Jupiter's pull on the asteroid is particularly strong, and over time it pulls the asteroid into an orbit further from the Sun. This periodic gravitational influence, called a *resonance* (see Chapter 7), may even eject the asteroid from the belt.

According to Kepler's third law, a period of 5.93 years corresponds to a semimajor axis of 3.28 AU. Because of Jupiter, no asteroids orbit the Sun at this average distance. Similarly, there is a gap corresponding to an orbital period of one-third Jupiter's period, or 3.95 years. Additional gaps exist for other simple relationships between the orbital periods of asteroids and Jupiter.

Although it is likely that the asteroid belt contains millions of asteroids, their average separation is a staggering 10 million kilometers. This is quite unlike the image that has been created by innumerable science fiction movies of asteroids so close together that you must dodge them as you fly past.

100 km across. The vast majority of asteroids are less than 1 km across.

You may have heard the common belief that the asteroids were once all a single planet that was somehow destroyed. It is much more likely that Jupiter's gravitational force pulling on the planetesimals in the region of the asteroid belt by differing amounts made it virtually impossible for large numbers of asteroids to meet, coalesce, and form a single object in that region. (We consider Jupiter's gravitational effect in more detail in the next section.)

In fact, if all the asteroids had once been part of a single body, it would have had a diameter of only 1500 km, or 12% of the Earth's diameter. This is less than two-thirds the diameter of Pluto and half the diameter of our Moon. Because Pluto just barely makes it into the category of a planet, a single body containing the entire asteroid mass would not qualify as a planet.

## 8-2 Jupiter's gravity creates gaps in the asteroid belt

In 1867 the American astronomer Daniel Kirkwood called attention to gaps in the asteroid belt. These features, called **Kirkwood gaps,** show the influence of Jupiter's gravitational attraction, as best seen in a graph of asteroid orbital

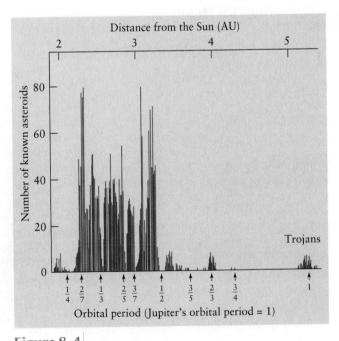

**Figure 8-4** **The Kirkwood Gaps** This graph displays the number of asteroids at various distances from the Sun. Notice that few orbital periods of asteroids correspond to such simple fractions as ⅓, ⅖, 3/7, and ½ of Jupiter's orbital period. Repeated alignments with Jupiter have deflected asteroids away from these orbits. The Trojan asteroids accompany Jupiter as it orbits the Sun.

**Insight into science    Get real** Scientists continually apply the "laws of nature" to new situations, problems, observations, and experiments, even to popular culture. For example, applying Newton's law of gravity to the asteroids reveals that they could never swarm, as science fiction movies suggest. At those close quarters, their gravity would cause them either to collide or to pass so close together that they would actually fly rapidly and permanently apart.

Nevertheless, over the 4.6 billion years that the solar system has existed, the gravitational influences of Mars and Jupiter have surely sent some asteroids careening into each other. In 1918 the Japanese astronomer Kiyotsugu Hirayama drew attention to groups of asteroids that share nearly identical orbits. These are fragments of parent asteroids.

A collision between kilometer-sized asteroids must be an awesome event. Typical collision velocities are estimated to be 3600 to 18,000 km/h (2000 to 11,000 mph), which is more than sufficient to shatter rock. In some collisions, the resulting fragments may not have enough speed to escape from each other's gravitational attraction, and thus they reassemble. Alternatively, several large fragments may end up orbiting or in contact with each other. The asteroid Toutatis appears to be composed of two comparably sized pieces connected to each other, as does the asteroid Castalia.

In the early 1990s the Jupiter-bound *Galileo* spacecraft passed near two asteroids—Gaspra (see Figure II-10) and Ida (Figure 8-5)—and sent back close-up views. Both

R I V U X G

**Figure 8-5** **Ida and Its Moon** The 55-km-long rocky asteroid Ida, shown here with its moon Dactyl, is about twice the size of the younger asteroid Gaspra (see Figure II-10). **Inset:** Dactyl is also heavily cratered.

asteroids are probably fragments of larger parent bodies that were broken apart by catastrophic collisions. Because Ida's surface is more heavily cratered than Gaspra's, Ida is much older.

At least two asteroids have their own moons: Ida is accompanied by Dactyl, and the moon of Dionysus has yet to be named. Dactyl is a pockmarked asteroid some 1.5 km in diameter that orbits Ida at a distance of 100 km. Dionysus's moon appears to be 0.5 km across, orbiting only a few kilometers from the larger body.

Asteroids are varied in composition and characteristics. For example, the 1997 visit of the *Near Earth Asteroid Rendezvous (NEAR)* spacecraft with asteroid Mathilde (Figure 8-6) revealed it to be only 1.3 times denser than water and to have an albedo of 0.04, making it darker than charcoal. This heavily cratered, carbon-rich body therefore has only half the density of the other asteroids we have studied, such as Ida. Observations indicate that there are many gaps in Mathilde's interior, but we do not yet know whether these are due to collisions that broke the asteroid into bits that recoalesced or whether it formed with lots of cracks originally.

## 8-3 Asteroids exist outside the asteroid belt

While Jupiter's gravitational pull depletes certain orbits within the asteroid belt, it actually captures asteroids at two locations in the path of its own orbit. The gravitational forces of the Sun and Jupiter work together to hold asteroids in orbit at these locations, called **stable Lagrange points,** in honor of the French mathematician Joseph Lagrange, whose calculations explained them. One Lagrange point is located 60° ahead of Jupiter, and the other is 60° behind, as shown in Figure 8-7.

The asteroids trapped at Jupiter's Lagrange points are called **Trojan asteroids,** each named after a hero of the Trojan War. Approximately 460 Trojan asteroids have been catalogued so far. Closer to home, asteroid number 3753 has also been discovered at one of Earth's Lagrange points.

Some asteroids have highly elliptical orbits that bring them into the inner regions of the solar system (see Figure 8-2). Others have similarly elliptical orbits that extend from the asteroid belt out beyond the farthest reaches of Pluto's orbit. Others, the *Amor asteroids*, cross Mars's orbit, while still others, called **Apollo asteroids,** even cross Earth's orbit.

The Apollo asteroid Eros passed within 23 million kilometers of our planet in 1931 (Figure 8-8). On October 30, 1937, the asteroid Hermes passed within 900,000 km of Earth—only a little more than twice the distance to the Moon. On June 14, 1968, asteroid Icarus passed Earth at

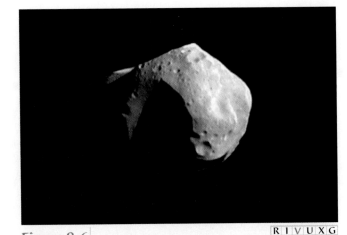

**Figure 8-6**    **Asteroid Mathilde**  Reflecting only half as much light as a charcoal briquette, Mathilde is only half as dense as typical, stone asteroids. Slightly larger than Ida, irregularly shaped Mathilde measures 66 km by 48 km by 46 km, rotates once every 17.4 d, and has a mass equivalent to 110 trillion tons. The part of the asteroid shown is about 59 by 47 km. The large crater in shadow is about 20 km across.

a distance of only 6 million kilometers. In 1972, space debris was observed to skip off Earth's atmosphere and retreat back into space.

There were at least two other close calls recently. On March 23, 1989, an asteroid called 1989 FC passed

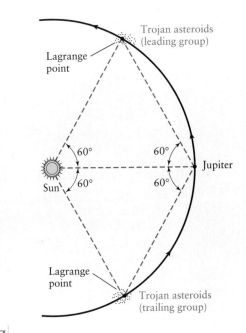

**Figure 8-7**    **The Trojan Asteroids**  Groups of asteroids orbit at the two stable Lagrange points along Jupiter's orbit, trapped by the combined gravitational forces of Jupiter and the Sun. Asteroids at these locations are named after Homeric heroes of the Trojan War.

**Figure 8-8**    **Eros** The *Near Earth Asteroid Rendevous* (*NEAR*) spacecraft photographed asteroid Eros in February 1999, taking images from as close as 3830 km (2375 mi). Eros's dimensions are 33 × 13 × 13 km (21 × 8 × 8 mi) and it rotates every 5¼ hours. Its density is 2700 kg/m$^3$, close to the average density of the Earth's crust and twice as dense as asteroid Mathilde. Eros's surface is peppered with impact craters.

within 800,000 km of the Earth, and on December 9, 1994, asteroid 1994 XM1 passed within 105,000 km. This latter asteroid is about 10 m across—the size of a small bus (Figure 8-9). Almost 300 *Earth-crossing* asteroids are known, and astronomers estimate more than 100,000 others are yet to be discovered.

During these close encounters, astronomers can examine the details of asteroids. For example, an asteroid's brightness often varies as it rotates because different surface features reflect different amounts of light. Such data show that typical rotation periods for asteroids are between 5 and 20 hours.

Most of the Apollo asteroids will eventually strike a moon or planet—even the Earth. Some asteroids will end up heading straight into the Sun. However, the chance of the Earth being hit by an asteroid in the next few thousand years is remote. We will examine this possibility at the end of the chapter.

# COMETS

While asteroids consist primarily of rock and metal, other pieces of space debris are composed of frozen water, along with rock, metal, and ices of other compounds. We have seen in earlier chapters that water ice, along with carbon dioxide, methane, and ammonia ices, was locked up in planets and moons. Ices in the young solar system also condensed with roughly equal amounts of small rocky and metallic debris into bodies that still remain in orbit around the Sun. These dirty icebergs in space are the **comets**. To better determine their composition, the *Stardust* probe is on its way to Comet Wild 2, where it will collect samples and return to Earth in 2006.

**Figure 8-9**    **Asteroid 1994 XM1** This image was obtained on December 9, 1994, shortly before the asteroid arrived in the Earth's vicinity. When it passed by the Earth just 12 hours later, asteroid 1994 XM1 was less than half the distance from the Earth to the Moon.

## 8-4 Comets lack tails until they enter the inner solar system

Comets first formed in the outer reaches of the solar system at the distances of Uranus and Neptune. In that re-

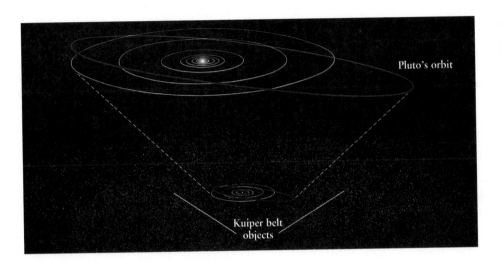

**Figure 8-10** **The Kuiper Belt**
The Kuiper belt of comets spreads from
Pluto out 500 AU from the Sun. Most of
the estimated 200 million belt comets
are believed to orbit in or near the plane
of the ecliptic. More than 70 of these
objects have been located; the largest is
320 km across. Some astronomers
believe that Pluto and Charon are
members of the belt.

gion, water was plentiful and the temperature low enough for the ices to condense into chunks several kilometers across. Then, gravitational tugs from Uranus and Neptune flung the comets in every direction.

The solar system contains two reservoirs of comets. Most comets that eventually return to the inner solar system and develop the long tails that we usually associate with them are believed to come from a doughnut-shaped region out beyond the orbit of Pluto. This **Kuiper belt** of comets, named after the American astronomer Gerard Kuiper, who first proposed its existence in 1951, is centered on the plane of the ecliptic and extends out some 500 AU from the Sun (Figure 8-10). More than 70 Kuiper belt objects have been observed (Figure 8-11). Given the locations of these objects, astronomers estimate that the Kuiper belt contains at least 200 million comets. The largest object so far observed in this region is about one-fifth as large as Pluto.

The vast majority of the several billion comets estimated to exist lie even farther from the Sun. Unlike the Kuiper belt comets and the rest of the solar system, these comets have a spherical distribution around the Sun called the **Oort cloud,** named after the Dutch astronomer Jan Oort, who first proposed its existence in the 1950s.

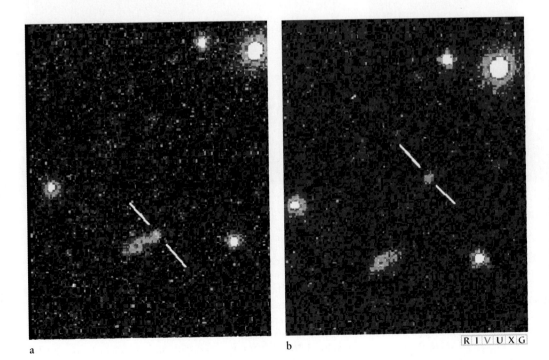

**Figure 8-11** **Kuiper Belt
Objects** These 1993 images
show the discovery (white lines)
of one of more than 70 known
Kuiper belt objects. These two
images of Kuiper belt object
1993 SC were taken 4.6 hours
apart, during which time the
object moved from **(a)** to **(b)**
against the background stars.

R I V U X G

Astronomers calculate that the Oort cloud extends out at least 50,000 AU, one-fifth the distance to the nearest stars. Most of these comets have orbits so nearly circular that they never even get as close as Pluto is to the Sun. However, occasionally a passing star's gravitational force nudges a distant comet toward the inner solar system. As a result of its inward plunge, comets from the Oort comet cloud, like Hale-Bopp and Hyakutake, have highly elliptical orbits. Often comets also have orbits that are highly tilted, even perpendicular, to the plane of the ecliptic.

Because the comets in the Kuiper belt and Oort cloud are far from the Sun, they are completely frozen. Solid comet bodies, called **nuclei** (*singular* **nucleus**), are typically 10 kilometers across. The first pictures of a comet's nucleus were obtained when a fleet of spacecraft flew past Comet Halley in 1986 (Figure 8-12). Halley's potato-shaped nucleus is darker than coal, probably because of carbon-rich compounds left behind after its ice evaporated.

As a comet nucleus comes within 20 AU of the Sun, solar radiation begins to vaporize the ices on its surface. The liberated gases form an atmosphere, or **coma**, around the nucleus. Because the coma scatters sunlight, it appears as a fuzzy, luminous ball. The largest coma ever measured was more than a million kilometers across—nearly as large as the Sun. Not visible to the human eye is the **hydrogen envelope**, a sphere of tenuous gas surrounding the comet's nucleus and measuring as much as 20 million kilometers in diameter (Figure 8-13).

Of course, the most visible and inspiring features of comets are their long, flowing, diaphanous **tails**, an awesome sight (Figure 8-14). Comet tails develop from coma gases and dust pushed outward from the Sun. This means that comet tails do not trail behind the nucleus, as the exhaust from a jet plane does in the Earth's atmosphere. Rather, *at the comet's nucleus, the tails always point away from the Sun* (Figure 8-15), regardless of the direction of the comet's motion. The implication that something from the Sun was "blowing" the comet's gases radially outward led Ludwig Biermann to predict the existence of the solar wind (see Chapters 5 and 9). This stream of particles from the Sun was actually discovered in 1962, a full decade later, by instruments on the spacecraft *Mariner 2*.

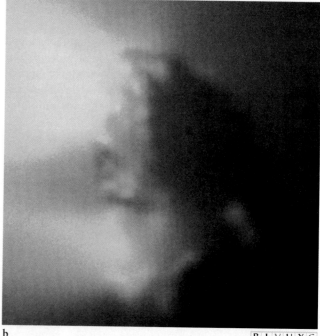

a    b    R I V U X G

## Figure 8-12

**Comet Halley** **(a)** The Head of Comet Halley. This photograph shows the bluish head of Comet Halley as it approached the Sun in 1985. Comet Halley orbits the Sun with an average period of 76 years along a highly elliptical path that stretches from just inside the Earth's orbit to slightly beyond the orbit of Neptune. This color photograph was constructed from three black-and-white photographs taken in rapid succession with red, blue, and green filters on the same telescope. Because of the comet's motion during the exposures, the images of background stars are elongated. **(b)** The Nucleus of Comet Halley. This close-up picture, taken by a camera on board the *Giotto* spacecraft, shows the potato-shaped nucleus of the comet. Its dark nucleus measures 15 km in its longest dimension and about 8 km in its shortest. The Sun illuminates the comet from the left. The numerous bright areas are thin jets of gas or icy outcroppings that reflect more sunlight than surrounding areas of the comet.

a     R I V U X G     b     R I V U X G

### Figure 8-13 Comet Kohoutek and Its Hydrogen Envelope
Comet Kohoutek as seen in visible light (a) and to the same scale at ultraviolet wavelengths (b). The ultraviolet picture reveals a huge hydrogen cloud surrounding the comet's nucleus.

R I V U X G

### Figure 8-14 Comet West
Astronomer Richard M. West first noticed this comet on a photograph taken with a telescope in 1975. After passing near the Sun, Comet West became one of the brightest comets of the 1970s. This photograph shows the comet in the predawn sky in March 1976.

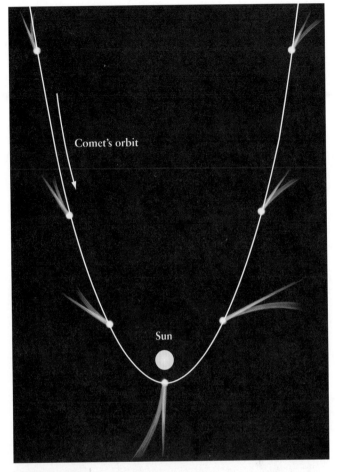

### Figure 8-15 The Orbit and Tails of a Comet
The solar wind and sunlight blow a comet's dust particles and ionized atoms away from the Sun. Consequently, the comet's tails always point away from the Sun.

**Insight into science  Working in reverse**
Science often advances by working backward from observations and experiments that require scientific explanations. For example, seeing comet tails and how they behave led Biermann to wonder what caused them. The simplest physical explanation is that matter from each comet body is being evaporated and pushed upon by something from the Sun, the solar wind. Remember Occam's razor.

The Sun's radiation usually produces two comet tails: a **gas** (or **ion**) **tail** and a **dust tail** (Figure 8-16). Positively charged ions (atoms missing one or more electrons) from the coma are swept directly away from the Sun by the solar wind to form the gas tail. This tail often appears blue, because of blue light emitted by carbon monoxide ions in

| August 22 | August 24 | August 26 |
|---|---|---|

R I V U X G

**Figure 8-16** **The Two Tails of Comet Mrkos** Comet Mrkos dominated the evening sky in August 1957. These three views, taken at two-day intervals, show dramatic changes in the comet's gas tail. In contrast, the slightly curved dust tail remained fuzzy and featureless.

it. Ions typically leave the coma at 1.4 million kilometers per hour. The relatively straight gas tail can change dramatically from night to night (see Figure 8-16).

The dust tail is formed when photons strike dust particles that have been freed from the comet's evaporating nucleus. Light exerts pressure on any object that absorbs

or scatters it. While this pressure, called **radiation pressure** or **photon pressure**, is quite weak, fine-grained dust particles in a comet's coma are sufficiently light to be blown away from the comet, thus producing a dust tail. The dust tail often is the color of sunlight. The dust particles are massive enough not to flow straight away from the Sun; rather, the dust tail arches in a path that lies between the gas tail and the direction from which the comet came (Figure 8-17). Comet tails can stretch to more than 150 million kilometers (1 AU) in length.

Figure 8-17 outlines the structure of a comet. Some, like the comet shown in Figure 8-18, have a large, bright coma but short, stubby tails. Others have an inconspicuous coma but one or more tails of astonishing length. The gas tail in Figure 8-19 is long enough to stretch all the way from the Earth to the Sun.

## 8-5 Comets don't last forever

Astronomers, many of them amateurs, typically discover at least a dozen new comets each year. Falling sunward from the Kuiper belt or Oort cloud, most are **long-period comets**, which move so fast that they leave the inner solar system after one pass by the Sun and take roughly 1 million to 30 million years to return.

However, sometimes a comet passes so close to a planet that the planet's gravitational force changes the comet's

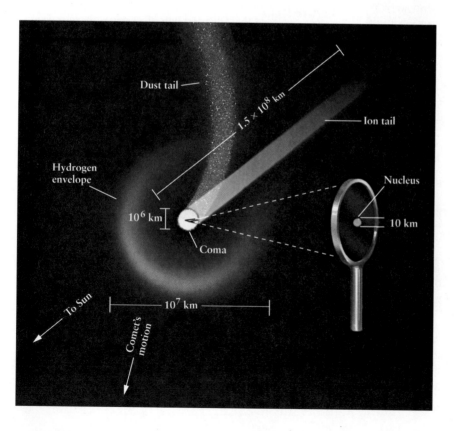

**Figure 8-17** **The Structure of a Comet** The solid part of a typical comet (the nucleus) is roughly 10 km in diameter. The coma can be as large as $10^5$ to $10^6$ km across, and the hydrogen envelope is typically $10^7$ km in diameter. A comet's tail can be enormous—as long as 1 AU. (This drawing is not to scale.)

**Figure 8-18** **The Head of Comet Brooks** This comet, named after its discoverer, had an exceptionally large, bright coma. It dominated the night skies in October 1911.

**Figure 8-19** **The Tail of Comet Ikeya-Seki** Named after its codiscoverers in Japan, this comet dominated the predawn skies in late October 1965. Although its coma was tiny, its tail spanned 1 AU.

orbit, slowing it down and trapping it in the inner solar system (Figure 8-20). The comet then becomes a **short-period comet**, orbiting the Sun in fewer than 200 years. Like Comet Halley, short-period comets appear again and again at predictable intervals. If you missed Halley's comet in its 1985 visit, your next chance to see it is in 2061.

Comets cannot survive an infinite number of passages near the Sun. A typical comet is estimated to lose between 1/60th and 1/100th of its mass at each perihelion (closest approach to the Sun). Therefore, a typical comet survives at most 100 close passes to the Sun before its ices evaporate completely. It is likely, then, that Halley's comet, which has been seen at least 27 times, will survive for fewer than 5600 more years. Figure 8-21 shows the nucleus of a comet breaking up at the distance of Mercury's orbit, shortly after it passed the Sun. Soon thereafter, its remaining dust and rock fragments spread out in a loose collection of debris that continues to circle the Sun along the comet's orbital path.

A comet can also be torn apart when it comes too close to a planet, or it can be destroyed completely by striking a planet, a moon, or the Sun. A spectacular example was Comet Shoemaker-Levy 9. As discussed in Chapter 7, that comet fragmented under the tidal force from Jupiter in 1992. Two years later, with the world's astronomers watching carefully, the pieces returned and struck the planet (see Figures 7-8 and 7-9).

The comet's disintegration provided an important clue to its structure. For Shoemaker-Levy 9 to break up in

the first place tells astronomers that its nucleus must have been held together very weakly. It may actually have been composed of separate pieces that had stuck together until

**Figure 8-20** **Transforming a Long-Period Comet into a Short-Period Comet** The gravitational force of a planet can change a comet's orbit. Initially, on highly elliptical orbits, comets are sometimes deflected into more circular paths that keep them in the inner solar system.

March 8          March 12          March 14          March 18          March 24

R I V U X G

**Figure 8-21** **The Fragmentation of Comet West** Shortly after passing near the Sun in 1976, the nucleus of Comet West broke into at least four pieces. This series of photographs clearly shows the disintegration of the comet's nucleus. Figure 8-14 shows a wide-angle view of this comet's brilliant tail.

R I V U X G

**Figure 8-22** **Comet Hale-Bopp** Discovered on July 23, 1995, this comet was at its breathtaking best in mid-1997. Did you see it? **Inset:** Jets of gas and debris were observed shooting out from Comet Hale-Bopp at least twice between September and October 1997. This image shows the comet nucleus (lower bright region), an ejected piece of the comet's surface (upper bright region), and a spiral tail. The ejected piece eventually disintegrated, following the same spiral pattern as the tail.

Jupiter's tidal force pulled them apart. Astronomers hope to learn whether this structure is typical of other comets as well.

Comets occasionally lose mass quickly by ejecting it in bursts. Astronomers observed this in September and October 1995, when Comet Hale-Bopp ejected 7 to 10 times more mass than usual (Figure 8-22). Astronomers believe that this event resulted from surface ice and dust being heated by the Sun and then being rapidly ejected from a local region on the comet's surface called a *vent*. Hale-Bopp rotates with a period of about 12 hours. The ejected matter therefore formed a pinwheel-shaped distribution spiraling from the comet's nucleus at a speed of 109 km/hr

(68 mph). While the ejected matter did not represent a significant fraction of the comet's mass, its light-reflecting dust made it look like a huge comet fragment, indeed.

# METEOROIDS, METEORS, AND METEORITES

Once formed, asteroids occasionally collide with each other, sending fragments into interplanetary space. These smaller pieces, along with rocky and metallic fragments from evaporating comets, and debris that never coalesced with larger bodies, are still strewn throughout the solar system.

## 8-6 Small rocky debris peppers the solar system

**Meteoroids** are the rocky and metallic debris smaller than asteroids scattered throughout the solar system. Although no official size standard separates the two, meteoroids are no more than a few tens of meters across, and the vast majority are smaller than a millimeter.

Meteoroids are often pulled by gravity into the Earth's atmosphere. Air friction generates so much heat that a meteoroid's outer layer begins to vaporize. (The same process creates the heat on the hull of the Space Shuttle as it reenters our atmosphere.) As the meteoroid penetrates further into the air, leaving behind a trail of dusty gas, it becomes a **meteor**. Common names for these dramatic streaks of light flashing across the sky include *shooting stars, fireballs,* and *bolides.* Fireballs are meteors at least as bright as Venus (Figure 8-23); bolides are bright meteors that explode in the air. It is worth noting that shooting stars are not stars of any kind, nor are they dying stars.

## 8-7 Impact craters and meteor showers mark remnants of space debris on Earth

Most meteors vaporize completely before they can strike the Earth. Their dust settles to the ground, often carried by raindrops. (This is not the source of acid rain, however, which comes from natural and human-made gases ejected into the atmosphere.) Any part of a meteor that survives its fiery descent to Earth may leave an **impact crater.** Weather and water erosion are wearing away all of the 200 or so impact craters now known on the Earth, and thousands more have long since been drawn into the Earth by the motion of its tectonic plates. Indeed, those known today are all less than 500 million years old because of these forces reshaping the Earth's surface.

RIVUXG

**Figure 8-23** **A Meteor** A streak of light is produced when a piece of interplanetary rock or dust strikes the Earth's atmosphere at high speed. Exceptionally bright meteors, such as the one shown in this exposure that tracked the meteor (notice the star trails), are usually called fireballs.

One of Earth's best-preserved impact craters is the famous Barringer (or Meteor) Crater near Winslow, Arizona (Figure 8-24). Measuring 1.2 km across and 200 m deep, it formed 50,000 years ago when an iron-rich meteoroid some 50 m across (about half the length of a football field) struck the ground at 40,000 km/h (25,000 mph). The blast was like the detonation of a 20-megaton hydrogen bomb.

**Figure 8-24** **The Barringer Crater** An iron meteoroid measuring 50 m across struck the ground in Arizona 50,000 years ago. The result was this beautifully symmetric impact crater.

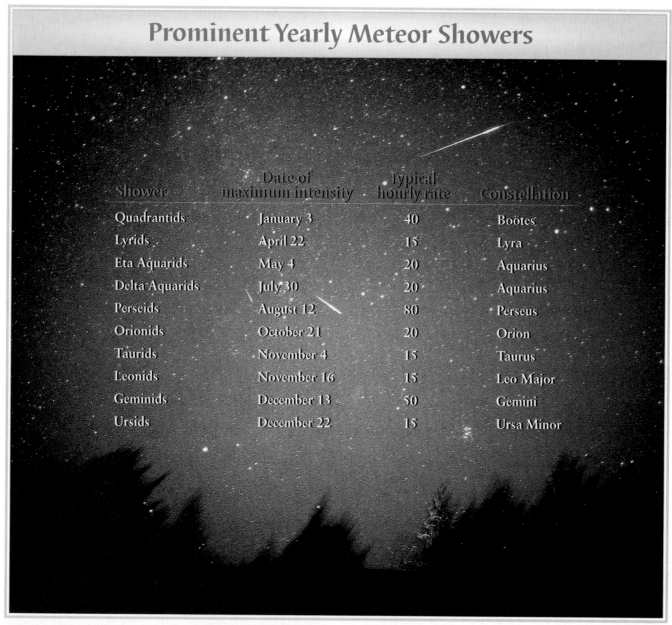

# Prominent Yearly Meteor Showers

| Shower | Date of maximum intensity | Typical hourly rate | Constellation |
| --- | --- | --- | --- |
| Quadrantids | January 3 | 40 | Boötes |
| Lyrids | April 22 | 15 | Lyra |
| Eta Aquarids | May 4 | 20 | Aquarius |
| Delta Aquarids | July 30 | 20 | Aquarius |
| Perseids | August 12 | 80 | Perseus |
| Orionids | October 21 | 20 | Orion |
| Taurids | November 4 | 15 | Taurus |
| Leonids | November 16 | 15 | Leo Major |
| Geminids | December 13 | 50 | Gemini |
| Ursids | December 22 | 15 | Ursa Minor |

R I V U X G

**Figure 8-25** **Meteor Streaks Seen During a Meteor Shower** This table lists highly active, yearly meteor showers, which last for several days. The time exposure behind the table shows meteors streaking away from one place in the sky, in the constellation Perseus, during the Perseid meteor shower.

On a typical clear night, you can expect to see a meteor about every 10 minutes. However, at predictable times throughout each year, the Earth is inundated with them. These **meteor showers** occur when the Earth moves through the orbit of remnant debris left from a comet.

Some 30 meteor showers can be seen each year. Because the meteors in each shower appear to come from a fixed region of the sky, they are named after the constellation from which the meteors appear to radiate (Figure 8-25). For example, meteors in the Perseid shower appear to originate in the constellation Perseus. More than one meteor can be seen each minute at the peaks of such prodigious meteor showers as the Perseids, which take place in the summertime and are among the best light shows in astronomy. Except for the Lyrids, these showers are best seen after midnight.

R I V U X G

**Figure 8-26**    **A Stony Meteorite** Most meteorites that fall to Earth are stones. Many freshly discovered specimens, like the one shown here, are coated with dark crusts. This stony meteorite fell in Texas.

R I V U X G

**Figure 8-27**    **A Cut and Polished Stone** Some stony meteorites contain tiny specks of iron, which can be seen when the stones are cut and polished. This specimen was discovered in California.

## 8-8 Meteorites are space debris that land intact

Although most meteors completely vaporize in the atmosphere, some reach the ground before totally disintegrating. Such debris are called **meteorites.** Most meteors, and therefore meteorites, are believed to have been broken off from asteroids by collisions. Some meteorites are debris that was never part of larger bodies. Still others come from the Moon, Mars, and comets. We saw in Chapter 6, for example, how SNC meteorites offer clues to the chemistry of Mars. Others tell us the age of the solar system.

People have been picking up debris from space for thousands of years. The descriptions of meteorites in historical Chinese, Indian, Islamic, Greek, and Roman literature show that early peoples placed special significance on these "rocks from heaven." They also have a practical "impact": Infalling space debris is increasing the Earth's mass by more than 300 tons per day on average.

Meteorites are classified as stones, stony-irons, and irons. Most **stony meteorites** look much like ordinary rocks, although some are covered with a dark crust. This crust is produced when the meteorite's outer layer melts during its fiery descent through the atmosphere (Figure 8-26). When a stony meteorite is cut in two and polished, tiny flecks of iron are sometimes found in the rock (Figure 8-27).

Meteorites with a high iron content can be located with a metal detector. They also look unusual and hence are more likely to be noticed. Consequently, the easily found iron and stony-iron meteorites dominate most museum collections, even though stony meteorites account for about 95% of all meteoritic material that falls on the Earth.

**Iron meteorites** (Figure 8-28) may also contain from 10% to 20% nickel by weight. Iron is moderately abundant in the universe as well as being one of the most common rock-forming elements, so it is not surprising that iron is an important constituent of asteroids and meteoroids. Another element, iridium, is common in the iron-rich minerals of meteorites but rare in ordinary rocks, because most iridium settled deep into the Earth eons ago. Measurements of iridium in the Earth's crust can thus tell us the rate at which meteoritic material has been deposited on the Earth over the ages.

In 1808 Count Alois von Widmanstätten, director of the Imperial Porcelain Works in Vienna, discovered a conclusive test for the most common type of iron meteorite. Most irons have a unique structure of long nickel-iron crystals called **Widmanstätten patterns,** which become visible when the meteorites are cut, polished, and briefly dipped into a dilute solution of acid (Figure 8-29). Because nickel-iron crystals can grow to lengths of several centimeters only if the molten metal cools slowly over many millions of years, Widmanstätten patterns are never found in counterfeit meteorites, or "meteorwrongs."

The final category of meteorites, the **stony-irons,** consist of roughly equal amounts of rock and iron. Figure

**Figure 8-28**    RIVUXG    **An Iron Meteorite** Irons are composed almost entirely of iron-nickel minerals. The surface of a typical iron is covered with thumbprintlike depressions created as the meteorite's outer layers melted away during its high-speed descent through the atmosphere. This specimen was found in Australia.

**Figure 8-29**    RIVUXG    **Widmanstätten Patterns** When cut, polished, and etched with a weak acid solution, most iron meteorites exhibit interlocking crystals in designs called Widmanstätten patterns. This meteorite was found in Australia.

8-30, for example, shows the greenish mineral olivine suspended in a matrix of iron.

To understand why different types of meteorites exist, we consider their formation. Most meteorites were once pieces of asteroids. Heat from the rapid decay of radioactive isotopes melted newly formed asteroid interiors. Over the next few million years, differentiation occurred, just as in the young Earth. Iron sank toward the asteroid's center, while lighter rock floated up to the asteroid's surface. Iron meteorites are fragments of an asteroid's core, and stones are samples of its crust. Stony-irons are believed to come from the boundary region between the iron core and stony crust.

A class of rare stony meteorites, **carbonaceous chondrites,** shows no evidence of ever having been melted as parts of asteroids. These rarities may therefore be primordial material from which our solar system was created. Carbonaceous chondrites contain complex carbon compounds and as much as 20% water bound into their minerals. The organic compounds would have been broken down and the water driven out if these meteorites had been significantly heated. Asteroid Mathilde, shown in Figure 8-6, has a very dark gray color and virtually the same spectrum as a carbonaceous chondrite meteorite, so it is likely composed of primordial material.

**Amino acids,** the building blocks of proteins upon which terrestrial life is based, are among the organic compounds occasionally found inside carbonaceous chondrites, although these may be contaminants acquired after the meteoroids entered the Earth's atmosphere. Nevertheless, some scientists suspect that carbonaceous chondrites may have played a role in the origin of life on Earth.

## 8-9 The Tunguska mystery and the Allende meteorite provide evidence of catastrophic collisions

At 7:14 AM, local time on June 30, 1908, a spectacular explosion occurred over the Tunguska region of Siberia. The blast, comparable to a nuclear detonation of several megatons, knocked a man off his porch some 60 km away and was audible more than 1000 km away. Millions of tons of dust were injected into the atmosphere, darkening the air as far away as California.

Preoccupied with wars, along with political and economic upheaval, neither Russia nor its successor, the former Soviet Union, sent a scientific expedition to the site until 1927. At that time, Soviet researchers found that trees had been seared and felled radially outward in an area about 30 km in diameter (Figure 8-31). There was no clear evidence of a crater. In fact, the trees at "ground zero" were left standing upright, although they were completely stripped of branches and leaves. Because no significant meteorite samples were found, for many years scientists assumed that a small comet had struck the Earth.

Recently, however, several teams of astronomers have argued that a small comet breaks up too high in the atmosphere to cause significant damage on the ground. They argue that the Tunguska explosion was actually caused by a small asteroid or large meteoroid traveling at supersonic speed. The Tunguska event is consistent with an explosion in the air of an asteroid about 80 m (260 ft) in diameter entering the Earth's atmosphere at 79,000 km/h (50,000 mph) and exploding in the air as a result of becoming exceedingly hot.

**Figure 8-30** **A Stony-Iron Meteorite** Stony-irons account for about 1% of all meteorites that fall to Earth. This specimen, a variety of stony-iron called a pallasite, was found in Chile.

A chance to study debris immediately after impact came shortly after midnight on February 8, 1969, when a brilliant, blue-white light moved across the night sky around Chihuahua, Mexico. Hundreds of people witnessed the dazzling display. The light disappeared in a spectacular, noisy explosion that dropped thousands of rocks and pebbles over the terrified onlookers. Within hours, teams of scientists were on their way to collect

specimens of carbonaceous chondrites, collectively named the *Allende meteorite,* after the locality in which they fell (Figure 8-32).

One of the most significant discoveries to come from the Allende meteorite was evidence of the detonation of a nearby supernova 4.6 billion years ago. Among nature's most violent and spectacular phenomena, a *supernova explosion* occurs when a massive star dies. A massive star blows apart in a cataclysm that hurls matter outward at tremendous speeds, as we will see in Chapter 12. During this detonation, violent collisions between atomic nuclei produce a host of radioactive elements, including a short-lived radioactive isotope of aluminum. Based on its decay products, scientists found unmistakable evidence that this isotope once lay within the Allende meteorite. Some astronomers interpret this as evidence for a supernova in our vicinity at about the time the Sun was born. Indeed, by compressing interstellar gas and dust, the supernova's shock wave may have helped stimulate the birth of our solar system.

## 8-10 An asteroid's impact with Earth apparently killed off the dinosaurs

In the late 1970s, the geologist Walter Alvarez and his father, physicist Luis Alvarez, discovered a different sort of shock wave closer to home. Working at a site of exposed marine limestone in the Apennine Mountains in Italy that had been on the Earth's surface 65 million years ago, the

**Figure 8-31** **Aftermath of the Tunguska Event** In 1908 a stony asteroid traveling at supersonic speed struck the Earth's atmosphere and exploded over the Tunguska region of Siberia. Trees were blown down for many kilometers in all directions from the impact site.

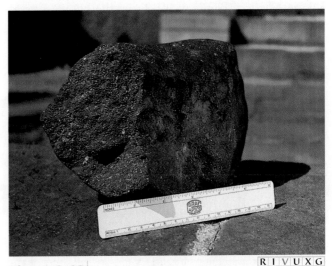

**Figure 8-32** **A Piece of the Allende Meteorite** This carbonaceous chondrite fell near Chihuahua, Mexico, in February 1969. Note the meteorite's dark color, caused by a high abundance of carbon. Geologists believe that this meteorite is a specimen of primitive planetary material. The ruler is 15 cm long.

Alvarez team discovered an exceptionally high abundance of iridium in a dark-colored layer of clay between limestone strata (Figure 8-33).

Since this discovery was announced in 1979, a comparable layer of iridium-rich material has been uncovered at numerous sites around the world. In every case, geologic dating reveals that this apparently worldwide iridium-rich layer was deposited about 65 million years ago. Paleontologists were quick to realize the significance of this date, because it was 65 million years ago when all the dinosaurs rather suddenly became extinct. In fact, two-thirds of all the species on Earth disappeared within a brief span of time back then.

The Alvarez discovery suggests a startling explanation for the dramatic extinction of so much of the life that once inhabited our planet—an asteroid impact. An asteroid 10 km in diameter slamming into the Earth at high speed could have thrown enough dust into the atmosphere to block out sunlight for several years. As the temperature dropped drastically and plants died for lack of sunshine, the dinosaurs would have perished, along with many other creatures in the food chain that were highly dependent on vegetation. The dust eventually settled, depositing an iridium-rich layer around the world. Tiny, rodentlike creatures capable of ferreting out seeds and nuts were among the animals that managed to survive this holocaust, setting the stage for the rise of mammals and, consequently, the evolution of humans.

In 1992 a team of geologists suggested that the hypothesized asteroid crashed into a site in Mexico. They based this conclusion on glassy debris and violently shocked grains of rock ejected from the multiringed, 195-km-diameter Chicxulub Crater buried under the Yucatán Peninsula in Mexico (Figure 8-34). From the known rate at which radioactive potassium decays, the scientists have pinpointed the date when the asteroid struck—64.98 million years ago. In 1998 geologists digging on the Pacific Ocean floor discovered a piece of meteoritic debris with precisely the same age, apparently a piece of the offending asteroid. Other geologists and paleontologists are not yet convinced that an asteroid impact led to the extinction of the dinosaurs, but most agree that this hypothesis fits the available evidence better than any other explanation that has been offered so far.

Could such an extinction-level event happen again? There are so many asteroids whose paths cross the Earth's path that scientists agree that it's a matter of "when," rather than "if." The good news is that studies of craters show that larger asteroids strike the Earth significantly less often than do smaller ones. While asteroids large enough to create the Barringer Crater strike the Earth about once every 10,000 years, killer asteroids, like the one that struck the Yucatán, collide with Earth only once every 100 million years. The threat of a catastrophic impact by an asteroid or comet in our lifetime, thankfully, is remote.

## WHAT DID YOU THINK?

1. *Are the asteroids a planet that was somehow destroyed?* No. The gravitational pull from Jupiter prevented a planet from ever forming in the asteroid belt. Also, the total mass of the asteroids is much less than even the mass of tiny Pluto, the smallest planet.

2. *How far apart are the asteroids on average?* The distance between asteroids averages 10 million kilometers.

3. *Why do comets have tails?* Gas and dust that evaporate from the comet nucleus are pushed away from the Sun by sunlight and the solar wind. We see them by the sunlight they scatter in our direction.

**Figure 8-33** **Iridium-Rich Layer of Clay** This photograph of strata in the Apennine Mountains of Italy shows a dark-colored layer of iridium-rich clay sandwiched between white limestone (below) from the late Mesozoic era and grayish limestone (above) from the early Cenozoic era. The coin is the size of a U.S. quarter.

R I V U X G

RIVUXG

Chicxulub crater

Meteorite
fragment
discovered

**Figure 8-34**    **Confirming an Extinction-Level Impact Site**  By measuring slight
variations in the gravitational attraction of different materials under the Earth's surface, geologists
create images of underground features. **(Right inset)** Concentric rings of the underground
Chicxulub Crater, shown here, lie under a portion of the Yucatán Peninsula. This crater has been
dated to 65 million years ago and is believed to be the site of the impact that led to the extinction
of the dinosaurs. **(Left inset)** A piece of 65-million-year-old meteorite discovered in the middle of
the Pacific Ocean in 1998 and believed to be a fragment of the meteorite that struck the Yucatán
Peninsula. The meteorite, about a tenth of an inch long, was cut into two pieces, shown here.

4  *In what direction does a comet tail point?* Comets' gas tails point directly away from the Sun; their dust tails make arcs pointing away from the Sun.

5  *What is a shooting star?* A shooting star is a piece of space debris plunging through the Earth's atmosphere—a meteor. It is not a star.

## Key Words

amino acid, 220

Apollo asteroid, 209

asteroid (minor planet), 205

asteroid belt, 206

belt asteroid, 206

carbonaceous chondrite, 220

coma (of a comet), 212

comet, 210

dust tail (of a comet), 213

gas (ion) tail, 213

hydrogen envelope, 212

impact crater, 217

iron meteorite, 219

Kirkwood gaps, 207

Kuiper belt, 211

long-period comet, 214

meteor, 217

meteor shower, 218

meteorite, 219

meteoroid, 217

nucleus (of a comet), 212

Oort cloud, 211

radiation (photon) pressure, 214

short-period comet, 215

stable Lagrange points, 209

stony meteorite, 219

stony-iron meteorite, 219

tail (of a comet), 212

Trojan asteroid, 209

Widmanstätten patterns, 219

## Key Ideas

### Asteroids

∗  Thousands of belt asteroids with diameters larger than a kilometer are known to orbit the Sun between the orbits of Mars and Jupiter. The gravitational attraction of Jupiter depletes certain orbits within the asteroid belt. The resulting gaps, called Kirkwood gaps, occur at simple fractions of Jupiter's orbital period.

∗  Jupiter's and the Sun's gravity combine to capture Trojan asteroids in two locations, called stable Lagrange points, along Jupiter's orbit.

∗  The Apollo asteroids move in highly elliptical orbits that cross the orbits of Mars and Earth. Many of these asteroids will eventually strike one of the inner planets.

### Comets

∗  Comets are fragments of ice and rock that generally move in highly elliptical orbits about the Sun often at a great inclination to the plane of the ecliptic.

∗  As a comet approaches the Sun, its icy nucleus develops a luminous coma surrounded by a vast hydrogen envelope. A gas (or ion) tail and a dust tail extend from the comet, pushed away from the Sun by the solar wind and radiation pressure.

∗  Many comets orbit the Sun in the Kuiper belt, a doughnut-shaped region beyond Pluto. Billions of cometary nuclei also exist in the spherical Oort cloud located far beyond Pluto.

### Meteoroids and Meteorites

∗  Boulders and smaller rocks in space are called meteoroids. When a meteoroid enters the Earth's atmosphere, it produces a fiery trail, and it is then called a meteor. If part of the object survives the fall, the fragment that reaches the Earth's surface is called a meteorite.

∗  Meteorites are grouped in three major classes according to their composition: iron, stony-iron, and stony meteorites. Rare stony meteorites called carbonaceous chondrites may be relatively unmodified material from the primitive solar nebula. These meteorites often contain organic hydrocarbon compounds, including amino acids.

∗  Fragments of rock from "burned-out" comets produce meteor showers.

∗  An analysis of the Allende meteorite suggests that a nearby supernova explosion may have been involved in the formation of the solar system some 4.6 billion years ago.

∗  An asteroid that struck the Earth 65 million years ago probably contributed to the extinction of the dinosaurs and many other species. Such devastating impacts occur on average every 100 million years.

## Review Questions

1 Why are asteroids, meteoroids, and comets of special interest to astronomers who want to understand the early history of the solar system?

2 Describe the asteroid belt.

3 Why do you suppose that there are many small asteroids but only a few very large ones?

4 Do comets always have tails? Explain.

5 What are Kirkwood gaps and what causes them?

6 What are the Trojan asteroids, and where are they located?

7 Describe the three main classifications of meteorites. How do astronomers believe these different types of meteorites originated?

8 Suppose you found a rock that you suspect to be a meteorite. Describe some of the things you could do to see if it were a meteorite or a "meteorwrong."

9 Why do astronomers believe that meteoroids come from asteroids, whereas meteor showers are related to comets?

10 Make a drawing that describes the structure of a comet.

11 Why is the phrase "dirty snowball" an appropriate characterization of a comet's nucleus?

12 What is the Kuiper belt, and how might it be related to debris left over from the formation of the solar system?

13 Why do scientists think the Tunguska event was caused by a large meteoroid and not a comet?

14 Which tail in Figure 8-22 is the gas tail and which is the dust tail? Justify your answers.

## Advanced Questions

The answers to all computational problems, which are preceded by an asterisk (*), appear at the end of the book.

15 Why do you suppose comets are generally brighter after passing perihelion (closest approach to the Sun) than before reaching perihelion?

16 Can you think of another place in the solar system where a phenomenon similar to the Kirkwood gaps in the asteroid belt exists? Explain.

17 Where on Earth might you find large numbers of stony meteorites that have not been significantly changed by weathering?

*18 Assuming a constant rate of meteor infall, how much mass has the Earth gained in the past 4.6 billion years?

## Discussion Questions

19 Suppose it was discovered that the asteroid Hermes had been perturbed in such a way as to put it on a collision course with Earth. Describe what you would do to counter such a catastrophe using present technology.

20 From the abundance of craters on the Moon and Mercury, we know that numerous asteroids and meteoroids struck the inner planets during the early history of the solar system. Is it reasonable to suppose that numerous comets also pelted the planets 4.0 to 4.5 billion years ago? Speculate about the effects of such a cometary bombardment, especially with regard to the evolution of the primordial atmospheres of the terrestrial planets.

## What If ...

21 An ocean on Earth were struck by a comet nucleus several kilometers across? What physical effects to the Earth would occur?

22 We passed through the tail of a comet? What would happen to the Earth and life on it?

23 As some astronomers have recently argued, passage of the solar system through an interstellar cloud of gas could perturb the Oort cloud, causing many comets to deviate slightly from their original orbits? What might be the consequences for Earth?

24 All the space debris (asteroids, meteoroids, and comets) had been cleared out of the solar system 3.8 billion years ago? What would have been different in the history of the Earth and in the history of life on the Earth?

# ⚹ Observing Projects

**25** Make arrangements to view an asteroid. At opposition, some of the largest asteroids are bright enough to be seen through a modest telescope. Check the "Minor Planets" section of the current issue of the *Astronomical Almanac* to see if any bright asteroids are near opposition. If so, check the current issue as well as the most recent January issue of *Sky & Telescope* for a star chart showing the asteroid's path among the constellations. You will need such a chart to distinguish the asteroid from background stars. Observe the asteroid on at least two occasions separated by a few days. On each night, draw a star chart of the objects in your telescope's field of view. Has the position of one starlike object shifted between observing sessions? Does the position of the moving object agree with the path plotted on published star charts? Do you feel confident that you have in fact observed an asteroid?

**26** Make arrangements to view a comet through a telescope. Because astronomers discover roughly a dozen comets each year, a comet usually is visible somewhere in the sky. Unfortunately, because most comets are quite dim, you will need access to a moderately large telescope. Consult recent issues of the *IAU Circular,* published by the International Astronomical Union's Central Bureau for Astronomical Telegrams, which contains predicted positions and the anticipated brightness of comets in the sky. Also, if there is an especially bright comet in the sky, the latest issue of *Sky & Telescope* might contain useful information. Is a comet visible and do you have a telescope at your disposal? If so, can you distinguish the comet from background stars? Can you see its coma? Can you see a tail?

**27** Make arrangements to view a meteor shower. The date of maximum intensity in Figure 8-25 is the best time to observe a particular shower, although good displays can often be seen a day or two before and after the maximum. In order to see a fine meteor display, you need a clear, moonless sky. The Moon's presence above the horizon can significantly detract from the number of faint meteors you will be able to see.

**28** View Comet Hale-Bopp on your computer using the Starry Night software. Set the date on 3/22/97. Turn off the daylight, planets, and horizon, go to ±90 latitude (North or South Pole), and lock on Hale-Bopp. Predict what direction the Sun is located relative to the comet. To verify that you are correct, turn on the planets (which control whether the sun is visible), and zoom back out. Return "Home." Now center on either Halley's Comet or Comet Hyakutake, but do not zoom in on either. Grabbing the sky with the little hand, move around until you locate the Sun. Predict the direction of your comet's tail on the sky. Lock back in on your comet, zoom in, and confirm that you are correct.

9

RIVUXG

**The X-ray Sun**
Our momentary glances toward the Sun from
day to day give the impression of a blindingly
bright, uniform body. This sequence of X-ray
images of the Sun taken by the *Yohkoh*
spacecraft shows another, darker aspect.
X rays are emitted near the solar surface and
appear here in shades of yellow, orange, and
red. The Sun's X-ray emissions are far less
uniform than the visible light it emits and
sometimes are so intense as to be extremely
dangerous to astronauts. By studying the
locations and intensities of X rays from the
Sun, astronomers gain an understanding of
the energy transfer from the Sun's interior
through its atmosphere. This sequence shows
how X-ray emissions from the Sun's outer
layer change by a factor of about 100 over an
11-year cycle.

# The Sun

## *In this chapter you will discover*

✳  why the Sun is a typical star

✳  how today's technology has led to new understanding of
solar phenomena, from sunspots to powerful ejections
of matter that sometimes enter our atmosphere

✳  that some features of the Sun, generated by the
Sun's varying magnetic field, occur in cycles

✳  how the Sun generates the energy that makes it shine

✳  the new insights provided by solar neutrinos

## WHAT DO YOU THINK?

1. What fraction of the solar system's mass is in the Sun?

2. Does the Sun have a solid and liquid interior, like the Earth?

3. What is the surface of the Sun like?

4. Does the Sun rotate?

5. What makes the Sun shine?

Without the Sun's energy, we on Earth could not exist. The Sun's ideal balance of heat, visible light, and ultraviolet radiation enabled life to form and flourish here. No wonder our star has been revered for millennia. No wonder, too, that astronomers have studied it intensively since Galileo first trained his telescope on the drama of sunspots.

We now complete our exploration of the solar system by studying the Sun. Today the Sun holds the key to our understanding stellar evolution because, for all its grandeur, the Sun is in fact a typical star. Figure 9-1 lists the Sun's essential properties. Comparing the mass of the Sun here to the total mass of all the planets in Appendix Table A-2, you can see that the Sun contains more than 99.85% of the total mass of the solar system.

## THE SUN'S ATMOSPHERE

Although astronomers often speak of the solar "surface," the Sun is so hot that it has neither liquid nor solid matter anywhere inside it. Moving down through the Sun, one continually encounters ever denser and hotter gases.

### 9-1 The photosphere is the visible layer of the Sun

The Sun appears to have a surface only because most of its visible light comes from one specific gas layer (see Figure 9-1). This region, which is about 400 km thick, is appropriately called the **photosphere** ("sphere of light"). The density of the photosphere's gas is low by Earth standards,

## The Sun's Vital Statistics

| | |
|---|---|
| Mass | $1.99 \times 10^{30}$ kg ($3.33 \times 10^5$ M$_\oplus$) |
| Visual radius | $6.96 \times 10^5$ km (109 R$_\oplus$) |
| Mean density | 1410 kg/m$^3$ |
| Luminosity (rate of energy emission) | $3.90 \times 10^{26}$ J/s |
| Surface temperature | 5800 K |
| Central temperature | $15.5 \times 10^6$ K |
| Equatorial rotation period | 25 days |

R I V U X G

**Figure 9-1    Our Star, the Sun** The Sun emits most of its visible light from a thin layer of gas called the photosphere, shown here. Although it has no solid or even liquid region, we see the bottom of the photosphere as the Sun's "surface." Astronomers think of the photosphere as the lowest level of the Sun's atmosphere. Astronomers always take great care when viewing the Sun by using extremely dark filters or by projecting the Sun's image onto a screen. **Never look at the Sun directly. Doing so will destroy the retinas of your eyes.**

**Figure 9-2**  **Solar Granulation**  High-resolution photographs of the Sun's surface reveal a blotchy pattern called granulation. Granules, which measure about 1000 km across, are convection cells in the Sun's photosphere. **Inset:** Gas rising upward produces bright granules. Cooler gas sinks downward along the darker, cooler boundaries between granules. This convective motion transports energy from the Sun's interior outward to the solar atmosphere.

about 0.01% as thick as the air we breathe. The photosphere has a nearly perfect blackbody spectrum corresponding to an average temperature of 5800 K (review Figure 4-3).

The photosphere is the lowest of the three layers comprising the Sun's atmosphere. Because the upper two layers are transparent to most wavelengths of visible light, we see through them down to the photosphere. We cannot, however, see through the shimmering gases of the photosphere, and so everything below the photosphere is called the Sun's interior.

The photosphere appears darkest toward the edge, or **limb,** of the solar disk, a phenomenon called **limb darkening** (examine Figure 9-1). This effect arises because we always see through the same amount of gas in the Sun's atmosphere. At the limb, that distance takes us only as far as higher, cooler, and, therefore, darker gases. In contrast, looking straight down at the center of the Sun's disk, we see the photosphere farther into the Sun's atmosphere, where it is hotter and brighter than at the limb.

Under good observing conditions with a telescope and using special dark filters for protection, you can see a blotchy pattern called *granulation* on the photosphere (Figure 9-2). The lightly colored **granules** measure about 1000 km across and are surrounded by darkish boundaries. Time-lapse photography shows that granules form, disappear, and then reform in cycles lasting several minutes. At any single moment, several million granules cover the solar surface.

At the end of Chapter 4 we learned that radial motion of a light source affects its spectral lines through the Doppler effect (see Figures 4-17 and 4-18 and An Astronomer's Toolbox 4-3). By carefully measuring the wavelengths of spectral lines in various parts of individual solar granules, astronomers have determined that hot gas rises upward in the center of each granule. The gas radiates its energy out into space, thereby cooling. We see this energy as visible light and other electromagnetic radiation. The cooled gas then spills over the edges of the granule and plunges back down into the Sun along the boundaries between granules (see Figure 9-2 inset). We saw in Chapter 7 that convection also helps form the belts and zones on the giant planets.

According to the Stefan-Boltzmann law (see Chapter 4), hotter regions emit more photons per square meter than do cooler regions. Note in Figure 9-2 that the centers of granules appear brighter than their edges. Measurements indicate that a granule's center is typically 100 K hotter than its edge, thus explaining the observed brightness difference.

## 9-2 The chromosphere is characterized by spikes of gas called spicules

Immediately above the photosphere is a dim layer of less dense stellar gas called the **chromosphere** ("sphere of color"). This unfortunate name suggests that it is the layer we normally see, but for centuries it was visible only when the photosphere was blocked, during a total solar eclipse.

During an eclipse, the chromosphere is visible as a pinkish strip some 2000 km thick around the edge of the dark Moon (Figure 9-3). Today, astronomers can also study the chromosphere through filters that pass light with specific wavelengths strongly emitted by it—but not by the photosphere—or through telescopes sensitive to nonvisible wavelengths that the chromosphere emits intensely.

High-resolution images of the chromosphere reveal numerous spikes, which are jets of gas called **spicules** (see Figure 9-3 inset). A typical spicule rises for several minutes at the rate of 72,000 km/h (45,000 mph) to a height of nearly 10,000 km (Figure 9-4). Then it collapses and fades away. At any one time, roughly a third of a million spicules cover a few percent of the Sun's chromosphere.

Spicules are generally located on the boundaries of enormous regions of rising and falling chromospheric gas called **supergranules** (Figure 9-5). A typical supergranule

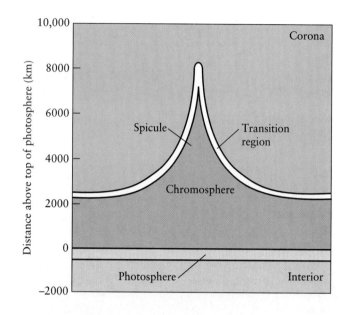

**Figure 9-4**    **The Solar Atmosphere** This schematic diagram shows the solar atmosphere's three layers. The photosphere is about 400 km thick. The chromosphere above it extends to an altitude of about 2000 km, with spicules jutting up to nearly 10,000 km above the photosphere. The outermost layer, the corona (discussed in Section 9-3), extends millions of kilometers above the photosphere.

has a diameter slightly larger than the Earth's and contains about 900 granules.

## 9-3 The corona ejects some of its mass into space as the solar wind

The outermost region of the Sun's atmosphere, the **corona,** extends several million kilometers from the top of the chromosphere (see Figure 9-4). It seems plausible that the temperature should fall as one rises through the Sun's atmosphere. After all, by moving upward, one moves farther from the apparent heat source. Indeed, starting in the photosphere at 5800 K, the temperature drops to around 4000 K in the lower chromosphere. Surprisingly, however, the temperature then begins to rise much higher as one ascends through the corona.

The unexpected increase in temperature was discovered around 1940, when astronomers realized that the spectrum of the Sun's corona contains the emission lines of a number of highly ionized elements. For example, a prominent green line caused by the presence of Fe XIV, an iron atom stripped of 13 electrons, appears in the coronal spectrum. (Recall that Fe I is a neutral iron atom.) Because extremely high temperatures are required to strip that many electrons from atoms, it is clear that the corona

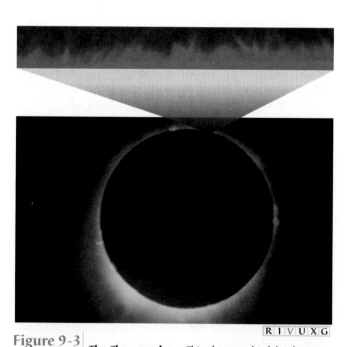

R I V U X G

**Figure 9-3**    **The Chromosphere** This photograph of the chromosphere was taken during an eclipse. It appears pinkish because the gas in the chromosphere emits only certain colors (wavelengths), among which the red ones dominate. **Inset:** Spicules are seen in this photograph of the Sun's chromosphere, which was taken through an $H_\alpha$ filter. The spicules are jets of gas that surge upward into the Sun's outer atmosphere.

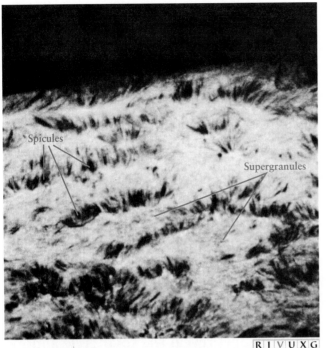

Figure 9-5  **Supergranules**  Surrounded by spicules, super-granules are regions of rising and falling gas in the chromosphere. Each supergranule spans hundreds of granules in the photosphere below.

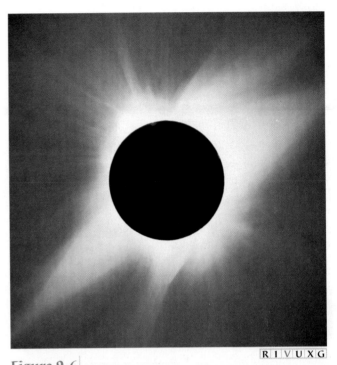

Figure 9-6  **The Solar Corona**  This extraordinary photograph was taken during the total solar eclipse of July 11, 1991. Numerous streamers are visible, extending millions of kilometers above the solar surface.

must be very hot. It is now known that coronal temperatures are typically in the range of 1 to 2 million Kelvins. Some parts may even be 10 times higher!

Although the temperature is extremely high, the density of gas in the corona is very low, about 10 trillion times less dense than the air at sea level on Earth. The low density partly accounts for the dimness of the corona, which otherwise would outshine the photosphere. Astronomers have mounting evidence that the corona is heated by energy carried aloft and released there by the Sun's complex magnetic fields, discussed in Sections 9-5 and 9-6.

The total amount of visible light we receive from the solar corona is comparable to the brightness of the full Moon—or only about one-millionth as bright as the photosphere. As with the chromosphere, the corona can be seen only when the photosphere is blocked out or through special filters or at nonvisible wavelengths (such as ultra-violet and X-ray), at which the corona is especially bright compared to the photosphere. Blocking the photosphere occurs naturally during a total eclipse or artificially in a specially designed telescope called a *coronagraph*. Figure 9-6 is a photograph of the corona taken during a total eclipse. The photo that opens this chapter shows stunning X-ray images of the corona.

Just as the Earth's gravity prevents most of our atmosphere from escaping into space, so, too, does the Sun's gravity keep most of its outer layers from leaving. However, some of the gas in the corona is moving fast enough—around a million kilometers per hour—to escape forever into space. As we saw in Chapter 5 in discussing Earth's magnetosphere and in Chapter 8 in discussing comets, such outflowing gas is called the **solar wind.**

The Sun ejects around a million tons of matter each second in the solar wind. Even at this rate, this process will use up only a few tenths of a percent of the Sun's total mass throughout its lifetime. While electrons and hydrogen and helium nuclei comprise 99.9% of the solar wind, silicon, sulfur, calcium, chromium, nickel, neon, and argon ions have also been detected in it.

## THE ACTIVE SUN

Granules, supergranules, spicules, and the solar wind occur continuously. These are features of the *quiet* Sun. But the Sun's atmosphere is periodically disrupted by magnetic fields that stir things up, creating the *active* Sun. The Sun's most obvious transient features are **sunspots,** regions of the photosphere that appear dark because they are cooler than the rest of the Sun's lower atmosphere. Sometimes sunspots occur in isolation (Figure 9-7a), but often they arise in clusters called *sunspot groups* (Figure 9-7b).

a

b

R I V U X G

**Figure 9-7**    **Sunspots**  (a) This dark region on the Sun is a typical isolated sunspot. Granulation is visible in the surrounding, undisturbed photosphere. (b) This high-resolution photograph shows a sunspot group in which several sunspots overlap and others are nearby.

🚀 **Insight into science**  **Perception versus reality**
Our senses (and often our technology) are limited, and so we must be careful how we interpret what we perceive. For example, the sunspots in Figure 9-7 certainly look like black spots. However, they appear black only in contrast to the bright light around them. As we will discuss shortly, sunspots are actually red and orange.

## 9-4  Sunspots reveal the solar cycle and the Sun's rotation

Like other transient features of the active Sun, the average number and location of sunspots vary in predictable cycles. As shown in Figure 9-8a, the average sunspot cycle lasts approximately 11 years. Within an 11-year cycle, most sunspots appear at a **sunspot maximum** (Figure 9-8b). Sunspot maxima occurred most recently in 1979, 1989, and 2000.

a

b

c

R I V U X G

**Figure 9-8**    **(a) The Sunspot Cycle**  The number of sunspots on the Sun varies with a period of about 11 years. The most recent sunspot maximum occurred in 2000, and the most recent sunspot minimum occurred in 1996. **(b)** The active Sun has many sunspots and other features (this photo was taken in 1989). **(c)** The quiet Sun is devoid of such features (this photo was taken in 1986).

During a **sunspot minimum,** the Sun is almost devoid of sunspots, as it was in 1976, 1986, and 1996 (Figure 9-8c). The next trough in the cycle is projected to occur in 2007.

A typical sunspot is 10,000 km across and lasts between a few hours and a few months. Each sunspot has two parts: a dark, central region, called the *umbra,* and a brighter ring surrounding the umbra, called the *penumbra,* visible in Figures 9-7a and 9-7b. While these are the same names as the blocked areas of an eclipse (see Chapter 1), we will see in the next section that their causes are completely different—another good example of how words can have different meanings in different contexts.

Seen without the surrounding, brilliant granules to outshine it, a sunspot's umbra appears red and its penumbra orange. From Wien's law (see Chapter 4), these colors indicate that the umbra is typically 4300 K and the penumbra 5000 K, both cooler than the normal photosphere.

On rare occasions a sunspot group is so large that it can be seen with the unaided eye. Chinese astronomers recorded such sightings 2000 years ago, and a huge sunspot group visible to the naked eye was seen in 1979. **(Always use special dark filters or other means to protect your eyes when viewing the Sun. Looking directly at the Sun for more than a few moments can cause blindness!)** Of course, a telescope gives a much better view, and so it was not until Galileo that anyone examined sunspots in detail.

By following sunspots as they moved across the solar disk (Figure 9-9), Galileo discovered that *the Sun rotates once in about four weeks.* Because a typical sunspot group lasts about two months, it can be followed for two solar rotations. Sunspot activity also lets us see the Sun's *differential rotation* (see Chapter 7): The equatorial regions rotate more rapidly than the polar regions. A sunspot near the solar equator takes 25 days to go once around the Sun, but a sunspot at 30° north or south of the equator takes about 27 days. The rotation period at 75° north or south of the equator is about 33 days, and near the poles it may be as long as 35 days.

The average latitude at which new sunspots appear changes throughout the sunspot cycle. At the beginning of each cycle, the sunspots appear mostly at about 30° north and south latitudes. Ones that form later in the cycle typi-

cally occur closer to the equator. Figure 9-10 shows that the average latitude at which sunspots are located varies at the same 11-year rate as does the number of sunspots. Satellite measurements reveal that the Sun emits about 0.1% more energy at the peak of the sunspot cycle than at its minimum.

| | |
|---|---|
| 1947 March 3 | March 30 |
| Overcast March 4 | 1947 April 1 |
| March 5 | April 2 |
| March 6 | April 3 |
| March 7 | April 4 |
| March 8 | April 5 |
| March 9 | April 6 |
| March 10 | April 7 |
| March 11 | April 8 |
| March 12 | April 9 |
| March 13 | April 10 |
| March 14 | April 11 |
| March 15 | April 12 |
| March 16 | April 13 |

R I V U X G

**Figure 9-9** **The Sun's Rotation** This series of photographs shows the same sunspot group over 1½ solar rotations. Note how the sunspot groups have changed over this time. By observing the same group of sunspots from one day to the next in this same manner, Galileo found that the Sun rotates once in about four weeks. Sunspot activity also reveals the Sun's differential rotation: The equatorial regions rotate faster than the polar regions.

**Figure 9-10** **Average Latitude of Sunspots Throughout the Sunspot Cycle** This graph shows that at the beginning of each sunspot cycle, most sunspots are at moderate latitudes, around 30° north or south. Sunspots arising later in each cycle typically form closer and closer to the Sun's equator.

## 9-5 The Sun's magnetic fields create sunspots

In 1908 the American astronomer George Ellery Hale discovered that sunspots are directly linked to intense magnetic fields on the Sun. When Hale focused a spectroscope on sunlight coming from a sunspot (Figure 9-11a), he found that each spectral line in the normal solar spectrum is flanked by additional, closely spaced spectral lines not usually observed (Figure 9-11b). This "splitting" of a single spectral line into two or more lines is called the **Zeeman effect**. The Dutch physicist Pieter Zeeman, who first observed it in the laboratory in 1896, showed that an intense magnetic field splits the spectral lines of a light source inside the field. The more intense the magnetic field, the more the split lines are separated.

Observationally, sunspots are areas where concentrated magnetic fields project through the hot gases of the photosphere. But how do the fields create the spots? The answer lies in the interaction between the magnetic fields and the photosphere's gases. Because of the photosphere's high temperature, many atoms in it are *ionized:* One or more of their electrons have been stripped off by high-energy photons there. As a result, the photosphere is a mixture of electrically charged ions and electrons called a **plasma**.

Plasmas are extremely good conductors of electricity and are repelled from regions of high magnetic field. Therefore, the magnetic field protruding through the photosphere prevents hot, ionized gases inside the Sun from rising to the surface as they normally do. Thus, such regions of the photosphere are left relatively devoid of hot gas and are therefore cooler and darker than the surrounding solar surface. We observe these darker, cooler regions as sunspots.

Recall from Chapter 5 that magnetic fields form complete loops, with each magnetic field having a north and a south pole. Likewise, when the Sun's magnetic field emerges through one sunspot or sunspot group, it forms a loop that reenters the Sun at another sunspot or sunspot group (Figure 9-12 insets). We associate the names north pole or south pole with each sunspot or sunspot group depending on whether the magnetic field there is leaving (a south pole) or entering (a north pole) the Sun. Sunspots and sunspot groups in each hemisphere are connected in

a

b

R I V U X G

**Figure 9-11** | **Zeeman Splitting by a Sunspot's Magnetic Field** (a) The black line drawn across the sunspot indicates the location toward which the slit of the spectroscope was aimed. (b) In the resulting spectrogram, one line in the middle of the normal solar spectrum is split into three components. The separation between the three lines corresponds to a magnetic field roughly 5000 times stronger than the Earth's magnetic field.

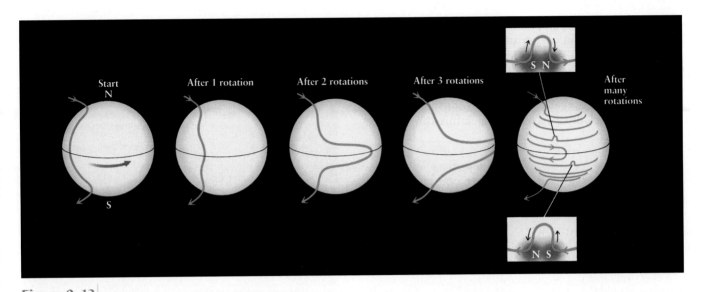

**Figure 9-12** | **Babcock's Magnetic Dynamo** In a possible partial explanation for the sunspot cycle, differential rotation wraps a magnetic field around the Sun. Convection under the photosphere tangles the field, which becomes buoyant and rises through the photosphere, creating sunspots and sunspot groups. (**Insets**) In each group, the sunspot that appears first has the same polarity as the Sun's magnetic pole in that hemisphere. (In this drawing, N in the upper hemisphere and S in the lower hemisphere.)

pairs, one where the magnetic field leaves the Sun, the other where it returns.

During his observations, Hale found that on one hemisphere of the Sun, sunspots with a magnetic north pole always come into view before the corresponding sunspots with a magnetic south pole. On the other hemisphere, the order is reversed (see Figure 9-12 insets). Hale also found that this pattern reverses itself every 11 years. The hemisphere where north magnetic poles come first during one 11-year cycle has south magnetic poles coming first during the next. Astronomers therefore speak of the 22-year **solar cycle**, the time it takes the magnetic fields to return to the same orientation.

In 1960 another American astronomer, Horace Babcock, proposed the **magnetic dynamo** model to explain the 22-year solar cycle. The Sun's rotation generates a magnetic field via the movement of its electrically charged gases. This magnetic field normally lies just below the surface. As shown in Figure 9-12, the Sun's differential rotation causes the field to become increasingly stretched.

This magnetic field becomes even more complicated as convection of the gases under the photosphere causes the fields to move vertically and horizontally, thereby developing tangles. These jumbled regions of magnetic field trap gases and expand, becoming buoyant and floating up through the solar surface. This first occurs at high latitudes; over the next 11 years it occurs closer and closer to the Sun's equator.

Sunspots, with magnetic fields typically 1000 times stronger than the Earth's magnetic field, form where the loops of solar magnetic field leave and reenter the Sun. The gas trapped in these loops eventually leaks out, and the fields interact with other parts of the Sun's magnetic field, eventually disappearing along with their sunspots. Because of how these fields vanish, every 11 years the Sun's entire magnetic field is reversed—the Sun's north magnetic pole becomes its south magnetic pole and vice versa. After another 11-year cycle, the field is back to its original orientation. This is why the solar cycle is 22 years long.

Notable irregularities occur in the solar cycle. For example, the overall reversal of the Sun's magnetic field is often piecemeal and haphazard: One pole may reverse polarity long before the other, so that for weeks or months on end the Sun may exhibit two north poles but no south pole. The corresponding south poles remain under the surface during this time.

More intriguing still is the strong historical evidence that all traces of sunspots and the sunspot cycle have vanished for decades at a time. For example, in 1893 the British astronomer E. Walter Maunder used historical observations to conclude that virtually no sunspots occurred from 1645 through 1715. This period, called the *Maunder minimum*, coincides with a period of cold in Europe so extreme that it was called the Little Ice Age. At the same time, western North America was subject to severe drought.

Similar sunspot-free periods apparently occurred at irregular intervals in earlier times as well. Conversely, periods of increased sunspot activity in the eleventh and twelfth centuries coincided with periods of warmer than average temperatures. It remains to be seen, however, if scientists can prove that the number of sunspots actually *caused* these periods of extreme temperatures on Earth.

> **Insight into science   Cause and effect** While extremes of sunspot activity often occur at the same times as weather extremes on Earth, models of these phenomena must also take into account the possibility of other, terrestrial causes for these effects. Remember, just because one event follows another does not mean that the first *causes* the second.

Theoretical calculations suggest that sunspots should last only for a matter of days before the magnetic fields collapse back into the Sun. Astronomers are just now beginning to understand how sunspots can persist for weeks or longer. Observations indicate that the magnetic fields on the Sun's surface are replenished about once every two days by a mechanism in which the magnetic field lines are broken apart and reconnected (Figure 9-13). Current theoretical work in this area is providing many new insights into the behavior of the Sun's surface.

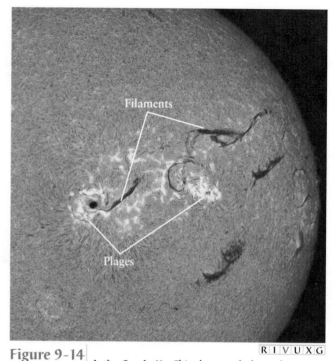

**Figure 9-14**     R I V U X G     **Active Sun in H$_\alpha$** This photograph shows the chromosphere and corona during a solar maximum, when sunspots are abundant. The image was taken through a filter allowing only light from H$_\alpha$ emission to pass through. The hot upper layers of the Sun's atmosphere are strong emitters of H$_\alpha$ photons. Besides sunspots, several other features do not appear at the solar minimum, such as the snakelike features shown here called filaments and the bright areas called plages.

**Figure 9-13**     **Magnetic Fields of the Sun** Based on data from the Solar and Heliospheric Observatory (SOHO), this image shows the magnetic fields extending from inside the Sun into the corona. The black and white spots at the bases of the magnetic loops indicate different polarity (that is, north and south poles). These fields are replenished every two days.

## 9-6 Solar magnetic fields also create other atmospheric phenomena

Figure 9-14 shows the active Sun's chromosphere and corona. The bright areas in this photograph are called **plages** (pronounced "plahj," from the French word for "beaches"). Best seen in the light emitted by calcium or hydrogen atoms, they are hotter, and therefore brighter, than the surrounding chromosphere. Plages, which often appear just before nearby sunspots form, are believed to be created by magnetic fields under the photosphere crowding upward just before they emerge through the photosphere. In pushing upward, the fields compress the gases of the upper Sun, causing this gas to become hotter and therefore to glow more brightly.

The dark streaks in Figure 9-14 are features in the corona called **filaments**. These huge volumes of gas are lofted upward from the photosphere by the Sun's magnetic field. When viewed from the side rather than from above, filaments form gigantic loops or arches called **prominences** (Figure 9-15a). The temperature of gas in prominences can

**Figure 9-15** **Prominences** (a) A huge prominence arches above the solar surface in this Skylab photograph taken in 1973. The radiation that exposed this picture is from singly ionized helium at a wavelength of 30.4 nm, corresponding to a temperature of about 50,000 K. (b) The gas in this prominence was so energetic that it broke free from the magnetic fields that shaped and confined it. This eruptive prominence occurred in 1997 and did not strike the Earth.

reach 50,000 K. These features are almost always associated with sunspots. Some prominences last only for a few hours, while others persist for months. The most energetic prominences escape the magnetic fields that confine them and surge out into space (Figure 9-15b).

X-ray photographs also reveal numerous coronal bright and dark spots that are hotter or colder, respectively, than the surrounding corona (Figure 9-16). Temperatures in the bright regions occasionally reach 4 million K. Many of the bright coronal hot spots are located over sunspots. The darker, cooler **coronal holes** act as conduits for gases to flow out of the Sun. Therefore, when a coronal hole on the rotating Sun faces Earth, the solar wind in our direction increases dramatically.

Violent, eruptive events on the Sun, called **solar flares,** also send out vast quantities of high-energy particles, as well as X rays and ultraviolet radiation from the Sun (Figure 9-17). At the maximum of the sunspot cycle, there are about 1100 flares per year. Most flares last for less than an hour, but during that time, temperatures soar to 5 million Kelvin. By following the paths of particles

**Figure 9-16** **A Coronal Hole** This X-ray picture of the Sun's corona was taken by the *Yohkoh* satellite in 1992. A huge coronal hole dominates the top part of the corona. Numerous bright points are also visible. The brightest of these are remnants of solar flares.

**Figure 9-17** **A Flare** Solar flares, which are associated with sunspot groups, produce energetic emission of particles from the Sun.

emitted by flares as they are guided outward by the Sun's magnetic field, astronomers have recently begun mapping that field as it extends out beyond the Earth (Figure 9-18).

> **Insight into science** **Accept coincidences** As the word "serendipitous" suggests, purely chance occurrences imply no preplanning, natural, or supernatural influences. Literally as I was typing the last paragraph, NASA flashed news on the Internet that the Sun was seen to vibrate as a recent solar flare occurred (more about this in Section 9-8). All too often, people believe that simple coincidences have cause-and-effect relationships. Pseudosciences like astrology thrive on this sort of misconception.

Astronomers recently discovered that huge, balloon-shaped volumes of high-energy gas are ejected from the corona. These **coronal mass ejections** typically expel 2 trillion tons of matter at 400 km/s, and each one lasts for up to a few hours (Figure 9-19). Coronal mass ejections have enough energy to break through the Sun's magnetic fields that normally contain them or even to carry fields outward, altering the Sun's magnetic field. Coronal mass ejections are often, but not always, associated with flares. The connection between the two is not yet understood.

Some coronal mass ejections, solar flares, and prominences head toward Earth. It takes about 8 minutes for their electromagnetic radiation to get here; their high-energy particles arrive a few days later. At times when these surges of particles are *not* coming to us, the normal solar wind particles are trapped by the Earth's magnetic fields in the Van Allen belts (Section 5-4). The additional particles from a coronal mass ejection or other solar event overwhelm the Van Allen belts, enabling matter in them to cascade earthward.

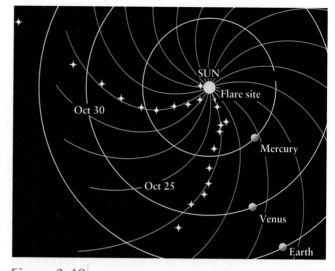

**Figure 9-18** **A Snapshot of the Sun's Global Magnetic Field**
By following the paths of particles emitted by a solar flare, astronomers have begun mapping the solar magnetic field. The field guides the outflowing particles, which in turn emit radio waves that indicate the position of the field. This data was collected by the *Ulysses* spacecraft in 1994.

a    b    c    R I V U X G

**Figure 9-19** **A Coronal Mass Ejection** Visible light photographs of a coronal mass ejection from the Sun taken by the Solar Maximum Mission. The Sun is blocked so that the ejection is visible. Figure **(a)** shows a relatively quiet corona. The time interval between the coronal mass ejection in images **(b)** and **(c)** is only 16 minutes. When coronal mass ejections head toward the Earth, the particles often penetrate the Van Allen belts, causing aurorae, disrupting radio communications and electrical power transmission, damaging satellites, and ejecting some of Earth's atmosphere into interplanetary space.

This inflow of high-energy particles damages satellites, disrupts radio communications, shorts out power grids, produces intense aurorae in the Earth's atmosphere, and causes some of the Earth's atmosphere to gush into interplanetary space. Because we now understand these solar phenomena sufficiently well, we can use our technology to monitor the Sun, watch for the radiation from high-energy events, and thereby forecast several days in advance the arrival of dangerous levels of solar particles. This enables us to protect satellites and, if necessary, bring astronauts back to Earth.

The numbers of plages, prominences, flares, and coronal mass ejections vary with the same 11-year cycle as the number of sunspots. Coronal mass ejections, the major source of hazardous particles from the Sun, occur with varying frequency throughout the sunspot cycle, but they never completely cease.

## THE SUN'S INTERIOR

During the nineteenth century, geologists and biologists found convincing evidence that the Earth must have existed in more or less its present form for at least hundreds of millions of years. This fact posed severe problems for astrophysicists, because at that time it seemed impossible to explain how the Sun could continue to shine for so long, radiating immense amounts of energy into space. If the Sun was shining by burning coal or gas, for example, it would be ablaze for only 5000 years before consuming all its fuel.

Everyday experience tells us the Sun is the source of an enormous amount of energy. Formal observations reveal hot gas, intense magnetic fields, and a variety of continuous and transient features on the Sun's surface. But the energy doesn't come from the surface gas or the magnetic fields—they have no mechanisms to create it. The answer lay deep within the Sun. To understand why the Sun shines, we must understand where its energy originates and how that energy is transported to its surface.

## 9-7 Thermonuclear reactions in the core of the Sun produce its energy

In 1905, Albert Einstein provided an important key to the source of the Sun's energy with his special theory of relativity. One of the implications of this theory is that matter and energy are related by the simple equation:

$$E = mc^2$$

In other words, a mass ($m$) can be converted into an amount of energy ($E$) equivalent to $mc^2$, where $c$ is the speed of light. Because $c^2$ ($c \times c$) is huge, $9 \times 10^{10}$ km$^2$/s$^2$, a small amount of matter can be converted into an awesome amount of energy.

Inspired by Einstein's work, astrophysicists began to wonder if the Sun's energy output might come from the conversion of matter into energy. But how? The answer came in the 1920s from the British astrophysicist Arthur Eddington. Eddington speculated that temperatures at the center of the Sun, its **core**, must be much greater than had

# AN ASTRONOMER'S TOOLBOX 9-1

## Thermonuclear Fusion

What drives the thermonuclear fusion that powers the Sun? For nuclei to fuse, they must be brought together at incredibly high temperatures and pressures. That is exactly what occurs in the Sun's core, where the entire mass of the Sun compresses inward. The core's temperature is 15.5 million K, its pressure is about $3.4 \times 10^{11}$ atm, and its density is 160 times greater than that of water.

Under less severe conditions, nuclei could not penetrate each other because the positive electric charge on each proton prevents nearby protons from coming together. (Remember that like charges repel each other.) But in the extreme heat and pressure of the Sun's center, the protons move so fast that they can stick, or fuse, together.

The nuclear transformations inside the Sun follow several routes, but each begins with the simplest atom, hydrogen (H). Most hydrogen nuclei consist of a single proton. The outcome of fusion is creation of the nucleus of the next simplest atom, helium (He), consisting of two protons and two neutrons. The fusion of hydrogen into helium takes several steps. The accompanying diagram shows one common path for the reaction. This particular sequence, called the proton-proton, or PP, chain, is the most common fusion process occurring in the Sun.

Note that the proton-proton fusion releases positively charged electrons, e⁺, called **positrons.** When these positrons encounter regular electrons in the Sun's core, both particles are annihilated and their mass is converted into energy in the form of gamma-ray photons. In addition, neutral, nearly massless particles called **neutrinos,** ν, are also emitted. The final step, in which the helium forms, returns two protons, which are then available to fuse again.

Because the PP chain produces neutrinos and the Sun's energy, we can summarize hydrogen fusion this way:

$$4H \rightarrow He + neutrinos + energy$$

*Example:* From our summary equation, we can easily calculate the energy released during a fusion reaction. We simply look at how much mass is converted into energy:

$$
\begin{array}{r}
\text{Mass of 4 hydrogen atoms} = 6.693 \times 10^{-27} \text{ kg} \\
- \quad \text{Mass of 1 helium atom} = 6.645 \times 10^{-27} \text{ kg} \\
\hline
\text{Mass lost} = 0.048 \times 10^{-27} \text{ kg}
\end{array}
$$

ever been imagined. Calculations reveal that the temperature in the Sun's core is about $15.5 \times 10^6$ K. At these high temperatures, the nuclei of the Sun's hydrogen fuse together to produce nuclei of the element helium. In this reaction a tiny amount of mass is lost: It is transformed into a very large amount of energy—the energy of the Sun.

The process of fusing nuclei at such extreme temperatures is called **thermonuclear fusion.** In particular, conversion of hydrogen into helium is called **hydrogen fusion,** and the same process provides the devastating energy of a hydrogen bomb. (We will encounter the thermonuclear fusion of helium and other elements in later chapters.) The energy generated by hydrogen fusion in the Sun's core eventually escapes through the photosphere into space and that energy makes the Sun shine.

You may have heard comments that mass is always conserved or that energy is always conserved in a reaction. We know that both of these concepts are inaccurate, because mass can be converted into energy and vice versa.

What is true, however, is that the total amount of mass *plus* energy is conserved. So, the destruction of mass by the Sun does not violate any laws of nature.

Hydrogen fusion is also called *hydrogen burning,* even though nothing is actually burned in the conventional sense. The ordinary burning of wood, coal, or any flammable substance is a chemical process involving only the electrons orbiting the nuclei of the atoms. Thermonuclear fusion is a far more energetic process that involves violent collisions between the atomic nuclei themselves. An Astronomer's Toolbox 9-1 shows you just *how* energetic.

## 9-8 Solar models describe how energy escapes from the Sun's core

A scientific description of the Sun's interior, called a **solar model,** explains how the energy from nuclear fusion gets to the photosphere. The model begins with the inward

Thus, a small fraction (0.7%) of the mass of the hydrogen going into the nuclear reactions does not show up in the mass of the helium. This last mass is converted into energy, as predicted from Einstein's famous equation,

$$E = mc^2$$
$$= (0.048 \times 10^{-27} \text{ kg}) \times (3 \times 10^8 \text{ m/s})^2$$
$$= 4.3 \times 10^{-12} \text{ J}$$

*Compare!* This energy is released from the formation of a single helium atom. It would light a 10-watt lightbulb for almost one-half a trillionth of a second.

Now let's add up the energy emitted from the entire Sun. The Sun's mass, usually designated 1 M$_\odot$, is equal to 333,000 M$_\oplus$. Its total energy output per second, called the **solar luminosity** and denoted L$_\odot$, is $3.9 \times 10^{26}$ W. To produce this luminosity, the Sun converts 600 million metric tons of hydrogen into helium within its core each second. This prodigious rate is possible because the Sun contains a vast supply of hydrogen—enough to continue the present rate of energy output for another five billion years.

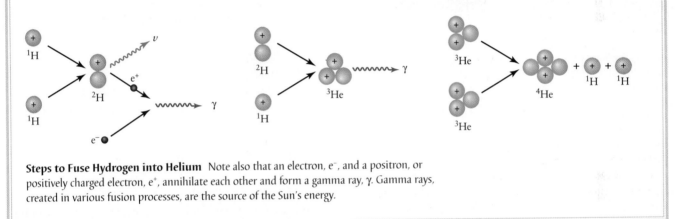

**Steps to Fuse Hydrogen into Helium**  Note also that an electron, e⁻, and a positron, or positively charged electron, e⁺, annihilate each other and form a gamma ray, γ. Gamma rays, created in various fusion processes, are the source of the Sun's energy.

force of the Sun's gravity. This force raises the pressure and temperature in the Sun's core. However, because the Sun is not shrinking today, an outward force must counter the inward force of gravity. That outward force is produced by the gamma-ray photons created during fusion.

These photons slam into nearby ions and electrons in the solar core. These particles absorb the energy of the photons and rebound at very high speeds. Because they are densely packed together, they do not travel far before striking other particles. In each collision, the particles exert forces on each other. They then rebound, emitting photons and colliding with other particles. The result of these frequent interactions is an outward force sufficient to counterbalance gravity. The balance between the inward force of gravity and the outward force from the motion of the hot gas is called **hydrostatic equilibrium.**

The photons emitted after particles collide in the Sun's core have slightly less energy than the photons that these particles had previously absorbed. The energy the photons lose provides the outward force keeping the Sun in equilibrium. Photons collide and new ones are emitted. Photon energies slowly diminish as the photons travel outward from the Sun's core to the photosphere, an odyssey that typically takes 170,000 years.

The outward transportation of energy by photons hitting particles, which then bounce off other particles and thereby reemit photons, is called *radiative transport,* because individual photons are responsible for carrying energy from collision to collision. Calculations show that radiative transport is the dominant means of outward energy flow in the **radiative zone,** extending from the core to 80% of the way out to the photosphere.

Near the photosphere, bulk motion of the hot gas, rather than the flying, energetic photons, carries most of the energy the remaining distance to the surface. As we discussed earlier, this upward motion of gas blobs is called convection. We thus say that the Sun has a **convective zone.** Once the gas reaches the photosphere, it emits photons into

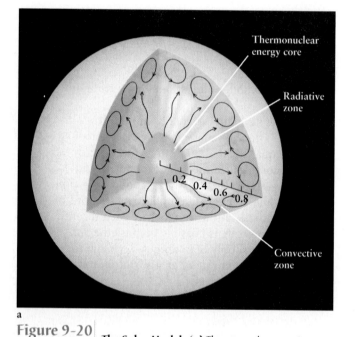

a

**Figure 9-20** | **The Solar Model** (a) Thermonuclear reactions occur in the Sun's core, which extends to a distance of 0.25 solar radius from the center. In this model, energy from the core radiates outward to a distance of 0.8 solar radius. Convection is responsible for energy transport in the Sun's outer layers. (b) The Sun's internal structure is displayed here with graphs that show how the luminosity, mass, temperature, and density vary with the distance from the Sun's center. A solar radius (the distance from the Sun's center to the photosphere) equals 696,000 km.

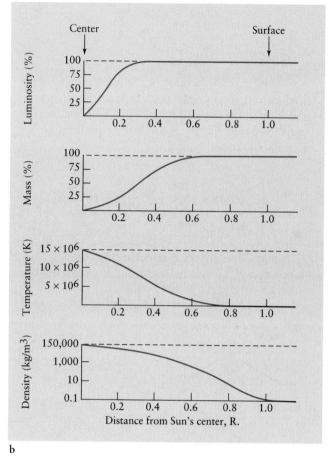

b

space, cools, and settles back into the Sun. This convective flow is the origin of solar granules, and the departing energy is the light that the Sun emits into space. Figure 9-20a sketches our model of the internal structure of the Sun.

Different gamma rays created in the Sun's core lose different amounts of energy as they and their successors travel upward through the Sun. Therefore, the photons emitted from the photosphere have a wide range of energies and, hence, wavelengths. The most intense emission is in the visible part of the electromagnetic spectrum. This is the origin of the blackbody nature of the photosphere's spectrum.

The solar model carefully charts the Sun's internal characteristics, such as its pressure, temperature, and density at various depths. Our model of the Sun's interior is expressed as a set of mathematical equations called the *equations of stellar structure*. These describe the conditions required inside *any* star to keep it stable, meaning that it neither expands nor collapses. Because the equations are so complex, astrophysicists today use computers to solve them. Figure 9-20b presents their results graphically. The four graphs show how the Sun's luminosity, mass, temperature, and density vary from the Sun's center to its surface. For example, the upper graph gives the per-

centage of the Sun's luminosity created within that radius. The luminosity rises to 100% at about one-quarter of the way from the Sun's center to its surface. This tells us that all the Sun's energy is produced within a volume extending out to ¼ $R_\odot$, where $R_\odot$ is the radius of the Sun.

The mass curve rises to nearly 100% at about 0.6 $R_\odot$ from the Sun's center. Almost all of the Sun's mass is therefore confined to a volume extending only 60% of the distance from the Sun's center to its surface, which helps explain why the density of the gas in the photosphere is $10^4$ times lower than the density of the air we breathe.

To learn more about the Sun's interior, astronomers record its vibrations—a study called **helioseismology**—just as geologists use earthquakes to study the Earth's interior structure. Although there are no true sunquakes, the Sun does vibrate at a variety of frequencies, somewhat like a ringing bell. These vibrations, first noticed in 1960, can be detected with sensitive Doppler shift measurements. They have shown that portions of the Sun's surface move up and down by about 10 km every 5 minutes (Figure 9-21a).

Slower vibrations with periods ranging from 20 minutes to nearly an hour were discovered in the 1970s. More recently, oscillations lasting several days have been detected.

25 Days                                            35 Days

**a**

**Figure 9-21**  **Helioseismology** (a) This computer-generated image shows one of the millions of ways that the Sun vibrates because of sound waves resonating in its interior. The regions that are moving outward are colored blue; those moving inward are red. The cutaway shows how deep these oscillations are believed to extend. (b) This cutaway picture of the Sun shows how the rate of solar rotation varies with depth and latitude. Red and yellow denote faster-than-average motion; blue regions move more slowly than average. The pattern of differential surface rotation, which varies from 25 days at the equator to 35 days near the poles, persists at least 19,000 km down into the Sun's convective envelope. Sunspots preferentially occur on the boundaries between different rotating regions. Earthlike jetstreams and other wind patterns have also been discovered in the Sun's atmosphere.

One important discovery from helioseismology is that the convective zone is twice as thick as prior solar models predicted. Another is that below the convective zone the Sun apparently rotates like a rigid body (Figure 9-21b).

Although the Sun is much more nearly spherical than the Earth, Doppler measurements reveal that it has "dimples" one-half kilometer deep, like a giant golf ball. Also, in 1998, astronomers first detected ripples in the Sun's atmosphere during solar flares. These rings of vibrating gas appear to be just like those you create when you throw a rock into a lake.

## 9-9  The mystery of the missing neutrinos inspired research into the fundamental nature of matter

As explained in An Astronomer's Toolbox 9-1, for every proton that changes into a neutron during thermonuclear fusion, a *neutrino* is released. Neutrinos have no electric charge, and they are extraordinarily difficult to detect because they rarely interact with ordinary matter. In fact, neutrinos interact so infrequently with matter that they can easily pass through the entire Earth as if it were not there. The Sun is also largely transparent to neutrinos, allowing these particles to stream outward, unimpeded, from its core.

According to our model of fusion in the Sun, nearly $10^{38}$ *solar neutrinos* are produced at the Sun's center each second. This output is so huge that, here on Earth, roughly 100 billion solar neutrinos pass through every square centimeter of your body every second!

On very rare occasions a solar neutrino strikes a neutron and converts it into a proton. If astronomers could detect even a few of these converted protons, it might be possible to build a "neutrino telescope" that could be used to "see" directly into the center of the Sun and the thermonuclear inferno there that is now hidden from ordinary view.

Inspired by such possibilities, Raymond Davis of the Brookhaven National Laboratory designed and built a large neutrino detector. This device consists of a huge tank containing 100,000 gallons of perchloroethylene ($C_2Cl_4$, the fluid your local dry cleaner uses) buried deep in a mine in South Dakota. Because matter is virtually transparent to neutrinos, most of the solar neutrinos pass right through Davis's tank with no effect whatever. On rare occasions, however, a solar neutrino strikes the nucleus of one of the chlorine atoms in the cleaning fluid and converts one of its neutrons into a proton, creating a radioactive atom of argon. The rate at which argon is produced is therefore correlated with the number of solar neutrinos arriving at the Earth.

On average, solar neutrinos create one radioactive argon atom every three days in Davis's tank. To the consternation of astronomers, this rate corresponds to only one-third of the neutrinos predicted from the "standard" solar model presented in Figure 9-20b. This experiment, which began in the mid-1960s, has been repeated with extreme care by other researchers around the world during the past four decades using a variety of different detection techniques. There are three types of neutrinos, each associated with different nuclear reactions, but until recently all of the experiments were designed to detect just solar neutrinos (that is, those made when neutrons transform into protons). All these traditional neutrino detectors get comparable, low results.

Most astronomers believe that the solar neutrino experiments are sound, implying that there really are fewer solar neutrinos reaching the Earth than theory predicts. This implies, of course, that the theory needs revision. Attention became focused on the two other types of neutrinos that exist and are created by different nuclear reactions than those commonly occurring in the Sun. By revising the theory with the assumption that neutrinos *have* a tiny amount of mass, the new equations predict that neutrinos can *change* from one type into another.

This is the key to resolving the dilemma of the low neutrino numbers previously detected. If neutrinos have mass, calculations predict that many of the neutrinos emitted by the Sun change to the other types of neutrinos by the time they reach the Earth. Astronomers set about to build neutrino detectors sensitive to these other neutrinos. The first such experiment to go online is in Japan (Figure 9-22) and uses a water-based neutrino detector sensitive to the alternative types of neutrinos. It began discovering them in 1998.

To understand how this new generation of neutrino detector works, note that water contains many protons, namely, the nuclei of hydrogen atoms. When one of the alternative types of neutrino strikes a proton, it produces a *positron*, which is identical to an electron, except that it

**Figure 9-22**    **The Solar Neutrino Experiment** Located in the Kamioka Mozumi mine in Japan, the Superkamiokande solar neutrino detector is centered around a tank containing 50,000 tons of ultrapure water. Occasionally, a neutrino entering the tank interacts with a proton, producing a positron, a positively charged electron. The positron moves so fast that it gives off a light flash called Cerenkov radiation. Some 13,000 light detectors (seen from below in the inset) detect this light.

has a positive charge. The positron in turn emits a flash of light called **Cerenkov radiation.** As the Russian physicist Pavel A. Cerenkov first observed, the flash occurs whenever a particle moves through water faster than light can. Such motion does not violate the tenet that the speed of light *in a vacuum* ($3 \times 10^5$ km/s) is the ultimate speed limit in the universe. Light is slowed considerably as it passes through water, and high-energy particles can exceed this reduced speed without violating the laws of physics.

Thus, scientists detect neutrinos indirectly by observing Cerenkov radiation flashes with light-sensitive devices called *photomultipliers* mounted in the water (see Figure 9- 22). This evidence that neutrinos have mass not only explains the earlier low rate of solar neutrino observations, but also has a profound implication about the fate of the universe, as we will see in Chapter 17.

## WHAT DID YOU THINK?

1. *What fraction of the solar system's mass is in the Sun?* The Sun contains about 99.85% of the solar system's mass.

2. *Does the Sun have a solid and liquid interior, like the Earth?* No. The entire Sun is composed of hot gases.

3. *What is the surface of the Sun like?* The photosphere is composed of hot, churning gases.

4. *Does the Sun rotate?* The Sun's surface rotates differentially, varying between once every 25 days near its poles and once every 35 days at its equator.

5. *What makes the Sun shine?* Thermonuclear fusion at the Sun's core is the source of the Sun's energy.

## Key Words

Cerenkov radiation, 245
chromosphere, 230
convective zone, 241
core (of the Sun), 239
corona, 230
coronal hole, 237
coronal mass ejection, 238
filament, 236
granules, 229

helioseismology, 242
hydrogen fusion, 240
hydrostatic equilibrium, 241
limb (of the Sun), 229
limb darkening, 229
magnetic dynamo, 235
neutrino, 240
photosphere, 228
plage, 236

plasma, 234
positron, 240
prominence, 236
radiative zone, 241
solar cycle, 235
solar flare, 237
solar luminosity ($L_\odot$), 241
solar model, 240
solar wind, 231

spicule, 230
sunspot, 231
sunspot maximum, 232
sunspot minimum, 233
supergranule, 230
thermonuclear fusion, 240
Zeeman effect, 234

## Key Ideas

### The Sun's Atmosphere

✳ The visible surface of the Sun is a layer at the bottom of the solar atmosphere called the photosphere. The gases in this layer shine with almost perfect blackbody radiation. Convection of this gas from below the surface produces features there called granules.

✳ Above the photosphere is a layer of hotter but less dense gas called the chromosphere. Jets of gas called spicules rise up into the chromosphere along the boundaries of supergranules.

✳ The outermost layer of thin gases in the solar atmosphere is called the corona, which blends into the solar wind at great distances from the Sun. The gases of the corona are very hot but at low density.

### The Active Sun

✳ Some surface features on the Sun vary periodically in an 11-year cycle, with the magnetic fields that cause these changes actually varying over a 22-year cycle.

✳ Sunspots are relatively cool regions produced by local concentrations of the Sun's magnetic field. The average number of sunspots and their average latitude increases and decreases in an 11-year cycle.

✳ A prominence is gas lifted into the Sun's corona by magnetic fields. A solar flare is a brief, but violent, eruption of hot, ionized gases from a sunspot group. Coronal mass ejections send out large quantities of gas from the Sun. Coronal mass ejections and flares that head our way affect satellites, communication, electrical power, and cause aurorae.

✳ The magnetic dynamo model suggests that many transient features of the solar cycle are caused by the effects of differential rotation and convection on the Sun's magnetic field.

### The Sun's Interior

✳ The Sun's energy is produced by the thermonuclear process called hydrogen fusion, in which four hydrogen nuclei release energy when they fuse to produce a single helium nucleus.

The energy released in a thermonuclear reaction comes from the conversion of matter into energy according to Einstein's equation $E = mc^2$.

✳   A stellar model is a theoretical description of a star's interior derived from calculations based on the laws of physics. The solar model suggests that hydrogen fusion occurs in a core that extends from the Sun's center to about 0.25 solar radius.

✳   Throughout most of the Sun's interior, energy moves outward from the core by radiative diffusion. In the Sun's outer layers, energy is transported to the Sun's surface by convection.

✳   Neutrinos generated and emitted by the Sun are detected at a lower rate than is predicted by our standard model of thermonuclear fusion. This apparently occurs because neutrinos actually have mass; therefore, some of them change into other forms of neutrinos before they reach Earth. These alternative forms are now being detected.

## Review Questions

The answers to all computational problems, which are preceded by an asterisk (*), appear at the end of the book.

1   Describe the features of the Sun's atmosphere that are always present.

2   Describe the three main layers in the solar atmosphere and how you would best observe them.

3   Name and describe seven features of the active Sun. Which two are the same, seen from different angles? Which have direct impact on the Earth? Explain.

4   Describe the three main layers of the Sun's interior.

*5   When will the next sunspot minimum and maximum occur after the maximum in 2000 and the minimum in 2007? Explain your reasoning.

6   Why is the solar cycle said to have a period of 22 years, even though the sunspot cycle is only 11 years long?

7   How do astronomers detect the presence of a magnetic field in hot gases, such as those in the solar photosphere?

8   Describe the dangers in attempting to observe the Sun. How have astronomers learned to circumvent these hazards?

9   Give an everyday example of hydrostatic equilibrium.

10   Give some everyday examples of heat transfer by convection and radiative diffusion.

11   What do astronomers mean by "a model of the Sun"?

12   Why do thermonuclear reactions in the Sun take place only in its core?

13   What is hydrogen fusion? This process is sometimes called "hydrogen burning." How is hydrogen burning fundamentally unlike the burning of a log in a fireplace?

14   Describe the Sun's interior, including the main physical processes that occur at various depths within the Sun.

15   What is a neutrino, and why are astronomers so interested in detecting neutrinos from the Sun?

## Advanced Questions

*16   Using the mass and size of the Sun, calculate the Sun's average density. Compare your answer to the average densities of the outer planets. Hint: The volume of a sphere of radius $r$ is $(4/3)\pi r^3$.

*17   Assuming that the current rate of hydrogen fusion in the Sun remains constant, what fraction of the Sun's mass will be converted into helium over the next five billion years? How will this affect the chemical composition of the Sun?

*18   Calculate the wavelengths at which the photosphere, chromosphere, and corona emit the most radiation. Explain how the results of your calculations suggest the best way to observe these regions of the solar atmosphere. Hint: Use Wien's law and assume that the average temperatures of the photosphere, chromosphere, and corona are 5800 K, 50,000 K, and $1.5 \times 10^6$ K, respectively.

19   When we are near a sunspot maximum, the Hubble Space Telescope must be put in a higher orbit. Why? Hint: Think about how the increased solar energy affects the Earth's atmosphere.

## Discussion Questions

20  Discuss the extent to which cultures around the world have worshiped the Sun as a deity throughout history. Why do you suppose our star inspires such widespread veneration?

21  Discuss some of the difficulties of correlating solar activity with changes in the terrestrial climate.

22  Describe some advantages and disadvantages of observing the Sun (a) from space and (b) from Earth's south pole. What kinds of phenomena and issues do solar astronomers want to explore from both Earth-orbiting and Antarctic observatories?

## What If ...

23  The Sun was not rotating? What about it would be different?

24  The typical solar wind was much stronger (say 100 times stronger) than it presently is? What differences would there be in the solar system?

25  The typical solar wind was much weaker (say 100 times weaker) than it presently is? What differences would there be in the solar system?

## Observing Projects

26  Use a telescope to view the Sun by projecting the Sun's image onto a screen or sheet of white paper. **DO NOT LOOK DIRECTLY AT THE SUN! Looking at the Sun causes blindness.** Do you see any sunspots? If so, sketch their appearance. Can you distinguish between the umbrae and penumbrae of the sunspots? Can you see limb darkening? Can you see granulation?

27  If you have access to an $H_\alpha$ filter attached to a telescope especially designed for viewing the Sun safely, use this instrument to examine the solar surface. How does the filtered appearance of the Sun differ from that in white light (see Observing Project 26)? What do sunspots look like in $H_\alpha$? Can you see any prominences? Can you see any filaments?

Are the filaments in the $H_\alpha$ image near any sunspots seen in white light?

28  Use Starry Night software to determine the Sun's rotation rate. Lock on the Sun and turn off the horizon. Zoom in on the Sun until the sunspots are clearly visible. Note the darkening of the Sun on the image. Set the timestep to 1 sidereal day and use the single timestep button (on the far right of the time window) to observe the Sun's rotation. Step through enough times to determine the rotation rate for the Sun. The rotation period portrayed by the program is closest to the rotation rate where on the Sun? (Note that the program does not show the Sun's differential rotation.)

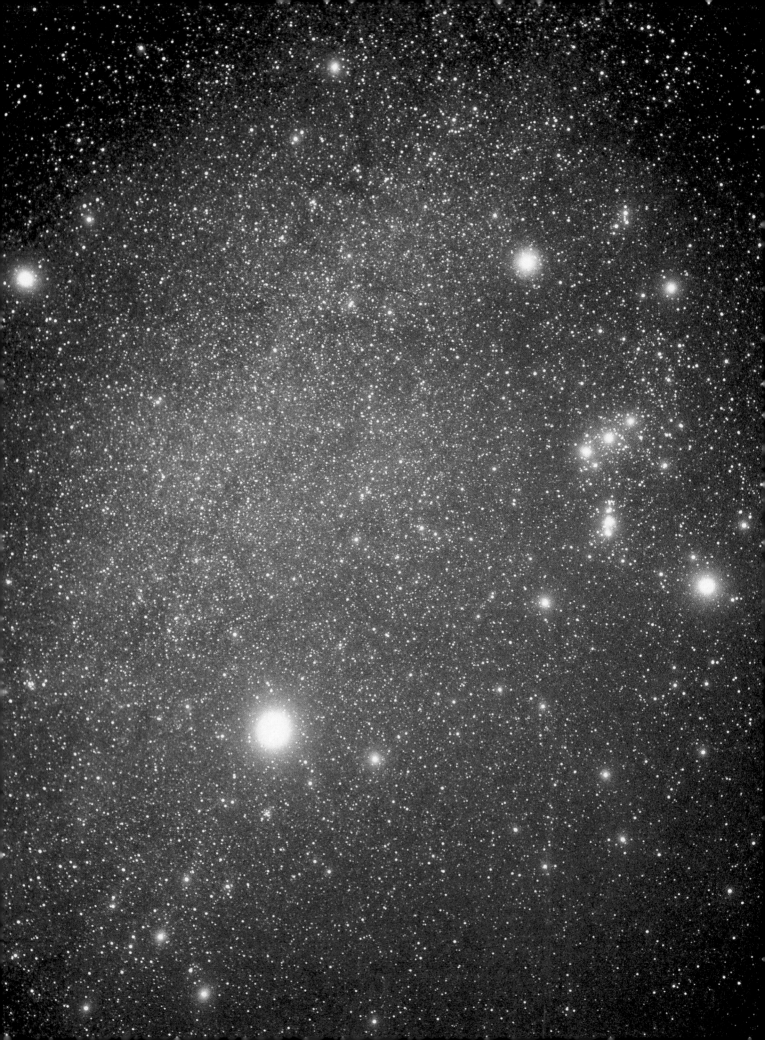

# III

# BUILD YOUR FOUNDATION

## The Stars

### *In the chapters ahead you will discover*

* ✳ the nature of stars and just how special—
and how ordinary—the Sun really is

* ✳ the observational properties of stars on which astronomers
base their models of stellar evolution

* ✳ that the distances to many nearby stars can be
measured directly, while the distances to
farther ones are determined indirectly

* ✳ that stellar luminosities and spectra are clues
to understanding stellar evolution

* ✳ that stars with different masses evolve differently

RIVUXG

**Science by Starlight** By analyzing starlight, an astronomer can determine stellar temperatures, chemistries, and masses; how far away the stars are; and how fast they are moving toward or away from us. This image shows some of the stars in Orion and adjoining constellations. These stars span virtually the entire range of star types. The brightest star in the night sky, Sirius (part of Canis Major), is toward the lower left. It is only 8.6 ly away, making it one of the stars closest to us. Most stars in this photograph are hundreds, or even thousands, of light-years away. The blue haze is starlight scattering off interstellar gas and dust.

## WHAT DO YOU THINK?

1  Do stars last forever?

2  How near is the closest star other than the Sun?

3  How luminous is the Sun compared with other stars?

An ancient legend tells of a race of remarkable insects, the Ephemera, who inhabited a great forest. These noble creatures were blessed with great intelligence but were cursed with tragically short life spans. To the Ephemera, the forest seemed eternal and unchanging. Members of each generation lived out their brief lives without ever noticing any changes in their leafy world. Nevertheless, careful observations and reasoning led some Ephemera to postulate that the forest was not static. They began to suspect that small green shoots grew to become huge trees and that mature trees eventually died, toppled over, and littered the forest with rotting logs, enriching the soil for future trees. Although unable to witness the transformation personally, the Ephemera predicted the existence of life cycles stretching over the awesome periods of many years.

1    To us, the heavens, too, seem eternal and unchanging; the views that greet us every night are virtually indistinguishable from those seen by our ancestors. This permanence, however, is an illusion. We see the stars because they emit vast amounts of radiation, created at the expense of converting one element into another in their interiors. And the change isn't limited to chemistry. We will soon see that over the course of their existence, stars also change in size, luminosity (total energy emitted each second), temperature, and color. Astronomers call this cycle **stellar evolution.**

We do have one advantage over the Ephemera, who were so small that they could see only a few trees, leaves, shoots, and rocks. Astronomers are witnesses to literally billions of stars in various stages of evolution. Therefore, although we cannot see any one star go through more than a tiny fraction of the cycle, we see every stage of stellar evolution. By understanding the changes that stars undergo, we gain insight into our place in the cosmos. We begin our journey to the stars by learning how far across interstellar space we have to go to reach the nearest ones.

## III-1  Distances to nearby stars are determined by stellar parallax

Nothing in the human experience prepares us for understanding the distances between the stars. On Earth, the greatest distance we encounter in our daily lives is at most a few thousand kilometers. From observing the Sun, Moon, and planets, we have some comprehension of hundreds of thousands or even millions of kilometers.

2    How far away do you think the nearest stars are? Ask ten people and you'll probably hear answers ranging from thousands to billions of kilometers or miles. In fact, the closest star other than the Sun, Proxima Centauri, in the constellation Centaurus, is about 40 trillion kilometers (24 trillion miles) away. It takes light about four years to get here from there. Most of the stars you see in the night sky are hundreds of times farther away than Proxima Centauri. (Centaurus, by the way, is not visible from most of the northern hemisphere.) How do we measure the distance to a nearby star?

Because Proxima Centauri is closer to us than other stars, its position among the background stars changes as the Earth orbits the Sun. This is precisely the same effect that we see when, for example, Mars appears to have retrograde motion as we pass between it and the Sun (see Figures 2-1 and 2-2). The motion of nearby stars among the background of more distant stars due to the Earth's motion around the Sun is called **stellar parallax.**

Parallax is actually an everyday phenomenon. We experience it when nearby objects appear to shift their positions against a distant background as we move (review Figure 2-4). We also experience it continuously when we are awake, because our eyes change angle when looking at objects at different distances. As you look at a tree 10 m away your eyeballs cross only slightly. Looking at something closer, say this book, your eyes cross much more in order for both eyes to focus on the same word. The parallax angle formed between your eyes and the tree or the book lets your brain judge just how close they are to you. Working out distances of nearby stars is done similarly, but it requires painstaking angle measurements and a little geometry.

As the Earth moves from one side of its orbit around the Sun to the other, a nearby star's apparent position shifts among the more distant stars. Referring to Figure III-1a, the parallax angle, $p$, is half the angle by which it shifts, measured in arcseconds. The difference in parallax angles for stars at different distances can be seen by comparing Figures III-1a and b.

Measuring parallaxes yields the most reliable distance measurements to the stars (see An Astronomer's Toolbox III-1). Naturally, astronomers want to measure the distances to as many stars as possible using the parallax technique. However, there is a catch—stellar parallax angles are extremely tiny. So astronomers must rely on complex

a

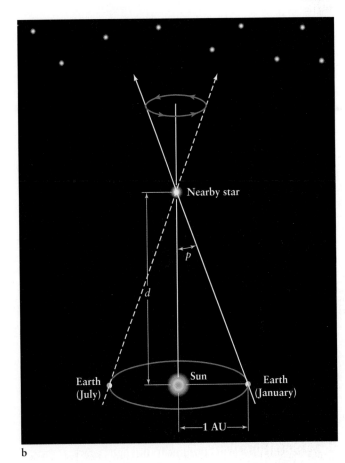

b

**Figure III-1** **Using Parallax to Determine Distance** **(a)** As the Earth orbits the Sun, a nearby star appears to shift its position against the background of distant stars. The star's parallax angle ($p$) is equal to the angle between the Sun and Earth as seen from the star. The stars on the scale of this drawing are shown much closer than they are in reality. If drawn to the correct scale, the closest star, other than the Sun, would be about 5 km (3.2 mi) away. **(b)** The farther the star is from us, the smaller the parallax angle $p$. The distance to the star (in parsecs) is found by taking the inverse of the parallax angle $p$ (in arcseconds), $d = 1/p$.

physics and incomplete knowledge about the cosmos in determining non-parallax-based distances. We will discuss other distance-measuring methods in later chapters.

The first parallax measurement was made by Friedrich Wilhelm Bessel, a German astronomer and mathematician. He found the parallax angle of the star 61 Cygni to be 1/3 arcsec, and so its distance is about 3 pc. Because parallax angles smaller than about 0.01 arcsec are difficult to measure from Earth-based observatories, the stellar parallax method using Earth-based telescopes gives stellar distances only up to about 100 pc. The precision of stellar parallax measurements is limited by the angular resolution of the telescope, as discussed in Chapter 3.

Telescopes in space are unhampered by our atmosphere and therefore have higher resolutions than Earth-based telescopes. Parallax measurements made in space

thus enable astronomers to determine the distances to stars well beyond the reach of ground-based observations.

In 1989 the European Space Agency (ESA) launched a satellite called *Hipparcos* (an acronym for *hi*gh *p*recision *par*allax *c*ollecting *s*atellite, named for Hipparchus, an astronomer in ancient Greece who created an early classification system for stars). Although the satellite failed to achieve its proper orbit, astronomers have used it to measure the distances to more than one million of the nearest stars up to 150 pc (500 ly) away. The success of *Hipparcos* has led to plans for better satellites to collect parallax data from stars even farther away.

Despite the information gained from stellar parallax, astronomers need to know the distances to more remote stars for which parallax cannot yet be measured. Several of these methods of determining ever greater distances will be

# AN ASTRONOMER'S TOOLBOX III-1

## Distances to Nearby Stars

Recall from An Astronomer's Toolbox I-2 that 1 parsec (1 pc) is the distance at which two objects 1 AU apart appear 1 arcsecond apart. This distance is $3.09 \times 10^{13}$ km, or 206,265 AU. The name parsec (from *parallax sec*ond) originates in the use of parallax to measure distance. Using parsecs, we can write down an especially simple equation for the distances to stars:

$$\text{distance to a star in parsecs} = \frac{1}{\text{parallax angle of that star in arcseconds}}$$

or

$$d_p = 1/p$$

where $d_p$ is the distance to the star in parsecs and $p$ is the parallax angle of that star in arcseconds.

The equation is only this simple in these units, which is one of the main reasons that many astronomers discuss cosmic distances in parsecs rather than light-years. We will continue to use light-years

primarily throughout this book, however, as they are more intuitive. In these latter units, the same equation becomes approximately

$$\text{distance to a star in light years} = \frac{326}{\text{parallax angle of that star in arcseconds}}$$

or

$$d_{ly} = 3.26/p$$

where $d_{ly}$ is the distance to a star in light years.

*Example:* The nearest star, Proxima Centauri, has a parallax angle of 0.77 arcsec, and so its distance is 1/0.77, or approximately 1.3 pc. Equivalently, Proxima Centauri is 4.24 ly away. The parallax of Proxima Centauri is comparable to the angular diameter of a dime seen from a distance of 3 km.

*Compare!* The average distance from the Earth to the Sun is $5 \times 10^{-6}$ parsecs.

*Self-Test:* Rigel is 773 ly away. How far away is it in parsecs? What is its parallax angle in arcseconds?

introduced in Chapters 10, 11, 12, and 15. Having established the fact that different stars are at different distances from Earth, we now turn to considering the brightnesses that stars appear to have as seen from here. This will prepare us to calculate how much light stars actually emit.

## III-2 Apparent magnitude measures the brightness of stars as seen from Earth

Greek astronomers, from Hipparchus in the second century B.C. to Ptolemy in the second century A.D., undertook the classification of stars strictly by evaluating how bright they appear to be compared to each other. This made sense because the stars were assumed to all be at the same distance from us, and, therefore, differences in brightness due to different stellar distances from Earth were not expected. These brightnesses are called **apparent magnitudes.**

The brightest stars were originally said to be of first magnitude, and their apparent magnitudes were designated as +1. (Greek astronomers did not try to classify the Sun's

dazzling brightness in this scheme.) Those stars about half as bright as first-magnitude stars were said to be second-magnitude stars (designated +2), and so forth, down to sixth-magnitude stars, the dimmest ones still visible to the unaided eye. This scale was created before accurate measurements could be made and it has since been refined.

A first-magnitude star is now defined to be exactly 100 times brighter than a sixth-magnitude star. Therefore, a magnitude difference of 1 between two stars means that we receive 2.512 times more light from the brighter of the two. (The factor 2.512 originates in the fact that going from a sixth-magnitude to a first-magnitude star leads to an increase in brightness of $2.512 \times 2.512 \times 2.512 \times 2.512 \times 2.512 = 100$ times, consistent with the definition above.) It would take 100 stars of apparent magnitude +6 to provide as much light as we receive from a single star of apparent magnitude +1. Similarly, it takes 2.512 third-magnitude stars to provide as much light as we receive from a single second-magnitude star.

Rigorous measurements of stellar brightness reveal that the brightest stars are actually brighter than apparent magnitude +1.0 (first magnitude). Therefore, astronomers resort to *negative* numbers for the magnitude of the very *brightest* objects. Sirius, for example, the brightest star in the night sky, shown in the figure that opens this Foundation, has an apparent magnitude of −1.44. With this convention we can describe other bright objects in the sky, such as the Sun, Moon, comets, and planets. At its brightest, Venus shines with an apparent magnitude of −4.4, the full Moon has an apparent magnitude of −12.6, and the Sun has an apparent magnitude of −26.7. Remember: *The smaller, or more negative, the apparent magnitude number for an astronomical body, the brighter that body is as seen from Earth.*

> **Insight into science    Diminishing returns**
> Numbering as well as naming schemes may be counterintuitive in science. One would expect brighter stars to have larger, more positive numbers than dimmer stars, but the apparent magnitude scheme is just the opposite. Similarly, in Chapter 10 you will discover that on the standard plot of stars used by astronomers, the Hertzsprung-Russell diagram, the hottest stars fall on the *left* and the coolest stars on the *right*.

Astronomers also extended the magnitude scale to larger positive numbers in order to describe dimmer stars visible only through their telescopes. For example, the dimmest stars visible through a pair of binoculars have an apparent magnitude of about +10. Time-exposure photographs reveal even dimmer stars. Through telescopes such as the Keck telescopes and the Hubble Space Telescope, stars nearly as dim as magnitude +30 can be observed. Similarly, apparent magnitudes from entire groups of stars, such as distant galaxies, can be measured. Figure III-2 illustrates the modern apparent magnitude scale.

The apparent magnitudes do not directly reveal fundamental stellar properties, because the brightnesses we measure are affected by the differing distances to the stars. Clearly, the closer of two identical stars appears brighter to us (has a smaller apparent magnitude) than the farther star. We now introduce another magnitude scale, which takes into account the different distances of stars.

## III-3    Absolute magnitudes do not depend on distance

Suppose we observe two identical stars, one twice as far away as the other. How much dimmer will the farther one appear to be as seen from Earth? As light moves outward from a source, it spreads out over increasingly larger areas of space and its brightness decreases. Thus, the farther away a source of light is, the dimmer it appears. The **inverse-square law** provides a simple rule for just *how* quickly the brightnesses of objects change with distance.

Imagine a shell of light moving out from a star (Figure III-3). Start with a small square area of light on the shell when the shell has moved out a distance $d = 1$. The light in that square has a certain brightness. When the same shell has gone twice as far ($d = 2$), you can see that the light square has become 4 times larger. The light in each of the squares at $d = 2$ contains one-quarter of the photons that were in the single square at $d = 1$. Therefore, the small squares at $d = 2$ are one-quarter as bright as the same-sized square at $d = 1$. Similarly, when the shell has moved to $d = 3$, there are now nine small squares, each one-ninth as bright as the original square at $d = 1$.

The inverse-square law can be summarized mathematically as follows: Apparent brightness decreases inversely with the square of the distance between the source

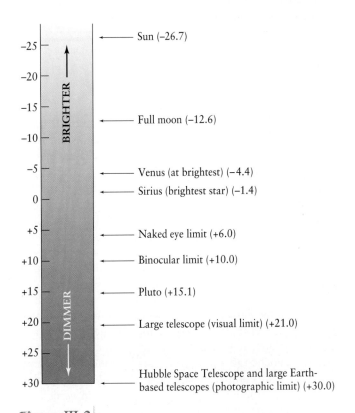

**Figure III-2**    **Apparent Magnitude Scale** Astronomers denote the brightnesses of objects in the sky by their apparent magnitudes. Stars visible to the naked eye have magnitudes between −1.44 (Sirius) and about +6. CCD (charge-coupled device) photography through the Hubble Space Telescope or a large Earth-based telescope can reveal stars and other objects nearly as faint as magnitude +30.

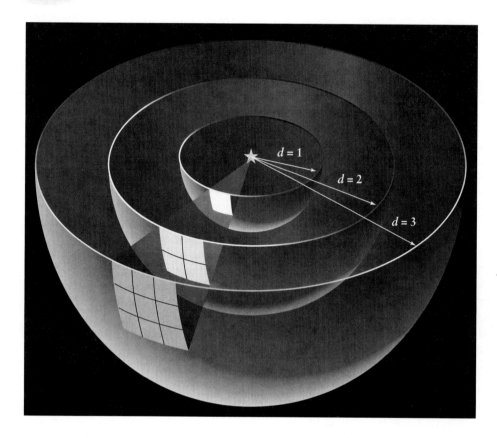

**Figure III-3** **The Inverse-Square Law** This drawing shows how the same amount of radiation from a light source must illuminate an ever-increasing area as the distance from the light source increases. Because the light spreads out as it moves away from the source, the apparent brightness of the source decreases. The decrease in brightness follows the inverse-square law, which means, for example, that doubling the distance decreases the brightness by a factor of 4.

and the observer. For example, consider two identical stars. The brightness of the one that is twice as far away decreases to $(\frac{1}{2})^2 = \frac{1}{4}$ the brightness of its closer twin by the time the starlight gets to us.

To determine the total energy emitted by stars, astronomers need to remove the effects of distance by calculating the brightnesses stars would have if they were all at the same distance from Earth. The correction for distance is made by calculating the apparent magnitude every star would have if it were 10 pc from the Earth. The resulting number is a star's **absolute magnitude.**

For example, if the Sun was moved to a distance of 10 pc from the Earth, it would have an apparent magnitude of +4.8. Therefore, the absolute magnitude of the Sun is +4.8. Absolute magnitude tells astronomers how much electromagnetic radiation stars actually emit and also how bright they are compared with each other. This information is used to evaluate models of stellar evolution, which we discuss in the next few chapters. Absolute magnitudes range from roughly –10 for the brightest stars to +17 for the dimmest.

Absolute magnitudes are always determined by calculations based on apparent magnitudes. Of course, we cannot just move a star to 10 pc distance and then remeasure

its apparent magnitude. However, we can easily calculate the absolute magnitude of a nearby star. To do this, we first measure its apparent magnitude, and then we find its distance by measuring its parallax angle. An Astronomer's Toolbox III-2 shows how these two numbers lead to the star's absolute magnitude.

Recall from Chapter 4 that a star is very nearly a blackbody, and, thus, its brightness (and, hence, absolute magnitude) is directly related to the total amount of energy radiating from its surface each second, called its **luminosity.** The smaller or more negative a star's absolute magnitude, the greater its luminosity. For convenience, stellar luminosities are expressed in multiples of the Sun's luminosity, denoted $L_\odot$, which is roughly $3.83 \times 10^{26}$ W. The intrinsically brightest stars (absolute magnitude of –10) have luminosities of $10^6$ $L_\odot$. In other words, each of these stars has the energy output of a million Suns. The dimmest stars (absolute magnitude of +17) have luminosities of $10^{-5}$ $L_\odot$.

The Sun is a wonderfully typical star. Its luminosity is about in the middle of the range for all stars, indicating that the Sun's energy output is average. We will see in Chapter 12 that this is extremely fortunate for life on Earth: If the Sun was too much brighter (and hotter), it would not be able to shine long enough for life to have

## AN ASTRONOMER'S TOOLBOX III-2

### The Distance-Magnitude Relationship

The closer a star, the brighter it appears. The inverse-square law leads to a simple equation for absolute magnitude $M$. Suppose a star's apparent magnitude is $m$, and its distance from the Earth is $d$ (measured in parsecs). Then

$$M = m - 5 \log d/10$$

where log stands for the common logarithm.

*Example:* Consider Proxima Centauri, the nearest star in the night sky. By measuring its parallax angle, we know this star is at a distance from the Earth of $d = 1.3$ pc. Its apparent magnitude is $m = +11.1$. Therefore, its absolute magnitude is

$$M = 11.1 - 5 \log 1.3/10 = 11.1 - (-4.4) = 15.5$$

*Compare!* The Sun is an average star with $M = 4.8$, so Proxima Centauri is an intrinsically dim star. The distance-magnitude relation can be rewritten as

$$m - M = 5 \log d - 5$$

If you know any two of $d$, $m$, and $M$, you can calculate the third variable. For example, if we know a star's absolute and apparent magnitude, the equation can be used to determine its distance.

*Self-Test:* A star is observed to have an apparent magnitude $m = 0.77$ and an absolute magnitude $M = +2.2$. How far from Earth is the star in parsecs and light-years? You can check your results and identify the star by referring to Appendix Table A-5.

---

evolved here. And if the Sun were too much dimmer (and cooler), the Earth would not get nearly the energy needed for our life to flourish.

### III-4 The brightnesses of stars help astronomers classify them and study their evolution

Armed with knowledge of stellar luminosities, astronomers early in the twentieth century searched for ways to use it to discover the nature of stars. In the following four chapters, we will learn how astronomers found a crucial relationship between stellar luminosities (or, equivalently, their absolute magnitudes) and their surface temperatures. Just as biologists graph the relationship between height and weight of each species of animal, stars lie in only certain regions of a graph of absolute magnitude versus surface temperature.

The information provided by the luminosity-surface temperature relationship gave astrophysicists insights into stellar evolution and led to development of the mathematical models of stars and stellar evolution that we explore in Chapter 10. These models are continually being tested and refined as they are compared to ever improving observations.

You will see in Chapters 11, 12, and 13 that stars with different masses evolve differently. Some, including the Sun, cease fusion after their cores are converted first from hydrogen into helium and then into carbon and oxygen. Other stars eventually convert their core hydrogen into neutrons; still others end up with their cores collapsing to become black holes.

### WHAT DID YOU THINK?

1 *Do stars last forever?* No, all stars go through evolutionary processes and eventually stop emitting radiation.

2 *How near is the closest star other than the Sun?* Proxima Centauri is 40 trillion kilometers (24 trillion miles) away. It takes light about 4 years to reach the Earth from there.

3 *How luminous is the Sun compared with other stars?* The Sun is roughly in the middle of the range of stellar brightnesses.

## Key Words

## Key Ideas

\* Stars differ in size, luminosity, temperature, color, and chemical composition.

\* Determining stellar distances from Earth is the first step to understanding the nature of the stars. Distances to the nearer stars can be determined by stellar parallax, the apparent shift of a star's location against the background stars while the Earth moves along its orbit.

\* The apparent magnitude of a star is a measure of how bright the star appears to Earth-based observers. The absolute magnitude of a star is a measure of the star's true brightness and is directly related to the star's energy output, or luminosity.

\* The absolute magnitude of a star is the apparent magnitude it would have if viewed from a distance of 10 pc. Absolute magnitudes can be calculated from the star's apparent magnitude and distance.

\* The luminosity of a star is the amount of energy emitted by it each second.

\* There is a correlation between absolute magnitudes and surface temperatures of stars.

## Review Questions

The answers to all computational problems, which are preceded by an asterisk (\*), appear at the end of the book.

1 Describe how the parallax method of finding a star's distance is similar to the binocular (two-eye) vision of animals.

2 What is stellar parallax?

3 How do astronomers use stellar parallax to measure the distances to stars?

4 Why do stellar parallax measurements work only with nearby stars?

5 What is the difference between apparent magnitude and absolute magnitude?

6 Briefly describe how you would determine the absolute magnitude of a nearby star.

7 What does a star's luminosity measure?

8 Why is the magnitude scale "backward" from what common sense dictates?

9 Does the star Betelgeuse, whose apparent magnitude is 0.5, look brighter or dimmer than the star Pollux, whose apparent magnitude is 1.1?

\* 10 Consider two identical stars, with one star 5 times farther away than the other. How much brighter will the closer star appear than the more distant one?

## Advanced Questions

11 What is the inverse-square law? Use it to explain why a headlight on a car can appear brighter than a star, even though the headlight emits far less light energy per second.

\* 12 Van Maanen's star, named after the Dutch astronomer who discovered it, has a parallax angle of 0.232 arcsec. How far away is the star?

13 Explain how sailors on a ship traveling parallel to a coastline at a known speed can use parallax angle measurements to determine the distance to the shore.

\* 14 Suppose that a dim star were located 2 million AU from the Sun. Find (a) the distance to the star in parsecs and (b) the parallax angle of the star.

\* 15 How many times brighter is a star of apparent magnitude −1 than a star of apparent magnitude +7?

## Discussion Question

16  Discuss the advantages and disadvantages of measuring stellar parallax from a space telescope in a large solar orbit, say, at the distance of Jupiter from the Sun.

## What If ...

17  All the stars in our Milky Way Galaxy actually were the same distance from Earth, say, 10 pc? Describe what the night sky might look like. How about the daytime sky? *Hint:* If you are doing this quantitatively, assume for simplicity that all the stars have the same absolute magnitudes as the Sun and that there are roughly 200 billion stars in the galaxy.

## Observing Projects

18  Use your Starry Night software to locate the stars listed in Appendix Table A-5. You can do this by using the Edit and then Find buttons and then typing the name of the object. Make a list of which stars are up tonight, including their apparent magnitudes, and arrange to observe them. You may find it helpful to print out star charts showing them (see Discover for Yourself: The Constellations, in Chapter 1). By comparing stars of similar apparent magnitude, such as Antares and Spica, determine the difference in apparent magnitude for which you can tell stars have different apparent brightnesses. If you bring out a star chart, identify and observe other stars, estimating their apparent magnitudes based on the ones on your list. Your instructor may have a copy of Starry Night Deluxe, with which you can confirm your apparent magnitude estimates.

# AN ASTRONOMER'S ALMANAC

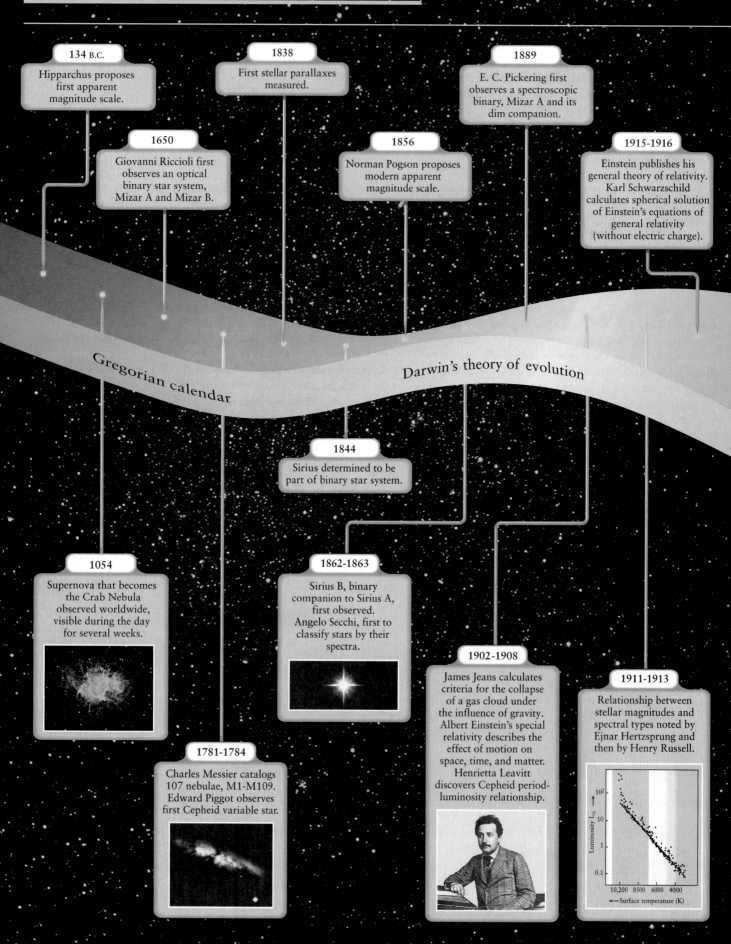

**134 B.C.**
Hipparchus proposes first apparent magnitude scale.

**1650**
Giovanni Riccioli first observes an optical binary star system, Mizar A and Mizar B.

**1838**
First stellar parallaxes measured.

**1856**
Norman Pogson proposes modern apparent magnitude scale.

**1889**
E. C. Pickering first observes a spectroscopic binary, Mizar A and its dim companion.

**1915-1916**
Einstein publishes his general theory of relativity. Karl Schwarzschild calculates spherical solution of Einstein's equations of general relativity (without electric charge).

Gregorian calendar

Darwin's theory of evolution

**1844**
Sirius determined to be part of binary star system.

**1054**
Supernova that becomes the Crab Nebula observed worldwide, visible during the day for several weeks.

**1862-1863**
Sirius B, binary companion to Sirius A, first observed. Angelo Secchi, first to classify stars by their spectra.

**1902-1908**
James Jeans calculates criteria for the collapse of a gas cloud under the influence of gravity. Albert Einstein's special relativity describes the effect of motion on space, time, and matter. Henrietta Leavitt discovers Cepheid period-luminosity relationship.

**1911-1913**
Relationship between stellar magnitudes and spectral types noted by Ejnar Hertzsprung and then by Henry Russell.

**1781-1784**
Charles Messier catalogs 107 nebulae, M1-M109. Edward Piggot observes first Cepheid variable star.

# THE STARS

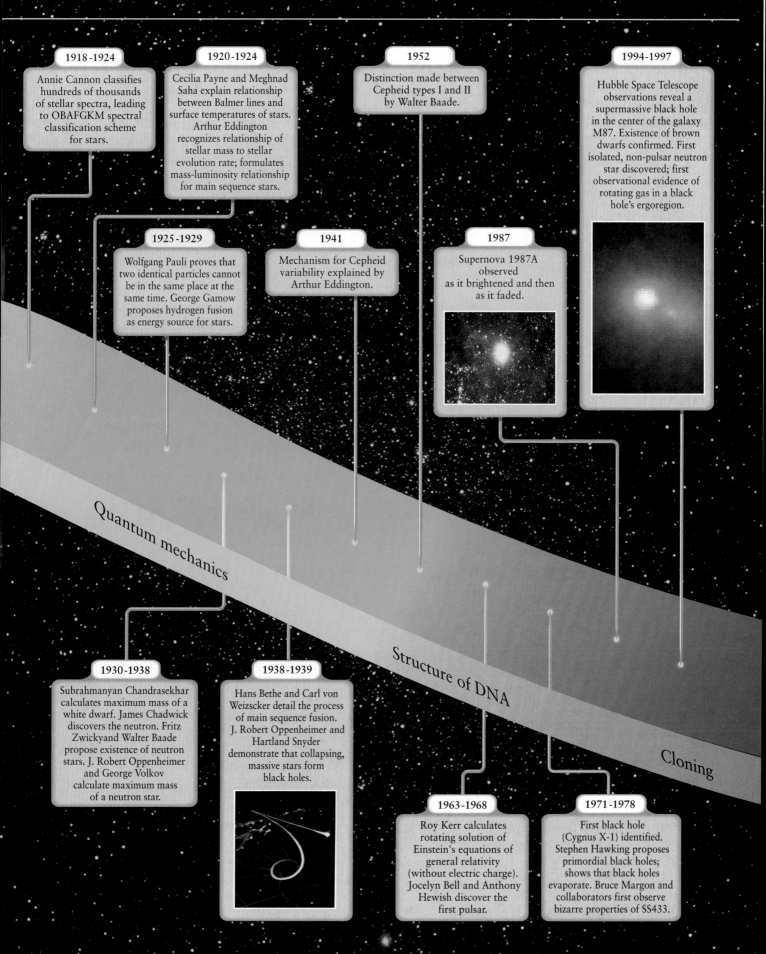

**1918-1924**
Annie Cannon classifies hundreds of thousands of stellar spectra, leading to OBAFGKM spectral classification scheme for stars.

**1920-1924**
Cecilia Payne and Meghnad Saha explain relationship between Balmer lines and surface temperatures of stars. Arthur Eddington recognizes relationship of stellar mass to stellar evolution rate; formulates mass-luminosity relationship for main sequence stars.

**1952**
Distinction made between Cepheid types I and II by Walter Baade.

**1994-1997**
Hubble Space Telescope observations reveal a supermassive black hole in the center of the galaxy M87. Existence of brown dwarfs confirmed. First isolated, non-pulsar neutron star discovered; first observational evidence of rotating gas in a black hole's ergoregion.

**1925-1929**
Wolfgang Pauli proves that two identical particles cannot be in the same place at the same time. George Gamow proposes hydrogen fusion as energy source for stars.

**1941**
Mechanism for Cepheid variability explained by Arthur Eddington.

**1987**
Supernova 1987A observed as it brightened and then as it faded.

Quantum mechanics

Structure of DNA

Cloning

**1930-1938**
Subrahmanyan Chandrasekhar calculates maximum mass of a white dwarf. James Chadwick discovers the neutron. Fritz Zwicky and Walter Baade propose existence of neutron stars. J. Robert Oppenheimer and George Volkov calculate maximum mass of a neutron star.

**1938-1939**
Hans Bethe and Carl von Weizscker detail the process of main sequence fusion. J. Robert Oppenheimer and Hartland Snyder demonstrate that collapsing, massive stars form black holes.

**1963-1968**
Roy Kerr calculates rotating solution of Einstein's equations of general relativity (without electric charge). Jocelyn Bell and Anthony Hewish discover the first pulsar.

**1971-1978**
First black hole (Cygnus X-1) identified. Stephen Hawking proposes primordial black holes; shows that black holes evaporate. Bruce Margon and collaborators first observe bizarre properties of SS433.

RIVUXG

## The Nature of Stars

### *In this chapter you will discover*

* some physical properties of stars

* how astronomers analyze starlight to determine a star's temperature and chemical composition

* how stellar luminosities and surface temperatures are related

* different classes of stars

* the variety and importance of binary star systems

* how astronomers calculate stellar masses

**Structure by Starlight**
By analyzing starlight, astronomers determine details about stars, such as their surface temperatures, chemical compositions, luminosities, and motion toward or away from us. This image of the Jewel Box, a cluster of stars also labeled NGC 4755 in the southern constellation Crux, shows stars of many colors. These colors reveal the stars' surface temperatures. Reddish stars are comparatively cool, with surface temperatures around 3000 K. Blue-white stars have much higher temperatures (15,000–30,000 K).

## WHAT DO YOU THINK?

1. What colors are stars?

2. Are brighter stars hotter than cooler stars?

3. What sizes are stars?

4. Are most stars isolated from other stars, as the Sun is?

Our understanding of the Sun has been aided immeasurably by the high-resolution images and electromagnetic spectra we can take of it and by the bits of solar matter that directly affect us via the solar wind, coronal mass ejections, flares, and neutrinos. This wealth of knowledge helps us build a sophisticated scientific model of why the Sun shines.

Now we are beginning to explore the rest of the stars, and our goal is to explain how they form, evolve, and finally cease their activity. However, there are roughly 200 billion stars in our Milky Way Galaxy alone, and trying to collect data from so many stars and create models of each is clearly a hopeless, and hopefully unnecessary, task. Rather, we want to assemble observations of stars so that we can explain with just one or a few models how they all

work. We begin organizing that data in this chapter. The insights we gain will then be applied in the next three chapters to testing stellar models.

## THE TEMPERATURE OF STARS

Before models of stellar activity could be developed, astronomers needed to take the data available from stars and organize it to provide the maximum physical insight possible. The question was whether there is a way to group stars on a graph so that physically similar ones will be near each other. The patterns thus provided would hopefully lead to useful models. This search begins with a fact we might easily overlook: Stars are not all the same color.

### 10-1 A star's color reveals its surface temperature

Even with the naked eye, you can see that stars have different colors, most commonly red, orange, yellow, white, and blue-white, as the photo that opens this chapter reveals. Because a star behaves very nearly like a perfect blackbody, its color reveals its surface temperature. Recall from Chapter 4 that associated with the temperatures of such blackbodies are different blackbody curves. These curves differ in both their intensities and the locations of their peaks, as described by Wien's law (see Figure 4-2).

Three typical blackbody curves are presented in Figure 10-1. The intensity of light from a relatively cool star peaks at long wavelengths, and so the star looks red

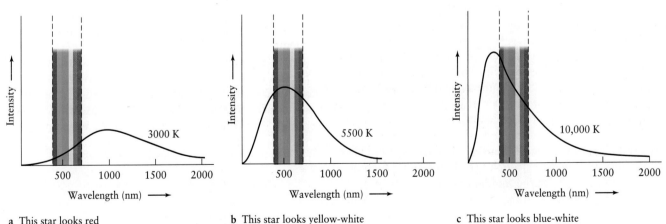

a  This star looks red          b  This star looks yellow-white          c  This star looks blue-white

**Figure 10-1**  **Temperature and Color** This diagram shows the relationship between the color of a star and its surface temperature. The intensity of light emitted by three hypothetical stars is plotted against wavelength (compare Figure 4-2). The range of visible wavelengths is indicated. Where the peak of a star's intensity curve lies relative to the visible light band determines the apparent color of its visible light.

(Figure 10-1a). The intensity of light from a very hot star peaks at shorter wavelengths, making it look blue-white (Figure 10-1c). The maximum intensity of a star of intermediate temperature, such as the Sun, is found near the middle of the visible spectrum (Figure 10-1b).

To accurately determine the peaks of their blackbody spectra, and, hence, their surface temperatures, astronomers need to know which blackbody curve most accurately describes each star. **Photometry**, in which a telescope collects starlight that is then passed through one of a set of colored filters and recorded on a light-sensitive device, such as a CCD (recall Figure 3-24), provides this information. At least three photometric images are taken of each star through filters passing different wavelengths of light. The intensity of a star's image is different at different wavelengths, and photometric data can thus be used to locate the peak of the star's blackbody radiation and its surface temperature.

If a star's surface is very hot, for example, 10,000 K, its radiation is skewed toward the short-wavelength ultraviolet, which makes the star bright through an ultraviolet-passing filter, dimmer through a blue filter, and dimmer still through a yellow filter. Regulus, in Leo, is such a star. Alternatively, if star is cool, say, 3000 K, its radiation peaks at long visible or even infrared wavelengths, making the star bright through a red filter, dimmer through a yellow filter, and dimmer still through a blue filter. Aldebaran and Betelgeuse are examples of such cool stars.

## 10-2  A star's spectrum also reveals its surface temperature

Another way to determine a star's surface temperature is from a study of its spectrum, a technique called **stellar spectroscopy** (see Chapter 4). Recall that spectral lines result from the absorption and scattering of starlight by gases in the star's atmosphere, in interstellar space, and in the Earth's atmosphere. By identifying and discarding spectral lines due to interstellar gas and Earth's atmosphere, astronomers are able to determine the stellar spectrum and, hence, the temperature of the star's atmosphere.

Recall from Chapter 4 that the Sun's absorption lines were first observed by Joseph von Fraunhofer early in nineteenth century. By the middle of that century, Italian astronomer Angelo Secchi had discovered spectral lines in the spectra of many stars. At first glance, stellar spectra seem to come in a bewildering variety, some of which are shown in Figure 10-2. Some stellar spectra show prominent absorption lines of hydrogen. Some exhibit many absorption lines of calcium and iron. Still others are dominated by broad absorption lines created by molecules such as titanium oxide.

Fortunately, this remarkable variety of spectra holds clues to each star's nature and, in particular, its surface temperature. Consider the abundant hydrogen gas found in every star's atmosphere. Although this gas accounts for about three-quarters of the mass of a typical star, strong (meaning dark) hydrogen absorption lines do not show up in every star's spectrum. The strength of the absorption lines depends on the star's temperature, and in the late 1920s Harvard astronomer Cecilia Payne and physicist Meghnad Saha succeeded in explaining how a star's visible hydrogen spectrum is affected by its surface temperature.

Niels Bohr's model of the hydrogen atom (recall Figure 4-14) explains why. As Bohr proposed in the early 1900s, the visible hydrogen lines, called Balmer lines, are produced when photons excite electrons in the second, $n = 2$, energy level of hydrogen to a higher energy level. Because hydrogen electrons are most strongly excited at a temperature of 10,000 K, a star with that surface temperature produces the strongest Balmer lines.

Why do hotter or cooler stars produce weaker (less dark) Balmer absorption lines? Suppose first that the star is much hotter than 10,000 K. High-energy photons streaming through the photosphere then completely strip away (ionize) electrons from most of the hydrogen atoms there. Because an ionized hydrogen atom cannot produce spectral lines, a very hot star has very dim hydrogen Balmer lines, even though it contains great quantities of hydrogen. Conversely, if a star is much cooler than 10,000 K, most of the photons escaping from it possess too little energy even to boost many electrons from the $n = 1$ (ground state) to the $n = 2$ (first excited state) of the hydrogen atoms, much less into more excited states. These stars, too, therefore produce dim Balmer lines. In summary, to produce strong Balmer lines, a star must be hot enough to excite electrons out of the $n = 2$ state but not hot enough to ionize a significant fraction of the atoms. Thus, we know its temperature.

At temperatures cooler and hotter than 10,000 K, the spectral lines of other elements dominate a star's spectrum. For example, the spectral lines of neutral helium are pronounced at around 25,000 K, because photons have enough energy to excite helium atoms without tearing away the electrons. Conversely, the spectral lines of neutral iron are especially strong at around 3500 K.

## 10-3  Stars are classified by their spectra

We have seen that astronomers can determine a star's surface temperature from either the peak of its blackbody or the strength of its various spectral lines. To make use of the diverse stellar spectra, astronomers since Secchi have grouped similar spectra into classes, or **spectral types**.

**Figure 10-2** | **Principal Types of Stellar Spectra** This figure shows the spectra for stars with different surface temperatures. The corresponding spectral types are indicated on the right side of each spectrum. The hydrogen Balmer lines are strongest in stars with surface temperatures of about 10,000 K (called A type stars; see Section 10-3). Cooler stars (G and K type stars) exhibit numerous atomic lines caused by various elements indicating temperatures from 4000 to 6000 K. The broad, dark bands in the spectrum of the coolest stars (M type stars) are caused by titanium oxide (TiO) molecules, which can exist only if the temperature is below about 3500 K. The Roman numeral I after a chemical symbol means that the absorption line is caused by a neutral atom; a numeral II means that the absorption is caused by atoms that have each lost one electron.

According to one classification scheme popular in the late 1800s, a star was assigned a letter from A through P, depending on the strength of the Balmer hydrogen lines in the star's spectrum. It was assumed that the strength of these lines was directly related to the star's surface temperature, but, as we have seen in the preceding section, this belief is unjustified.

In the early 1900s Annie Cannon and her colleagues at Harvard Observatory set up the spectral classification scheme we use today. Many of the early A through P categories were dropped because the Balmer lines in a star's spectrum can be weak whether the star is very cool or very hot. The remaining Balmer-based spectral types were thus reordered by stellar surface temperature into the sequence **OBAFGKM.** This sequence is most easily memorized with a mnemonic, such as: "Oh, Be A Fine Guy, Kiss Me!" or "Oh, Be A Fine Girl, Kiss Me!"

The hottest stars are the O type, with surface temperatures of more than 35,000 K; their spectra are dominated by He II and Si IV (triply ionized silicon). M stars are the coolest type, with surface temperatures around 3000 K. Table 10-1 includes representative examples of each spectral type.

Astronomers have found it useful to subdivide the OBAFGKM temperature sequence further. Each spectral type is broken up into ten temperature subranges. These

### Table 10-1
### The Spectral Sequence

| Spectral class | Color | Temperature (K) | Spectral lines | Examples |
| --- | --- | --- | --- | --- |
| O | Blue-violet | 30,000–50,000 | Ionized atoms, especially helium | Naos (ζ Puppis), Mintaka (δ Orionis) |
| B | Blue-white | 11,000–30,000 | Neutral helium, some hydrogen | Spica (α Virginis), Rigel (β Orionis) |
| A | White | 7500–11,000 | Strong hydrogen, some ionized metals | Sirius (α Canis Majoris), Vega (α Lyrae) |
| F | Yellow-white | 5900–7500 | Hydrogen and ionized metals such as calcium and iron | Canopus (α Carinae), Procyon (α Canis Minoris) |
| G | Yellow | 5200–5900 | Both neutral and ionized metals, especially ionized calcium | Sun, Capella (α Aurigae) |
| K | Orange | 3900-5200 | Neutral metals | Arcturus (α Boötis), Aldebaran (α Tauri) |
| M | Red-orange | 2500-3900 | Strong titanium oxide and some neutral calcium | Antares (α Scorpii), Betelgeuse (α Orionis) |

ten finer steps are indicated by adding an integer from 0 (hottest) through 9 (coolest). Thus, an A8 star is hotter than an A9 star, which is hotter than an F0 star, which is hotter than an F1 star, and so on. Test yourself on this: What class of star is just slightly cooler than a K9? (The caption for Figure 10-3 has the answer.) The Sun, whose spectrum is dominated by singly ionized metals (especially Fe II and Ca II), is a G2 star.

**Insight into science   Tolerate idiosyncracies**
In astronomy, the word "metal" applies to all elements other than hydrogen or helium. In chemistry, sodium and iron are metals, while carbon and oxygen are not. In astronomy, all these elements are metals.

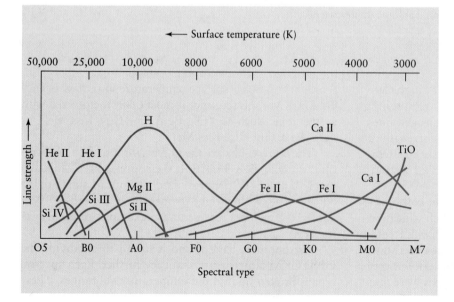

**Figure 10-3** | **Spectral Type and Temperature** The more radiation absorbed at a given wavelength, the stronger the corresponding spectral absorption line will be. The strengths of the absorption lines of various elements are directly related to the temperature of the star's outer layers, where the absorption lines are created. For example, the Sun's spectrum has strong lines of singly ionized iron and calcium (Fe II and Ca II), corresponding to a spectral type of G2 and a surface temperature of about 5800 K. Note that hydrogen lines are strongest in A stars, whereas stars cooler than about 3500 K show strong absorption caused by titanium oxide molecules. Roman numeral I indicates a neutral element. Numerals II, III, and IV indicate absorption by atoms with one, two, or three electrons missing, respectively. *Answer to text question:* An M0 star is the next coolest after a K9.

The modern spectral classification scheme is summarized in Figure 10-3, which plots the strength of absorption lines against spectral type. Astronomers use this information to deduce the surface temperature and corresponding spectral type of a star from the intensity of the lines in its spectrum. For example, a star exhibiting especially strong Ca II and Fe I lines in its spectrum is a K3 star with a surface temperature around 4500 K.

## TYPES OF STARS

Astronomers began observing stellar spectra in the early 1800s and classifying spectral types by mid-century. The first accurate measurements of stellar parallaxes were also being made at about the same time. Since then, observing techniques have vastly improved, and the spectral types and absolute magnitudes of millions of stars have been catalogued.

### 10-4 The Hertzsprung-Russell diagram identifies distinct groups of stars

Around 1911 the Danish astronomer Ejnar Hertzsprung noticed that patterns emerge when the luminosities of stars (or their equivalent absolute magnitudes) are plotted against their surface temperatures or spectral types. Within two years, the American astronomer Henry Norris Russell independently discovered the same result. Graphs of stellar brightness (luminosity or absolute magnitude) against surface temperature (or, equivalently, spectral type) are now known as **Hertzsprung-Russell diagrams,** or **H-R diagrams.**

The H-R diagram is valuable because it shows that stars do not have random surface temperatures and brightnesses; the two factors are correlated. Figure 10-4 is a typical Hertzsprung-Russell diagram. Each dot represents a star whose luminosity and spectral type have been determined. To help you see the equivalence, the surface temperatures are plotted along the top of the figure and the absolute magnitudes along the right side.

Bright stars are near the top of the diagram; dim stars are near the bottom. Contrary to intuition, hot (O and B) stars are toward the left side of the graph and cool (M) stars are toward the right. Hertzsprung and Russell made this choice because of the standard sequence OBAFGKM.

**Insight into science  From patterns to models**
Scientists look at patterns of behavior in related objects as valuable clues to underlying properties and their causes. Guided by this data, they then create theoretical models and make fresh predictions. For example, astronomers analyze the relationship between the luminosities and surface temperatures of stars as displayed on an H-R diagram to gain insight into the internal activities of stars.

## DISCOVER FOR YOURSELF

Most of the stars never received fanciful names such as Betelgeuse or Aldebaran. Indeed, most are so dim that they have been observed only through telescopes in the last two centuries. To study the stars yourself, you must be able to keep track of them. Astronomers have created a system of labels for all of them.

The 24 brightest stars in each constellation are assigned Greek lowercase letters:

| | | | | | |
|---|---|---|---|---|---|
| α | alpha | ι | iota | ρ | rho |
| β | beta | κ | kappa | σ | sigma |
| γ | gamma | λ | lambda | τ | tau |
| δ | delta | μ | mu | υ | upsilon |
| ε | epsilon | ν | nu | φ | phi |
| ζ | zeta | ξ | xi | χ | chi |
| η | eta | ο | omicron | ψ | psi |
| θ | theta | π | pi | ω | omega |

A *bright* star's name is a Greek letter together with its constellation. We use the Latin possessive form of the constellations, for example:

## Star Names

| Constellation | Possessive | Constellation | Possessive |
|---|---|---|---|
| Aries | Arietis | Scorpius | Scorpii |
| Taurus | Tauri | Ophiuchus | Ophiuchi |
| Gemini | Geminorum | Sagittarius | Sagittarii |
| Cancer | Cancri | Capricornus | Capricorni |
| Leo | Leonis | Aquarius | Aquarii |
| Virgo | Virginis | Pisces | Piscium |
| Libra | Librae | | |

In most cases the brightest star in the constellation is α, the second brightest is β, the third is γ, and so on. For example, the brightest star in the constellation Gemini is called α Geminorum. This name is more informative than its common name, Pollux.

For the millions of stars extending beyond the 24 brightest, a variety of catalogs list the stars numerically. For example, HDE 226868 is a bright, blue star, the 226,868th star in the *Henry Draper Extended Catalogue* of stars.

The band of stars in Figure 10-4 stretching diagonally across the H-R diagram and on which a red curve is superimposed represents most of the stars we see in the nighttime sky. This band, called the **main sequence**, extends from the hot, bright, bluish stars in the upper left corner of the diagram down to the cool, dim, reddish stars in the lower right corner. Each star on this band is called a **main-sequence star.** About 90% of the stars surrounding the solar system fall into this category.

Observations reveal that the number of main-sequence stars decreases with increasing surface temperature. There-

fore, along the main sequence, the cooler M, K, and G stars are the most common ones and the hot O stars are the rarest. The Sun (spectral type G2, absolute magnitude +4.8) is a main-sequence star.

To the right of the main sequence on the H-R diagram is a second major grouping of stars. These stars are bright but cool. From the Stefan-Boltzmann law (see An Astronomer's Toolbox 4-1), we know that a cool object radiates much less light from each unit of surface area than does a hot object. In order to be so bright, these cool stars must therefore be huge compared to main-sequence stars of the same temperature, and so they are called **giants.** By contrast, main-sequence stars are often called **dwarfs.**

Giants are typically 10 to 100 times the radius of the Sun and have surface temperatures between 3000 and 20,000 K. The cooler members of this class of stars (those with surface temperatures between 3000 and 4500 K) are often called **red giants** because they appear reddish in the nighttime sky. Aldebaran in the constellation Taurus and Arcturus in Boötes are examples of red giants that you can easily see with the naked eye.

A few rare stars are considerably bigger and brighter than typical giants. Located along the top of the H-R diagram, these superluminous stars are appropriately called **supergiants.** Betelgeuse in Orion and Antares in Scorpius are two examples that are visible in the nighttime sky. Together, giants and supergiants comprise less than 1% of the stars in our vicinity.

The remaining 9% or so of stars in our neighborhood of space fall in a final grouping toward the lower left and bottom of the Hertzsprung-Russell diagram. As their placement on the diagram shows, these stars are hot, dim, and tiny compared to the Sun. Called **white dwarfs,** we will see that they are actually remnants of stars. White dwarfs are roughly the same size as the Earth, and because of their great distances, they can be seen only with the aid of a telescope.

These results are summarized in Figure 10-5, where the dashed lines indicate the radii of stars. Notice that most main-sequence stars are roughly the same size as the Sun.

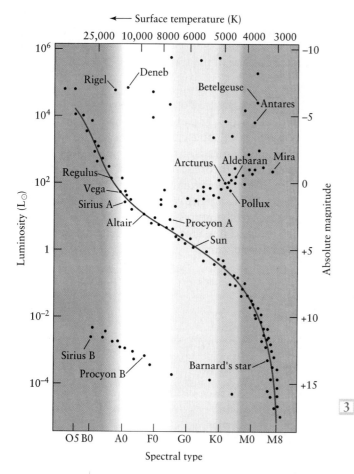

**Figure 10-4**

**A Hertzsprung-Russell Diagram** On an H-R diagram, the luminosities of stars are plotted against their spectral types. Each dot on this graph represents a star whose luminosity and spectral type have been determined. Some well-known stars are identified. The data points are grouped in just a few regions of the diagram, revealing that luminosity and spectral type are correlated: Main-sequence stars fall along the red curve, giants are to the right, supergiants are on the top, and white dwarfs are below the main sequence. The absolute magnitudes and surface temperatures are listed at the right and top of the graph, respectively. These are sometimes used on H-R diagrams instead of luminosities and spectral types.

## 10-5 Luminosity classes

We have seen that a star's surface temperature largely determines which lines are prominent in its spectrum. Therefore, classifying stars by spectral type is essentially the same as categorizing them according to surface temperature. However, as we saw in Figure 10-4, stars of the same surface temperature can have different luminosities. As an example, a star with a surface temperature of 5800 K could be either a white dwarf, a main-sequence star, a giant, or a supergiant.

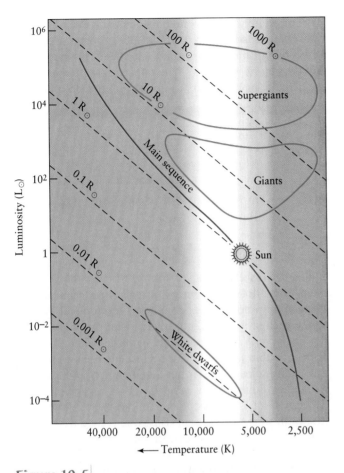

**Figure 10-5** **Determining the Sizes of Stars from an H-R Diagram** On this H-R diagram, done to the same scale as Figure 10-4, stellar luminosities are graphed against the surface temperatures of stars. The dashed diagonal lines indicate stellar radii. For a given stellar radius, as the surface temperature increases (corresponding to moving from right to left on the H-R diagram), the star glows more intensely and the luminosity increases (corresponding to moving upward on the diagram). While individual stars are not plotted, we show the regions of the diagram in which main-sequence, giant, supergiant, and white dwarf stars are found. Note that the Sun's size is intermediate in luminosity, surface temperature, and radius; it is very much a middle-of-the-road star.

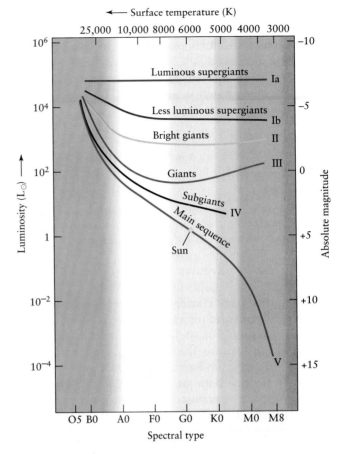

**Figure 10-6** **Luminosity Classes** It is convenient to divide the H-R diagram into regions called luminosity classes. This subdivision permits finer distinctions between giants and supergiants. Luminosity classes Ia and Ib encompass the supergiants. Luminosity classes II, III, and IV indicate giants of different sizes. Luminosity class V are the main-sequence stars. White dwarfs do not have their own luminosity class.

By studying the absorption lines in detail, however, astronomers can determine to which category a star belongs. This is possible primarily because absorption lines are also affected by the density and pressure of the gas in a star's atmosphere, both of which depend on whether the star is a white dwarf, main-sequence star, giant, or supergiant.

Based upon these differences in spectra, W. W. Morgan and P. C. Keenan of the Yerkes Observatory developed a system of **luminosity classes** in the 1930s. Luminosity classes Ia and Ib include all the supergiants, and luminosity

class V includes the main-sequence stars. The intermediate classes distinguish giants of various luminosities, as indicated in Figure 10-6. We will see in Chapter 11 that the different luminosity classes correspond to different stages of stellar evolution.

Plotting luminosity classes on the H-R diagram (see Figure 10-6) provides a useful subdivision of star types. In fact, astronomers commonly describe a star by both its spectral type and its luminosity class. The Sun, for example, is called a G2 V star. This notation supplies a great deal of information about the star, because its spectral type is correlated with its surface temperature. Thus, an astronomer knows immediately that a G2 V star is a main-sequence star with a luminosity of 1 $L_\odot$ and a surface temperature of around 5800 K. Similarly, knowing that Aldebaran is a K5 III star tells an astronomer that it is a red giant with a luminosity of around 370 $L_\odot$ and a surface temperature of about 4000 K (see Figure 10-4).

## 10-6 A star's spectral type and luminosity class provide a second distance-measuring technique

A star's spectral type and luminosity class, combined with information on the H-R diagram, enable astronomers to estimate stellar distances from the Earth. Consider, for example, the star Regulus in the constellation Leo. Its spectrum reveals Regulus to be a B7 V star (a hot, blue, main-sequence star). Placing it on the H-R diagram (Figure 10-4), we can read off its luminosity as 140 $L_\odot$ and an absolute magnitude of –0.52. Given the star's apparent magnitude, we can use the inverse-square law to determine its distance from Earth (see An Astronomer's Toolbox III-1). Because both the spectral type and luminosity class are obtained spectroscopically, this method of determining distances is called **spectroscopic parallax**. The name is misleading, because no parallax angle is involved.

Spectroscopic parallax is limited in accuracy because of the spread of stars in each luminosity class—the stars do not fall on a single line. It is also limited in that spectra of distant stars become increasingly hard to determine. As a result, errors of 10% in distance are common using spectroscopic parallax. Accepting the larger errors, the power of this method is that it can be used for stars at much greater distances than those determined by stellar parallax. Indeed, it even provides distances to stars in other galaxies tens of millions of light-years away.

## BINARY STARS AND STELLAR MASS

There is one more physical property of stars, their mass, that might provide insight into why stars only occur in limited places on the H-R diagram. The problem is that there is no way of determining stellar mass directly by examining isolated stars. The mass of a star must be determined by its gravitational effects on other bodies using Newton's law of gravity. As we saw in Chapter 2, when Newton derived the formula for Kepler's third law, he discovered that for any two objects in orbit, such as a planet around the Sun, a moon around a planet, or two stars around each other, the period of their orbits is related to the sum of their masses.

Fortunately for astronomers, about two-thirds of the stars near our solar system are members of star systems in which two or more stars orbit each other. This means that half the objects we see as single stars with our unaided eyes are actually pairs of stars in orbit around each other that are so distant they appear to us as one. Using telescopes to observe the periods of the orbits and the distances between stars, astronomers can determine the sum of the stellar masses of the pair.

## 10-7 Binary stars provide information about stellar masses

A pair of stars located at nearly the same position in the night sky is called a *double star*. Between 1782 and 1838, William Herschel and his son John catalogued thousands of them. Some double stars are not really near each other in space and do not orbit each other. These **optical doubles** just happen to lie in the same direction as seen from Earth. For example, δ Herculis, visible with the naked eye, is an optical double with a dim star. Through a telescope these stars appear close together but that is an optical illusion.

Other double stars are true **binary stars**—pairs in which two stars orbit each other. In the case of **visual binaries**, both stars can be seen, using a telescope if necessary (Figure 10-7). Astronomers can plot the orbit of one star about the other in a visual binary.

To see how astronomers determine the masses of stars, consider a visual binary. We rewrite Kepler's third law as a relation between the masses of the stars (in solar masses), their orbital period around each other in Earth years (this period is the same for both so it doesn't matter which star we assume orbits the other), and the average separation between the stars in AU (which equals the semimajor axis of the elliptical orbit):

$$\text{The sum of the masses} = \frac{\text{the cube of the semimajor axis}}{\text{the square of the orbital period}}$$

So, by observing the separation between a pair of stars and how long one of them takes to complete its orbit, we can calculate the sum of the masses of the two stars. The details are presented in An Astronomer's Toolbox 10-1.

In many cases we can also determine the individual masses of the stars. To do this, we need to know the distances of the stars from the **center of mass** of the pair. Each of the stars in a binary system actually moves in an elliptical orbit around the same point between them. This point, called the center of mass, is determined by the masses of the stars and can be understood by analogy to two children sitting on a seesaw. To keep the seesaw horizontal, the more massive child has to sit closer to the pivot point, or fulcrum, than the less massive child (Figure 10-8a). Just as

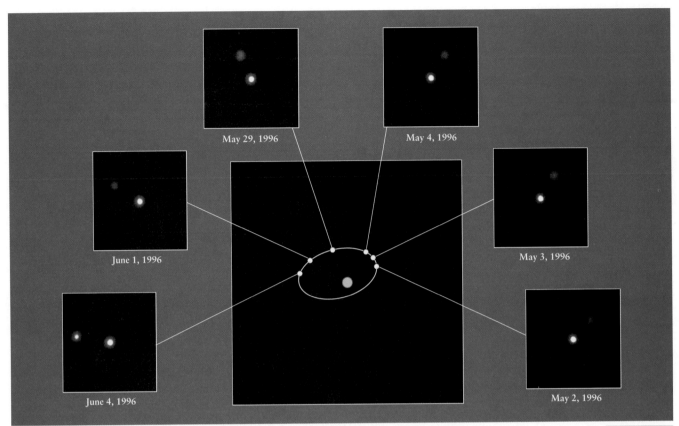

**Figure 10-7**  **A Binary Star System**  About one-half of the visible "stars" are actually double stars. ζ (zeta) Urase Majoris in Ursa Major is a binary system with stars separated by only about 0.01 arcseconds. The images surrounding this diagram show the relative positions of the two stars over half of their orbital period. The orbital motion of the two binary stars about each other is evident: Either star can be considered fixed in making such plots.

## AN ASTRONOMER'S TOOLBOX 10-1

### Kepler's Third Law and Stellar Masses

The same gravitational force that holds the Earth or a satellite in orbit also keeps pairs of stars in orbit about each other. For *any* pair of orbiting bodies, the orbits are ellipses, and Kepler's third law can be written as:

$$M_1 + M_2 = a^3/P^2$$

Here $M_1$ and $M_2$ are the two masses (expressed in solar masses), $a$ is the length of the semimajor axis of the ellipse (in astronomical units), and $P$ is the orbital period (in years). Note that $a$ is also the average separation between the two bodies. Thus, the sum of the masses can be found once the orbital period and average separation are known.

*Example:* Suppose two stars make up a double system, and one star follows an ellipse with an average separation of 4 AU. We find that it takes 2.5 years to complete one orbit. Then the sum of the stars' masses is

$$M_1 + M_2 = (4)^3/(2.5)^2 = 10.2 \ M_\odot$$

The *total* mass of the system is 10.2 $M_\odot$.

*Self-Test:* As a simple test of what you have just learned, recall from Chapter 2 that in these units we write Kepler's third law for a planet orbiting the Sun as

$$P^2 = a^3$$

Do you see why? *Hint:* For the Sun and a planet, the combined mass is extremely close to the Sun's mass alone.

the seesaw naturally balances at its center of mass, the two stars in a binary system naturally orbit around their center of mass (Figure 10-8b).

The center of mass of a visual binary is located by plotting the separate orbits of the two stars, as in Figure 10-7, using the background stars as reference points. The center of mass lies at the common focus of the two elliptical orbits. Comparing the relative sizes of the two orbits around the center of mass yields the ratio of the two stars' masses, $M_1/M_2$. To get the relative sizes of the orbits, however, we must be able to determine the plane of orbit of the two stars. This is not always possible, which is why we cannot always determine the individual masses of the stars. When we can determine $M_1/M_2$, we can find the individual masses, because we already know the sum $M_1 + M_2$ from Kepler's third law. The individual masses are determined by combining the ratio of the masses and the sum of the masses. The range of stellar masses thus determined extends from .08 $M_\odot$ to about 100 $M_\odot$.

## 10-8    There is a relationship between mass and luminosity for main-sequence stars

As binary star data accumulated, an important trend began to emerge. *On the main sequence, the more luminous the star, the more massive it is.* This **mass-luminosity relation** can be conveniently displayed graphically (Figure 10-9a). The Sun's mass lies in the middle of this range. The mass-luminosity relation demonstrates that the main sequence on the H-R diagram is a progression in mass as well as in luminosity and surface temperature. The hot, bright, bluish stars in the upper left corner of the H-R diagram (see Figure 10-9b) are the most massive main-sequence stars. As we move down the sequence, stellar masses decrease until we reach the dim, cool, reddish stars in the lower right corner of the H-R diagram. It is clear from how dramatically the luminosity changes with mass that a star's energy production is closely linked to its mass, an important clue for our upcoming model of stellar evolution.

## 10-9    The orbital motion of binary stars affects the wavelengths of their spectral lines

Many binary stars are scattered throughout our Milky Way Galaxy, but only those that are nearby or are widely separated can be distinguished as visual binaries. A remote binary often presents the appearance of a single star because telescopes cannot resolve the images of its individual stars. These binary systems can still be detected, however, through spectroscopy.

Spectral analysis yields incongruous spectral lines for some stars. For example, the spectrum of what appears at first to be a single star may include strong absorption lines for both hydrogen (indicating a hot, type A star) and also titanium oxide (indicating a cool, type M star). Because a single star cannot display both types of absorption lines prominently, such an observation must be a binary system.

Spectroscopy can also detect the movements of stars orbiting each other because of the Doppler shift in spectral lines. As we saw in Chapter 4, an approaching source of

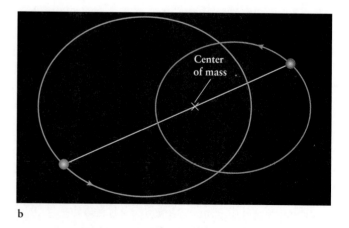

a

### Figure 10-8 | Center of Mass of a Binary Star System  (a) For a seesaw to balance, the center of mass of the two children must be at the fulcrum. The center of mass is closer to the more massive of the two children. (b) Similarly, the center of mass of a binary star system

b

is always nearer the more massive of the two stars, in this case, the reddish star. Both stars move in elliptical orbits around their common center of mass. Although the orbits cross each other, the two stars are always on opposite sides of the center of mass and thus never collide.

---

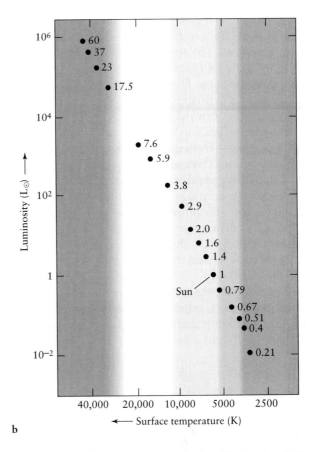

**Figure 10-9** | **The Mass-Luminosity Relation** (a) For main-sequence stars, mass and luminosity are directly correlated—the more massive a star, the more luminous it is. A main-sequence star of mass 10 $M_\odot$ has roughly 3000 times the Sun's luminosity (3000 $L_\odot$); one with 0.1 $M_\odot$ has a luminosity of only about 0.001 $L_\odot$. To fit them on the page,

the luminosities and masses are plotted using logarithmic scales. **(b)** On this H-R diagram, each dot represents a main-sequence star. The number next to each dot is the mass of that star in solar masses ($M_\odot$). As you move up the main sequence from the lower right to the upper left, the mass, luminosity, and surface temperature of main-sequence stars all increase.

light has shorter wavelengths than if the same source were stationary, and a receding source has longer wavelengths. The amount of the shift is proportional to the speed of the light source: The greater the speed, the greater the shift.

If the two stars in a binary are orbiting at more than a few kilometers per second, they will produce spectral lines that shift back and forth in a regular fashion. Such stars are called **spectroscopic binaries.** The motions of the stars revolving about their center of mass produce the periodic shifting of the spectral lines. In many spectroscopic binaries, one of the stars is so dim that its spectral lines cannot be detected. Instead, a single set of spectral lines from the star that we can see shifts regularly back and forth. Such a *single-line spectroscopic binary* yields less information about its two stars than does a *double-line spectroscopic binary,* in which the lines from both stars are visible.

Figure 10-10 shows two spectra of the spectroscopic binary system κ Arietis taken a few days apart. In

**Figure 10-10** | **A Double-Line Spectroscopic Binary** The spectrum of the double-line spectroscopic binary κ (kappa) Arietis has spectral lines that shift back and forth as the two stars revolve about each other. **(a)** The stars are moving parallel to the line of sight with one star approaching Earth, the other star receding as in stages 1 or 3 of Figure 10-11. These motions produce two sets of shifted spectral lines. **(b)** Both stars are moving perpendicular to our line of sight as in stages 2 or 4 of Figure 10-11. As a result, the spectral lines of the two stars have merged.

Figure 10-10a, two sets of spectral lines are visible, each slightly offset in opposite directions from the normal positions of these lines. The spectral lines of the star moving toward the Earth are blueshifted; those of the other star (moving away from the Earth) are redshifted. A few days later, the stars have progressed along their orbits so that one star is moving toward the left and the other toward the right as seen from Earth. Because neither star is moving toward or away from us, their spectral lines return to their usual positions (Figure 10-10b).

The shifts in spectral lines can yield significant information about the orbital velocities of the stars in a spectroscopic binary. This information is best displayed as a **radial-velocity curve,** in which radial velocity is graphed over time (Figure 10-11). Recall that radial velocity is the portion of a star's motion that is directed along our line of sight to the star.

In Figure 10-11, the wavy pattern repeats with a period of about 15 days, which is the orbital period of the binary. This pattern is displaced upward from the zero-velocity line by about 12 km/s, which is the overall motion of the binary system away from the Earth. Superimposed on this overall recessional motion are the periodic approaches and recessions of the two stars as they orbit around each other.

## 10-10 Some binary stars eclipse each other

Some binary systems are oriented so that the two stars periodically eclipse each other as seen from Earth. Such **eclipsing binaries** can be detected even when the stars cannot be resolved as two distinct images in a telescope. The apparent magnitude of the image of the binary dims each time one star blocks out part of the other. An astronomer can measure light intensity from binaries very accurately as a function of time. From these **light curves,** such as those shown in Figure 10-12, we can see at a glance whether the eclipse is partial, creating a V-shaped trough in the light curve (Figure 10-12a), or total, creating a U-shaped trough (Figure 10-12b).

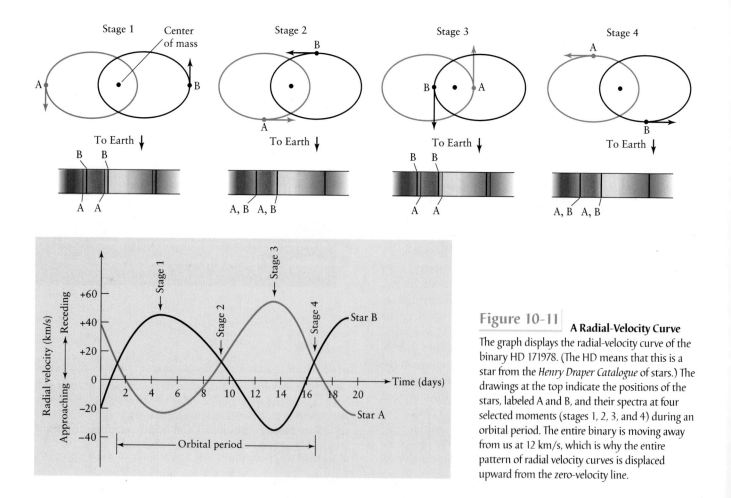

**Figure 10-11**    **A Radial-Velocity Curve**
The graph displays the radial-velocity curve of the binary HD 171978. (The HD means that this is a star from the *Henry Draper Catalogue* of stars.) The drawings at the top indicate the positions of the stars, labeled A and B, and their spectra at four selected moments (stages 1, 2, 3, and 4) during an orbital period. The entire binary is moving away from us at 12 km/s, which is why the entire pattern of radial velocity curves is displaced upward from the zero-velocity line.

**Figure 10-12** **Representative Light Curves of Eclipsing Binaries** The shape of the light curve (in blue) reveals details about the two stars that make up an eclipsing binary. Illustrated here are **(a)** a partial eclipse and **(b)** a total eclipse.

a  Partial eclipse

b  Total eclipse

The light curve of an eclipsing binary can yield other useful information as well. For example, if an eclipsing binary is also a double-line spectroscopic binary, astronomers can calculate the mass and diameter of each star, brightnesses, speeds, and stellar separation from the light curves and the radial-velocity curves. These stars are rare, however, because the orbital planes of most spectroscopic binaries are tilted so that eclipses do not occur as seen from Earth.

Light curves can also reveal information about stellar atmospheres. Suppose that one star of a binary is a white dwarf and the other is a bloated giant. By observing exactly how the light from the bright white dwarf is gradually cut off as it begins to move behind the edge of the giant during an eclipse, astronomers can infer the pressure and density in the upper atmosphere of the giant. Such information is invaluable in testing models of stellar structure.

Many binary stars are separated by several AU or more. Other than orbiting each other, these stars essentially behave as though they are isolated. That is, the models we develop in the coming chapters for the evolution of isolated stars apply to them as well. However, there are also **close binary** systems, with only a few stellar diameters separating the stars. Such stars are so close together that the gravity of each one dramatically affects the appearance and evolution of the other. If one member of a close binary is a giant, some of the gas of its outer layers is pulled onto its more compact companion—mass is transferred from one star to the other. We will explore more about such systems in Chapter 11.

Because each star only has a finite amount of fuel to fuse in its core, it is plausible that the different types of stars plotted on the H-R diagram represent different stages of stellar evolution. Perhaps each stage represents different fuel being fused. If so, stellar models should explain how the stars change from each stage to the next.

This chapter has provided us with observational data that serves, in part, as the basis for the models of stellar activity and evolution presented in the next three chapters. The models make predictions, of course, that lead to new observations and refined models. We will take up some further observations thus driven as we proceed through the models.

## WHAT DID YOU THINK?

1. *What colors are stars?* Stars are found in a wide range of colors, from red through violet as well as white.

2. *Are brighter stars hotter than cooler stars?* Not necessarily. Many brighter stars are cooler than dimmer stars.

3. *What sizes are stars?* Stars range from more than 1000 times the Sun's diameter to less than 1/100 the Sun's diameter.

4. *Are most stars isolated from other stars, as the Sun is?* In the vicinity of the Sun, two-thirds of the stars are found in pairs or larger groups.

## Key Words

## Key Ideas

✳ Stars vary in their physical properties—color, temperature, mass, and size—a fact that helps astronomers understand stellar structure and evolution.

### The Temperatures of Stars

✳ Stellar temperatures can be determined from stars' colors or stellar spectra.

✳ Stars are classified into spectral types (O, B, A, F, G, K, and M) based on their spectra. The spectral type of a star is directly related to its surface temperature.

### Types of Stars

✳ The Hertzsprung-Russell (H-R) diagram is a graph on which luminosities of stars are plotted against their spectral types (or, equivalently, their absolute magnitudes are plotted against surface temperatures). The H-R diagram reveals the existence of four major groupings of stars: main-sequence stars, giants, supergiants, and white dwarfs.

✳ The mass-luminosity relation expresses a direct correlation between a main-sequence star's mass and the total energy it emits.

✳ Distances to stars can be determined using their spectral type and luminosity class.

### Binary Stars and Stellar Masses

✳ Binary stars are surprisingly common. Those that can be resolved into two distinct star images by an Earth-based telescope are called visual binaries.

✳ The masses of the two stars in a binary system can be computed from measurements of the orbital period and orbital dimensions of the system.

✳ Some binaries can be detected and analyzed, even though the system may be so distant (or the two stars so close together) that the two star images cannot be resolved with Earth-based telescopes.

✳ A spectroscopic binary is a system detected from the periodic shift of its spectral lines. This shift is caused by the Doppler effect as the orbits of the stars carry them alternately toward and away from the Earth.

✳ An eclipsing binary is a system whose orbits are viewed nearly edge-on from the Earth, so that one star periodically eclipses the other. Detailed information about the stars in an eclipsing binary can be obtained by studying its light curve.

✳ Mass transfer occurs between binary stars that are close together.

## Review Questions

1 How and why is the spectrum of a star related to its surface temperature?

2 What is the primary chemical component of most stars?

3 A star of which spectral type has the strongest H absorption line? At approximately what wavelength is this line normally found?

4 Why does an M2 star have many more absorption lines than an F0 star?

5 Draw an H-R diagram and sketch the regions occupied by main-sequence stars, giants, supergiants, and white dwarfs.

Briefly discuss the different ways you could have labeled the axes of your graph.

6 How can observations of a visual binary lead to information about the masses of its stars?

7 What is a radial-velocity curve? What kinds of stellar systems exhibit such curves?

8 What is the difference between a single-line and a double-line spectroscopic binary?

9 What is meant by the light curve of an eclipsing binary? What sorts of information can be determined from such a light curve?

10  What is the mass-luminosity relation? To what kind of stars does it apply?

11  Refer to Figure 10-4: (a) What are the hottest and coolest named stars on the diagram? (b) What are the brightest and dimmest named stars on the diagram? (c) What are the hottest and coolest named main-sequence stars on the diagram? (d) What named stars are white dwarfs? Giants? Supergiants?

## Advanced Questions

12  Referring to Figure 10-3, why does the TiO curve stop at roughly 3500 K?

13  Sketch the radial-velocity curve of a binary whose stars are moving in nearly circular orbits that are (a) perpendicular and (b) parallel to our line of sight.

14  Sketch the light curve of an eclipsing binary whose stars are moving along highly elongated orbits with the major axes of the orbits (a) pointed toward the Earth and (b) perpendicular to our line of sight.

15  (a) Estimate the mass of a main-sequence star that is 10,000 times as luminous as the Sun. (b) What is the luminosity of a main-sequence star whose mass is one-tenth that of the Sun?

16  Estimate the surface temperature of a main-sequence star with a luminosity 100 times as bright as the Sun.

17  What temperature and classification would you assign to a star with equal H and Ca II line strengths?

## Discussion Question

18  How does a star's rotation affect the appearance of its spectral lines?

## What If ...

19  The Sun was part of a binary star system? (a) Describe qualitatively what the Earth would be like if it orbited around one of the stars, while the other star just changed the Earth's distance of orbit slightly. You might want to discuss such things as climate, tides, impacts from meteoroids and asteroids, and inhabitability. (b) Describe qualitatively what the Earth would be like if its orbit periodically took it from orbiting one star to the other and back. You might want to again discuss such things as climate, tides, impacts from meteoroids and asteroids, and inhabitability.

## Observing Projects

20  Locate the star Betelgeuse in Orion. Observe it both by eye and through a small telescope. What color is it in each viewing? It helps to compare its color to those of its neighbors.

21  Locate one or more of the following double star systems: Regulus in Leo, Algieba in Leo, Mirak in Boötes, Ras Algethi in Hercules, Albireo in Cygnus, Vega in Lyra, Polaris in Ursa Minor, Rigel in Orion, Antares in Scorpius, Sirius in Canis Major, and Schedar in Cassiopeia. You can find their locations and print star charts using your Starry Night software. View them with and without a telescope. What are their colors?

22  Observe the eclipsing binary β Algol (Persei) using nearby stars to judge its brightness during the course of an eclipse. You can find its location and print a star chart using your Starry Night software. Algol has an orbital period of 2 days, 20 hours, and 53 minutes. When the eclipse begins, its apparent magnitude drops from 2.1 to 3.4. It remains this faint for about 2 hours. The entire eclipse, from start to finish, takes about 10 hours. Consult the "Celestial Calendar" section of the current issue of Sky & Telescope for the predicted dates and times of minimum brightness of Algol. Note that the schedule is given in Universal Time (the same as Greenwich Mean Time), so you will have to convert to your own time zone. Algol is normally the second brightest star in the constellation Perseus. Because of its northerly position (right ascension = $3^h$ 08.2$^m$, declination = +40° 57′), Algol is readily visible from northern latitudes during the fall and winter months.

R I V U X G

**Reflection and Emission Nebulae** Newly formed stars are found in and around interstellar clouds called nebulae. The two main bluish objects, NGC 6589 (top) and NGC 6590 (below), are reflection nebulae surrounding young, hot, main-sequence stars. Interstellar dust around these stars efficiently reflects their bluish light. Several smaller reflection nebulae are scattered around the large reddish patch of ionized hydrogen gas called IC 1283-4. Dust mixed with the gas dilutes the intense red emission of the hydrogen atoms with a soft blue haze.

# The Life Cycles of Stars

*In this chapter you will discover*

&#42; how stars form

&#42; how astronomers use the physical properties of stars to learn about stellar life cycles

&#42; the remarkable transformations of older stars into giants and supergiants

&#42; that some dying stars eject material that creates new generations of stars, while others act as beacons that enable astronomers to pinpoint distant galaxies

&#42; that the H-R diagram is your guide to the stellar life cycle

## WHAT DO YOU THINK?

1  How do stars form?

2  Are stars forming today?

3  Do stars with greater or lesser mass shine longer?

Stars emit huge amounts of radiation. As we saw for the Sun in Chapter 9, the price for radiating energy is change: Stars lose mass over time and their chemical makeup evolves. Major stages in the life of a star can last for millions or even billions of years.

The stars seem unchanging to us only because of the colossal time scales over which these cycles of change occur. You saw in Build Your Foundation III that by observing stars with different temperatures and brightnesses, astronomers have discovered groupings of stars with similar properties. We now begin to explore the models of **stellar evolution** that are the basis for our understanding of how stars form and how they "mature," "grow old," and "die." We will also expand our observational database to provide evidence for the validity of our models.

**Insight into science   Beware poetic license** Anthropomorphism—assigning human attributes to nonhuman creatures or even to nonliving objects like stars—is descriptive but not scientific. In fact, scientists avoid this practice where possible because it creates unjustified expectations. All too often, the names and descriptions we use for nonliving objects may appear to give them human qualities. For example, we use "birth" line to understand stellar "evolution"; we probe the "lives" and "deaths" of stars. This is poetic license of a sort: It piques our interest but muddles our science. As you study stellar evolution, always be mindful of that.

## PROTOSTARS AND PRE–MAIN-SEQUENCE STARS

We saw in Build Your Foundation II that the solar system is believed to have formed from a collapsing, rotating cloud of gas and dust some 4.6 billion years ago. This model requires that interstellar matter once existed. Tak-

ing this reasoning one step further, if such matter exists today, perhaps star formation continues.

## 11-1  Stars condense from clouds of gas and dust lying between existing stars

Radio telescopes have indeed revealed the existence of matter between the stars. We call it the **interstellar medium,** and it contains about 10% of all the known mass in our Galaxy. From painstaking observations of the spectra of the interstellar medium, astronomers have determined that it is composed of atoms, molecules, and tiny pieces of dust, similar to the dust you see when sunlight streams into a room.

Ninety percent of the particles comprising the interstellar medium are hydrogen atoms and molecules, 9% are helium atoms, and the remaining 1% are all the other naturally forming elements, often in the form of molecules or dust. Astronomers often speak of the mass of different elements in the interstellar medium, rather than the numbers of particles. About 74% of the interstellar medium's mass is hydrogen, 25% is helium, and all the other naturally forming elements make up the remaining 1%.

Many types of molecules are found in interstellar space, including molecular hydrogen ($H_2$), carbon monoxide (CO), water ($H_2O$), ammonia ($NH_3$), and formaldehyde ($H_2CO$). These molecules can be identified by their unique spectral emissions, just as we can identify elements from atomic spectra.

Working from the belief that new stars form from the gas and dust of the interstellar medium, astronomers map this matter to identify places to look for newly forming stars. Like clouds in our atmosphere, gas and dust in space scatters visible light from nearby stars. Therefore, we can see the gas and dust that is close to us in space. However, that same nearby interstellar medium scatters visible light passing through it from more distant regions, preventing us from seeing this latter material. This problem of seeing distant regions of the interstellar medium in the visible part of the spectrum is compounded by the fact that most interstellar gas and dust is so cold that it emits very little visible light of its own.

Interstellar dust and its dramatic effects are clearly visible in the Pleiades star cluster (Figure 11-1). Easily visible to the unaided eye, the Pleiades is in the constellation Taurus. Note the distinctly bluish color of the nebulosity around these stars. This haze, called a **reflection nebula,** is caused by fine grains of interstellar dust that efficiently scatter and reflect blue light from surrounding stars. Indeed, reflection nebulae are blue for the same reason that Earth's sky is blue: Particles scatter short-wavelength light much more efficiently than longer-wavelength radiation. Blue light is therefore scattered back toward us much more intensely than is light of any other color.

a

R I V U X G

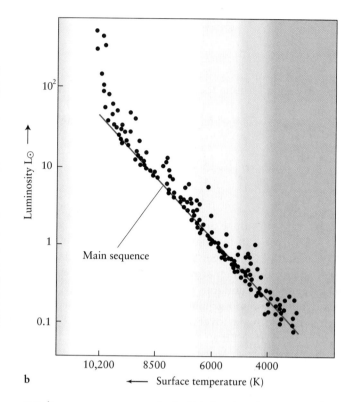

b

## Figure 11-1

**Dating the Pleiades** (a) The open cluster called the Pleiades can easily be seen with the naked eye in the constellation Taurus, the Bull. It lies about 375 ly (116 pc) from Earth. The stars are not shedding mass as are those in Figure 11-10. Their blue glow is a reflection nebula created as some of the stars' radiation scatters off preexisting dust grains in their vicinity. (b) Each dot plotted on the

H-R diagram represents a star in the Pleiades whose luminosity and surface temperature have been determined. Note that most of the cool, low-mass stars have arrived at the main sequence, indicating that hydrogen fusion has begun in their cores. The cluster has a diameter of about 5 ly, is about 100 million years old, and contains about 500 stars.

However, many interstellar atoms and molecules emit radio photons, which are scattered relatively little by intervening gas and dust. Therefore, we search for the interstellar medium primarily in the radio part of the electromagnetic spectrum. Because hydrogen is so common in the Galaxy and because the Sun is primarily hydrogen, we hypothesize that stars are composed mostly of hydrogen. However, star-forming molecular hydrogen is hard to detect in space. Therefore, radio astronomers often search instead for carbon monoxide, which emits copious photons at a wavelength of 2.6 mm. Calculations based on the known abundances of elements reveal that there are about 10,000 hydrogen molecules ($H_2$) for every CO molecule in a typical interstellar cloud. Consequently, wherever astronomers detect strong emission of CO, they know that an enormous amount of hydrogen gas must also be present.

In mapping the locations of CO emission, astronomers came to realize that vast amounts of interstellar gas are concentrated in **giant molecular clouds**. In some cases these regions appear as dark areas silhouetted against a glowing background light, such as Orion's famous Horsehead Nebula (Figure 11-2). In other cases, the clouds appear as dark blobs that obscure the background stars (Figure 11-3). Some 6000 of these molecular clouds are known, with masses ranging from $10^5$ to $2 \times 10^6$ $M_\odot$ and diameters ranging from 50 to 300 ly. The density inside one of these clouds ranges from $10^2$ to $10^5$ hydrogen molecules per cubic centimeter, several thousand times greater than the average density of the gas and dust dispersed throughout interstellar space, but some $10^{15}$ times less dense than the air you breathe.

## 11-2 Supernova explosions in cold, dark nebulae trigger the birth of stars

We will see in detail in Chapter 12 that a *supernova* is a violent detonation that ends the life cycle of a massive star. In a matter of seconds the core of the doomed star collapses, releasing vast quantities of particles and energy that blow the star apart. The star's outer layers are blasted into space at speeds of several thousand kilometers per second.

Astronomers find remains of many such dead stars scattered across the sky. These **nebulae** or **supernova rem-**

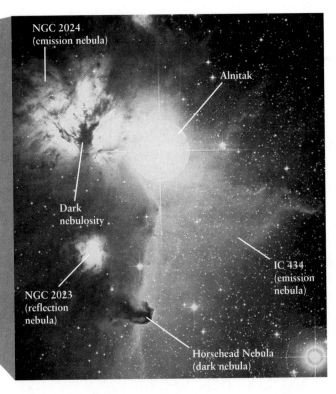

R I V U X G

**Figure 11-2** **The Horsehead**
**Nebula** A variety of nebulae appear in the sky
around Alnitak, also called ζ (zeta) Orionis, the
easternmost star in the belt of Orion. To the left of
Alnitak is a bright, red emission nebula called NGC 2024.
The glowing gases in emission nebulae are excited by
ultraviolet radiation from young, massive stars. Dust grains
obscure part of NGC 2024, giving the appearance of black streaks,
while the distinctively shaped dust cloud called the Horsehead Nebula
blocks the light from the background nebula IC 434. The Horsehead is
part of a larger complex of dark interstellar matter, seen in the lower left of
this image. Above and to the left of the Horsehead Nebula is the *reflection
nebula* NGC 2023, whose dust grains scatter blue light from stars between
us and it more effectively than any other color. All this nebulosity lies about
1600 ly from Earth, while the star Alnitak is only 135 ly away from us.

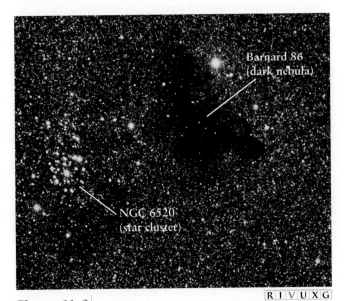

R I V U X G

**Figure 11-3** **A Dark Nebula** This dark nebula, Barnard 86, is
located in Sagittarius. It is visible in this photograph simply because it
blocks out light from the stars beyond it. The cluster of bluish stars to
the left of the dark nebula is a star cluster called NGC 6520.

nants, like the Cygnus Loop shown in Figure 11-5, have a
distinctly arched appearance, as would be expected for a
shell of gas expanding at supersonic speeds. As it passes
through the surrounding interstellar medium, the supernova remnant excites the atoms and molecules there,
causing the gases to glow. If the expanding shell of a supernova remnant rams into a giant molecular cloud, it can
cause the cloud to contract, thus stimulating star birth. As
we learned in Foundation II, there is evidence that such an
event occurred around the time the solar system formed.

A simple collision between two interstellar clouds can
also create new stars, because compression must occur at
the boundary between the clouds. Radiation from an especially bright star or group of stars may also compress
the surrounding interstellar medium enough to give birth
to stars (Figure 11-6).

Once a giant molecular cloud contracts and cools
enough, gravitational attraction causes small regions of
gas and dust in it to collapse until they become stars. This
gas must be cool, because the higher the temperature, the
faster its atoms and molecules move. As they move, they
collide with other particles, driving them apart. In other

The Nebulae

Binoculars and the naked eye are enough to let you "get your hands dirty" exploring star dust. Distant nebulae—clusters of stars and glowing gases—are among the most impressive objects in the night sky.

You can observe the Great Nebula of Orion (M42) during winter in the northern hemisphere even with the naked eye. In all likelihood, you've seen it dozens of times without knowing it. To locate the Great Nebula, find the constellation Orion (using, for example, the star charts at the end of the book). Locate Orion's belt. Due south of the belt are three stars in a row making up Orion's sword. Examine the sword very carefully with your naked eye. Do any of the stars in it look at all odd? Now look at them through a pair of binoculars. Which one is different from the others? That one is the Great Nebula in Orion, not a star at all! How does what you see compare with Figure 11-12?

The North American Nebula and the Pelican Nebula in Cygnus are best spotted in the fall. Pick a dark, moonless night and use binoculars rather than a telescope. Higher magnification reveals too small a region of the sky for you to see the entirety of these vast, dim nebulae. To find them, first locate the bright star Deneb on the tail of Cygnus (using, for example, the star charts at the end of the book). The North American Nebula is located 3° east of Deneb, while the Pelican Nebula is located 2° southeast of it. These are both small angles, so sweep around the sky east of Deneb. If your binoculars are powerful enough, you should be able to see the outlines that give these nebulae their names.

The constellations of Orion and Monoceros encompass one of the most accessible regions of the sky for studying star formation and the interaction of young stars with the interstellar medium. Figure 11-4 shows a map of this region made with a radio telescope tuned to a wavelength of 2.6 mm, which is emitted by CO. Note the ex-

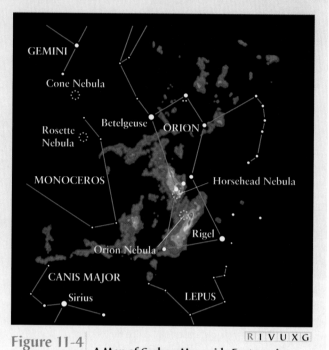

**Figure 11-4**    RIVUXG    **A Map of Carbon Monoxide Features in Orion** This color-coded radio map of a large section of the sky shows the extent of giant molecular clouds in Orion and Monoceros. The intensity of carbon monoxide (CO) emission is displayed by colors in the order of the rainbow, from violet for the weakest to red for the strongest. Black indicates no detectable emission. The locations of four prominent star-forming nebulae are indicated on the star chart overlay. Note that the Orion and Horsehead nebulae are sites of intense CO emission, indicating that stars are forming in these regions.

tensive areas covered by giant molecular clouds. Such comprehensive maps of CO emission help astronomers understand how the large-scale structure of the interstellar medium is related to the formation of stars.

words, the collisions between atoms and molecules create the pressure in the cloud. If the temperature is too high, the pressure overcomes any gravitational attraction between the particles, preventing them from drawing close enough together to form stars.

When a region of a cloud is sufficiently cold and dense, the gravitational attraction of the matter in this region overwhelms the pressure there and pulls the gas together to form a new star. This collapse is called a **Jeans instability**, after the British physicist James Jeans, who in 1902 calculated the conditions for it to occur.

Infrared observations show compact regions called **dense cores** inside many interstellar clouds. Their temperatures, around 10 K, are so low that the dense cores are destined to form stars. Often a giant molecular cloud has several hundred or even thousands of dense cores. In this case, hundreds or thousands of stars form together. Such stellar nurseries will become **open clusters** of stars like the Pleiades, shown in Figure 11-1.

At first, a collapsing dense core is just a cool, dusty region thousands of times larger than our solar system. The dense core actually collapses from the inside out. The

a

b

R I V U X G

R I V U X G

**Figure 11-5** **A Supernova Remnant** **(a)** X-ray image of the Cygnus Loop, the remnant of a supernova that occurred nearly 20,000 years ago. The expanding spherical shell of gas now has a diameter of about 120 ly. The entire Cygnus Loop has an angular diameter in our sky 6 times wider than the Moon. **(b)** This visible-light Hubble Space Telescope image of part of the Cygnus Loop shows emission from different atoms: blue from oxygen, red from sulfur, and green from hydrogen.

inner region falls in rapidly, leaving the outer layers of the dense core to drift in at a more leisurely rate. This process of increasing mass in the central region is called *accretion*, and the newly forming object at the center is called a **protostar.** Although fusion has not begun, a protostar glows from the heat generated by the compression of the gas it contains.

Figure 11-7 shows views of an interstellar cloud in which stars are forming. The view at visible wavelengths (Figure 11-7a) is a familiar sight to many telescope observers, but it tells only part of the story. Visible light from the

protostar never reaches us because it is absorbed by the surrounding shell of infalling dust. Only the infrared picture (Figure 11-7b) shows the hundreds of protostars.

If a dense core is not spinning, it collapses into a sphere, which ultimately becomes an isolated star. If it is spinning, it collapses into a disk, which may then condense into two or three stars. Or, if the disk has a low enough mass, it may become a single star with orbiting protoplanets. One such disk of gas and dust has been observed around β Pictoris (see Figure II-12). It was the first sign of planetary development outside the solar system.

R I V U X G

**Figure 11-6** **The Core of the Rosette Nebula** The large, circular Rosette Nebula (NGC 2237) is near one end of a sprawling giant molecular cloud in the constellation Monoceros, the Unicorn. Radiation from young, hot stars has blown gas away from the center of this nebula. Some of this gas has become clumped in dark globules that appear silhouetted against the glowing background gases. New star formation is taking place within these globules. The entire Rosette Nebula has an angular diameter on the sky nearly 3 times that of the Moon, and it lies some 3000 ly from Earth.

a    R I V U X G

b    R I V U X G

**Figure 11-7**    **Newborn Stars in the Swan Nebula** **(a)** This image at visible wavelengths shows an H II region called the Swan (or Omega) Nebula, so named because of its characteristic shape. It is 5500 ly from Earth in the constellation Sagittarius. **(b)** This infrared view, constructed from images taken at wavelengths of 1.2, 1.6, and 2.2 μm, reveals hundreds of stars that do not appear in (a). A comparison of the visible and infrared views demonstrates that circumstellar and interstellar dust obscures many of the stars in this star-forming region, which contains around 800 $M_\odot$.

## 11-3 When a protostar ceases to accumulate mass, it becomes a pre–main-sequence star

Models of star formation predict that after about $10^5$ years of mass accretion, the protostar builds up to about the mass of the Sun. A protostar of 1 $M_\odot$ is about 5 times larger in diameter than the Sun. Its large size makes it brighter than the Sun, and it can be observed as an intense point of infrared radiation. Much matter is still slowly falling inward from the dense core's outer shell. However, the radiation and particles flowing off the protostar exert outward forces on this remaining gas and dust, preventing it from ever reaching the protostar. Now, mass accretion stops, and the protostar becomes a **pre–main-sequence star.**

A pre–main-sequence star contracts slowly, unlike the rapid collapse of a protostar. When the temperature at its core reaches $10^7$ K, hydrogen fusion begins there. As we saw in Chapter 9, this thermonuclear process releases enormous amounts of energy. The outpouring of energy from hydrogen fusion creates pressure inside the pre–main-sequence star sufficient to finally halt its contraction.

The more massive a pre–main-sequence star is, the more rapidly it begins hydrogen fusion in its core. For

example, calculations indicate that a 5-$M_\odot$ pre–main-sequence star starts fusing only $10^5$ years after it first forms from a protostar, whereas a 1-$M_\odot$ pre–main-sequence star takes a few tens of millions of years to do the same.

In the final stages of pre–main-sequence evolution, the outer shell of gas and dust finally dissipates. For the first time, the star is directly revealed to the outside universe.

## 11-4 The evolutionary track of a pre–main-sequence star depends on its mass

Astrophysicists use computers and the equations of stellar structure (described in Chapter 9) to model the evolution of a pre–main-sequence star. By calculating changes in the energy that the contracting star emits, computer models can follow its changing position on a Hertzsprung-Russell diagram (Figure 11-8). Keep in mind that such an **evolutionary track** represents changes in a star's temperature and luminosity, not its motion in space.

Pre–main-sequence stars are relatively cool when they begin to shine at visible wavelengths, in agreement with Wien's displacement law. Hence, the evolutionary tracks of

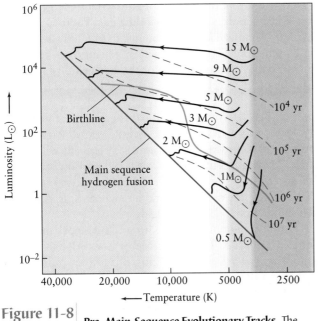

**Figure 11-8** **Pre–Main-Sequence Evolutionary Tracks** The evolutionary tracks based on models of seven stars having different masses are shown in this H-R diagram. The dashed lines indicate the stage reached after the indicated number of years of evolution. The birth line, shown in blue, is the location where each protostar stops accreting matter and becomes a pre-main-sequence star. Stars more massive than about 7 $M_\odot$ start fusing so rapidly that they go directly from protostars to the main sequence. Note that all tracks terminate on the main sequence at points agreeing with the mass-luminosity relation.

pre–main-sequence stars begin near the right side of the H-R diagram along a curve called the **birth line** (see Figure 11-8). A star's exact location on this curve depends primarily on its mass and, to a much smaller extent, on the amount of metal it contains. As a pre–main-sequence star of less than 2 $M_\odot$ contracts, its diminishing surface area causes its luminosity to drop significantly. On the H-R diagram, therefore, the track of the star drops below the birth line. Eventually, its surface temperature increases, and the star's track moves to the left in the diagram.

Pre-main-sequence stars more massive than 2 $M_\odot$ become hotter without much change in overall luminosity. The evolutionary tracks of these pre–main-sequence stars thus traverse the H-R diagram horizontally, from right to left. A star more massive than about 7 $M_\odot$ has no pre–main-sequence phase at all. Its gravitational compression is so great that it begins to fuse hydrogen in its protostellar phase.

Calculations of the temperature required to start hydrogen fusion in the core of a star indicate that pre–main-sequence stars less massive than about 0.08 $M_\odot$ do not have enough gravitational force compressing and heating their cores to initiate fusion. Instead, these small bodies contract to become planetlike orbs of hydrogen and heli-um, sometimes called **brown dwarfs.** As we saw in Chapter 7, Jupiter has the same chemistry as most stars but 75 times too little mass for fusion to begin in its core. There is no universal agreement on the precise mass that represents the boundary between brown dwarfs and planets. We will use one number that has been proposed, namely that a brown dwarf has at least 13 times the mass of Jupiter. Below that limit, the object is classified as a planet.

At least 11 brown dwarfs have now been located, beginning with Gliese 229B, a brown dwarf in the constellation Lepus about 18 ly (6 pc) from Earth. It was first observed in 1994 from Mount Palomar Observatory, and a year later the Hubble Space Telescope confirmed that it is indeed a brown dwarf (Figure 11-9). Even more exciting, astronomers have confirmed the existence of at least a dozen planets outside our solar system. These *extrasolar planets* are detected by their gravitational effects on the stars they orbit.

R I V U X G

**Figure 11-9**    **A Brown Dwarf** Located 18 ly (6 pc) from Earth in the constellation Lepus (the Hare), Gliese 229B (the smaller object in the image) is the first confirmed brown dwarf ever observed. With a surface temperature of about 1000 K, its spectrum is like that of Jupiter. Gliese 229B is in a binary star system. The overexposed image of part of its companion, Gliese 229, is seen on the left. The two stars are separated by about 43 AU. Gliese 229B has from 20 to 50 times the mass of Jupiter, but the brown dwarf is compressed to the same size as our giant planet. The spike of light was produced when Gliese 229 overloaded part of the Hubble Space Telescope's electronics.

The upper limit to stellar mass is around 100 $M_\odot$. This limit is the result of mass loss by extremely massive stars in the early stages of their development. Protostars with greater than about 100 $M_\odot$ rapidly develop extremely high temperatures—so high that their outer layers are expelled into interstellar space, thereby lowering their mass (Figure 11-10). One of the brightest stars in our Galaxy, the Pistol Star, may have started its life with as much as 200 $M_\odot$. Observations reveal that every few thousand years it expels shells of gas, and it may have less than 10 $M_\odot$ left when the expulsion of matter stops. The entire process of mass loss from very massive stars takes only a few million years.

## 11-5 H II regions harbor young star clusters

We can detect the formation of a cluster of stars as a magnificent glow in the nebula. Figures 11-2 and 11-11 show these **emission nebulae**. Because these nebulae are predominantly ionized hydrogen, they are also called **H II regions**. To see why we can observe H II regions, remember that most massive pre–main-sequence stars, those of spectral types O and B, are exceptionally hot. Because their surface temperatures are typically 15,000 to 35,000 K, they emit

Figure 11-10    **Mass Loss from a Supermassive Star** Named the Pistol and located 25,000 ly from Earth, astronomers calculate that the star in the middle of this pistol-shaped gas cloud originally had 100–200 $M_\odot$. The structure of the gas cloud suggests the star ejected the gas we see in two episodes 6000 and 4000 years ago. The gas from previous ejections is so thinly spread now that we cannot see it. The nebula shown here is more than 4 ly (1.25 pc) across—it would stretch from the Sun nearly to the closest star, Proxima Centauri.

vast quantities of ultraviolet radiation. This energetic radiation easily ionizes any surrounding gas. Photons from an O5 star can ionize hydrogen atoms up to 500 ly away.

While some hydrogen atoms are being knocked apart by ultraviolet photons, some of the free protons and electrons manage to get back together. As these new hydrogen atoms assemble, their electrons return to their ground state ($n = 1$). This downward cascade through each atom's energy levels is what makes the nebula glow. Particularly prominent is the transition from $n = 3$ to $n = 2$, which produces $H_\alpha$ photons at 656 nm in the red portion of the visible spectrum (review the emission line spectrum in Figure 4-11). Thus, the nebula around a newborn star cluster often shines with a distinctive reddish hue.

An H II region is a small, bright "hot spot" in a giant molecular cloud. The collection of hot, bright O and B stars that produces the ionizing ultraviolet radiation is called an **OB association**. The famous Orion Nebula (Figure 11-12) is an example. Four O and B stars at the heart of the Orion Nebula are responsible for the ionizing radiation that causes the surrounding gases to glow. The Orion Nebula is embedded in a giant molecular cloud whose mass is estimated at 500,000 $M_\odot$.

The OB association creating the H II region also affects the rest of the giant molecular cloud (see Figure 11-12 insets). Detailed models propose that vigorous stellar winds, along with ionizing ultraviolet radiation from the O and B stars, carve out a cavity in the cloud. Where this outflow is supersonic, it creates a shock wave, like the sonic boom created by fast-flying aircraft. The shock wave forms along the outer edge of the expanding H II region, compressing hydrogen gas as it passes and thereby stimulating a new round of star birth. As more O and B stars form, they power the expansion of the H II region still farther into the giant molecular cloud. Meanwhile, the older O and B stars left behind begin to disperse (Figure 11-13). In this way, an OB association "eats into" a giant molecular cloud, creating stars in its wake.

To test the model of star birth just described, astronomers have looked for protostars adjacent to stars in an OB association. The inset in Figure 11-13 shows star formation around a single O star. Figure 11-12 shows star formation in the core of the Orion Nebula. Visible light observations easily reveal four O and B stars amid the glowing gas and dust, but infrared observations of the same region may be showing us cocoons of warm dust still enveloping pre–main-sequence stars.

## 11-6 Plotting a star cluster on an H-R diagram reveals its age

The young open clusters we have studied so far offer astronomers a rich source of information about stars in their infancy. By measuring each star's apparent magnitude,

**Figure 11-11** **An H II Region** This emission nebula, called the Eagle Nebula because of its shape, surrounds the star cluster called M16. Star formation is occurring in M16, which is located 7000 ly from Earth in the constellation of Serpens Cauda (Serpent's Tail). Several bright, hot O and B stars are responsible for the ionizing radiation that causes the gases to glow. **Inset:** Star formation is occurring inside these dark pillars of gas and dust. Intense ultraviolet radiation from existing massive stars off to the right of this image is evaporating the dense cores in the pillars, thereby prematurely terminating star formation there. Newly revealed stars are visible at the tips of the columns.

R I V U X G

Newly revealed stars

Protostars hidden within "fingers" of nebular material

R I V U X G

**Figure 11-12** **The Orion Nebula** As you have observed, the middle "star" in Orion's sword is actually the Orion Nebula, a part of a huge system of interstellar gas and dust in which new stars are now forming. The Orion Nebula is an H II region visible to the naked eye. It is 1600 ly (490 pc) from Earth and has a diameter of roughly 16 ly (5 pc). This nebula's mass is about 300 $M_\odot$. **Inset (left)** This view at visible wavelengths shows the inner regions of the Orion Nebula. At the lower left are four massive stars, called the Trapezium, which cause the nebula to glow. **Inset (right)** This view of exactly the same region shows that infrared radiation penetrates interstellar dust that absorbs visible photons. Numerous infrared objects, many of which are probably stars in the early stages of formation, can be seen, along with shock waves caused by matter flowing out of protostars faster than the speed of sound waves in the nebula. Shock waves from the Trapezium may have helped trigger the formation of the protostars in this view.

R I V U X G

Young stars

Shock waves

R I V U X G

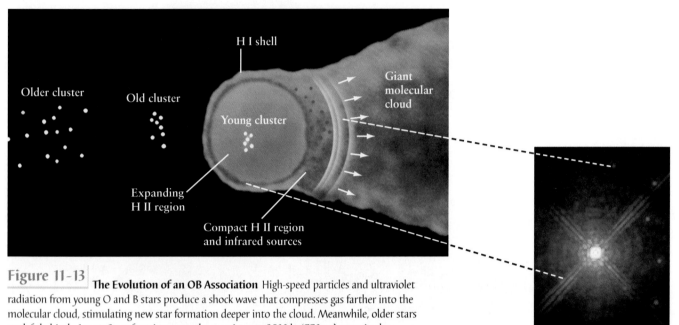

**Figure 11-13** **The Evolution of an OB Association** High-speed particles and ultraviolet radiation from young O and B stars produce a shock wave that compresses gas farther into the molecular cloud, stimulating new star formation deeper into the cloud. Meanwhile, older stars are left behind. **Inset:** Stars forming around a massive star 2500 ly (770 pc) away in the constellation Monoceros's Cone Nebula. The six stars (small dots on the right side of the inset) arrayed around the bright, massive central star are believed to have formed as a result of the central star compressing surrounding gas with high-speed particles and radiation. The younger stars are just 0.04 - 0.08 ly from the central star.

color, and distance, an astronomer can deduce its luminosity and surface temperature. The data for all the stars in the cluster can then be plotted on an H-R diagram, as shown for the cluster NGC 2264 in Figure 11-14. (NGC stands for *New General Catalogue*, a listing of more than 7000 objects other than stars published in 1888. The original *General Catalogue* was complied by William Herschel and son John during the first half of the nineteenth century.)

Stars with masses over the entire allowed range form in each cluster. Because all the stars in the cluster begin forming at essentially the same time and because stars with different masses arrive on the main sequence at different times, astronomers can use the H-R diagram to determine the age of the cluster. Note that the hottest stars in NGC 2264, with surface temperatures around 20,000 K, lie on the main sequence. These hot stars are extremely bright and massive: Their radiation causes the surrounding gases to glow. Most of the stars cooler than about 10,000 K have not yet arrived at the main sequence. These less massive stars, which are in the final stages of pre–main-sequence contraction, are just now beginning to ignite thermonuclear reactions at their centers.

We will discover in Section 11-7 that stars with different masses evolve at different rates and thus leave the main sequence at different times. The fact that none of the high-mass stars have left the main sequence yet, combined with the fact that many low-mass stars have not yet even

reached it, can be applied to models of stellar evolution to reveal that the open cluster in Figure 11-14 is roughly two million years old.

Spectroscopic observations of the cooler stars in NGC 2264 show that many are vigorously ejecting gas just before they reach the main sequence. Gas-ejecting stars in spectral classes G and cooler (that is, G, K, and M) are called **T Tauri stars,** after the first example discovered in the constellation of Taurus. Some astronomers propose that the onset of hydrogen fusion is preceded by vigorous chromospheric activity marked by enormous spicules and flares that propel the star's outermost layers back into space. In fact, an infant star going through its T Tauri stage can lose as much as 0.4 $M_\odot$ of matter and also shed its cocoon while still a pre–main-sequence star. Observations reveal that T Tauri stars are variable.

In contrast to the H-R diagram for NGC 2264, nearly all the stars in the Pleiades (see Figure 11-1) have completed their pre–main-sequence stage. The cluster's age is about 100 million years, which is how long it takes for the least massive stars to finally begin hydrogen fusion in their cores.

Open clusters, such as the Pleiades and NGC 2264, possess barely enough mass to hold themselves together. A star moving faster than the average speed for the cluster occasionally escapes. Astronomers predict that in a few hundred million years, after its stars have separated from

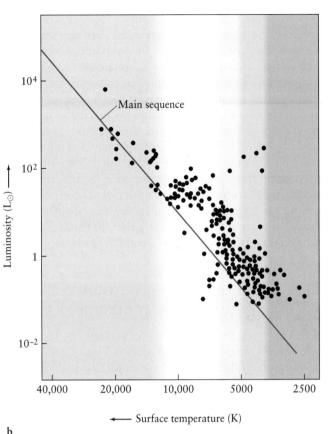

Main sequence

Surface temperature (K)

**Figure 11-14** **Plotting the Ages of Stars** **(a)** This photograph shows an H II region and the young star cluster NGC 2264 in the constellation Monoceros. The nebulosity is located about 2600 ly from Earth and contains numerous stars that are about to begin hydrogen fusion in their cores. **(b)** Each dot plotted on this H-R diagram represents a star in this cluster whose luminosity and surface temperature have been measured. Note that most of the cool, low-mass stars have not yet arrived at the main sequence. Models of stellar evolution indicate that this star cluster started forming about two million years ago.

each other and mixed with the rest of the stars in the galaxy, each open cluster ceases to exist.

# MAIN-SEQUENCE AND GIANT STARS

As a pre–main-sequence star evolves, fusion begins in its core. Taking the Sun as our model of main-sequence activity, we saw in Chapter 9 that thermal pressure then pushes outward from the core, balancing the inward force of gravity. We call that balance *hydrostatic equilibrium*. The pre–main-sequence star now ceases to collapse—and a star is born. It has reached the main sequence at last, as shown in Figure 11-8. *Main-sequence stars are those stars with nuclear reactions fusing hydrogen into helium in their cores at a constant rate.*

## 11-7 Stars spend most of their life cycles on the main sequence

The **zero-age main sequence (ZAMS)** is the location on the H-R diagram where the stellar model predicts that pre–main-sequence stars (fusing hydrogen in their cores) first become stable objects, neither shrinking nor expanding. This is the solid red line that appears on several of the H-R diagrams. As was the case with the birth line, the location of a ZAMS star on the main sequence depends primarily on its mass. Note that the evolutionary tracks in Figure 11-8 end at locations along the main sequence that agree with the mass–luminosity relation (recall Figure 10-9): The most massive main-sequence stars are the most luminous, while the least massive stars are the least luminous.

The equations of nuclear fusion predict that the more massive a star is, the faster it evolves and the less time it spends on the main sequence. Gravity presses down with greater force on a more massive star's core than on a less massive star's core. This tremendous pressure speeds up fusion so that some O and B stars consume all their core hydrogen in only a few million years, as shown in Table 11-1. Conversely, stars of very low mass take hundreds of billions of years to convert their cores from hydrogen into helium.

Whatever the rate of fusion, the equations reveal that the conversion of hydrogen into helium in a star's core takes a long time compared to any other stage of stellar evolution. (These stages are presented in the following sections.) That is why the vast majority of stars represented on an H-R diagram are located on the main sequence. Our model of solar evolution predicts that the Sun's total lifetime on the main sequence will be about ten billion years.

## 11-8 When core hydrogen fusion ceases, a main-sequence star becomes a giant

When the hydrogen in a star's core is completely converted into helium, fusion in the core ceases. This is because the temperature of the core is not hot enough to enable the helium to fuse into other elements. But the star continues to evolve. Let's see how the loss of the energy source can actually cause stars to *expand* in size to become *giants*.

Recall from our discussion of the Sun in Chapter 9 that while on the main sequence a star's outer layers are supported by thermal pressure, with energy supplied by fusion in the core. Without fusion, a star can no longer support the crushing weight of its outer layers, and so it begins compressing its helium core. We will return to the fate of the core shortly, but, in the meantime, consider the hydrogen-rich gas just outside it. Under the influence of gravity, this gas is compressed and heated enough to begin fusing into helium. This is called **shell hydrogen fusion**. Recall from An Astronomer's Toolbox 9-1 that the proton-proton chain is the major source of energy inside the Sun. This fusion process also occurs in shell hydrogen fusion.

Why then does the star expand to become a giant? In fact, we do not yet know! When our equations of stellar activity are used to create computer models of stars leaving the main sequence, the simulated stars swell up, just like real red giants. However, the equations are so complex that astrophysicists have been unable to isolate all the effects that go into creating red giants.

**Insight into science    More than the sum of its parts**  We have gotten to the point in science where some computer models that numerically solve complex sets of equations reproduce natural phenomena without illuminating it. The equations that describe stellar evolution are often in this category.

Far from the fusing shell, the surface gases cool and soon the temperature of the star's bloated surface falls to between 3000 and 6000 K, depending on the star's total mass. In about 5 billion years, our Sun will swell to a red giant with a diameter of about 1 AU, vaporizing Mercury and perhaps causing Venus to spiral into the Sun. The Earth will be scorched to a cinder.

Giant stars are so enormous that their bloated outer layers constantly leak gases into space. At times, this *mass loss* is significant (Figure 11-15), and it can be detected spectroscopically. The escaping gases exhibit narrow absorption lines, and the lines from gases coming toward us are slightly blueshifted, owing to the Doppler effect. This small shift corresponds to a speed of 10 km/s, typical of the expansion velocities with which gases leave the tenuous outer layers of giants. A typical rate of mass loss for a giant is roughly $10^{-7}$ $M_\odot$ per year. For comparison, in a main-sequence star such as the Sun, mass loss rates are only around $10^{-14}$ $M_\odot$ per year.

Although the surface of a giant is cooler than that of the main-sequence star from which it evolved, the giant is brighter: It can emit more photons each second because it

| Mass ($M_\odot$) | Surface temperature (K) | Luminosity ($L_\odot$) | Time on main sequence ($10^6$ years) | Spectral class |
|---|---|---|---|---|
| 25 | 35,000 | 80,000 | 3 | O |
| 15 | 30,000 | 10,000 | 15 | B |
| 3 | 11,000 | 60 | 500 | A |
| 1.5 | 7,000 | 5 | 3,000 | F |
| 1.0 (Sun) | 6,000 | 1 | 10,000 | G |
| 0.75 | 5,000 | 0.5 | 15,000 | K |
| 0.50 | 4,000 | 0.03 | 200,000 | M |

Table 11-1
Main-Sequence Lifetimes

has so much more surface area. As a full-fledged giant (Figure 11-16), our Sun will shine 2000 times brighter than it does today.

## 11-9  Helium fusion begins at the center of a giant

Consider now the helium core of the giant star. Helium is the ash of hydrogen fusion. When a star first becomes a giant, its hydrogen-fusing shell surrounds a small, compact core of almost pure helium. In a moderately low-mass giant, the dense helium core is about twice the size of the Earth.

At first, no thermonuclear reactions occur in the helium-rich core of a giant because the temperature there is too low to fuse helium nuclei. The hydrogen-fusing shell creates more helium, which becomes part of the core, while the hydrogen-fusing shell moves outward. The ever more massive helium core continues to contract and heat up.

When the central temperature reaches about 100 million K, **core helium fusion** begins at the giant's center. The aging star thus has a central energy source for the first time since it left the main sequence. The *triple-alpha process* provides the dominant path for helium in the core to fuse into carbon. First, two helium nuclei combine to create beryllium. Then, within $10^{-8}$ seconds a third helium

**Figure 11-15** **A Mass-Loss Star** This unusual star in the constellation Norma (the Carpenter's Square) is the brightest of a system of three stars orbiting each other. This extremely hot, massive star, named HD 148937, is continuously shedding its outer layers. Other vigorous outbursts in the past gave rise to the two symmetric shells of material, called NGC 6164 and NGC 6165, on either side of the star. These shells absorb ultraviolet radiation from the star, causing them to glow with the characteristic red color of excited hydrogen gas.

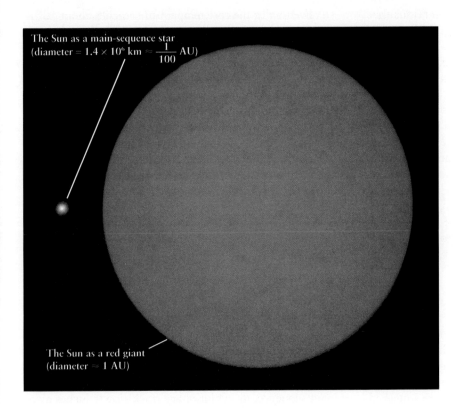

The Sun as a main-sequence star
(diameter = $1.4 \times 10^6$ km $\approx \frac{1}{100}$ AU)

The Sun as a red giant
(diameter $\approx$ 1 AU)

**Figure 11-16** **The Sun Today and as a Giant** In about five billion years, when the Sun expands to become a giant, its diameter will increase a hundredfold while its core becomes more compact. Today, the Sun's energy is produced in a hydrogen-fusing core whose diameter is about 200,000 km. When the Sun becomes a giant, it will draw its energy from a hydrogen-fusing shell surrounding a compact helium-rich core. The helium core will have a diameter of only 30,000 km. The Sun's diameter will be about 100 times bigger, and it will be about 2000 times more luminous as a giant than it is today.

nucleus is added to the short-lived beryllium to create carbon. This process also releases energy and it can be summarized as follows:

$$^4\text{He} + {}^4\text{He} + {}^4\text{He} \rightarrow {}^{12}\text{C} + \gamma$$

where $\gamma$ denotes energy emitted as photons. But the fusion in giants does not stop there. Some of the carbon created can then fuse with another helium nucleus to produce oxygen:

$$^{12}\text{C} + {}^4\text{He} \rightarrow {}^{16}\text{O} + \gamma$$

Again, energy is released as gamma rays.

The energy released in core helium fusion reestablishes hydrostatic equilibrium—but only briefly. A mature giant fuses helium in its core for about 20% of the time that it spent fusing hydrogen as a main-sequence star. While fusion is again occurring in the core, further gravitational contraction of the giant star ceases. Starting in the distant future, the Sun will consume the helium it is now creating in its core for about two billion years.

The models of stellar evolution reveal that how helium fusion begins at a giant's center depends on the mass of the star. In high-mass stars, helium fusion begins gradually as temperatures in the star's core approach 100 million Kelvin. In low-mass stars, helium fusion begins explosively and suddenly in an event called the **helium flash**. Depending on the model of stellar evolution you choose to believe, the upper limit to stars that undergo the helium flash is between 2 $M_\odot$ and 4 $M_\odot$. For convenience, we will adopt the limit of 2 $M_\odot$ in what follows. The helium flash is the result of unusual conditions that develop in the core of a low-mass star after it leaves the main sequence. To appreciate these conditions, we must first understand how an ordinary gas behaves, then explore how the densely packed electrons at the star's center alter this behavior.

When an ordinary gas is compressed, it heats up; when it expands, it cools down. The helium in the core of a giant star with mass greater than about 2 $M_\odot$ behaves the same way. If energy production overheats its core, the core expands, cooling the gases and slowing the rate of thermonuclear reactions. Conversely, if too little energy is being created to support the star's overlying layers, they move inward, compressing the core. The resulting increase in temperature speeds up the thermonuclear reactions and thus increases the energy output, which stops the contraction. Either way, the star has a "safety valve" to keep it from exploding.

A giant with less than 2 $M_\odot$ lacks that safety valve because densely packed electrons in its core change its behavior. The lower a giant's mass, the more it has to compress to drive temperatures high enough to ignite helium fusion. In fact, at the extreme pressures deep inside a giant with less than 2 $M_\odot$, the atoms are completely ionized into nuclei and electrons. Furthermore, the nuclei are squeezed

into a regular pattern, a crystal-like solid. The electrons, distributed between the nuclei, are so closely crowded together that another law of physics becomes important.

According to the **Pauli exclusion principle**, formulated in 1925 by the Austrian physicist Wolfgang Pauli, two identical particles cannot exist in the same place at the same time. As the electrons are pressed closer and closer together by the star's gravitational force, the exclusion principle prevents them from all just freezing in place. Instead, many of them must vibrate faster and faster so that they do not become "identical," here meaning moving with the same speeds, to adjacent electrons.

The faster the electrons vibrate, the more energy they have, and the more they can provide an outward pressure to slow the contraction of the core. Astronomers say that electrons in this state are degenerate and that the helium-rich core of a low-mass giant is supported by **electron degeneracy pressure**, which provides a greater outward force than did the normal pressure in the star before it became degenerate. Therefore, electron degeneracy pressure prevents the core from collapsing further. It sits there and heats up under the gravitational influence of the star's mass.

The equation of such degenerate matter as the electrons in this star predict that *degeneracy pressure, unlike the pressure of an ordinary gas, does not change with temperature*. This means that as the core's temperature grows, the pressure there does not increase. Without the "safety valve" of increasing pressure, the star's core cannot expand and cool.

Eventually, the core temperature reaches about $10^8$ K, the point where helium begins to fuse. This fusion creates energy and so the core's temperature goes up, but the degenerate electrons still do not increase the pressure. Therefore, the core does not immediately expand, and thus for several hours the core temperature and fusion rate rise dramatically. This is the *helium flash*.

In the helium flash, temperatures become so high (around $3.5 \times 10^8$ K) that the helium again becomes an ordinary gas. Suddenly, the usual safety valve operates once again. In a matter of hours, the star's core expands and cools, decreasing the fusion rate. After the helium flash subsides, the star's energy output declines, and so its outer layers again contract. A low-mass, helium-fusing giant is left smaller, dimmer, and hotter than it was before it began fusing helium.

## 11-10 As stars evolve, their positions on the H-R diagram shift

It is very enlightening to plot the theoretical evolutionary tracks of post-main-sequence stars on the Hertzsprung-Russell diagram (Figure 11-17). After each star reaches the main sequence, its evolutionary track slowly inches away

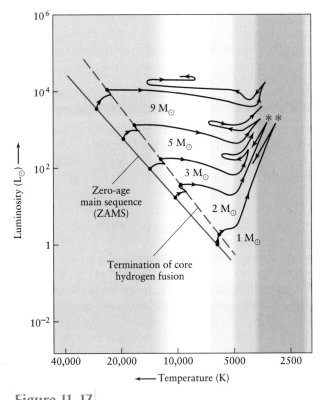

**Figure 11-17** **Post–Main-Sequence Evolution** Model-based evolutionary tracks of five stars are shown on this H-R diagram. In the high-mass stars, core helium fusion ignites smoothly where the evolutionary tracks make a sharp turn upward into the giant region of the diagram. The red asterisks on the 1-$M_\odot$ and 2-$M_\odot$ curves indicate where the helium flash in these stars occurs.

begins, however, the evolutionary tracks back away from these peak luminosities. The tracks then wander back and forth in the giant region while the stars readjust to their new energy sources.

We saw that after the helium flash, low-mass stars become dimmer but hotter. On the H-R diagram in Figure 11-17, note that the two low-mass stars move down and to the left.

## 11-11 Globular clusters are bound groups of old stars

Post–helium-flash stars are often found in old star clusters, called **globular clusters**, so named because of their spherical shapes. A typical globular cluster, like the one shown in Figure 11-18, may contain up to a million stars in a volume 300 ly across. Like open clusters, the stars in globular clusters all form at about the same time. Unlike open clusters of young stars, globular clusters are gravitationally bound groups of stars that do not disperse.

Astronomers know that globular clusters are old because they contain no high-mass main-sequence stars. If you measure the luminosity and surface temperature of many stars in a globular cluster and plot the data on a color-magnitude diagram, as shown in Figure 11-19, you will find that the upper half of the main sequence is missing. All the high-mass main-sequence stars evolved long ago into giants, leaving behind only lower-mass, slowly evolving stars still undergoing core hydrogen fusion.

from the ZAMS location as the hydrogen-fusing core grows to accommodate fresh fuel. The dashed line in Figure 11-17 shows the locations of the stars when all the core hydrogen has been consumed. As we saw in Section 11-7, it takes a few million years for the most massive main-sequence stars to reach this point, while a 1-$M_\odot$ star takes ten billion years. Calculations indicate that lower-mass stars take tens or even hundreds of billions of years to get there. Because, as we will see, the universe is roughly 15 billion years old, no main-sequence stars with masses less than about 0.75 $M_\odot$ have yet moved into the giant stage.

After core hydrogen fusion ceases, the points representing high-mass stars move rapidly from left to right across the H-R diagram. This is when each star's core begins to contract and its outer layers expand. Although the stars' surface temperatures are decreasing, their surface areas are increasing. Their overall luminosities remain roughly constant.

Just before core helium fusion begins, the evolutionary tracks of high-mass stars turn upward into the giant region of the H-R diagram. After the core helium fusion

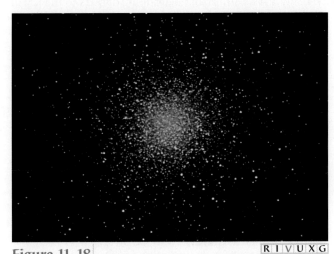

**Figure 11-18**  R I V U X G  **A Globular Cluster** A globular cluster is a spherical cluster that typically contains a few hundred thousand stars. This cluster, called M13, is located in the constellation Hercules, roughly 23,000 ly from Earth. The stars in M13 are all crowded within a diameter of only 150 ly, making the average density of stars there about 100 times greater than in the solar neighborhood.

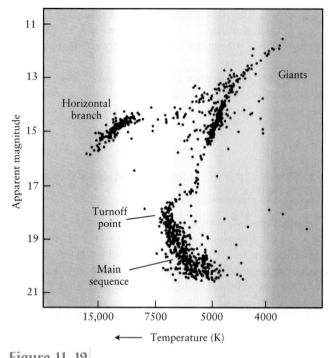

**Figure 11-19**    **An H-R Diagram of a Globular Cluster** Each dot on this graph represents the apparent magnitude and surface temperature of a star in the globular cluster called M55. Because all the stars are at virtually the same distance from Earth, their apparent magnitudes are measured rather than their absolute magnitudes. Note that the upper half of the main sequence is missing. The horizontal branch stars are low-mass stars that recently experienced the helium flash in their cores and now exhibit core helium fusion and shell hydrogen fusion.

The color-magnitude diagram of a globular cluster typically shows a horizontal grouping of stars to the left of the center portion of the diagram (see Figure 11-19). These stars, called **horizontal branch stars,** are post–helium-flash stars, and they have luminosities of about 50 $L_\odot$. Eventually, these stars will move back toward the giant region as core helium fusion and shell hydrogen fusion devour their fuel.

As noted earlier, an H-R diagram of a cluster can also be used to determine its age, assuming that all of the stars in the cluster formed at the same time. In the H-R diagram for a young open cluster (review Figure 11-14b), all the stars are on the main sequence. As a cluster gets older, like the Pleiades, stars begin to leave the main sequence (see Figure 11-1b). The high-mass, high-luminosity stars are the first to become giants as the main sequence starts to burn down like a candle.

Over time, the main sequence in a cluster gets shorter and shorter. The top of the surviving portion of the main sequence is called the *turnoff point* (see Figure 11-19). Stars at the turnoff point are just beginning to exhaust the hydrogen in their cores. The model of stellar evolu-

tion enables us to predict how long stars with different masses last on the main sequence before moving toward the giant stage. Therefore, the mass of the stars presently at the turnoff point is used to determine the age of the cluster (recall Table 11-1).

Consider the plot for the globular cluster M55 shown in Figure 11-19 (so named because it was fifty-fifth in the *Messier Catalogue* of astronomical objects). In M55, 0.8-$M_\odot$ stars have just left the main sequence, and so, according to Table 11-1, the cluster's age is approximately 15 billion years.

Using data collected from the Hipparcos satellite, Figure 11-20a is a composite diagram of isolated and cluster stars, reflecting the overall structure of the H-R diagram. Using data from stellar evolution theory as presented in Table 11-1, the ages of star clusters can be estimated from the turnoff points, such as those depicted in Figure 11-20b.

The youngest star clusters in the Milky Way (those with their main sequences still intact) are found in open clusters. These clusters exist in the plane of the Galaxy, that is, along the band of light sweeping majestically across the night sky. Stars in these young clusters are said to be *metal-rich,* because their spectra contain many prominent spectral lines of heavy elements. (Recall from Chapter 10 that all elements other than hydrogen and helium are considered metals by astronomers.) This material originally came from stars that exploded long ago, enriching the interstellar gases with the heavy elements formed in their cores. The young clusters are therefore formed from the debris of older generations of stars. The Sun is an example of a young, metal-rich star that was probably formed in an open cluster. Such stars are also called **Population I stars.**

Globular clusters contain the oldest stars and are generally located above or below the plane of our Galaxy. Because their spectra show only weak lines of heavy elements, these ancient stars are said to be *metal-poor.* They were created long ago from interstellar gases that had not yet been substantially enriched with heavy elements. They are also called **Population II stars.** Figure 11-21 compares the spectra from a Population II star and the Sun.

## VARIABLE STARS

After core helium fusion begins, the evolutionary tracks of mature stars move across the middle of the H-R diagram. In Figure 11-17, we saw the evolutionary tracks of post-main-sequence stars. During these excursions across the H-R diagram, a star can become unstable and pulsate. In fact, the region on the H-R diagram between the main sequence and the giant branch is called the **instability strip** (Figure 11-22). When a star's evolutionary track carries it

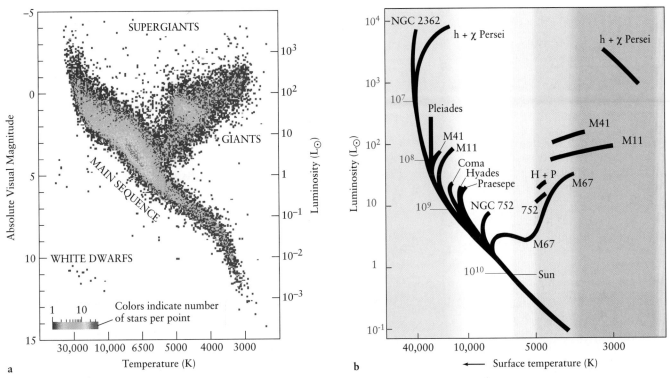

**Figure 11-20** **Structure of the H-R Diagram** **(a)** Data taken by the Hipparcos satellite place 20,853 stars more precisely on the H-R diagram than any previous observations. This figure shows the overall structure of the H-R diagram. The thickness of the main sequence is due in large part to stars of different ages turning off the main sequence at different places, as shown in (b). **(b)** The black bands indicate where data from various star clusters fall on the H-R diagram. The ages of turnoff points (in years) are listed in red alongside the main sequence. The age of a cluster can be estimated from the location of the turnoff point, where the cluster's most massive stars are just now leaving the main sequence.

through this region, the star slowly pulsates. As it does so, its brightness varies periodically. These so-called **variable stars** can be easily identified by their changes in brightness amid a field of stars of constant luminosity.

Low-mass, post–helium-flash stars pass through the lower end of the instability strip as they move in the horizontal branch along their evolutionary tracks. These stars

become **RR Lyrae variables,** named after the prototype in the constellation of Lyra (the Harp). RR Lyrae variables all have periods shorter than one day, and all have roughly the same average brightness as stars on the horizontal branch. High-mass stars pass back and forth through the upper end of the instability strip on the H-R diagram. These stars become **Cepheid variables,** often simply called Cepheids.

**Figure 11-21** **Spectra of a Metal-Poor and a Metal-Rich Star** These spectra compare **(a)** a metal-poor (Population II) and **(b)** a metal-rich (Population I) star (the Sun) of the same surface temperature. Numerous spectral lines prominent in the solar spectrum are caused by elements heavier than hydrogen and helium. Note that corresponding lines in the metal-poor star's spectrum are weak or absent. Both spectra cover a wavelength range that includes two strong hydrogen absorption lines labeled $H_\gamma$ (434 nm) and $H_\delta$ (410 nm).

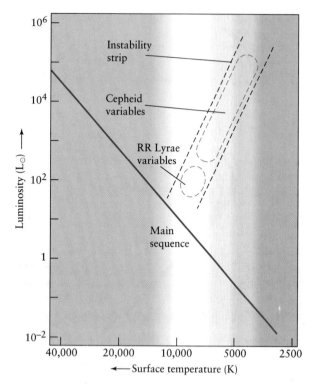

**Figure 11-22** **The Instability Strip** The instability strip occupies a region between the main sequence and the giant branch on the H-R diagram. A star passing through this region along its evolutionary track becomes unstable and pulsates.

## 11-12 A Cepheid pulsates because it is alternately expanding and contracting

A Cepheid variable is characterized by the way in which its light output varies—rapid brightening followed by gradual dimming. A Cepheid variable brightens and fades because the star's outer layers cyclically expand and contract. Lines in the spectrum of δ (delta) Cephei shift back and forth with the same 5.4-day period that characterizes its variations in magnitude. According to the Doppler effect, these shifts mean that the star's surface is alternately approaching and receding from us.

When a Cepheid variable pulsates, the star's surface oscillates up and down like a spring. Consequently, the star's gases alternately heat up and cool down; the surface temperature changes from about 6300 K to about 5000 K and back. Thus, the characteristic light curve of a Cepheid variable results from changes in both size and surface temperature.

Just as a bouncing ball eventually comes to rest, a pulsating star would soon stop pulsating without some mechanism to keep its oscillations going. In 1941 the British astronomer Arthur Eddington explained that a Cepheid

variable feeds energy into its pulsations by a valvelike action involving the periodic ionization and recombination of gas in the star's outer layers. According to this theory, a star is more opaque, or "light tight," when compressed than when expanded. When the star compresses, trapped heat pushes the star's surface outward. When the star expands, the heat escapes, and the star's surface, which is no longer supported, falls inward.

## 11-13 Cepheids enable astronomers to estimate vast distances

Cepheids are very important to astronomers, because there is a direct relationship between a Cepheid's period of pulsation and its average luminosity. This relationship is called, appropriately enough, the **period-luminosity relation**. Dim Cepheid variables pulsate rapidly with periods of one to two days and have average brightnesses of a few hundred Suns. The most luminous Cepheids have the longest periods of all Cepheids, with variations occurring over 100 days and average brightnesses equaling 10,000 $L_\odot$. Because the changes in brightness of Cepheids can be seen even in distant galaxies where many other techniques for measuring distance fail, the period-luminosity relation plays an important role in determining the overall size and structure of the universe, as we will see in Chapter 15.

The details of a Cepheid's pulsation depend on the abundance of heavy elements in its atmosphere. The average luminosity of metal-rich Cepheids is roughly 4 times greater than the average luminosity of metal-poor Cepheids having the same period. Thus, there are two classes: **Type I Cepheids** (also called δ *Cephei* stars), which are the brighter, metal-rich stars, and **Type II Cepheids** (also called *W Virginis* stars), which are the dimmer, metal-poor stars. The period-luminosity relation for both types of variables is shown in Figure 11-23.

**Insight into science** **Firm foundations?** To understand concepts far from our everyday size and time scales, scientists often must base their explanations of observations on one or perhaps several complex theories even before all the testing and refinement of the theories are complete. An error anywhere along the chain of ideas leads to incorrect results. For example, using the Cepheid variable stars to measure distances requires combining several intermediate concepts, such as knowing the period-luminosity relationship, knowing that there are two different types of Cepheids, knowing the luminosity of Cepheids, and knowing the relationship between luminosity and distance.

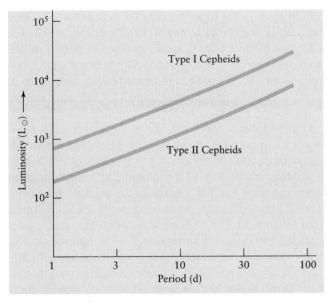

**Figure 11-23** **The Period-Luminosity Relation** The period of a Cepheid variable is directly related to its average luminosity: The more luminous the Cepheid, the longer its period and the slower its pulsations. Type I Cepheids (*δ Cephei* stars) are metal-rich Population I stars. They are brighter than the Type II Cepheids (*W Virginis* stars), which are metal-poor Population II stars.

In rare cases, stellar pulsations can be quite substantial. In some instances, the expansion velocity is so high that the star's outer layers are ejected completely. There are several other ways that stars shed matter. We end this chapter by discovering how the gravitational attraction of binary stars in near proximity to each other can cause stars to lose mass, and in the next chapter we explore the mass shedding that occurs at the ends of stellar life cycles.

## 11-14 Mass transfer in close binary systems can produce unusual double stars

In the mid-1800s, the French mathematician Edward Roche pointed out that the atmospheres of two stars in a binary system must remain within a pair of teardrop-shaped regions surrounding the stars. Otherwise, the gas escapes from the star of its origin, either transferring to the other star, where the teardrops make contact, or escaping completely from the binary in the opposite direction. Cut through, these **Roche lobes** take on a figure-eight shape (dashed lines in Figure 11-24). The more massive star is always located inside the larger Roche lobe.

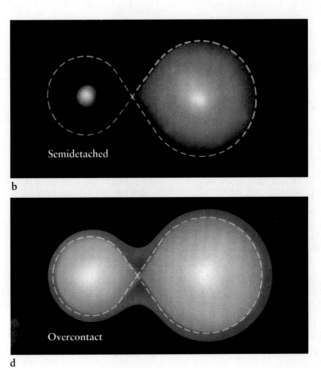

**Figure 11-24** **Detached, Semidetached, Contact, and Overcontact Binaries** These figures show the various types of binary star systems. **(a)** In a detached binary, neither star fills its Roche lobe. **(b)** If one star fills its Roche lobe, the binary is semidetached. Mass transfer is often observed in semidetached binaries. **(c)** In a contact binary, both stars fill their Roche lobes. **(d)** The two stars in an overcontact binary both overfill their Roche lobes. The two stars actually share the same outer atmosphere.

The stars in many binaries are so far apart that even during their giant stages the stars' surfaces remain well inside their Roche lobes. Other than orbiting one another, each star in such systems lives out its life cycle as if it were single and isolated, and the system is referred to as a **detached binary** (Figure 11-24a). If two stars are relatively close together, however, one star may fill or overflow its Roche lobe as it expands into the giant phase. This system is called a **semidetached binary,** and gases then flow across the point where the two Roche lobes touch and fall onto the companion star (Figure 11-24b). When both stars completely fill their Roche lobes, the system is called a **contact binary** because the two stars actually touch and exchange gas (Figure 11-24c). It is quite unlikely, however, that both stars will exactly fill their Roche lobes at the same time. It is more likely that they overflow their lobes, giving rise to a common atmospheric envelope. Such a system is called an **overcontact binary** (Figure 11-24d).

Semidetached and contact binaries are easiest to detect if they are also eclipsing binaries (recall Figure 10-12).

Their light curves have a distinctly rounded appearance caused by these tidally distorted egg-shaped stars. The eclipsing binary called β Persei, or Algol (from an Arabic term for "demon"), is a semidetached binary that can easily be seen with the naked eye in the constellation Perseus. From β Persei's light curve (Figure 11-25a), astronomers have determined that the binary contains a star that fills its Roche lobe. Sometime in the past, as it expanded and became a giant, this star dumped a significant amount of gas onto its companion.

Mass transfer is still occurring in a semidetached eclipsing binary called β Lyrae in the constellation Lyra. Like β Persei, β Lyrae contains a giant that fills its Roche lobe (Figure 11-25b). For many years astronomers were puzzled by the fact that the detached companion star in β Lyrae is severely underluminous, contributing virtually no light at all to the visible radiation coming from the system. Furthermore, the spectrum of β Lyrae contains unusual features, some of which are consistent with gas flowing between the stars and around the system as a whole.

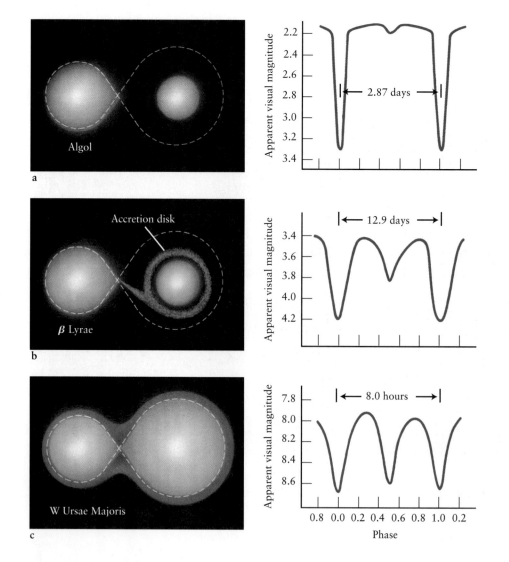

**Figure 11-25** **Three Close Binaries** Light curves for and sketches of three eclipsing binaries are shown. The phase denotes the fraction of the orbital period from one primary minimum to the next. **(a)** Algol, also known as β Persei, is a semidetached binary. The deep eclipse occurs when the giant star (on the left) blocks the light from the smaller, but more luminous, main-sequence star. **(b)** β Lyrae is a semidetached binary in which mass transfer has produced an accretion disk surrounding the detached star. This disk is so thick and opaque that it renders the secondary star almost invisible. **(c)** W Ursae Majoris is an overcontact binary. Both stars therefore share their outer atmospheres. The period of this binary is extremely short, indicating that the stars are very close to each other.

The β Lyrae system was explained in 1963, when Su-Shu Huang proposed that the underluminous star in β Lyrae is enveloped in a huge **accretion disk** of gas captured from its bloated companion. The disk is so large and thick that it completely shrouds the secondary star, making it impossible to observe at visible wavelengths. The primary star is overflowing its Roche lobe, with gases streaming onto the disk at the rate of 1 M☉ per hundred thousand years.

The fate of a semidetached system like β Persei or β Lyrae depends primarily on how fast its stars evolve. If the detached star expands to fill its Roche lobe while the companion star fills its own Roche lobe, then the result is a contact binary. An example is W Ursae Majoris, in which two stars share the same photosphere (Figure 11-25c).

Models of the evolution of binary systems that transfer mass reveal that some remarkable transformations are possible. This is seen in the case of φ (phi) Persei. As shown in Figure 11-26, the more massive star in a binary can actually be transformed into the lower-mass star and vice versa! In Chapters 12 and 13 we will see that mass transfer produces some of the most extraordinary objects in the sky.

## WHAT DID YOU THINK?

1 *How do stars form?* Stars form from gas and dust inside giant molecular clouds.

2 *Are stars forming today?* Yes. Astronomers have seen stars that have just arrived on the main sequence, as well as infrared images of gas and dust clouds in the process of forming stars.

3 *Do stars with greater or lesser mass shine longer?* Lower-mass stars last longer because the lower gravitational force inside them causes fusion to take place at slower rates compared to the fusion inside higher-mass stars.

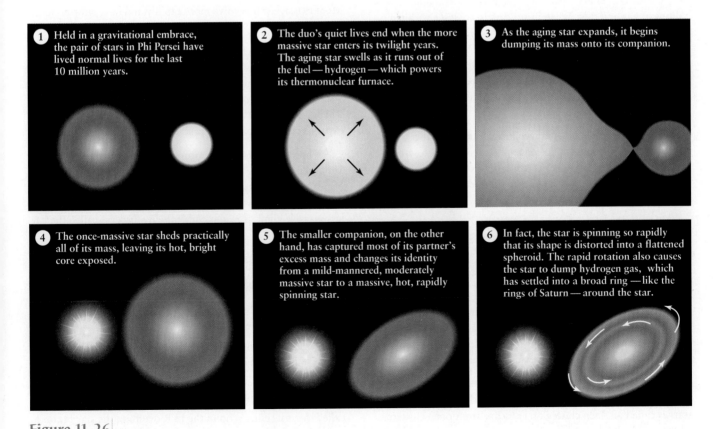

1 Held in a gravitational embrace, the pair of stars in Phi Persei have lived normal lives for the last 10 million years.

2 The duo's quiet lives end when the more massive star enters its twilight years. The aging star swells as it runs out of the fuel—hydrogen—which powers its thermonuclear furnace.

3 As the aging star expands, it begins dumping its mass onto its companion.

4 The once-massive star sheds practically all of its mass, leaving its hot, bright core exposed.

5 The smaller companion, on the other hand, has captured most of its partner's excess mass and changes its identity from a mild-mannered, moderately massive star to a massive, hot, rapidly spinning star.

6 In fact, the star is spinning so rapidly that its shape is distorted into a flattened spheroid. The rapid rotation also causes the star to dump hydrogen gas, which has settled into a broad ring—like the rings of Saturn—around the star.

**Figure 11-26** **Mass Exchange Between Close Binary Stars** This sequence of drawings shows how close binary stars can initially be isolated but, as they age, grow and exchange mass. Such mass exchange leads to very different fates than if the same stars had evolved in isolation.

## Chapter 11

## Key Words

accretion disk, 297

birth line, 283

brown dwarf, 283

Cepheid variable, 293

contact binary, 296

core helium fusion, 289

dense core, 280

detached binary, 296

electron degeneracy pressure, 290

emission nebula, 284

evolutionary track, 282

giant molecular cloud, 278

globular cluster, 291

H II region, 284

helium flash, 290

horizontal branch star, 292

instability strip, 292

interstellar medium, 277

Jeans instability, 280

nebulae (*singular* nebula), 278

OB association, 284

open cluster, 280

overcontact binary, 296

Pauli exclusion principle, 290

period-luminosity relation, 294

Population I star, 292

Population II star, 292

pre–main-sequence star, 282

protostar, 281

reflection nebula, 277

Roche lobe, 295

RR Lyrae variable, 293

semidetached binary, 296

shell hydrogen fusion, 288

stellar evolution, 277

supernova remnant, 278-279

T Tauri stars, 286

Type I Cepheid, 294

Type II Cepheid, 294

variable stars, 293

zero-age main sequence (ZAMS), 287

## Key Ideas

### Protostars and Pre–Main-Sequence Stars

✳   Enormous, cold clouds of gas and dust, called giant molecular clouds, are scattered about the Galaxy.

✳   Star formation begins when gravitational attraction causes clumps of gas and dust called protostars to coalesce within a giant molecular cloud. As a protostar contracts, its matter begins to glow. When the contraction slows down, the protostar becomes a pre–main-sequence star. When the pre–main-sequence star's core temperature becomes high enough to begin hydrogen fusion and stop contracting, it becomes a main-sequence star.

✳   The most massive pre–main-sequence stars take the shortest time to become main-sequence stars (O and B stars). They emit strong ultraviolet radiation that ionizes hydrogen in the surrounding molecular cloud, creating reddish emission nebulae called H II regions.

✳   In the final stages of pre–main-sequence contraction, when hydrogen fusion is about to begin in the core, the pre–main-sequence star may undergo vigorous chromospheric activity that ejects large amounts of matter into space. Such gas-ejecting stars are called T Tauri stars.

✳   Ultraviolet radiation and stellar winds from the OB association at the core of an H II region create shock waves that compress the gas cloud, triggering the formation of more protostars. Supernova explosions also compress gas clouds and trigger star formation.

✳   A collection of a few hundred or a few thousand newborn stars is called an open cluster. Stars escape from open clusters, which eventually dissipate.

### Main-Sequence and Giant Stars

✳   The Sun has been a main-sequence star for 4.6 billion years and should remain so for about another 5 billion years. Less massive stars evolve more slowly and have longer lifetimes. More massive stars evolve more rapidly and have shorter main-sequence lifetimes.

✳   Core hydrogen fusion ceases when hydrogen is exhausted in the core of a main-sequence star, leaving a core of nearly pure helium surrounded by a shell where hydrogen fusion continues. Shell hydrogen fusion adds more helium to the star's core, which contracts and becomes hotter. The outer atmosphere expands considerably, and the star becomes a giant.

✳   When the central temperature of a giant reaches about 100 million Kelvin, the thermonuclear process of helium fusion begins. This process converts helium to carbon, then to oxygen. In a massive giant, helium fusion begins gradually. In a less massive giant, it begins suddenly in a process called the helium flash.

✳   The age of a stellar cluster can be estimated by plotting its stars on an H-R diagram. The upper portion of the main sequence will be missing, because more massive main-sequence stars already will have become giants. Relatively young stars are metal-rich; ancient stars are metal-poor.

✳   Giants undergo extensive mass loss, sometimes producing circumstellar shells of ejected material.

### Variable Stars

✳   When a star's evolutionary track carries it through a region called the instability strip in the H-R diagram, the star becomes unstable and begins to pulsate.

* RR Lyrae variables are low-mass, pulsating variables with short periods. Cepheid variables are high-mass, pulsating variables exhibiting a regular relationship between the period of pulsation and luminosity.

* Mass can be transferred from one star to another in close binary systems. When this occurs the evolution of the two stars changes.

## Review Questions

1  What is a giant molecular cloud, and what role do these clouds play in star formation?

2  Why are low temperatures necessary for dense cores to form and contract into protostars?

3  What is an H II region?

4  Why don't thermonuclear reactions occur on the surface of a main-sequence star?

5  What is an evolutionary track, and how can such tracks help us interpret the H-R diagram?

6  Why are most stars we see in the sky main-sequence stars?

7  Draw the pre–main-sequence evolutionary track of the Sun on an H-R diagram. Briefly describe what was occurring throughout the solar system at various stages along this track. (You may find reviewing Foundation II useful.)

8  On what grounds are astronomers able to say that the Sun has about five billion years remaining in its main-sequence stage?

9  What will happen inside the Sun five billion years from now when it begins to evolve into a giant?

10  Draw the post-main-sequence evolutionary track of the Sun on an H-R diagram up to the point when the Sun becomes a

helium-burning giant. Briefly describe what might occur throughout the solar system as the Sun undergoes this transition.

11  What does it mean when an astronomer says that a star "moves" from one place to another on an H-R diagram?

12  What is the helium flash and what causes it?

13  Explain how and why the turnoff point on the H-R diagram of a cluster is related to the cluster's age.

14  Why do astronomers believe that globular clusters are made of old stars?

15  What are Cepheid variables and how are they related to the instability strip?

16  Which are older stars: Type I or Type II Cepheids? Justify your answer.

17  What are RR Lyrae variables and how are they related to the instability strip?

18  What is a Roche lobe and what is its significance in close binary systems?

19  What is the difference between detached, semidetached, contact, and overcontact binaries?

## Advanced Questions

The answers to all computational problems, which are preceded by an asterisk (*), appear at the end of the book.

20  How is a degenerate gas different from an ordinary gas?

21  If you took a spectrum of a reflection nebula, would you see absorption lines, emission lines, or no lines? Explain your answer.

22  Why is it useful to plot the apparent magnitudes of stars in a single cluster on an H-R diagram rather than their absolute magnitudes?

23  What observations would you make of a star to determine whether its primary source of energy was fusing hydrogen or helium?

24  What might happen to the massive outer planets when the Sun becomes a giant?

* 25  How many 1.5-$M_\odot$ main-sequence stars would it take to equal the luminosity of one 15-$M_\odot$ star?

* 26  How many times longer does a 1.5-$M_\odot$ star fuse hydrogen at its core than does a 15-$M_\odot$ star?

27  Why do you expect that a shock wave from a supernova seems to produce relatively few high-mass O and B stars compared to the lower-mass A, F, G, and K stars.

28  How would you distinguish a newly formed protostar from a giant, given that they occupy the same location on the H-R diagram?

29  What observational consequences would we find in H-R diagrams for star clusters if the universe had a finite age? Could we use these consequences to establish constraints on the possible age of the universe? Explain.

## Discussion Questions

30 Speculate about the possibility of life-forms and biological processes occurring in giant molecular clouds. In what ways might conditions favor or hinder biological evolution?

31 Is there any evidence that the Earth has ever passed through a star-forming region in space?

## What If ...

32 The solar system passed through a giant molecular cloud? How would this encounter affect the Earth and life on it?

33 The Sun had already entered its giant phase? What would happen to the Earth and to life on it?

34 The Earth was orbiting a 0.5-$M_\odot$ star at a distance of 1 AU? What would be different for the Earth and life on it? What effects would moving the Earth closer to the lower-mass Sun have?

## Observing Projects

35 Use a telescope to observe at least two of the following interstellar gas clouds: M42 (Orion), M43, M20 (Trifid), M8 (Lagoon), M17 (Omega). You can easily locate them with the aid of your Starry Night program, star charts published during the summer months in such magazines as *Astronomy* and *Sky and Telescope*, or the list of coordinates below. In each case, can you identify the stars responsible for the ionizing radiation that causes the nebula to glow? Draw a picture of what you see through the telescope and compare it with a photograph of the object. Which portions of the nebula are not visible through your telescope? Why are they obscured?

36 Several star clusters can be seen quite well with a good pair of binoculars. You can easily locate many of them with the aid of your Starry Night program, star charts published during the summer months in such magazines as *Astronomy* and *Sky & Telescope*, or the list of coordinates below. Observe as many of these clusters as you can. Look at them through a telescope, if available. Note the overall distribution of stars in each cluster. Can you see any of these clusters with the naked eye? What difference do you note between binocular and telescopic images of individual clusters?

| Nebula | Right ascension | Declination |
|---|---|---|
| M42 (Orion) | $5^h\ 35.4^m$ | $-5°\ 27'$ |
| M43 | $5^h\ 35.6^m$ | $-5°\ 16'$ |
| M20 (Trifid) | $18^h\ 02.6^m$ | $-23°\ 02'$ |
| M8 (Lagoon) | $18^h\ 03.8^m$ | $-24°\ 23'$ |
| M17 (Omega) | $18^h\ 20.8^m$ | $-16°\ 11'$ |

| Nebula | Right ascension | Declination |
|---|---|---|
| M45 (Pleiades) | $3^h\ 47.0^m$ | $+24°\ 07'$ |
| Hyades | $4^h\ 27.0^m$ | $+16°\ 00'$ |
| Praesepe | $8^h\ 40.1^m$ | $+19°\ 59'$ |
| Coma | $12^h\ 25.0^m$ | $+26°\ 00'$ |
| M11 | $18^h\ 51.1^m$ | $-06°\ 16'$ |

# The Earth Orbited a 1.5M$_\odot$ Sun?

The Earth is at a perfect distance from a wonderful star. The Sun provides just enough heat so that liquid water, necessary to sustain life, can exist here. Would our planet still be suitable for the evolution of life if the Sun were 1.5 M$_\odot$ rather than 1.0 M$_\odot$? To begin with, we need to know how far from the new Sun, let's call it Sol II, to place the Earth.

**A Very Sunny Day** Sol II's surface temperature would be 8400 K and it would appear blue-white in our sky. Sol II would give off 7 times as much energy per second as our present Sun, due to the combination of a higher temperature and a 20% larger radius than the Sun. The effect of Sol II's increased energy emission would require that Earth be located much farther away from Sol II than our present distance from the Sun. To understand why, consider the increase in infrared (heat) output from Sol II. That extra heat would have the initial effect of raising the average global temperature on Earth at our present distance by about 10 K (about 20°F). This does not seem like a lot, but the impact of that slight increase would actually boost the atmospheric temperature much higher.

The extra heat from Sol II would cause more ocean water to evaporate into the atmosphere. Because water is a greenhouse gas (that is, it traps infrared radiation), the air temperature would rise, causing even more water to evaporate from the oceans, which, in turn, would cause the air to heat even more. This vicious cycle, called a *runaway greenhouse effect*, would make the Earth's surface so hot and dry that it would be uninhabitable.

By moving the Earth about 2.6 times farther away from Sol II, the temperature would become suitable for life. At that distance, the year would be 1249 days long. While moving away from the heat is one thing, moving away from the ultraviolet radiation emitted by Sol II is something else altogether.

**Ultraviolet Excess** Just by increasing the Sun's mass by 50%, the ultraviolet radiation emitted would be several *thousand* times stronger. This is because the energy output of stars with different surface temperatures varies with wavelength. While the output of visible light would change slightly, the output of ultraviolet radiation would be vastly greater. Therefore, even though the Earth's surface temperature would be suitable farther from Sol II, the flood of ultraviolet radiation would be so strong that the ozone layer would be overwhelmed, and the level of ultraviolet radiation at the Earth's surface would be much higher than it is today. This would be so even with the greater concentration of ozone created by the increased ultraviolet from Sol II.

Life would have to evolve greater protection from ultraviolet radiation than it has today. And if this were not a great enough challenge, suppose intelligent life-forms were evolving on the Earth orbiting Sol II 4.6 billion years after the solar system formed. They would discover that their star had evolved so rapidly that it was just about to expand into the giant phase!

**12**

R I V U X G

**Eta Carinae** This spectacular image is Eta Carinae, believed to be a binary star system, with each star having more than 70 $M_{\odot}$. This Hubble Space Telescope image details the remnants of an ejection that was first seen in 1841 and that temporarily made Eta Carinae the second brightest object in the night sky. Despite losing more than 1 $M_{\odot}$ in this event, the stars survived. Because of their high masses, these stars are expected to self-destruct as supernovae within the next few hundred million years. Eta Carinae is 2450 pc (8000 ly) from Earth. Among its other mysteries is the mechanism by which it has created the two lobes of expanding gas. We will encounter other hourglass-shaped distributions later in this chapter.

# The Deaths of Stars

*In this chapter you will discover*

✳ what happens to stars after helium fusion in their cores ceases

✳ how heavy elements are created

✳ the end of stellar evolution, when stars eject matter back into space

✳ that lower-mass stars expel their outer layers relatively gently

✳ that higher-mass stars explode violently

✳ how astronomers have observed neutron stars

## WHAT DO YOU THINK?

1   Will the Sun end its existence? If so, how?

2   What is a nova?

3   What is the origin of the carbon, silicon, oxygen, iron, uranium, and other heavy elements on Earth?

4   What is a cosmic ray?

5   What is a pulsar?

Change is the underlying theme of stellar evolution. We have seen how gas and dust come together to form stars; how hydrogen fuses into helium; how post–main-sequence stars expand, contract, and expand again; and how helium transforms into carbon and oxygen. Does this spiral of changing elements and changing stellar properties continue indefinitely, or are there limits to what can be created inside stars? In this chapter and Chapter 13, we learn that there are indeed limits to stellar fusion processes, limits that are determined by stellar masses. Furthermore, we investigate the often spectacular finales to stellar evolution.

# LOW-MASS STARS AND PLANETARY NEBULAE

In Chapter 11 we discovered what happens when shell hydrogen fusion first begins: The outpouring of energy causes a star to expand and become a giant. In this process, low-mass stars (under 8 $M_\odot$) move over to, and ascend, the giant branch on the H-R diagram for the first time (Figure 12-1a). This period is accompanied by the expulsion of mass into space in the form of stellar winds, which reduce the masses of these stars. Then comes helium fusion, with stars of less than 2 $M_\odot$ undergoing a core helium flash in the giant stage. After fusion begins, these stars shrink and move onto the *horizontal branch* (Figure 12-1b). Eventually, the core is converted into carbon and oxygen. Helium fusion there ceases and these stars undergo another stage that closely parallels the end of core hydrogen fusion.

## 12-1 Low-mass stars expand into the supergiant phase before expanding into planetary nebulae

Calculations indicate that carbon and oxygen require a temperature of at least 600 million K to fuse. Because the core of a low-mass giant on the horizontal branch is only about 200 million K, fusion of these elements in the core does not occur. Photon production therefore drops off, and the inner regions of the star again contract, compressing

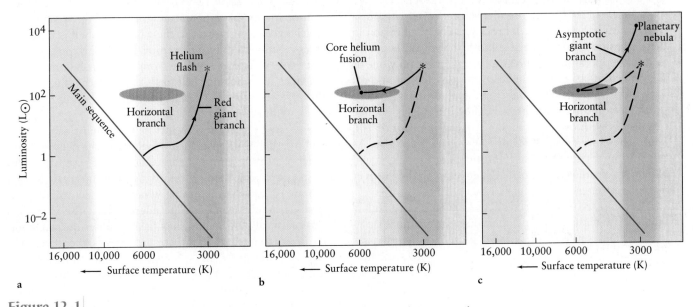

**Figure 12-1**   **Post–Main-Sequence Evolution of Low-Mass Stars**   (a) The H-R track as a star makes the transition from the main sequence to the giant phase. The asterisk shows the helium flash occurring in a low-mass star. (b) After the helium flash, the star converts its helium core into carbon and oxygen. While doing so, its core reexpands, decreasing shell fusion. As a result, the star's outer layers recontract. (c) After the helium core is completely transformed into carbon and oxygen, the core recollapses, and the outer layers reexpand, powered up the asymptotic giant branch by shell helium fusion.

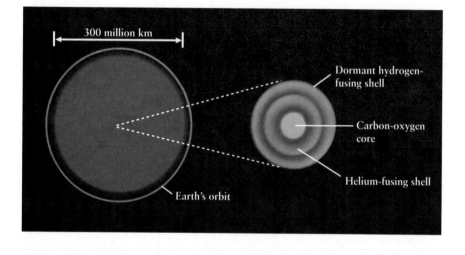

**Figure 12-2** **The Structure of an Old Low-Mass Star** Near the end of its life, a low-mass star becomes a supergiant, with a diameter almost as large as the diameter of the Earth's orbit. The star's core, the dormant hydrogen-fusing shell, and an active helium-fusing shell are contained within a volume roughly the size of the Earth.

and heating the shell of helium-rich gas just outside the core. As a result, **shell helium fusion** begins outside the core; this shell is itself surrounded by a hydrogen-fusing shell. All this takes place within a volume roughly the size of the Earth.

Once shell helium fusion commences, the new outpouring of energy pushes the outer envelope of the star out again. A low-mass star thus leaves the horizontal branch and ascends the giant branch for a second time. Powered by the shell fusion, it becomes brighter than it ever was before. Such **asymptotic giant branch stars**, or **AGB stars**, typically have diameters as big as the orbit of Mars and shine with the brightness of $10^4$ L$_\odot$. These stars are now low-temperature, red supergiants (Figure 12-1c).

Models of AGB stars reveal a variety of activities destined to lead to the stars' destruction. The shell fusion in these stars does not proceed smoothly. Sometimes conditions of temperature and pressure are right for only one or the other shell to operate, and so they switch on and off with increasing ferocity. Therefore, as these stars proceed up the asymptotic giant branch, they irregularly change brightness with periods ranging from about 100 to 700 days. These are called *Mira* or *long-period variables*. As the Sun moves along the asymptotic giant branch, it will expand to a radius of about 1 AU, enveloping the Earth in its tenuous outer layers (Figure 12-2).

Giant stars of all spectral types have strong stellar winds. Low-mass stars expel their outer layers more slowly than higher-mass stars, but they all reduce their masses significantly from what they were on the main sequence. During their initial ascent up the giant branch, stars can lose as much as 30% of their masses. On the asymptotic giant branch they lose even more, often surrounding themselves in thickening cocoons of gas and dust. A star of the Sun's mass loses about $10^{-5}$ M$_\odot$ per year at this stage, eventually dumping half of its mass back into space.

This mass loss limits the amount of gravity available to compress the star's core and regions of shell fusion. A

fine balance is struck: The core of our low-mass AGB star is compressed until it again becomes degenerate, meaning that the electrons in the core provide a growing repulsive force that stops its contraction. However, even for the most massive of these low-mass stars (now much less than their original 8 M$_\odot$ due to their stellar winds), the core temperature cannot reach the 600 million K necessary to fuse carbon or oxygen. Therefore, no further core fusion occurs.

The final act in the drama of a low-mass star begins with a "thermal runaway" (meaning a rapid rise in temperature) in the helium shell, like the helium flash in its core earlier in its life. This occurs because the triple alpha process (see Section 11-9) is extremely sensitive to temperature, and as the temperature goes up slightly, the fusion rate skyrockets. However, the gas in the shell is not compressed enough to be degenerate: The shell is too thin. The increase in energy output from the thin helium-fusing shell, called a **helium shell flash**, expands the star, thereby briefly decreasing the temperature in the shell and slowing the rate of fusion.

A star undergoes several helium shell flashes as its helium shell thickens. During each flash, the helium shell's energy output jumps a thousandfold. These brief outbursts are separated by relatively quiet intervals lasting about 100,000 years, during which the helium shell gradually becomes thicker. The energy from the increased fusion during the helium shell flashes causes the star's outer layers to expand and, therefore, cool further. Eventually, the outer gases are sufficiently cool so that electrons and ions there can recombine. This is exactly the opposite of ionization (discussed in Section 4-5). Whereas ionization requires an electron to absorb a photon, recombination forces an electron to emit a photon.

The pressure provided by the recombination photons combined with the pressure provided by the photons from the helium flashes generates enough energy to eject more and more of the star's outer layers into space. As the ejected material expands and cools, some of it condenses to form dust grains that are propelled outward by radiation pressure from the star's hot, burned-out core. The expanding dust and gases are called a **planetary nebula.** As the outer layers of the star are shed, an increasingly hot interior is revealed, and the star moves to the left across the H-R diagram (Figure 12-3).

Planetary nebulae have *nothing* to do with planets. This unfortunate term was coined by the astronomer William Herschel in the eighteenth century. When viewed through the small telescopes of the day, these glowing objects, often green in appearance, looked like planets, such as Uranus. Including the mass lost prior to the planetary nebula phase, low-mass stars lose as much as 80% of their masses by shedding their outer layers.

Planetary nebulae come in a breathtaking variety of shapes (Figures 12-4 and 12-5). The outer rings of planetary nebula NGC 7027 (Figure 12-4c) indicate that it initially emitted spherical shells of gas and dust, followed by a final period in which the material flowed outward more irregularly. The Hourglass Nebula (Figure 12-5) appears initially to have shed mass in a doughnut shape around itself. In its final death throes, this gas and dust forced the final outflow to go in two directions perpendicular to the plane of the doughnut, creating what is called a *bipolar planetary nebula.*

Our theories about why the gas and dust in planetary nebulae emerge in such a variety of patterns are incomplete, but evidence that many of the nonspherical planetary nebulae are created in binary star systems is growing. Astronomers believe that the companion star, along with

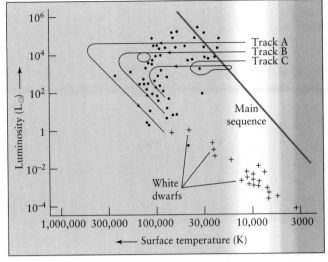

| Evolutionary track | Mass ($M_\odot$) | | |
|---|---|---|---|
| | Red supergiant | Ejected nebula | White dwarf |
| A | 3.0 | 1.8 | 1.2 |
| B | 1.5 | 0.7 | 0.8 |
| C | 0.8 | 0.2 | 0.6 |

**Figure 12-3**    **Evolution from Supergiants to White Dwarfs**
The evolutionary tracks of three low-mass, red supergiants are shown as they eject planetary nebulae. The table gives their mass as red supergiants, the amount of mass they lose as planetary nebulae, and the remaining (white dwarf) mass. The dots on this graph represent the central stars of planetary nebulae whose surface temperatures and luminosities have been determined. The crosses are white dwarfs for which similar data exist.

a        b        c        R I V U X G

**Figure 12-4**    **Some Shapes of Planetary Nebulae**  The outer shells of dying low-mass stars are ejected in a wonderful variety of patterns. **(a)** NGC 7293, the Helix Nebula, is located in the constellation Aquarius (the Water Bearer). The star that ejected these gases is seen at the center of the glowing shell. This nebula, located about 215 pc (700 ly) from Earth, has an angular diameter equal to about half that of the full Moon. **(b)** NGC 6826 shows jets of gas (in red) whose origin is as yet unknown. **(c)** NGC 7027 shows circular rings (actually spherical shells) of gas and dust that are flowing outward irregularly. It is about 920 pc (3000 ly) from Earth in the constellation Cygnus.

a

b

c

d

RIVUXG

**Figure 12-5** **Formation of a Bipolar Planetary Nebula** Bipolar planetary nebulae may form in two steps. Astronomers hypothesize that **(a)** first, a doughnut-shaped cloud of gas and dust is emitted, **(b, c)** followed by outflow that is channeled by the original gas to squirt out perpendicular to the plane of the doughnut. **(d)** The Hourglass Nebula appears to be a classic example of such a system. The bright ring is believed to be the doughnut-shaped region of gas, lit by energy from the planetary nebula. The Hourglass is located about 2500 pc (8000 ly) from Earth.

stellar rotation and stellar magnetic fields, propels the outgoing mass into the patterns we see.

Planetary nebulae are quite common in our Galaxy. More than 1000 have been identified, and astronomers estimate that 20,000 to 100,000 exist in the Milky Way. Indeed, such will be the fate of our Sun.

Spectroscopic observations of planetary nebulae show bright emission lines of hydrogen, carbon, neon, magnesium, oxygen, and nitrogen. From the Doppler shifts of these lines, we conclude that the expanding shell of gas is moving outward from the dying low-mass star at speeds of 10 to 30 km/s. A typical planetary nebula has a diameter of roughly 1 ly, which means that it began expanding about 10,000 years ago.

By astronomical standards, a planetary nebula is short-lived. After about 50,000 years, the nebula spreads over distances so far from the cooling central star that its nebulosity simply fades from view. The gases then mingle and mix with the surrounding interstellar medium, enriching it with moderately heavy elements, such as carbon. Astronomers estimate that planetary nebulae return a total of 5 $M_\odot$ to the interstellar medium of our Galaxy

# IV

# BUILD YOUR FOUNDATION

## The Universe

### In the chapters ahead you will discover

* the Milky Way Galaxy—billions of stars along with gas and dust bound together by mutual gravitational attraction

* how astronomers determined that myriad galaxies lie beyond the Milky Way

* how the period–luminosity law for Cepheid variables and the light curves of supernovae enable astronomers to calculate distances to other galaxies

* other distant objects, the large-scale structure of the universe, and its ultimate fate

* a way to estimate whether advanced life may have formed on planets orbiting distant stars

R I V U X G
**The Realm of the Galaxies**
Beyond the solar system, beyond the Milky Way, lies the universe of galaxies. Just as stars orbit in their respective galaxies, so, too, do many galaxies orbit each other. This image shows hundreds of galaxies grouped together some 325 million light-years from Earth. This computer-enhanced photograph shows only a tiny sliver of the universe, yet it provides a valuable glimpse at the variety and staggering numbers of galaxies that exist. This region of the universe is part of the so-called Great Wall of galaxies stretching across more than one-third of the sky.

each year. This amounts to about 15% of all matter expelled by all types of stars each year. Planetary nebulae therefore play an important role in the chemical evolution of the Galaxy as a whole.

## 12-2 The burned-out core of a low-mass star becomes a white dwarf

Our models predict that main-sequence stars less massive than about 8 $M_\odot$ never develop the central pressures or temperatures necessary to ignite thermonuclear reactions of the carbon or oxygen in their cores. Instead, as we have seen, the process of mass ejection strips away the stars' outer layers and exposes the carbon-oxygen cores, which simply cool off. Such burned-out hulks are called **white dwarfs.** This is to be the fate of the Sun's core.

Theoretical models predict that degenerate matter, such as the electrons in a white dwarf, can withstand only a finite amount of pressure, above which they cannot exert enough outward force to prevent the core from collapsing further. This limits the mass of a white dwarf to less than 1.4 $M_\odot$. This mass is called the **Chandrasekhar limit,** after the astrophysicist Subrahmanyan Chandrasekhar, who received a Nobel Prize for his pioneering theoretical studies of white dwarfs in the 1940s.

Theoretical evolutionary tracks of three burned-out stellar cores are shown in Figure 12-3. These particular white dwarfs evolved from main-sequence stars of between 0.8 and 3.0 $M_\odot$. During the ejection phase, the appearance of these stars changes rapidly. These objects appear to race along their evolutionary tracks on the H-R diagram, sometimes executing loops corresponding to thermal pulses (tracks B and C in Figure 12-3). Finally, as the ejected planetary nebulae fade and the stellar cores cool, the evolutionary tracks of the dying stars take a sharp turn downward toward the white dwarf region of the diagram.

As noted above, the temperature inside a white dwarf is too low to ignite the carbon or oxygen. Electron degeneracy pressure (discussed in Section 11-9) prevents the crushing weight of the carbon and oxygen from contracting the remnant core further, thereby creating a stable white dwarf roughly the size of the Earth. The density of matter in one of these stellar "corpses" is typically $10^9$ kg/m$^3$. In other words, a teaspoonful of white dwarf matter brought to Earth would weigh 5 tons.

As billions of years pass, an isolated white dwarf radiates its stored energy into space, cooling and decreasing in luminosity. It moves down and to the right on the H-R diagram. The Sun will become a cold, dark, dense sphere of degenerate gases rich in oxygen and carbon, about the size of the Earth. We can actually follow the theory a little further. It predicts that after billions of years of radiating away their heat, the interior temperatures of white dwarfs

will decrease to about 4000 K, and the carbon and oxygen in them will solidify, transforming them into giant crystals.

Many white dwarfs are found in our solar neighborhood, but all are too faint to be seen with the naked eye. The first white dwarf to be discovered is a companion to the bright star Sirius. The binary nature of Sirius was first deduced in 1844 by the German astronomer Friedrich Bessel, who noticed that the star was moving back and forth, as if orbited by an unseen object. This companion, called Sirius B, was first glimpsed in 1862 (Figure 12-6). Recent satellite observations at ultraviolet wavelengths—where white dwarfs emit most of their light—demonstrate that the surface temperature of Sirius B is about 30,000 K. Recalling our discussion of binary star systems in Section 11-14, the fact that Sirius B is already a white dwarf while its companion is not means that Sirius B must have originally been more massive than Sirius A. Sirius A is the more massive of the two today. Because most stars occur in binary star systems, it is important to take into account any mass transfer that occurs when determining the evolution of these stars.

## 12-3 White dwarfs in close binary systems can create powerful explosions

Occasionally, a star in the sky suddenly becomes between ten thousand and a million times brighter. This phenomenon is called a **nova.** Novae are fairly common. Their abrupt rise in brightness is followed by a gradual decline that may stretch over several months or more (Figures 12-7 and 12-8).

R I V U X G

**Figure 12-6**    **Sirius and Its White Dwarf Companion** Sirius, the brightest-appearing star in the night sky, is actually a double star. The secondary star is a white dwarf, seen here at the five o'clock position in the glare of Sirius. The spikes and rays around Sirius are created by optical effects within the telescope.

a                                                                                           b                                    R I V U X G

## Figure 12-7

**Nova Herculis 1934** These two pictures show a nova (**a**) shortly after peak brightness as a magnitude –3 star and (**b**) two months later, when it had faded to magnitude +12. Novae are named after the constellation and year in which they appear.

**2** Painstaking observations strongly suggest that novae occur in close binary systems containing a white dwarf. The ordinary companion star presumably fills its Roche lobe, so it gradually deposits fresh hydrogen onto the white dwarf. This new mass becomes a dense layer covering the hot surface of the white dwarf. As more gas is deposited and compressed, the temperature in the hydrogen layer continues to increase. Finally, at about $10^7$ K, hydrogen fusion ignites throughout the layer, blowing it into inter-

stellar space. This explosion is the nova. After a nova, fusion ceases. The companion star, however, may retain enough mass to supply a new layer of surface hydrogen, enabling some novae to reoccur.

# HIGH-MASS STARS AND SUPERNOVAE

In high-mass main-sequence stars (greater than 8 $M_\odot$) sufficient gravitational force compresses the core to heat it enough to enable the carbon and oxygen there to fuse. This leads to the creation of a host of other elements that are eventually ejected into interstellar space.

## 12-4 A series of fusion reactions in high-mass stars leads to luminous supergiants

When helium fusion ends in the core of a massive star, gravitational compression again collapses the core, driving the star's central temperature beyond 600 million K. Now, carbon fusion begins, producing such elements as neon and magnesium. When a star compresses itself enough to drive its central temperature to 1.2 billion K, neon fusion occurs. When the central temperature of the star then reaches 1.5 billion K, oxygen fusion begins. As the star consumes increasingly heavier nuclei, it produces

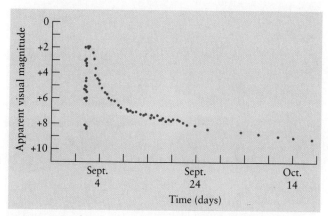

## Figure 12-8

**The Light Curve of a Nova** This graph shows the history of Cygni 1975, a nova that was observed to blaze forth in the constellation of Cygnus in September 1975. The rapid rise in magnitude followed by a gradual decline is characteristic of many novae, while some oscillate in intensity as they become dimmer.

sulfur and isotopes of silicon and phosphorus, among other elements.

Each successive thermonuclear reaction occurs more rapidly than the last. For example, detailed calculations for a star that was initially 25 $M_\odot$ on the main sequence demonstrate that carbon fusion occurs for 600 years, neon fusion for 1 year, and oxygen fusion for only 6 months. After half a year of core oxygen fusion, gravitational compression forces the central temperature up to 2.7 billion K and silicon fusion begins. This thermonuclear process proceeds so furiously that the entire core supply of silicon in a 25-$M_\odot$ star is used up in 1 day.

Each stage of fusion adds a new shell of matter outside the core (Figure 12-9), creating a structure resembling the layers of an enormous onion. Together, the tremendous numbers of fusion-generated photons created in all these shells, as well as in the core, push the outer layers of the star outward. It expands to become a luminous supergiant almost as wide across as Jupiter's orbit around the Sun.

Luminous supergiant stars, which are brighter than $10^5$ $L_\odot$, emit winds throughout most of their existence, with mass-loss rates exceeding those of giants (see Chapter 11). Figure 12-10 shows a supergiant star losing mass. Betelgeuse (see the figure that opens Build Your Foundation III), 470 ly away in the constellation of Orion, is another good example of a supergiant experiencing mass loss. Spectroscopic observations show that Betelgeuse is losing mass at the rate of $1.7 \times 10^{-7}$ $M_\odot$ per year and is surrounded by ejected gas, its *circumstellar shell*. This huge shell is expanding at 10 km/s, and escaping gases have been detected at distances of 10,000 AU from the

star. The expanding circumstellar shell has an overall diameter of 1/3 ly.

Silicon fusion in high-mass stars involves many hundreds of nuclear reactions, but its major final product is iron. Eventually, the core is converted entirely into iron, which is surrounded by layers of shell fusion that consume the star's remaining reserves of fuel (see Figure 12-9). Whereas the star's enormously bloated atmosphere is nearly as big as the orbit of Jupiter, its entire energy-producing region is again contained in a volume the size of the Earth. The buildup of an inert, iron-rich core signals the impending violent death of a massive star.

## 12-5 High-mass stars violently blow apart in supernova explosions

Unlike lighter elements, iron cannot fuel further thermonuclear reactions. The protons and neutrons inside iron nuclei are already so tightly bound together that no further energy can be extracted by fusing still more nuclei together. The sequence of fusion stages in the cores of high-mass stars therefore ends.

Because iron atoms do not fuse, the electrons in the core must now support the star's outer layers by the strength of electron degeneracy pressure alone. Soon, however, the continued deposition of fresh iron from the silicon-fusing shell causes the core's mass to exceed the Chandrasekhar limit. Electron degeneracy suddenly fails to support the star's enormous weight, and the core collapses.

Any isolated main-sequence star of greater than 8 $M_\odot$ develops an iron core as a luminous supergiant. When the

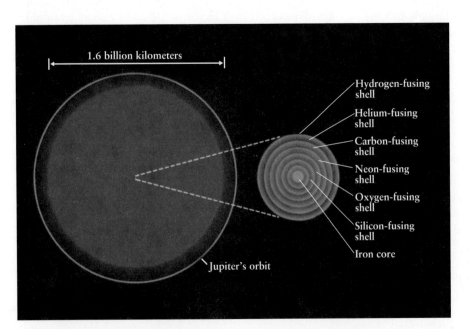

**Figure 12-9** **The Structure of an Old High-Mass Star** Near the end of its life, a high-mass star becomes a supergiant with a diameter almost as wide as the orbit of Jupiter. The star's energy comes from six concentric fusing shells, all contained within a volume roughly the same size as the Earth.

**Figure 12-10** R I V U X G
**Mass Loss from a Supergiant Star** Beautiful nebulosity surrounds a supergiant star (~100 $M_\odot$) that is experiencing significant mass loss. The ejected material collides and interacts with the surrounding interstellar gas and dust, thereby producing the cosmic bubble seen here. This supergiant and its nebulosity are located in the constellation of Canis Major.

core collapses, a rapid series of cataclysms is triggered that tears the star apart in a few seconds. Let us see how this happens in the death of a 25-$M_\odot$ star according to supercomputer simulations (Table 12-1).

In a 25-$M_\odot$ star, electron degeneracy pressure fails when the density is sufficiently high inside the iron core. The core, only some 3000 km in diameter, then collapses immediately. In roughly 1/10 second, the central temperature exceeds 5 billion K. Gamma-ray photons associated with this intense heat have so much energy that they begin to break apart the iron nuclei, a process called **photodisintegration**. Although millions of years passed from the time the star arrived on the main sequence with a hydrogen-

and helium-filled core until its core became iron, it takes less than a second to convert the core back into elemental protons, neutrons, and electrons.

Within another 1/10 second, as the density continues to climb, the electrons in the core are forced to combine with protons there to produce neutrons, and the process releases a flood of neutrinos. As we saw in Chapter 9, neutrinos have no electric charge, very little mass, and most pass through the Earth or Sun without interaction. Nevertheless, the matter deep inside a collapsing high-mass star is so fantastically dense that the newly created neutrinos there cannot immediately escape from the star's core.

According to the theory, about 1/4 second after the collapse begins, the density of the entire core reaches $4 \times 10^{17}$ kg/m$^3$, which is *nuclear density*, the density at which neutrons and protons are normally packed together inside atomic nuclei. (Compare this with the density of water, $10^3$ kg/m$^3$.) At nuclear density, matter is virtually incompressible. Thus, when the neutron-rich material of the core reaches nuclear density, it suddenly stiffens. The collapse of the core halts so abruptly that it rebounds and begins rushing back out in a process called *core bounce*.

The model predicts that during this critical stage, the star's unsupported layers of shell-fusing matter are plunging inward at up to 15% of the speed of light. The outward flowing neutrinos and rebounding core slam into this matter, causing it to reverse course. In just a fraction of a second, a tremendous volume of matter begins to move back up toward the star's surface. This matter accelerates rapidly as it encounters less and less resistance, and soon it forms an outgoing shock wave. After a few hours, this shock wave reaches the star's surface, lifting the star's outer layers away from the core in a mighty blast. The star becomes a **supernova**.

As the supernova expands, the star's luminosity suddenly increases by a factor of $10^8$. The energy emitted by a supernova is truly staggering—as much as all 200 billion stars in the Milky Way Galaxy combined! In other

| Table 12-1 | | | |
|---|---|---|---|
| Evolutionary Stages of a 25-$M_\odot$ Star | | | |
| Stage | Central temperature (K) | Central density (kg/m$^3$) | Duration of stage |
| Hydrogen fusion | $4 \times 10^7$ | $5 \times 10^3$ | $7 \times 10^6$ yr |
| Helium fusion | $2 \times 10^8$ | $7 \times 10^5$ | $5 \times 10^5$ yr |
| Carbon fusion | $6 \times 10^8$ | $2 \times 10^8$ | 600 yr |
| Neon fusion | $1.2 \times 10^9$ | $4 \times 10^9$ | 1 yr |
| Oxygen fusion | $1.5 \times 10^9$ | $1 \times 10^{10}$ | 6 mo |
| Silicon fusion | $2.7 \times 10^9$ | $3 \times 10^{10}$ | 1 d |
| Core collapse | $5.4 \times 10^9$ | $3 \times 10^{12}$ | 0.2 s |
| Core bounce | $2.3 \times 10^{10}$ | $4 \times 10^{17}$ | milliseconds |
| Supernova explosion | about $10^9$ | varies | 10 s |

words, for a few days following the explosion, a supernova may shine as brightly as an entire galaxy. Over its lifetime, our model 25-$M_\odot$ star ejects more than 20 $M_\odot$ of its mass back into space via stellar winds, gas ejected during unstable periods, and during its supernova phase. Less massive stars return proportionately less mass to the interstellar medium.

3  As the outer layers of a massive dying star are blasted into space, they are compressed so much by the neutrinos and the shock wave that fusion actually occurs during the explosion, creating a broad assortment of elements, including the heaviest elements that exist naturally in the universe. Indeed, the products of shell fusion and this final burst of fusion during supernovae are where most of the elements on the Earth and other terrestrial planets come from, including many of the atoms in our bodies.

**Insight into science**   **The process of science**
The best scientific theories and models also provide explanations about matters that they were not initially designed to study. In modeling stellar evolution and supernovae, the processes that create and eject metals (that is, elements heavier than hydrogen and helium) from stars also explain how the elements necessary for life are both created and made available.

## 12-6 Remnants of supernova explosions can be detected for millennia afterward

Astronomers find the debris of supernova explosions scattered across the sky. A beautiful example is the Cygnus Loop (see Figure 11-4). The doomed star's outer layers were blasted into space 20,000 years ago so violently that they are still traveling at supersonic speeds. As this expanding shell of gas plows through the interstellar medium, it collides with atoms and molecules, making the gases glow.

Many supernova remnants cover sizable fractions of the sky. The largest is the Gum Nebula, with an angular diameter of 60° (Figure 12-11). This nebula looks so big because it is so close: Its near side is only about 300 ly from Earth. Studies of the Gum Nebula's expansion rate suggest that the supernova exploded about 11,000 years ago. At maximum brilliance, the exploding star probably was as bright as the Moon at first quarter.

Many supernova remnants can be detected only at nonvisible wavelengths, ranging from X rays through radio waves. For example, Figure 12-12 shows both X-ray and radio images of the supernova remnant Cassiopeia A. Visible-light photographs of this part of the sky reveal

R I V U X G

**Figure 12-11**   **The Gum Nebula** The Gum Nebula is the largest known supernova remnant, spanning 60° across the sky, centered roughly on the southern constellation of Vela. The nearest portions of this expanding nebula are only 300 ly from the Earth. The supernova explosion occurred about 11,000 years ago, and its remnant now has a diameter of about 2300 ly. Only the central regions of the nebula are shown here.

only a few small, faint wisps. Thus, radio searches for supernova remnants are more fruitful than visual searches. Only two dozen supernova remnants have been found in photographs, but more than 100 remnants have been discovered by radio astronomers.

From the expansion rate of the nebulosity in Cassiopeia A, astronomers calculate that the supernova explosion first reached Earth about 300 years ago. Although telescopes were in wide use by the late 1600s, no record of the event is known. In fact, the last supernova of a star near its maximum brightness in our Galaxy was seen by Johannes Kepler in 1604. In 1572 Tycho Brahe also recorded the sudden appearance of an exceptionally bright star in the sky. To find previous accounts of supernova explosions, we must delve into astronomical records that are almost a thousand years old.

4  Besides the gas and dust enriched with heavy elements, supernovae may leave a heritage long after their clouds have dissipated. Astronomers have detected particles flying through space at speeds exceeding 90% the speed of light. These high-speed particles are the **cosmic rays**. *Cosmic rays are not rays at all!* They are primarily hydrogen nuclei, along with nuclei of more massive elements like helium and carbon, and a few percent electrons and positrons (positively charged electrons). While there is considerable controversy over the origins of cosmic rays, it appears that many of them are emitted by supernovae. These explosions are the best-known source of the energy necessary to give these particles their tremendous speeds.

a    R I V U X G

b    R I V U X G

**Figure 12-12**    **Cassiopeia A** Supernova remnants such as Cassiopeia A are typically strong sources of X rays and radio waves. **(a)** An X-ray picture of Cassiopeia A taken from the *Exosat* satellite, a European Space Agency mission. **(b)** A corresponding radio image produced by the Very Large Array (VLA). Radiation from the supernova that produced this nebula first reached Earth 300 years ago. The explosion occurred about 10,000 ly from Earth.

Most cosmic rays coming in our direction collide with gas in the atmosphere some 15 km above the Earth's surface. This is fortunate because each cosmic ray packs an enormous amount of energy that potentially could harm living tissue. The collision in the air divides a cosmic ray's energy among several gas particles, which are shoved earthward. These, in turn, often hit other gas atoms, creating a cascade of lower-energy particles that eventually reach the Earth's surface as a **cosmic ray shower**. These cosmic rays created from atoms in our atmosphere are called **secondary cosmic rays**.

## 12-7 Supernova 1987A offers a close-up look at a massive star's death

Supernovae are often seen in remote galaxies with the aid of a telescope. In 1885 a supernova in the Andromeda Galaxy was just barely visible to the unaided eye. On February 23, 1987, a supernova was observed in the Large Magellanic Cloud, a galaxy near our own Milky Way. The supernova, designated SN 1987A, was the first to be observed that year (Figure 12-13), and it was so bright that it could easily be seen with the naked eye. What made SN 1987A such an exciting discovery was that it gave astronomers a rare opportunity to study the death of a nearby massive star using modern equipment and thereby provide data with which to test the theory of supernovae described above.

At first, SN 1987A reached only a tenth of the luminosity predicted for an exploding massive star. For the next 85 days, it gradually brightened before slowly dimming, as

is characteristic of an ordinary supernova. Fortunately, the doomed star had been observed before it became a supernova. The Large Magellanic Cloud is about 160,000 ly from Earth—near enough to us that many of its stars have been individually observed and catalogued. The doomed star had been identified as a B3 I supergiant (Figure 12-14). The theory of the evolution of such stars provided an explanation for SN 1987A's unusually slow brightening.

**Figure 12-13**    R I V U X G

**Supernova 1987A** A supernova was discovered in a nearby galaxy called the Large Magellanic Cloud (LMC) in 1987. This photograph shows a portion of the LMC that includes the supernova and a huge H II region called the Tarantula Nebula. At its maximum brightness, observers at southern latitudes saw the supernova without a telescope.

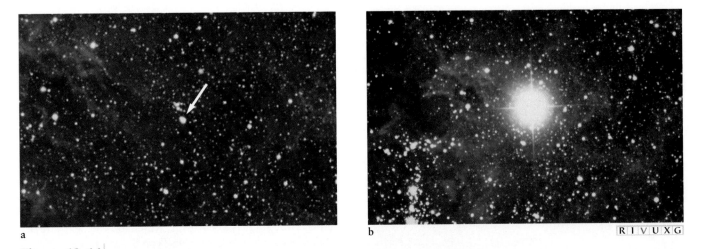

Figure 12-14  **The Doomed Star and SN 1987A** (a) A small section of the Large Magellanic Cloud before the outburst of SN 1987A. The doomed star, a blue supergiant, is identified by an arrow. (b) The supernova a few days after the explosion was first observed.

When this star was on the main sequence, its mass was about 20 $M_\odot$, although by the time it exploded it had shed many solar masses. The evolutionary track for an aging 20-$M_\odot$ star wanders back and forth across the top of the H-R diagram, and so the star alternates between being a hot (blue) supergiant and cool (red) supergiant. The star's size changes significantly as its surface temperature changes. A blue supergiant is only 10 times larger in diameter than the Sun, but a red supergiant of the same luminosity is 1000 times larger. Because the doomed star was relatively small when it exploded, it reached only a tenth of the brightness that it would have attained had it been a red supergiant at that time.

Ordinary telescopic observations of a supernova explosion can show us only the expanding outer layers of the dying star. Even X-ray or radio observations fail to see through the hot gases being blasted into space. Thus, using ordinary techniques we cannot observe the extraordinary events occurring in and around the doomed star's core. However, most of the energy of a supernova explosion is carried away in a characteristic form: Neutrinos are produced in great profusion as the star's core collapses. By detecting these neutrinos and measuring their properties, astronomers can learn many details about the star's collapsing core, especially about core bounce. As we saw in Chapter 9, however, neutrinos seldom interact with ordinary matter, making them difficult to detect. Fortunately, several neutrino detectors were operating when the neutrinos from SN 1987A reached Earth.

Nearly a day before SN 1987A was first observed in the sky, teams of scientists at neutrino detectors in Japan and the United States reported finding Cerenkov flashes (see Section 9-9) from a burst of neutrinos. The Kamiokande II detector (see Figure 9-22) in Japan detected 12 neutrinos at about the same time that 8 were found by the IMB (Irvine-Michigan-Brookhaven) detector in a salt mine under Lake Erie. Neutrinos preceded the visible supernova outburst because they escaped from the dying star before the shock wave from the collapsing core reached the star's surface. They were detected in the Earth's northern hemisphere, where the supernova is always below the horizon, after having passed through the Earth.

About three and a half years after SN 1987A's detonation was first seen here, astronomers using the Hubble Space Telescope obtained a picture showing several rings of glowing gas around the exploded star (Figure 12-15). This gas was emitted by the star 20,000 years before the supernova, when a hydrogen-rich stellar envelope was ejected by stellar winds from the doomed star, as discussed earlier in this chapter. This gas expanded in an hourglass shape, because it was blocked from expanding around the star's equator either by a preexisting ring of gas there or by the orbit of an as-yet-unseen companion star.

While emitting these gases, the doomed star was a red supergiant. In its final blue giant phase, the same star emitted higher-speed gas that compressed the hourglass gas into three rings—a narrow one around the equator and wider ones at the top and bottom of the hourglass. This relic gas is now being illuminated by photons from the supernova. Astronomers predict that the gas ejected from the supernova itself will strike the circumstellar shell sometime between now and 2004. This collision will cause the shell of gas to brighten considerably, eventually illuminating the entire hourglass.

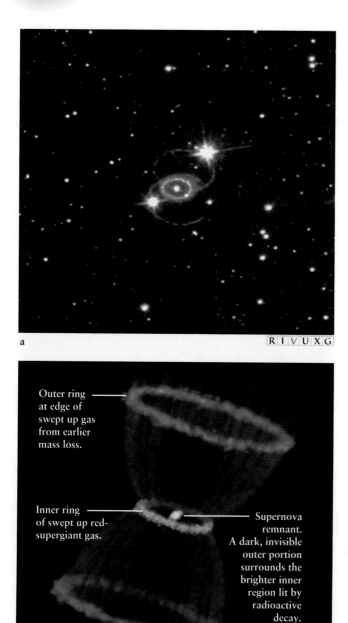

a

R I V U X G

Outer ring
at edge of
swept up gas
from earlier
mass loss.

Inner ring
of swept up red-
supergiant gas.

Supernova
remnant.
A dark, invisible
outer portion
surrounds the
brighter inner
region lit by
radioactive
decay.

b

**Figure 12-15** Shells of Gas Around SN 1987A **(a)** Intense
radiation from the supernova explosion caused three rings of gas
surrounding SN 1987A to glow in this Hubble Space Telescope image.
This gas was ejected from the star 20,000 years before it detonated. All
three rings lie in parallel planes. The inner ring is about 1.3 ly across.
The white spots are unrelated stars. **(b)** When the progenitor star of
SN 1987A was still a red supergiant, a slowly moving wind from the star
filled the surrounding space with a thin gas. When the star contracted
into a blue supergiant, it produced a faster-moving stellar wind. The
interaction between the fast and slow winds somehow caused gases to
pile up along an hourglass-shaped shell surrounding the star. The burst
of ultraviolet radiation from the supernova ionized the gas in the rings,
causing the rings to glow. The supernova itself, at the center of the
hourglass, glows because of energy released from radioactive decay.

SN 1987A provided astronomers with invaluable
information, in part because the doomed star had been
studied before it exploded and its distance from Earth was
known. The supernova was located in an unobscured part
of the sky, and neutrino detectors happened to be operating at the time of the outburst. Astronomers will be moni-
toring the progress of this supernova for years to come.

## 12-8 Accreting white dwarfs in close binary systems can also explode as supernovae

Supernovae such as SN 1987A, which are the death throes
of massive stars, are known as **Type II supernovae**, but not
every supernova originated as a high-mass star. Observations reveal that others, known as **Type Ia supernovae**,
begin as white dwarfs.

A supernova's spectrum indicates its type: Hydrogen
lines are prominent in Type II supernovae but absent in
Type Ia. Both types begin with a sudden rise in brightness
(Figure 12-16). A Type Ia supernova typically reaches an
absolute magnitude of −19 at peak brightness, while a
Type II supernova usually peaks at −17. Type Ia supernovae then decline gradually for more than a year, whereas Type II supernovae alternate between periods of steep
and gradual declines in brightness. Type II light curves
therefore have a steplike appearance.

A Type Ia supernova begins with a carbon–oxygen-rich
white dwarf in a close, semidetached binary system. As
we saw in Chapters 10 and 11, mass transfer can occur in
a close binary if one star overflows its Roche lobe (recall
Figures 11-24 and 11-26). To trigger a Type Ia supernova,
a swollen giant companion star dumps gas onto a white
dwarf. When the white dwarf's mass gets close to the
Chandrasekhar limit, the increased core pressure created
by the additional mass enables carbon fusion to begin. In
a catastrophic runaway process reminiscent of the helium
flash, the rate of carbon fusion skyrockets, and the star
blows up.

A Type Ia supernova is powered by nuclear energy.
What we see is simply the fallout from a gigantic thermonuclear explosion, which produces a wide array of radioactive isotopes. Especially abundant is an unstable isotope
of nickel that decays into a radioactive isotope of cobalt.
Most of the electromagnetic display of a Type Ia supernova, including the smooth decline of its light curve, results
directly from the radioactive decay of this nickel and
cobalt. The spectrum lacks hydrogen because the star that
explodes consumes it all.

Astronomers have seen more than 600 supernovae in
distant galaxies. These observations suggest that in a typical galaxy like the Milky Way, Type Ia supernovae occur
approximately once every 36 years, while Type II supernovae occur about once every 44 years. Thus, there should

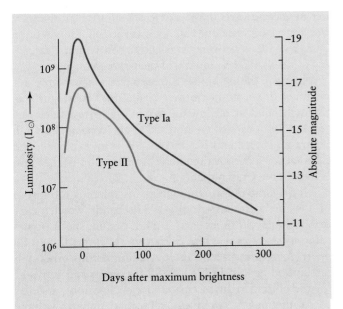

**Figure 12-16** **Supernova Light Curves** A Type Ia supernova, which gradually declines in brightness, is caused by an exploding white dwarf in a close binary system. A Type II supernova is caused by the explosive death of a massive star and usually has alternating intervals of steep and gradual declines in brightness.

be about five supernovae exploding in our Milky Way Galaxy each century. Where are they?

Vigorous stellar evolution in our Galaxy occurs primarily in the Galaxy's disk, where the giant molecular clouds are located. Our Galaxy's disk is therefore the place where massive stars are born and supernovae explode. However, this region of our Galaxy is so filled with interstellar gas and dust that we simply cannot see very far into it. Supernovae probably do erupt every few decades in remote parts of our Galaxy, but their detonations are hidden by interstellar debris.

# NEUTRON STARS AND PULSARS

When main-sequence stars of between approximately 8 $M_\odot$ and 25 $M_\odot$ explode as supernovae, their remnant cores are highly compressed clumps of neutrons called **neutron stars.** Isolated neutron stars are stable stellar remnants with masses between 1.4 $M_\odot$ and about 3 $M_\odot$.

## 12-9 The cores of many Type II supernovae become neutron stars

The neutron was discovered during laboratory experiments in 1932. Within a year, two astronomers had pre-

dicted the existence of neutron stars. Inspired by the realization that white dwarfs are supported by electron degeneracy pressure, Fritz Zwicky and Walter Baade proposed that a highly compact ball of neutrons could similarly support a stellar corpse. They noted that the Pauli exclusion principle prevents neutrons with identical properties from packing too closely together. When neutrons are pressed together, many of them have to move rapidly (so as not to be identical to their neighbors). This motion provides a pressure, called **neutron degeneracy pressure**, that the equations predict should be even greater than electron degeneracy pressure (see Chapter 11) and thus allows stellar remnants with masses beyond the Chandrasekhar limit. Zwicky and Baade wrote, "We advance the view that supernovae represent the transition from ordinary stars into neutron stars, which in their final stages consist of extremely closely packed neutrons."

Most scientists ignored Zwicky and Baade's theory for years. After all, a neutron star would have to be a rather unlikely object. In order to transform protons and electrons into neutrons, the density in the star would have to be equal to nuclear density, about $10^{17}$ kg/m$^3$. Thus, a thimbleful of neutron star matter brought back to Earth would weigh 100 million tons. Furthermore, an object compacted to nuclear density would be very small. A 2-$M_\odot$ neutron star would have a diameter of only 8 km, making it only as large as Jupiter's smallest known moons. The surface gravity on one of these neutron stars would be so strong that the escape velocity would equal one-half the speed of light. All these conditions seemed so outrageous that few astronomers paid any serious attention to the subject of neutron stars—until 1968.

As a young graduate student at Cambridge University, Jocelyn Bell spent many months helping to construct an array of radio antennas covering 4½ acres in the English countryside. By the fall of 1967 the instrument was completed, and Bell and her colleagues began detecting radio emissions from various celestial sources. In November, while scrutinizing data from the new telescope, Bell noticed that the antennas had detected regular beeps from one particular location in the sky. Careful repetition of the observations demonstrated that the radio pulses were arriving with a regular period of 1.3373011 seconds.

The regularity of this pulsating radio source was so striking that the Cambridge team suspected that they might be detecting signals from an advanced alien civilization. This possibility was soon discarded as several more of these pulsating radio sources, which soon came to be known as **pulsars,** were discovered across the sky. In all cases, the periods were extremely regular, ranging between 0.2 second and 1.5 seconds (Figure 12-17).

When the discovery of pulsars was officially announced in early 1968, astronomers around the world began proposing all sorts of explanations. Many of these theories were bizarre, and arguments raged for months.

**Figure 12-17**  **A Recording of a Pulsar**  This chart recording shows the intensity of radio emissions from one of the first pulsars to be discovered, PSR 0329+54. Note that some of the pulses are weak and others are strong. Nevertheless, the spacing between pulses is so regular (0.714 seconds) that it is more precise than most clocks on Earth.

We have already ruled out alien civilizations. The broad distribution of pulsars would require incredibly widespread civilizations, inconsistent with what we have found in our searches for alien intelligence (Chapter 18). Astronomers therefore looked for some recurring event in a star's life. We know that a star sheds matter as it swells to a giant. Perhaps some stars alternately expand and contract, emitting energy as they pulsate. However, this explanation also fails, because any star expanding and contracting as fast as a pulsar pulses would explode. Other astrophysicists suggested another kind of periodic event in a star's life: It might be spinning, giving off pulses with each revolution. But how?

Whatever the explanation, most astronomers doubted that pulsars were associated with dead stars. It was generally assumed that all dying stars somehow manage to eject enough matter so that their corpses do not exceed the Chandrasekhar limit. There seemed to be a sufficient number of white dwarfs in the sky to account for all the stars that have died since our Galaxy was formed. However, by late 1968, all controversy was laid to rest with the discovery of a pulsar in the middle of the Crab Nebula.

In A.D. 1054, Chinese astronomers recorded the appearance of a supernova (they called it a "guest star") in the constellation of Taurus. When we turn a telescope toward this location, we find the Crab Nebula, shown in Figure 12-18. It looks like the residue of an explosion and is, in fact, a supernova remnant. At the center of the Crab Nebula is the Crab pulsar.

If the Crab pulsar is spinning, it is spinning too fast to be a white dwarf. Its period is 0.033 seconds, which means that it rotates 30 times each second. At that speed, something as wide as a white dwarf would immediately fly apart. Astronomers have discovered numerous other pulsars, including PSR 1257+12 (PSR for pulsar and 1257+12 for the object's right ascension and declination), a stellar corpse that pulses 888 times each second. Staying with the belief that they are spinning, astronomers concluded that pulsars must be incredibly compact. Calculations revealed that if neutron stars exist, they have a sufficiently small diameter to remain intact rotating as fast as pulsars pulsate. Rejecting the belief that all stars end up as white dwarfs, the theory of neutron stars, as described above, was quickly developed.

## 12-10  A rotating magnetic field explains the pulses from a neutron star

We now consider how neutron stars can become pulsars. Based on observations of the Sun, Betelgeuse, and other nearby stars, astronomers theorize that many, perhaps

**Figure 12-18**  **The Crab Nebula and Pulsar**  **(a)** This beautiful nebula, named for the crablike appearance of its filamentary structure, is the remnant of a supernova seen in A.D. 1054. The distance to the nebula is about 6000 ly, and its present angular size (4 by 6 arcmin) corresponds to linear dimensions of about 7 by 10 ly. **(b)** This Hubble image reveals the pulsar at the heart of the nebula. **Insets:** The Crab pulsar in its "on" (top) and "off" (bottom) states. Both its radio pulses and visible flashes have identical periods of 0.033 second.

a    R I V U X G       b       R I V U X G          R I V U X G

most, stars on the main sequence have some angular momentum (rotation). The Sun, for example, takes nearly a full month to rotate once about its axis. But the rotation rate of collapsing stars increases, just as pirouetting ice skaters speed up when they pull in their arms. (Recall from Section 2-5 that the total amount of angular momentum in an isolated system always remains constant.) Even an ordinary star rotating once a month would spin faster than once a second if compressed to the size of a neutron star.

In addition to rapid rotation, most neutron stars are believed to have intense magnetic fields. In an average star like our Sun, the magnetic field is spread out over millions upon millions of square kilometers just under the star's surface (see Chapter 9). However, if a star of solar dimensions collapses down to a neutron star, its magnetic field becomes very concentrated. The strength of the magnetic field increases a billionfold.

The axis of rotation of a typical neutron star is not the same as the axis connecting its north and south magnetic poles (Figure 12-19), much as the magnetic and rotation axes of the planets and Sun are different axes that are inclined to each other. As the neutron star rotates, its powerful magnetic field therefore rapidly changes direction. Like a giant electric generator, the star creates intense electric fields, which act on protons and electrons near its surface. The powerful electric fields channel these charged particles, causing them to flow out from the neutron star's polar regions, as sketched in Figure 12-19. As the particles stream along the field, they accelerate and emit energy. The result is two very thin beams of radiation pouring out of the neutron star's north and south magnetic polar regions—a pulsar.

This explanation for pulsars is often called the **lighthouse model**. A rotating, magnetized neutron star is somewhat like a lighthouse beacon. As the star rotates, its beams of radiation sweep around the sky. If the Earth happens to be located in the right direction, a brief flash can be observed each time a beam whips past our line of sight.

The lighthouse model is accepted because it proved consistent with observations. Also, astrophysicists have been unable to identify any other object that can pulsate at the rate of a pulsar without flying apart. The center of the Crab Nebula is flashing on and off 30 times each second (see Figure 12-18).

The Crab pulsar is one of the youngest pulsars, its creation having been observed some 950 years ago. Also visibly flashing is the Vela pulsar at the core of the Gum Nebula (see Figure 12-11). The Vela pulsar, with a period of 0.089 second, is the slowest pulsar ever detected at visual wavelengths. Because they use some of their energy of rotation via the pulses they emit, *pulsars slow down as they get older.* Only the very youngest pulsars are energetic enough to emit visible flashes along with their radio pulses, and the Vela pulsar was created about 11,000 years

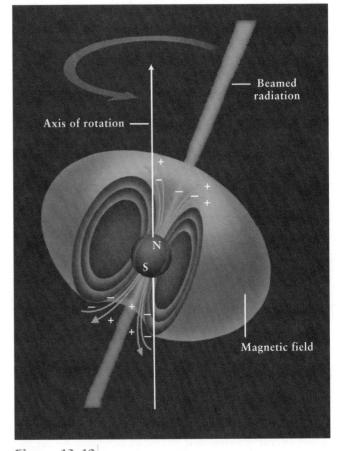

**Figure 12-19** **A Rotating, Magnetized Neutron Star** It is reasonable to assume that neutron stars rotate rapidly and possess powerful magnetic fields. Charged particles are accelerated near a star's magnetic poles and produce two oppositely directed beams of radiation. As the star rotates, the beams sweep around the sky. If the Earth happens to lie in the path of the beams, we see a pulsar.

ago. Astronomers are observing SN 1987A with the hope of seeing through the supernova's ejected gases to search for pulses that would indicate a rapidly spinning neutron star there.

**Insight into science** **Test and retest** New discoveries almost invariably allow for a variety of theories to explain them. Theories that are inconsistent with observations or make incorrect predictions are quickly discarded, while theories that explain what is observed are tested and retested until they break down. Such testing shows the range of situations over which the theory can be used reliably. Consider how the theories proposed above to explain pulsars were analyzed, tested, and, for the most part, rejected.

While most pulsars appear to be isolated bodies, two have been discovered with companions. One, labeled PSR 1957+20, but more commonly called the *binary pulsar,* has a massive companion, which is probably another neutron star. The two neutron stars orbit each other with a period of about 8 hours. Because they are so close together and so incredibly dense, these two bodies follow orbits predicted by Einstein's general theory of relativity (coming up in Chapter 13), rather than Newton's laws of motion (Chapter 2). Indeed, their behavior is a strong confirmation that general relativity is the more accurate model of the behavior of matter. The second pulsar with a companion, labeled PSR B1257+2, is even more unusual than the binary pulsar in that its Doppler shift reveals the presence of at least two planets in orbit around it.

## 12-11 Pulsating X-ray sources are neutron stars in close binary systems

Neutron stars in binary systems may also hold the key to another regular pulse from the sky—pulsating X-ray sources. During the 1960s astronomers obtained tantalizing X-ray views of the sky during short rocket and balloon flights that briefly lifted X-ray detectors above the Earth's atmosphere. Several strong X-ray sources were discovered, and each was named after the constellation in which it was located. For example, Scorpius X-1 is the first X-ray source found in the constellation of Scorpius.

Astronomers were so intrigued by these preliminary discoveries that NASA built and launched *Explorer 42,* an X-ray-detecting satellite that could make observations 24 hours a day. The satellite was launched from Kenya (to place it in an orbit above the Earth's equator) on the seventh anniversary of Kenyan independence in 1970. To commemorate the occasion, *Explorer 42* was renamed *Uhuru,* which means "freedom" in Swahili.

*Uhuru* gave us our first comprehensive look at the X-ray sky. As the satellite slowly rotated, its X-ray detectors swept across the heavens. Each time an X-ray source came into view, signals were transmitted to receiving stations on the ground. Before its battery and transmitter failed in early 1973, *Uhuru* had succeeded in locating 339 X-ray sources.

The discovery of pulsars was still fresh in everyone's mind when the *Uhuru* team discovered X-ray pulses coming from Centaurus X-3 in early 1971. Figure 12-20 shows data from one sweep of *Uhuru*'s detectors across Centaurus X-3. The pulses have a regular period of 4.84 seconds. A few months later, similar pulses were discovered coming from a source called Hercules X-1, which has a period of 1.24 seconds. Because the periods of these two X-ray sources are so short, astronomers began to suspect that they had found rapidly rotating neutron stars.

It soon became clear, however, that systems such as Centaurus X-3 and Hercules X-1 are not ordinary pulsars like the Crab or Vela pulsars. Centaurus X-3 completely turns off periodically. Every 2.087 days, it turns off for almost 12 hours. This fact suggests that Centaurus X-3 is an eclipsing binary and that the X-ray source takes nearly 12 hours to pass behind its companion star.

The case for the binary nature of Hercules X-1 is even more compelling. It has an off state corresponding to a 6-hour eclipse every 1.7 days, and careful timing of the X-ray pulses shows a periodic Doppler shift every 1.7 days. This information provides direct evidence of orbital motion about a companion star: When the X-ray source is approaching us, its pulses are separated by slightly less than 1.24 seconds. When the source is receding from us, slightly more than 1.24 seconds elapse between the pulses.

Careful visual searches around the location of Hercules X-1 soon revealed a dim star named HZ Herculis. The apparent magnitude of this star varies between +13 and +15, with a period of 1.7 days. Because this period is

**Figure 12-20** **X-Ray Pulses from Centaurus X-3** This graph shows the intensity of X rays detected by *Uhuru* as Centaurus X-3 moved across the satellite's field of view. The successive pulses are separated by 4.84 seconds. The variation in the height of the pulses was a result of the changing orientation of *Uhuru*'s X-ray detectors toward the source as the satellite rotated.

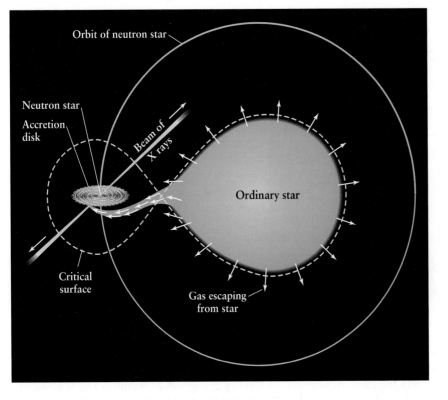

**Figure 12-21** **A Model of a Pulsating X-Ray Source** Gas transfers from an ordinary star to the neutron star. The infalling gas is funneled down onto the neutron star's magnetic poles, where it strikes the star with enough energy to create two X-ray–emitting hot spots. As the neutron star spins, beams of X rays from the hot spots sweep around the sky.

exactly the same as the orbital period of the X-ray source, astronomers conclude that HZ Herculis is the companion star around which Hercules X-1 orbits.

Astronomers now realize that systems such as Centaurus X-3 and Hercules X-1 are examples of binary stars in which one is a neutron star. Because all these binaries have very short orbital periods, the distance between the ordinary star and its neutron star must be small. This proximity enables the neutron star to capture gas escaping from the companion star.

To explain the pulsations of X-ray sources such as Centaurus X-3 or Hercules X-1, astronomers assume that the ordinary star either fills or nearly fills its Roche lobe. Either way, matter escapes from the star. If the star fills its lobe, as in the case of Hercules X-1, mass loss results from direct overflow through the Roche lobe. If the star's surface lies just inside its lobe, as with Centaurus X-3 (Figure 12-21), a stellar wind carries off the mass. A typical rate of mass loss for the ordinary star in such a pair is approximately $10^{-9}$ $M_\odot$ per year.

The neutron star in a pulsating X-ray source, like an ordinary pulsar, rotates rapidly and has a powerful magnetic field inclined to the axis of rotation (recall Figure 12-19). As the gas from the companion star falls toward the neutron star, the magnetic field funnels the incoming matter down onto its magnetic polar regions. The neutron star's gravity is so strong that the gas is traveling at nearly

half the speed of light by the time it crashes onto the star's surface. This violent impact creates hot spots at both poles, with temperatures of about $10^8$ K, that emit abundant X rays with a luminosity nearly 100,000 times brighter than the Sun. As the neutron star rotates, two beams of X rays from the polar caps sweep around the sky. If the Earth happens to be in the path of one of the beams, we can observe this pulsating X-ray source. The pulse period is thus equal to the neutron star's rotation period. For example, the neutron star in Hercules X-1 is spinning at the rate of once every 1.24 seconds.

## 12-12 Other neutron stars in binary systems emit powerful jets of gas

In a binary system, gas captured by a neutron star's gravity goes into orbit around the neutron star, as shown in Figure 12-21. Recall from Chapter 10 that the result is a revolving disk of material called an *accretion disk*. Accretion disks have been detected in many close binary systems where mass transfer is occurring (recall Figure 11-24b).

With pulsating X-ray sources like Hercules X-1, the rate at which gas falls onto the neutron star from the inner edge of the accretion disk is low enough to allow the resulting X rays to escape. If the companion star is dumping vast amounts of material onto the neutron star, however, the

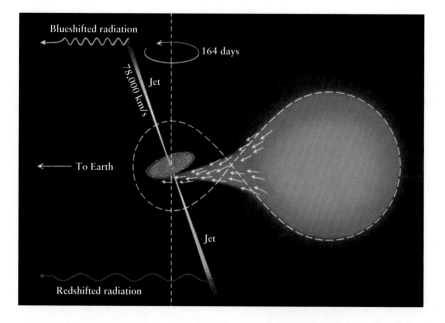

**Figure 12-22** **A Model of SS433** Gas from a normal star is captured into an accretion disk about a neutron star. Two high-speed, oppositely directed jets of gas are ejected from the faces of the disk. Because the disk is tilted, the gravitational pull of the normal star causes the jets to precess, with a period of 164 days.

resulting energy cannot escape easily. In the plane of the accretion disk, the newly arrived gases are constantly spiraling in toward the neutron star. X-rays striking the incoming gases create tremendous pressures in them. Some of this gas is thereby ejected perpendicular to the accretion disk. It squirts out as two powerful jets of high-velocity gas. That is just what was seen for the weird star known as SS433.

The spectrum of SS433 is composed of two sets of spectral lines, identical except that one set is very redshifted away from its usual wavelengths, while another set is comparably blueshifted. Somehow, SS433 is coming and going at the same time. Further study revealed that the system's redshifted and blueshifted lines are actually moving back and forth across the spectrum of SS433 with a period of 164 days. Astrophysicists were quick to point out that the two sets of spectral lines could be generated by two oppositely directed jets of gas, one tilted toward us and the other away from us. Furthermore, the 164-day variation could be explained by a precession, or "wobble," of the axis of revolution of the accretion disk and its two jets. As the two jets circle about the sky every 164 days, we see a periodic variation in the Doppler shift (Figure 12-22).

The large redshifts and blueshifts are explained by gas in the two oppositely directed jets, moving with a speed of 78,000 km/s, roughly one-quarter the speed of light. In addition, the accretion disk must be tilted with respect to the orbital plane of the two stars of the binary system. Just as the Earth's axis is tilted with respect to the plane of the ecliptic, causing the Earth to precess, the tilt of the accretion disk results in the 164-day precession of the two jets. Figure 12-23 shows four high-resolution radio views of SS433. Note the two oppositely directed appendages emerging from the central source. As we will see in Chap-

ter 16, many *quasars* have a similar radio structure, although on a much larger scale. Because most quasars are very far away, they are difficult to study. The real significance of SS433 may be that it gives us a miniature quasar-like object in our own celestial backyard.

In 1997, astronomers detected the first nonpulsar, nonbinary neutron star. This object is identified as a neutron star because of its physical characteristics. Its surface temperature is a sizzling $6.5 \times 10^5$ K, and it is less than 28 km across. Only a neutron star in the astronomical inventory is known to have these properties. This neutron star emits a strong, steady stream of X rays and is also seen in visible light telescopes (Figure 12-24).

## 12-13 Neutron stars in binary systems can also emit powerful isolated bursts of X rays

Neutron stars can acquire additional mass from a companion star and flare up. Beginning in late 1975, astronomers analyzing data from X-ray satellites detected sudden, powerful bursts of radiation. The record of a typical burst is shown in Figure 12-25. The source, called an **X-ray burster**, emits X rays at a constant low level; then, suddenly and without warning, there is an abrupt increase followed by a gradual decline. A typical burst lasts for 20 seconds. Several dozen X-ray bursters have been located, most lying in the plane of the Milky Way Galaxy. Their locations on the celestial sphere indicate that they are actually in the Milky Way.

X-ray bursters, like novae, are believed to arise from mass transfer in binary star systems. With a burster, however, the stellar corpse is a neutron star rather than a white dwarf. Gases escaping from the ordinary compan-

**Figure 12-23** **Four Views of SS433** These four radio views, taken in early 1981, show jets of gas extending out to one-sixth of a light-year on either side of SS433. Three-quarters of the radio emission come from SS433 itself (the red central blob), which is located at the center of a supernova remnant 13,000 ly from Earth.

ion star fall onto the neutron star. The energy released as this gas crashes down onto the neutron star's surface produces the low-level X rays that are continuously emitted by the burster.

Most of the gas falling onto the neutron star is hydrogen, which becomes compressed against the hot surface of the star by the star's powerful surface gravity. In fact, temperatures and pressures in this accreting layer are so high that the arriving hydrogen is promptly converted by fusion into helium. Constant hydrogen fusion soon produces a layer of helium that covers the entire neutron star.

When the helium layer is about 1 m thick, helium fusion ignites explosively, and we observe a sudden burst of X rays. In other words, whereas explosive hydrogen fusion on a white dwarf produces a nova, explosive helium fusion on a neutron star produces an X-ray burster. In both cases, the fusion is explosive, because the fuel is so strongly compressed against the star's surface that it is degenerate, like the star itself. As we saw with the helium flash inside giants, ignition of a degenerate thermonuclear fuel always involves a sudden thermal runaway, because

the usual safety valve between temperature and pressure is not operating.

In death as well as in life, the mass of a star determines its fate. Just as there is an upper limit to the mass of a white dwarf (the Chandrasekhar limit), there is also an upper limit to the mass of a neutron star, called the *Oppenheimer–Volkov limit*. Above this limit, neutron degeneracy pressure cannot support the overbearing weight of the star's matter pressing in from all sides. The Chandrasekhar limit for a white dwarf is 1.4 $M_\odot$, and the Oppenheimer–Volkov limit for a neutron star is about 3 $M_\odot$. What might happen if a dying massive star failed to eject enough matter to get below the Oppenheimer–Volkov limit?

The gravity associated with a neutron star is so strong that the escape velocity is roughly half the speed of light. With a stellar corpse greater than 3 $M_\odot$, there is so much matter crushed into such a small volume that the neutron star collapses and the escape velocity exceeds the speed of light. Because nothing can travel faster than light, nothing—not even light—can leave the dead star. The star therefore disappears, its powerful gravity leaving

**Figure 12-24** **Lone Neutron Star** This is the first visual image of an isolated neutron star, located within 400 pc (1300 ly) of Earth. It was discovered in 1997 by the X-ray satellite *ROSAT* because it is also a strong emitter of X rays.

**Figure 12-25** **X Rays from an X-Ray Burster** A burster emits X rays with a constant low intensity interspersed with occasional powerful bursts. This burst was recorded in September 1975 by an X-ray telescope that was pointed toward the globular cluster NGC 6624. About one-third of all known X-ray bursters are located in globular clusters.

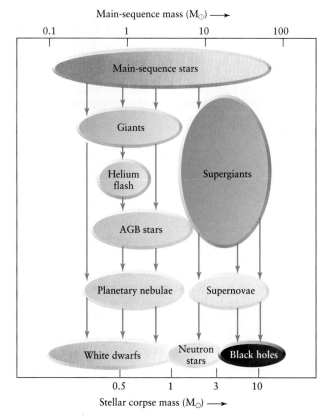

Main-sequence mass ($M_\odot$) ⟶

0.1       1       10       100

Main-sequence stars

Giants

Helium flash

Supergiants

AGB stars

Planetary nebulae

Supernovae

White dwarfs

Neutron stars

Black holes

0.5       1       3       10

Stellar corpse mass ($M_\odot$) ⟶

**Figure 12-26** **A Summary of Stellar Evolution** The evolution of an isolated star depends on its mass. The scale at the top indicates the mass of a star when it is on the main sequence. The scale on the bottom gives the mass of the resulting stellar corpse. Stars less massive than about 8 $M_\odot$ can eject enough mass to become white dwarfs. High-mass stars can produce Type II supernovae and become neutron stars or black holes.

a hole in the fabric of the universe. The discovery of neutron stars thus inspired astrophysicists to consider seriously one of the most bizarre and fantastic objects ever predicted by modern science—the black hole, which we examine in Chapter 13. Figure 12-26 summarizes the evolutionary history of stars.

## WHAT DID YOU THINK?

**1** *Will the Sun end its existence? If so, how?* The Sun will shed matter as a planetary nebula in about seven billion years and then cease fusing. Its remnant white dwarf will dim over the succeeding billions of years.

**2** *What is a nova?* A nova is a relatively gentle explosion of hydrogen gas on the surface of a white dwarf in a binary star system.

**3** *What is the origin of the carbon, silicon, oxygen, iron, uranium, and other heavy elements on Earth?* These elements are created during stellar evolution and by supernovae.

**4** *What is a cosmic ray?* Cosmic rays are high-speed particles (mostly hydrogen and other atomic nuclei) in space. Many of them are believed to have been created in supernovae.

**5** *What is a pulsar?* A pulsar is a rotating neutron star in which the magnetic field does not pass through the rotation axis. The beam of radiation it emits must sweep in our direction.

## Key Words

asymptotic giant branch (AGB) star, 304
Chandrasekhar limit, 307
cosmic rays, 311
cosmic ray shower, 312

helium shell flash, 304
lighthouse model, 317
neutron degeneracy pressure, 315
neutron star, 315
nova (*plural* novae), 307

photodisintegration, 310
planetary nebula, 305
pulsar, 315
secondary cosmic rays, 312
shell helium fusion, 304

supernova, 310
Type Ia supernova, 314
Type II supernova, 314
white dwarf, 307
X-ray burster, 320

## Key Ideas

✳ There are limits to the chemical changes due to fusion that can occur in stars.

✳ Stars lose mass via stellar winds throughout the giant and supergiant phases.

### Low-Mass Stars and Planetary Nebulae

✳ A low-mass main-sequence star becomes a giant when shell hydrogen fusion begins. It becomes a horizontal branch star when core helium fusion begins. It enters the asymptotic

giant branch and becomes a supergiant when shell helium fusion starts.

✳ Stellar winds during the thermal pulse phase can also eject mass from the star's outer layers.

✳ The burned-out core of a low-mass star becomes a dense carbon–oxygen sphere, called a white dwarf, with about the same diameter as the Earth. The maximum mass of a white dwarf (the Chandrasekhar limit) is 1.4 $M_\odot$.

✳ Explosive hydrogen fusion may occur in the surface layer of a white dwarf in a close binary system, producing the sudden increase in luminosity that we call a nova.

✳ An accreting white dwarf in a close binary system can also become a supernova when carbon fusion ignites explosively throughout such a degenerate star. Such a detonation is called a Type Ia supernova.

## High-Mass Stars and Supernovae
✳ After exhausting its central supply of hydrogen and helium, the core of a high-mass star undergoes a sequence of other thermonuclear reactions. These include carbon fusion, neon fusion, oxygen fusion, and silicon fusion. This last fusion eventually produces an iron core.

✳ A high-mass star dies in a supernova explosion that ejects most of the star's matter into space at very high speeds. This Type II supernova is triggered by the gravitational collapse of the doomed star's core.

✳ Neutrinos were detected from Supernova 1987A, which was visible to the naked eye.

## Neutron Stars and Pulsars
✳ A high-mass star's dead core becomes a neutron star or even a black hole. A neutron star is a very dense stellar corpse consisting of closely packed neutrons in a sphere roughly 8 km in diameter. The maximum mass of a neutron star, called the Oppenheimer–Volkov limit, is about 3 $M_\odot$.

✳ A pulsar is a rapidly rotating neutron star with a powerful magnetic field that makes it a source of periodic radio and other electromagnetic pulses. Energy pours out of the polar regions of the neutron star in intense beams that sweep across the sky.

✳ Some X-ray sources exhibit regular pulses. These objects are thought to be neutron stars in close binary systems with ordinary stars.

✳ Explosive helium fusion may occur in the surface layer of a companion neutron star, producing the sudden increase in X-ray radiation called an X-ray burster.

## Review Questions

1 What is the difference between a giant and a supergiant?

2 Why is knowing the temperature in a star's core so important in determining which nuclear reactions can occur there?

3 What determines the temperature in the core of a star?

4 What is a planetary nebula and how does it form?

5 What is a white dwarf?

6 What is the Chandrasekhar limit?

7 What is a neutron star?

8 Compare a white dwarf and a neutron star. Which of these stellar corpses is more common? Why?

9 What is the Oppenheimer–Volkov limit?

10 On an H-R diagram, sketch the evolutionary track that the Sun will follow between the time it leaves the main sequence and when it becomes a white dwarf. Approximately how much mass will the Sun have when it becomes a white dwarf? Where will the rest of the mass have gone?

11 Why do you suppose that all the white dwarfs known to astronomers are relatively close to the Sun?

12 Why do spherical planetary nebulae look like rings as seen from Earth?

13 Why have radio searches for supernova remnants been more fruitful than searches at visible wavelengths?

14 Why do astronomers believe that pulsars are rapidly rotating neutron stars?

15 What is the difference between Type Ia and Type II supernovae?

16 Compare a nova with a Type Ia supernova. What do they have in common? How are they different?

17 Compare a nova and an X-ray burster. What do they have in common? How are they different?

18 What is SS433? Describe its unusual features.

19 Describe what radio pulsars, X-ray pulsars, and X-ray bursters have in common. How are they different manifestations of the same type of astronomical object?

## Advanced Questions

The answers to all computational problems, which are preceded by an asterisk (*), appear at the end of the book.

20 What prevents thermonuclear reactions from occurring at the center of a white dwarf? If no thermonuclear reactions take place in its core, why doesn't the star collapse?

21 Suppose you wanted to determine the age of a planetary nebula. What observations would you make, and how would you use the resulting data?

22 What explains why the rate of expansion of the gas shell in a planetary nebula is often not uniform in all directions?

23 What kinds of stars would you monitor if you wished to observe a supernova explosion from its very beginning? Look up tabulated lists of the brightest and nearest stars. Which, if any, of these stars are possible supernova candidates? Explain.

24 To determine the period of a pulsar accurately, astronomers must take the Earth's orbital motion about the Sun into account. Explain why.

*25 The distance to the Crab Nebula is about 2000 pc. When did it actually explode?

## Discussion Questions

26 Suppose that you discover a small glowing disk of light while searching the sky with a telescope. How would you decide just from your observations if this object was a planetary nebula? Could your object be something else? Explain.

27 Immediately after the first pulsar was discovered, one explanation offered was that the pulses were signals from an extraterrestrial civilization. Why do you suppose astronomers discarded this idea?

## What If ...

28 The Earth passed through an old supernova remnant? What would happen to the Earth and life on it?

## Observing Projects

29 With the help of star maps or your Starry Night software, locate and observe as many planetary nebulae as possible on clear, moonless nights, using the largest telescope at your disposal. Some of the more notable ones include: Little Dumbbell (M76), NGC 1535, Eskimo, Ghost of Jupiter, Owl (M97), Ring (M57), Blinking Planetary, Dumbbell, Saturn Nebula, and NGC 7662. Note and compare the various shapes of the different nebulae.

30 Two supernova remnants can be seen through modest telescopes, one in the winter sky and the other in the summer sky. Both are quite faint, however, so you should schedule your observations for a moonless night. The winter sky contains the Crab Nebula, which is discussed in detail in this chapter. It is located near the star marking the eastern horn of Taurus (the Bull) and can be located using Starry Night software. Its coordinates are R.A. = $5^h$ $34.5^m$ and Dec. = +22° 00′. Whereas the entire Crab Nebula easily fits into the field of view of an eyepiece, the Veil (or Cirrus) Nebula in the summer sky is so vast that you can see only a small fraction of it at a time. The easiest way to find the Veil Nebula is to aim the telescope at the star 52 Cygni (R.A. = $20^h$ $45.7^m$ and Dec. = +30°43′), which lies on one of the brightest portions of the nebula. This star is also accessible in your Starry Night program. If you then move the telescope slightly north or south until 52 Cygni is just out of the field of view, you should see giant wisps of glowing gas.

# WHAT IF...

# A Supernova Exploded Near Earth?

In 1054 a new star appeared in the constellation Taurus, the Bull. Unlike normal stars, this one increased in brightness and within a week blazed so intensely as to be visible from Earth during the day. After a month of maximum radiance, the visitor to our skies began to fade. Two years later it was gone from sight. It took 900 years for astronomers to correctly explain this event. Today we know it was a supernova, the mightiest of all explosions in the universe. And astronomers have located the remnant of that long-ago explosion. Now called the Crab Nebula, it is an expanding cloud of gas and dust some 6000 ly from Earth.

**The Flash** Exotic as a nearby supernova and its consequences might sound, some estimates suggest that one might occur as close as 50 light-years of Earth as frequently as every 25 million years. Indeed, the events about to be described have probably occurred in the vicinity of the Earth at least once, and perhaps several times, since life emerged.

What would happen here if a supernova occurred only 50 ly away? (Considering the titanic forces supernovae release, it should come as no surprise that the high-energy electromagnetic radiation from such an explosion detonating much closer would immediately kill virtually all life on Earth.) Without warning, the doomed star would grow tremendously luminous, 50 times brighter than the Moon and only 8000 times less bright than the Sun. It would be brighter than the light from all the other stars in the night sky combined.

Neutrinos would foreshadow by a few hours the visible flash and pending flood of X rays and gamma rays from the supernova. These lethal radiations would be the first causes of death on Earth. While our atmosphere absorbs virtually all the X rays and gamma rays from the Sun, the radiation from the supernova would be so intense that much of it would reach the Earth's surface. Within days individuals from virtually all spe-

cies of plants and animals would begin dying of radiation poisoning. Entire radiation-sensitive species might be annihilated immediately. The radiation would also cause many cancers and other internal diseases over succeeding years, leading to many more plant and animal deaths. At the same time, ultraviolet radiation reaching the Earth's surface would cause an astronomical jump in the rates of skin cancer and cataracts.

The first blast of ultraviolet radiation from the supernova would destroy the Earth's ozone layer within a matter of days, transforming the ozone primarily into atomic oxygen. With the removal of this protective barrier, ultraviolet radiation from both the supernova and the Sun would saturate the Earth's surface. After the intensity of ultraviolet radiation from the supernova diminished, sunlight would begin repairing the ozone layer. As a result, both the ozone layer and the ultraviolet radiation level would eventually return to normal at the Earth's surface.

The brightness and emission of all the electromagnetic radiation from the supernova would peak after a month. It would then fade, but the supernova remnant would be visible for millennia as an expanding cloud of gas and dust in the night sky. Fortunately for life that survived the onslaught of high-energy radiation from the supernova, the remnant cloud would primarily emit harmless visible light.

**The Aftermath** The most energetic particles from the supernova, the so-called cosmic rays, would also affect the Earth's surface. Moving at 90% of the speed of light, cosmic rays would arrive here only five to ten years after the photons do. Cosmic ray energies are much higher than those of any particles normally existing on Earth. For example, a typical cosmic ray has enough energy to light a small light bulb for one second; it takes over 600,000 trillion electrons flowing through wiring in your house to do the same thing.

While the lower-energy cosmic rays would be trapped in the magnetic Van Allen belts surrounding the Earth, the more energetic particles would plunge into the upper atmosphere. Some would slam into nitrogen or oxygen molecules in the air, shattering the air molecules and creating the cosmic ray showers. The highest energy cosmic rays from the supernova would reach the Earth intact, as do the highest energy cosmic rays from other sources today. These would break up atoms of the objects they strike on the Earth's surface, enhancing the earlier biological damage caused by the supernova's electromagnetic radiation.

The bulk of the matter ejected into space by the supernova would travel at speeds of 16,000 km/sec (60 million km/hr), nearly 20 times slower than the emitted photons. Thus, the bulk of the supernova remnant would take least a thousand years longer to reach us than did its initial radiation. By the time the bulk of the blast wave reaches the solar system, it would have become so thin and diffuse that it would probably do little damage to the Earth's atmosphere. However, we could expect this material to deposit a thin but exotic mix of elements into the upper atmosphere. Some would be radioactive. All this material would eventually fall to Earth and alter the chemistry of both the ocean and the soil.

**The Outcome**  The damage done to life by both the electromagnetic radiation and cosmic ray particles would disrupt the global food chain. In the oceans large quantities of plankton and other microscopic organisms at the foundation of the chain would die off. As a result, many of the larger aquatic species that feed on these smaller organisms would starve. Similarly, on the surface most plant life would wither and many herbivorous animals would starve as a result. This would, of course, lead to the death and dislocation of animals throughout the food chain. Surviving plants and animals would undergo genetic mutations due to the supernova's radiation. Most of these genetic changes would lead to immediate death of the altered plant or animal, but a few changes would be beneficial and enable the mutated organisms to thrive.

How would the human race fare after the supernova? Our long-term survival would depend, in part, on whether surviving humans could coexist with surviving plants and animals, and with each other.

# Black Holes

*In this chapter you will discover*

✳ that Einstein's general theory of relativity predicts regions of space and time that are severely distorted by the extremely dense mass they contain

✳ how black holes arise

✳ the surprisingly simple theoretical properties of black holes

✳ the apparent bizarre fate of a black hole

**The Behavior of Light near a Black Hole** Black holes distort spacetime nearby. In this artist's rendition, two light beams fired in slightly different directions from the upper left travel on very different paths. They both move in straight lines, but the meaning of "straight" near a black hole defies our intuition. The upper beam, aimed slightly away from the black hole, is virtually unaffected by it. The lower beam, inclined toward the black hole, travels through space that is so distorted that it twists the light into a spiral, which is swept down into the black hole. Don't let this painting (or similar images in science fiction) mislead you into believing that black holes have a "top" or a "bottom." Theory predicts that they can be entered from any direction.

## WHAT DO YOU THINK?

1  Are black holes just holes in space?

2  What exists at the surface of a black hole?

3  What power or force enables black holes to draw things in?

4  Do black holes last forever?

The extreme gravitational force acting on a neutron star is hard to imagine. But for a neutron star smaller than 3 $M_\odot$ that force is counterbalanced by an equal, outward force from neutron degeneracy pressure. Prevented from collapsing further, a neutron star is incredibly compact, barely 8 km across. But not even neutron degeneracy pressure can stand up to masses greater than 3 $M_\odot$ and their inward gravitational force.

Once a massive star overcomes even neutron degeneracy pressure, what prevents it from collapsing until nothing—not even light—can manage to escape its gravitational attraction? The answer according to contemporary theories of physics is "Nothing"! It becomes a **black hole**. We must turn to Einstein's powerful theories of relativity in order to understand these exotic objects.

## THE EVIDENCE FOR BLACK HOLES

Black holes inspire awe, fear, and uncertainty. Many people harbor the belief that black holes are destined to "swallow up" all the matter in the universe. Happily, the equations describing them reveal that black holes are more benign than that, but they truly are strange.

## 13-1  Special relativity changes our conception of space and time

In 1905 Albert Einstein began a revolution in physics with his **special theory of relativity**, a description of how motion affects our measurements of distance, time, and mass. He was guided by two innovative ideas. While its implications proved revolutionary, the first notion seems simple: *Your description of physical reality is the same regardless of the constant velocity at which you move.* In other words, as long as you are moving in a straight line at a constant speed, you experience the same laws of physics as anyone else moving at any other constant speed and in any other direction. To illustrate, suppose you were inside a closed boxcar moving in a straight line at 100 km/h. Any scientific measurements you make in the moving boxcar would yield the same results as the measurements taken from the same experiment while the boxcar was at rest (or moving at any other constant speed).

The second idea seems more bizarre: *Regardless of your speed or direction, you always measure the speed of light to be the same.* Suppose that you are in a car moving toward a distant street lamp. Even if you are moving at 95% the speed of light, you will still measure the speed of the oncoming light photons to be the same as if you were standing by the roadside.

Einstein's special theory of relativity expresses these two assumptions mathematically, and the results of these assumptions have been confirmed in innumerable experiments. Here are three fundamental results of special relativity. The first is that the length of an object decreases as its speed increases; the faster an object moves past you, the shorter its dimension is in the direction of its motion. In other words, if a boxcar moves past you, it is actually shorter, from your perspective on the ground, than when it is at rest (Figure 13-1). However, if you measure the length of the moving boxcar while you are inside it and moving with it, your measurement of its length will be the same as measured on the ground when it was at rest. The word *relativity* emphasizes the importance of the relative speed between the observer and the measured object.

At rest                                    In motion

**Figure 13-1**    **Movement and Space**   According to the special theory of relativity, the faster an object moves, the shorter it becomes in its direction of motion, becoming infinitesimally small as its speed approaches the speed of light. The dimensions perpendicular to the object's motion are unchanged compared with the same dimensions when the object is at rest.

Einstein's second result is just as strange: Clocks passing by you run more slowly than do clocks at rest. This is called *time dilation*. Indeed, the faster a clock passes by, the slower it appears to tick from your perspective. For example, pilots and airline attendants actually age more slowly than they would if they didn't fly (although because of the relatively low speeds aircraft travel compared to the speed of light, this difference is imperceptibly small). People flying in aircraft do not feel time moving more slowly, however, because their biological activities slow down at the same rate as the clocks around them. Only an observer moving more slowly sees that the clock and the activities of the high-speed travelers have slowed. These connections between motion and clocks mean that space and time cannot be considered as two separate concepts. Relativity requires us to consider them as a single entity, thus creating the concept of **spacetime.**

Finally, Einstein showed that the mass of an object increases as it moves faster. A body approaching the speed of light becomes infinitely massive. To push mass faster than the speed of light would therefore require an infinite amount of energy. The speed of light is therefore the universal speed limit.

## 13-2 General relativity predicts black holes

First published in 1915, Einstein's **general theory of relativity** extends his special theory of relativity to include the effects of gravity. It describes how matter creates gravitational force by distorting the very fabric, or geometry, of space and time. He found that matter affects the nature of space and the rate at which time passes.

Newton's laws of motion and his universal law of gravitation are accurate only for objects with relatively small masses and low densities. We use Newton's laws to determine the force necessary to put the Space Shuttle in orbit around the Earth, for example. But Newton's laws are inadequate to predict Mercury's orbit around the Sun or the paths of objects near a collapsing neutron star. In such cases, general relativity correctly predicts how objects move.

In general relativity, unlike in Newtonian physics, the path of light is affected by nearby mass (Figure 13-2). As seen by observers far away, massive objects actually curve space and slow down the flow of time. The more massive or dense the object, the more it curves nearby space and the more it slows down time in its vicinity. Just by being near matter, clocks run slower than they do in empty space. Furthermore, photons leaving the vicinity of a star lose energy in climbing out of the star's gravitational field. They show this change by shifting to longer wavelengths, an effect we see in the spectra of some white dwarfs, whose light appears redder than it would if this effect did

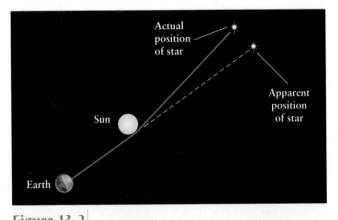

**Figure 13-2**    **Warped Space and the Curved Path of Light**
That the warping of space by matter deflects light was one of the first predictions of general relativity to be confirmed, in 1919. This confirmation came when stars near the Sun were observed during an eclipse. Starlight was deflected around the Sun by up to 1.75 arcsec, a small but measurable amount.

not occur. This shift of wavelengths leaving the vicinity of a massive object is called **gravitational redshift.**

Einstein's description of gravity is radically different from that of Newton. According to Newtonian mechanics, space is perfectly flat and extends infinitely far in all directions. Similarly, Newtonian clocks monotonously tick at an unchanging rate, never speeding up or slowing down. In this rigid, unalterable framework of space and time, gravity is described as a "force" that acts at a distance. The planets literally pull on each other across empty space with a strength described by Newton's universal law of gravitation.

The general theory of relativity does not treat gravity as a force at all. Instead, gravity causes space to become curved and time to slow down. A planet or spacecraft passing near the Sun is deflected from a straight-line path because space itself is curved.

General relativity has been confirmed again and again, as seen in these observations:

- Light is measurably deflected by the Sun's gravitational curving of space (see Figure 13-2).

- The perihelion position of Mercury as seen from the Sun shifts or precesses by 43 arcsec more than is predicted by Newtonian gravitational theory (Figure 13-3).

- Extremely accurate clocks run more slowly when being flown in airplanes than identical clocks remaining on the ground.

- The spectra of some stars are observed to have gravitational redshifts.

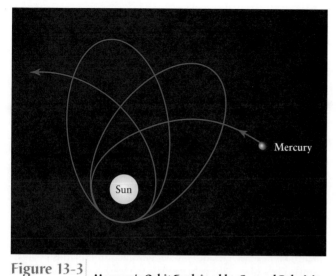

**Figure 13-3** **Mercury's Orbit Explained by General Relativity**
In this greatly exaggerated figure, the location of Mercury's perihelion (its position closest to the Sun) changes with each orbit because of the gravitational influences of the other planets as well as effects predicted by Einstein's general theory of relativity. The amount of this change is inconsistent with the predictions made by Newton's laws of gravity alone.

These observations are consistent with the general theory of relativity but not with the laws of Newtonian physics.

Neutron stars with more than 3 $M_\odot$ collapse because their gravitational force overcomes the repulsion created by neutron degeneracy pressure. Applying general relativity to collapsing neutron stars, we learn that sufficiently dense matter actually causes nearby space to curve so much that it closes in on itself (Figure 13-4). Photons flying outward at an angle from such a collapsing star arc back inward; photons flying straight outward lose all their energy, and they cease to exist.

If light, the fastest moving of all known things, cannot escape from the vicinity of such dense matter, then nothing can! Such regions out of which no matter or any form of electromagnetic radiation can escape are called black holes. As we will see next, their discovery has provided dramatic confirmation of Einstein's theories.

Plummeting in on itself, a collapsing neutron star with mass greater than 3 $M_\odot$ becomes so dense that it ceases to consist of neutrons. General relativity predicts that as a black hole, its matter compresses to infinite density, a state called a **singularity**. But when matter gets sufficiently dense, even general relativity ceases to be valid; a more comprehensive theory must still be developed to explain the nature of matter in this realm. By combining the theories of quantum mechanics and general relativity into a new, more comprehensive theory, astrophysicists hope to

one day explain the final form of matter in a black hole. All we can say with confidence is that the matter in black holes becomes incredibly dense and compact.

## 13-3 Several binary star systems contain black holes

Black holes are more than fine points of relativity theory: They are real, their presence has been observed by their effects on other objects, and more of them are being located all the time. To find evidence for black holes, we look first to binary star systems.

As we saw in Chapter 10, many stars occur in binary star systems. The model for detecting black holes formed from neutron stars is based on the interaction between the black hole and a binary companion. When one star in a close binary becomes a black hole, its gravitational attraction pulls off some of its companion's atmosphere. However, such black holes have diameters of only a few kilometers, and so there is not room for all this gas to fall straight in. Rather, the infalling gas swirls into the black hole like water going down a bathtub drain. The gas waiting to fall

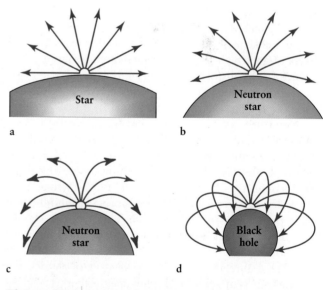

**Figure 13-4** **Trapping of Light by a Black Hole** **(a)** Light departing from a main-sequence, giant, or supergiant star is affected very little by the star's gravitational force. **(b)** However, as a neutron star collapses, its gravitational force increasingly curves the path of departing light until **(c)** some of the photons actually return to the star's surface. **(d)** When the neutron star becomes a black hole, all light leaving its surface remains trapped. Most photons curve back in, except those flying straight upward, which become infinitely redshifted, thereby disappearing.

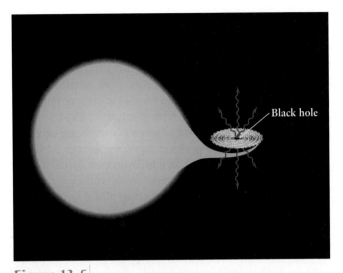

**X Rays Generated by Accretion of Matter near a Black Hole** Stellar-remnant black holes, such as Cygnus X-1, LMC X-3, V404 Cygni, and probably A0620-00, are detected in close binary star systems. This drawing of such a system shows how gas from the companion star transfers to the black hole, thereby creating an accretion disk. As the gas spirals inward, friction and compression heat it so much that the gas emits X rays, which astronomers can detect.

forth with a period of 5.6 days. This behavior is characteristic of a single-line spectroscopic binary (see Section 10-9); HDE 226868's companion is too dim to produce its own set of spectral lines. The clear implication is that HDE 226868 and Cygnus X-1 are the two components of a binary star system.

The B0 supergiant HDE 226868's mass is estimated at about 30 $M_\odot$, like other B0 supergiants. As a result, Cygnus X-1 must have about 7 $M_\odot$; otherwise, it would not exert enough gravitational pull to make the B0 star wobble by the amount deduced from the periodic Doppler shift of its spectral lines. Cygnus X-1 cannot be a white dwarf or a neutron star, because its mass is too large for either of these objects. The only remaining possibility is that it must be a fully collapsed neutron star—a black hole.

In the early 1980s, a similar binary system was identified in the nearby galaxy called the Large Magellanic Cloud. The X-ray source, called LMC X-3, exhibits rapid fluctuations, just like those of Cygnus X-1. LMC X-3 orbits a B3 main-sequence star every 1.7 days. From its orbital data, astronomers conclude that the mass of LMC X-3 is probably about 10 $M_\odot$, which would make it a black hole.

in forms an accretion disk (see Chapter 12). Calculations reveal that this gas is compressed and heats so much that it gives off X rays (Figure 13-5). Thus, if a visible star has a sufficiently tiny, sufficiently massive X-ray–emitting companion, we have located a black hole.

Shortly after the *Uhuru* X-ray satellite was launched in the early 1970s, astronomers found a promising black-hole candidate—an X-ray source called Cygnus X-1. This source is highly variable and irregular. Its strong X-ray emission flickers on time scales as short as a hundredth of a second. If different parts of an X-ray source grew bright at different times, its emission would be a continuous stream. In order for Cygnus X-1 to flicker, the entire star must brighten and dim as a unit. Therefore, light must have time to travel across Cygnus X-1 between each pulse. Because light travels 3000 km in a hundredth of a second, Cygnus X-1 must be smaller in diameter than the Earth.

Cygnus X-1 occasionally emits radio radiation, and in 1971 radio astronomers succeeded in associating these emissions with the visible star HDE 226868 (Figure 13-6). Spectroscopic observations revealed that HDE 226868 is a B0 supergiant with a surface temperature of about 31,000 K. Because such stars do not emit significant X rays, HDE 226868 alone cannot be the Cygnus X-1 X-ray source.

Further spectroscopic observations soon showed that the lines in the spectrum of HDE 226868 shift back and

R I V U X G

**HDE 226868** This star is the optical companion of the X-ray source Cygnus X-1. This binary system is located about 8000 ly from Earth and contains a 7-$M_\odot$ black hole in orbit with HDE 226868, a B0 blue supergiant star. The photograph was taken with the 200-in. telescope at Mount Palomar. The slightly dimmer star above is an optical double that is not part of the binary system.

Another black-hole candidate is a spectroscopic binary in the constellation Monoceros that contains the flickering X-ray source A0620-00. The visible companion of A0620-00 is an orange dwarf star of spectral type K that orbits the X-ray source every 7.75 hours. From orbital data, astronomers conclude that the mass of A0620-00 must be greater than 3.2 $M_\odot$, more probably about 9 $M_\odot$.

Perhaps the most convincing observational evidence for a black hole is the spectroscopic binary V404 Cygni, which consists of an X-ray source orbited by a low-mass G or K star. Doppler shift measurements reveal that the line-of-sight velocity of the visible star varies by more than 400 km/s as it orbits its unseen companion every 6.47 days. These data give a firm lower limit of at least 6.26 $M_\odot$ for the mass of the companion.

## 13-4 Supermassive black holes exist at the centers of most galaxies, and minuscule black holes may have formed in the early universe

Based on the equations of general relativity, the fate of massive neutron stars led to the idea of black holes as early as 1939. Could other mechanisms create black holes? Calculations support at least two other sources. **Supermassive black holes** could have formed from the clumping of mass at the centers of newborn galaxies. And the Big Bang itself may have compressed matter sufficiently to form **primordial black holes**.

As we will explore in more detail in Chapter 17, galaxies were created from condensing gas and stars in the early universe. When a galaxy was forming, some of the gas plunged straight inward, colliding with similar gas coming in from the other direction. As it piled up at the center, much like the mass in a protostar, this matter compressed itself into a supermassive black hole. In May 1994, the Hubble Space Telescope obtained compelling evidence for a black hole at the very center of the galaxy M87. At the heart of M87 is a tiny, bright source of light. Spectra showed that nearby gas and stars are orbiting it extremely rapidly. They can be held in place only if the bright object contains some three *billion* solar masses (Figure 13-7). Given that its size is only slightly larger than the solar system, the source can only be a black hole.

Since 1994, numerous black holes in the centers of galaxies have been identified by their gravitational effect on surrounding gas and stars. While spectroscopy reveals that the dust in Centaurus A orbits in an accretion disk seen edge-on, a distinct, hubcap-shaped accretion disk around the supermassive black hole in NGC 7052 was seen by the Hubble Space Telescope in 1998 (Figure 13-8). As we will see in the next chapter, a black hole containing several million solar masses has even been found lurking at the center of our Milky Way, only 25,000 ly from the Earth. Indeed, in a survey done by the Hubble Space Telescope of 27 nearby galaxies and in follow-up studies, evidence for central, supermassive black holes was found in most galaxies!

Even more exotic black holes may have formed along with the universe itself. British astrophysicist Stephen

b

### Figure 13-7  **Supermassive Black Holes**
Notice the bright regions in the centers of each of these galaxies, where stars and gas are held in tight orbits by black holes. **(a)** M87's bright nucleus (**center of inset**) is only about the size of the solar system and pulls on the nearby stars with so much force that astronomers believe it is a 3-billion-$M_\odot$ black hole. **(b)** The nucleus of the nearby spiral galaxy Andromeda (M31) harbors a black hole of a few million solar masses.

a

R I V U X G

**Figure 13-8** **Accretion Disk Around a Supermassive Black Hole** Swirling around a 300-million-$M_\odot$ black hole in the center of the galaxy NGC 7052, this disk of gas and dust is 3700 ly across. The gas is cascading into the black hole, which will consume it over the next few billion years. NGC 7052 is 191 million ly from Earth in the constellation of Vulpecula.

R I V U X G

Hawking has proposed that the Big Bang explosion from which most astronomers believe the universe emerged may have been chaotic and powerful enough to have compressed tiny knots of matter into primordial black holes. Their masses may have ranged from a few grams to greater than the mass of the Earth. Astronomers have not yet observed evidence of primordial black holes.

Black holes with masses between 100 and $10^4$ $M_\odot$ have also been detected. They may have been created by the merger of stellar black holes.

## INSIDE A BLACK HOLE

The formation of a black hole is complicated, but its nature is surprisingly simple. It has a boundary shaped like a sphere, it either rotates or it does not, and it slowly disappears.

### 13-5 Matter in a black hole becomes much simpler than elsewhere in the universe

A black hole is separated from the rest of the universe by a boundary called the **event horizon.** Within this shell, even light cannot escape the black hole's enormous gravitation-al attraction. The event horizon is not like the surface of a solid body. No matter exists at this location except for the instant it takes all infalling mass to cross the event horizon and enter the black hole.

We cannot look inside a black hole because no electromagnetic radiation escapes from it. Our understanding of its structure comes from the equations of general relativity. According to Einstein's theory, the event horizon is a sphere. The distance from the center of the black hole to its boundary is called the **Schwarzschild radius** (abbreviated $R_{Sch}$), after the German physicist Karl Schwarzschild, who first determined its properties. An Astronomer's Toolbox 13-1 shows how to calculate this distance, which depends only on the black hole's mass. The more massive the black hole, the larger its event horizon.

According to the equations of general relativity, when a stellar remnant collapses to a black hole, it loses its magnetic field. The field's energy radiates away in the form of **gravitational waves,** which are ripples in the very fabric of spacetime. These waves are also created when neutron stars or black holes collide or even when they are in close orbits around each other. The *binary pulsar* mentioned in Section 12-10 is a pair of neutron stars orbiting each other and emitting gravitational waves. The energy in these waves comes from the orbital energy of the neutron stars, and so they are spiraling toward each other. Although gravitational waves have not yet been detected directly, their effect on the orbits of the neutron stars is in complete

## AN ASTRONOMER'S TOOLBOX 13-1

### The Size of Black Holes

In 1918, Karl Schwarzschild, a German physicist, discovered the first solution of Einstein's equations. His solution describes the nature of nonrotating black holes. According to Einstein's general theory of relativity, the Schwarzschild radius $R_{Sch}$ of any black hole can be found from its mass $M$:

$$R_{Sch} = \frac{2GM}{c^2}$$

where $R_{Sch}$ is measured in meters; $M$ is the black hole's mass in kilograms; $c$ is the speed of light, $3 \times 10^8$ m/s$^2$; and $G$ is the gravitational constant, $6.67 \times 10^{-11}$ m$^3$/kg · s$^2$. This equation can be approximated by:

$$R_{Sch} \approx 3\,M'$$

where $M'$ is the black hole's mass in solar masses. It is also often convenient to convert $R_{Sch}$ from meters to kilometers.

*Example:* A 5-$M_\odot$ main-sequence star has a radius of $3 \times 10^6$ km, while the above equation reveals that a black hole with the same mass has a 15-km Schwarzschild radius.

*Compare!* Even a galactic black hole containing 3 billion solar masses has a Schwarzschild radius of only 60 AU (which is just twice the size of Neptune's orbit around the Sun).

*Self-Test:* A primordial black hole with the mass of Mount Everest would have a Schwarzschild radius of just $1.5 \times 10^{-15}$ m! What is Mount Everest's mass? (The answer is in this chapter.)

---

agreement with the predictions of general relativity for such a system. This agreement earned the Nobel Prize in 1993 for Joseph Taylor and Russell Hulse, who discovered the binary pulsar in 1974 and measured the changes in its neutron stars' orbits.

General relativity predicts that wherever a gravitational wave passes, spacetime itself becomes momentarily distorted. Scientists are now building instruments to detect these distortions directly. These devices will measure the minuscule changes in the separation between objects caused by passing gravitational waves.

Besides losing its magnetic field, matter within a black hole loses almost all traces of its composition and origin. It retains only three properties that it had before entering the black hole: its *mass,* its *angular momentum,* and its *electrical charge.* All other attributes of matter vanish inside a black hole, and familiar concepts, such as proton, neutron, electron, atom, and molecule, no longer apply. In addition, because few large bodies appear to have a net charge, it is also doubtful that black holes do. We therefore predict that there are only two different types of black holes: those that rotate and those that do not.

If the mass creating a black hole is not rotating, the black hole it forms does not rotate either. We call nonrotating black holes **Schwarzschild black holes** (Figure 13-9).

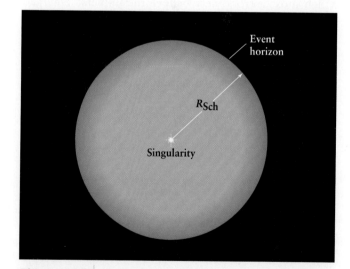

**Figure 13-9**     **Structure of a Schwarzschild (Nonrotating) Black Hole**   A nonrotating black hole has a very simple structure. In fact, it has only two notable parts: its singularity and its boundary. Its mass, called a singularity because it is so dense, collects at its center. The spherical boundary between the black hole and the outside universe is called the event horizon. The distance from the center to the event horizon is the Schwarzschild radius, $R_{Sch}$. There is no solid, liquid, or gas surface at the event horizon. In fact, except for its location at the boundary of the black hole, an event horizon lacks any features at all.

General relativity predicts that all the mass in such a black hole collapses to a point of infinite density at its center, the singularity mentioned earlier. The rest of the volume inside the Schwarzschild radius of a Schwarzschild black hole is empty space.

When the matter creating a black hole possesses angular momentum, that matter collapses to a ring-shaped singularity located inside the black hole between its center and the event horizon (Figure 13-10). Such rotating black holes are called **Kerr black holes** in honor of the New Zealand mathematician Roy Kerr, who first calculated their structure in 1963. Most Kerr black holes should be spinning thousands of times every second, even faster than the pulsars we studied in Chapter 12.

Unlike Schwarzschild black holes, the equations indicate that Kerr black holes possess a doughnut-shaped region directly *outside* their event horizons in which objects cannot remain at rest without falling into the black hole. Called **ergoregions**, they are regions of spacetime that the rotating black hole drags around, like so much batter in a blender. In 1997, matter orbiting a black hole was observed to behave in a fashion consistent with the existence of an ergoregion surrounding the hole. If it is moving fast enough, an object entering the ergoregion can fly out of it again; if it stops in the ergoregion, however, it will fall into the black hole.

## 13-6  Falling into a black hole is an infinite voyage

Imagine being in a spacecraft orbiting only 100 Schwarzschild radii (1500 km) from an isolated 5-M$_\odot$ black hole (see An Astronomer's Toolbox 13-1 for the equation of the Schwarzschild radius). You are held in orbit by the black hole's gravitational force. Even at that short distance, the only effect the black hole has on you is its gravitational attraction. It is only when you get very, very close to the event horizon that bizarre things begin to happen. To investigate these changes, you send a cube-shaped probe toward the black hole, with the same side of the cube always facing "downward" toward the black hole. The probe emits a blue glow so that you can follow its progress. What happens to the cube as it approaches the black hole?

From the time you eject it until it reaches about 10 Schwarzschild radii (150 km), you see the probe descend as if it were falling toward a planet or moon (Figure 13-11a). At 10 Schwarzschild radii, however, the probe begins to respond to a severe tidal effect from the black hole. The face of the probe closest to the event horizon receives more gravitational pull from the black hole than its farther-away parts, and it begins to stretch apart.

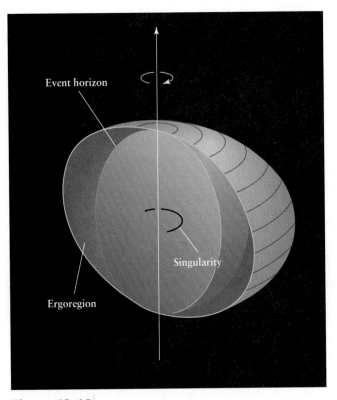

**Figure 13-10**  **Structure of a Kerr (Rotating) Black Hole**
Rotating black holes are only slightly more complex than nonrotating ones. The singularity of a Kerr black hole is located in an infinitely thin ring around the center of the hole. It appears as an arc in this cutaway drawing. The event horizon is again a spherical surface. There is also a doughnut-shaped region, called the ergoregion, just outside the event horizon, in which nothing can remain at rest. Space in the ergoregion is being curved or pulled around by the rotating black hole.

By the time the probe comes within a few Schwarzschild radii of the event horizon, the tidal forces on it are so great that it violently elongates. The part of the probe closest to the black hole accelerates downward and away from the rest of the probe. Furthermore, the sides of the probe are drawn together: They are falling in straight lines toward a common center. The net gravitational effect of moving close to the event horizon is for the probe to be pulled long and thin. From a practical perspective, this means that the probe would be violently torn apart, because it is not composed of perfectly elastic material.

As the probe nears the black hole, blue photons leaving it must give up more and more energy to escape the increasing gravitational force. However, unlike a projectile fired upward, photons cannot slow down. Rather, they lose energy by increasing their wavelengths (see An

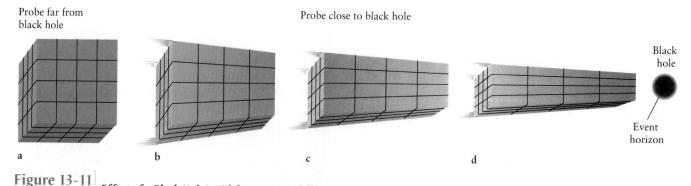

Probe far from black hole

Probe close to black hole

Black hole

Event horizon

a            b            c            d

**Figure 13-11**    **Effect of a Black Hole's Tidal Force on Infalling Matter**    **(a)** A cube-shaped probe intact at 1500 km from a 5-$M_\odot$ black hole. **(b, c, d)** Near the Schwarzschild radius, the probe is pulled long and thin by the difference in the gravitational forces felt by its different sides. This tidal effect is a greatly magnified version of the Moon's gravitational force on the Earth. The probe changes color as its photons undergo extreme gravitational redshift.

Astronomer's Toolbox 3-1). This is another example of the gravitational redshift predicted by general relativity. The closer the probe gets to the event horizon, the more its light is redshifted—first to green, then yellow, then orange, then red, then infrared, and finally radio waves (see Figures 13-11b, c, d).

Stranger still is the black hole's effect on time. General relativity predicts that when the probe approaches within a few Schwarzschild radii of the black hole, its infall rate will slow down as seen from far away. Also, signals from the probe show that its clocks are running more slowly than they did when it left your spacecraft. Time dilation is so great near the event horizon that the probe will appear to hover above it and its clocks will stop. That is what we observe from far away. Someone in the probe observes something else altogether: They see the probe actually cross the event horizon and continue falling toward the black hole's singularity. Pulled apart by tidal effects, it disintegrates as it falls inward. Contrary to the science fiction concept of traveling great distances quickly by passing through a black hole, calculations indicate that objects entering them could not survive passage through, even if there were a way to come out somewhere else.

Could a black hole be connected to another part of spacetime or even some other universe? General relativity predicts such connections, called **wormholes,** for Kerr black holes, but astrophysicists are skeptical. Their conviction is called **cosmic censorship:** Nothing can leave a region containing a singularity.

## 13-7  Black holes evaporate

In exploring the fate of black holes, astrophysicists find that, once again, these objects confound common sense.

With a black hole cloaked behind the event horizon and its mass presumably collapsed into a singularity, it must seem that there is no way of getting mass from the black hole back out into the universe again. No way, that is, until one recalls that mass and energy are two sides of the same coin. What if there was a way of converting the mass into a form of energy that *could* get out of the black hole, such as gravitational energy? The key to the conversion is Einstein's equation $E = mc^2$, and the conversion does occur, according to an idea first proposed by Stephen Hawking.

Black holes convert their mass into energy by a process called **virtual particle** production. Quantum mechanics (see Chapter 4) allows pairs of particles to form spontaneously. Each pair always consists of a particle and its antiparticle, such as a proton and an antiproton, an electron and a positron, or a pair of photons (because a photon is its own antiparticle). Normally, virtual particles form, come back together, and annihilate each other without trace, all within the incredibly short time of about $10^{-21}$ seconds. For example, an electron and its antiparticle, a positron, might form and destroy each other in this process. This is occurring everywhere in the universe, not just near black holes. Astronomers visualize the entire volume of the universe, even "empty space," as a seething foam of virtual particles spontaneously appearing and disappearing.

Virtual particles have actually been observed. This is done in high-energy particle accelerators called synchrotrons and linear accelerators. These machines accelerate electrons, protons, or other bodies to nearly the speed of light. Occasionally, one of these high-speed particles encounters a pair of virtual particles *before* the two virtual particles annihilate each other. The collision forces the virtual particles apart, preventing them from annihilating each other and making the two particles real and observable.

The virtual particles that form *just outside* the event horizon of a black hole feel an exceptionally powerful gravitational field. If one particle is created slightly farther from the hole than its companion, the two virtual particles feel a *tidal* force from the black hole's tremendous gravitational pull (recall the blue probe discussed earlier). When the gravitational tidal force is strong enough, it can pull the two virtual particles apart, making them real (Figure 13-12).

To conserve momentum, at least one of the newly formed real particles always falls into the black hole. But the other particle will sometimes have enough energy to escape from the vicinity of the black hole. The particle flies free into space, and the black hole has effectively emitted energy equal to $E = mc^2$, where $m$ is the mass of the freed particle. Where the virtual particles became real is a temporary void in space outside the event horizon. The black hole fills the void with energy that comes from its mass. Some of the black hole's mass is converted into gravitational energy. This energy is then transmitted *outside* the event horizon to replace the energy taken away by the escaped particle! This is called the **Hawking process,** named after its proposer.

The time it takes a black hole to completely evaporate by the Hawking process increases with the mass of the hole. More massive black holes have lower tidal forces at their event horizons than do less massive and, therefore, smaller-diameter, black holes. Therefore, the rate at which virtual particles are converted into real particles around bigger black holes is actually slower than it is around smaller ones. The faster the real particles are produced, the faster they wear down the mass of the black hole; thus, lower-mass black holes evaporate more rapidly.

A stellar black hole of 5 $M_\odot$ would take more than $10^{62}$ years to evaporate (much longer than the present age of the universe), while a $10^{10}$-kg primordial black hole (equivalent to the mass of Mount Everest) would take only about 15 billion years. Depending on the precise age of the universe, a black hole of this size and age should be in the final throes of evaporating. Astronomers are trying to identify events in space corresponding to the violent, particle-generating deaths of primordial black holes.

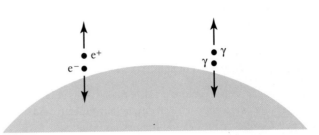

**Figure 13-12** | **Evaporation of a Black Hole** Throughout the universe, pairs of virtual particles spontaneously appear and disappear so quickly that they do not violate any laws of physics. The tidal force just outside the event horizon of a black hole is strong enough to tear apart two virtual particles that appear there before they destroy each other. The gravitational energy that goes into separating them makes them real and, therefore, permanent. Some of these newly created particles fall into the black hole, while some escape into interstellar space. Because the gravitational energy used to create the particles came from the black hole, the hole loses mass and shrinks, eventually evaporating completely. Here we see just a few particles in the making: an electron ($e^-$) and positron ($e^+$), and a pair of photons ($\gamma$).

## WHAT DID YOU THINK?

1. *Are black holes just holes in space?* No, black holes contain highly compressed matter—they are not empty.

2. *What exists at the surface of a black hole?* The surface of a black hole, called the event horizon, is empty space. No stationary matter exists there.

3. *What power or force enables black holes to draw things in?* The only force that pulls things in is the gravitational attraction of the matter in the black hole.

4. *Do black holes last forever?* No, black holes evaporate.

## Key Words

## Key Ideas

* A black hole is an object so dense that the escape velocity from it exceeds the speed of light.

* If a stellar corpse is more massive than 3 $M_\odot$, gravitational compression overcomes neutron degeneracy pressure and forces it to collapse further and become a black hole.

### The Evidence for Black Holes

* According to general relativity, mass causes space to curve and time to slow down. These effects are significant only near large masses or compact objects.

* Observations indicate that some binary star systems harbor black holes. In such systems, gases captured by the black hole from the companion star heat up and emit detectable X rays.

* Supermassive black holes originated in the cores of some galaxies. Other black holes may have formed at the beginning of the universe.

### Inside a Black Hole

* The event horizon of a black hole is a spherical boundary where the escape velocity equals the speed of light. No matter or electromagnetic radiation can escape from inside the event

horizon. The distance from the center of the black hole to the event horizon is called the Schwarzschild radius.

* The matter inside a black hole collapses to a singularity. The singularity for nonrotating matter is a point at the center of the black hole. For rotating matter, the singularity is a ring inside the event horizon.

* Matter inside a black hole has only three physical properties: mass, angular momentum, and electrical charge.

* Nonrotating black holes are called Schwarzschild black holes. Rotating black holes are called Kerr black holes. The event horizon of a Kerr black hole is surrounded by an ergoregion in which all matter must constantly move to avoid being pulled into the black hole.

* Matter approaching a black hole's singularity is torn by extreme tidal forces generated by the black hole, light from the matter is redshifted, and time appears to slow.

* Black holes can evaporate by the Hawking process, in which virtual particles near the black hole become real. This transition decreases the mass of a black hole.

## Review Questions

1 Under what conditions do all outward pressures on a collapsing star fail to stop its inward motion?

2 In what way is a black hole blacker than black ink or a black piece of paper?

3 If the Sun suddenly became a black hole, how would the Earth's orbit be affected?

4 What is cosmic censorship?

5 What are the differences between rotating and nonrotating black holes?

6 Why are all the black hole candidates that are stellar remnants members of close binary systems?

7 If light cannot escape from a black hole, how can we detect X rays from such objects?

## Advanced Questions

The answers to all computational problems, which are preceded by an asterisk (*), appear at the end of the book.

*8 What is the Schwarzschild radius of a black hole, measured in kilometers, containing 3 $M_\odot$? 30 $M_\odot$?

9 If more massive stars evolve and die before less massive ones, why do some black-hole candidates have lower masses than their normal stellar companions?

10 Under what circumstances might a white dwarf or a neutron star in a binary star system become a black hole?

11 Which type of black hole do science fiction writers (implicitly) use in sending spaceships from one place to another through the hole? Why would the other type not be suitable?

## What If ...

**12** A black hole of 5 $M_\odot$ passed by the Earth at, say, Pluto's average distance from the Sun? What would happen to our planet's orbit and to life here?

**13** A primordial black hole with the mass of our Moon approached, passed through, and exited the Earth? What would happen to our planet and to life here?

**14** A primordial black hole with the mass of Jupiter entered the solar system and became trapped in orbit around the Sun halfway between Mars and Jupiter (in the same way that comets are trapped in close orbits; see Chapter 8)? What would happen to the Earth and life on it?

## Observing Projects

**15** As you well know, you cannot see a black hole with a telescope. Nevertheless, you might want to observe the visible companion of Cygnus X-1. The coordinates of this ninth-magnitude star are R.A. = $19^h$ $58.4^m$ and Decl. = +35°12′. This location is quite near the bright star ε Cygni (the star in Cygnus, the Swan, that represents the swan's long neck). You can find it with your Starry Night software by going to edit/find/eta cygni. This constellation appears high in the night sky in the summer and in the shape of a cross. Compare what you see in the night sky and in Starry Night with Figure 13-6.

## WHAT DO YOU THINK?

1   Are all the stars located in the Milky Way?

2   How many galaxies exist?

You have discovered stars singly, in binary systems, and in the relatively small groups called open and globular star clusters. In binary systems and in globular clusters, the stars are *gravitationally bound*: They are held together by their mutual attraction. Now we will see that the same forces create far larger groupings of stars. This final part of the book explores how matter is distributed throughout the universe, what the universe is, where it came from, and where it is going. We start with the group of stars most familiar to us, the Milky Way, and the tools that astronomers used to establish its place in the vast universe.

## IV-1 The Milky Way is only one among billions of galaxies

For those of us fortunate enough to live away from bright outdoor lights, the Milky Way appears as a filmy white band overlaid with the glow of individual stars. As children we are all told—correctly, it turns out—that our solar system is part of the **Milky Way Galaxy,** an enormous assemblage of stars. But just what else is a part of it? Perhaps, we might imagine, all the stars in the universe belong to the Milky Way. If so, then "the galaxy" and "the universe" would mean the same thing. Indeed, that is precisely what most astronomers believed until the early decades of the twentieth century.

Today, we know that the universe contains myriad galaxies, of which the Milky Way is just one. Each **galaxy** is a grouping of millions, billions, or even trillions of stars, all gravitationally bound together. These stars are sometimes accompanied by huge quantities of interstellar gas and dust. In fact, as we saw in Chapter 10, new stars are forming in giant molecular clouds every day.

As early as 1755, the German philosopher Immanuel Kant suggested that vast collections of stars lie far beyond the confines of the Milky Way. Less than a century later, an Irish astronomer observed the structure of some of those "island universes" proposed by Kant. William Parsons was the third Earl of Rosse in Ireland. He was rich, he liked machines, and he was fascinated with astronomy. Accordingly, he set about building gigantic telescopes. In February 1845, his pièce de résistance was finished. This telescope's massive mirror measured 1.8 m (6 ft) in diameter and was mounted at one end of an 18-m (60-ft) tube controlled by cables, straps, pulleys, and cranes (Figure IV-1a). For many years, this triumph of nineteenth-century engineering enjoyed distinction as the largest telescope in the world.

With this new telescope, Lord Rosse examined many of the glowing interstellar clouds discovered and catalogued by William Herschel. Recall from your study of stars that Herschel and his son John, among others, discovered and

a                                                                              b

**Figure IV-1**   **High-Tech Telescope of the Mid-Nineteenth Century**   **(a)** Built in 1845, this structure housed the 1.8-m-diameter telescope that was the largest of its day. The improved resolution it provided over other telescopes is analogous to the improvement that the Hubble Space Telescope provides over most Earthbound optical instruments today. The telescope, as shown here, was restored to its original state during 1996–1998. **(b)** Using his telescope, Lord Rosse made this sketch of the spiral structure of M51.

recorded details of many astronomical objects, including double star systems and, as is relevant here, fuzzy-looking objects they called **nebulae** (singular **nebula**). Lord Rosse observed that some of these nebulae have a distinct spiral structure. Perhaps the best example is M51, also called the Whirlpool or NGC 5194.

Lacking photographic equipment, Lord Rosse made drawings of what he observed. Figure IV-1b shows his drawing of M51. Views like this inspired Lord Rosse to echo Kant's proposal of "island universes." Figure IV-2 shows a modern photograph of M51. It is interesting to note the differences between the perception and interpretation of astronomical objects on the one hand (Figure IV-1b) and the camera's recording of them on the other (Figure IV-2).

Most astronomers of Rosse's day did not agree with the notion of island universes outside our Galaxy. A considerable number of the "nebulae" listed in the *New General Catalogue* were, in fact, interstellar clouds and star clusters scattered throughout the Milky Way. It seemed likely that the intriguing spiral nebulae were also contained within our Galaxy.

The astronomical community became increasingly divided over the nature of spiral nebulae. Finally, in April 1920, a formal discussion, now known as the **Shapley–Curtis debate**, was held at the National Academy of Sciences in Washington, D.C. On one side was Harlow Shapley, an astronomer renowned for his determination of the size of the Milky Way Galaxy. Shapley believed the spiral nebulae to be relatively small, nearby objects scattered around our Galaxy like the globular clusters he had studied. Opposing Shapley was Heber D. Curtis of the Lick

Observatory near San Jose, California. Curtis championed the island universe theory, arguing that each of these spiral nebulae is a separate rotating system of stars much like our own Galaxy.

The Shapley–Curtis debate generated much heat but little light. While focusing scientific attention on the size of the universe, nothing was decided, because no one had any firm evidence to demonstrate exactly how far away the spiral nebulae are. Astronomers desperately needed to devise a way to measure the distances to them. Such a measurement was the first great achievement of a young teacher and former basketball player who moved from Kentucky to Chicago to study astronomy. His name was Edwin Hubble.

**Insight into science  Room for debate**  Lacking definitive data, competent scientists can develop and believe strikingly different explanations for the same observations. At the time of the Shapley–Curtis debate, the observations allowed both points of view. It was only with the advent of the distance measurement technique used by Hubble that the spiral nebulae were definitely shown to be outside our Galaxy.

## IV-2  Studies of Cepheid variable stars led to the discovery of the distances to other galaxies

Edwin Hubble joined the staff of the Mount Wilson Observatory in Pasadena, California, and in 1923 he took a historic photograph of the Andromeda Nebula, M31, one of the spiral nebulae around which controversy raged. Hubble carefully examined his photographic plate and discovered what he first thought to be a nova. Referring to previous plates of that region, he soon realized that the object was actually a Cepheid variable star. As we saw in Chapter 11, these pulsating stars vary in brightness periodically. Further scrutiny over the next several months revealed many other Cepheids. Figure IV-3 shows a Cepheid in the galaxy M100 at different stages of brightness.

Only a decade before, in 1912, the American astronomer Henrietta Leavitt had published an important study of Cepheid variables. Leavitt studied numerous Cepheids in the Small Magellanic Cloud, a galaxy very near the Milky Way. (On a clear night, this galaxy is visible to the naked eye in the southern hemisphere as a thin haze near the Large Magellanic Cloud.) As we learned in Chapter 11, there are two kinds of Cepheid variables: the metal-rich Type I Cepheids, which Leavitt studied, and the slightly dimmer, metal-poor Type II Cepheids. The latter were not discovered until the 1940s, when U.S. cities were blacked out during World War II and the sky was, therefore, especially dark.

**Figure IV-2**  **The Spiral Galaxy M51 (NGC 5194)**  This spiral galaxy in the constellation of Canes Venatici is known as the Whirlpool Galaxy because of its distinctive appearance. Its distance from Earth is about 20 million light-years. The blob on the right, at the end of one of its spiral arms, is a companion galaxy.

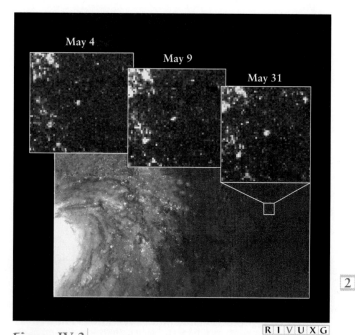

**Figure IV-3**   **M100**   One of the most reliable ways to determine the distance to remote galaxies is to locate Cepheid variable stars in them, as discussed in the text. At 17 Mpc (56 Mly), the galaxy M100 in the constellation Coma Berenices is among the most remote objects whose distances have been determined using Cepheids. **Insets:** The Cepheid in this view, one of 20 located to date in M100, is shown at different stages in its brightness cycle, which recurs over several weeks.

## IV-3  Discovering the universe helps us understand our place in it

Hubble's results, which he presented at the end of 1924, settled the Shapley–Curtis debate once and for all. The universe was recognized to be far larger and populated with far bigger objects than most astronomers had imagined. Hubble had discovered the realm of the galaxies.

In Chapter 14 we apply Hubble's method to find Earth's place in the Milky Way Galaxy. Recall that the Sun's proximity to us makes it our best understood star. It might seem that the nearness of the stars and clouds in the Milky Way would make it the best understood galaxy. However, we will see that the clouds of gas and dust surrounding the solar system make it very challenging for astronomers to survey the Galaxy completely.

Moving out beyond the Milky Way, observations of distant galaxies by the Hubble Space Telescope led astronomers to estimate that some 50 billion galaxies exist throughout the universe. In Chapter 15 we learn about their different types. By looking at neighboring galaxies not concealed by clouds, we learn about the activity in their centers. We also discover that galaxies are themselves often found in groups, and the large-scale structure of the universe does not stop there. Most groups of galaxies are themselves parts of larger groups, distributed as if on the surfaces of enormous, empty bubbles. These groupings and

Leavitt's study led her to the period–luminosity law for Type I Cepheids. As we saw in Figure 11-23, a direct relationship exists between a Cepheid's luminosity (or absolute magnitude) and its period of oscillation. Suppose you find a Cepheid variable, measure its period, and use a graph such as Figure IV-4 to determine the star's average luminosity. The star's brightness can be expressed as an absolute magnitude. Meanwhile, you observe the star's apparent magnitude. Because you now know both the apparent and the absolute magnitudes, you can calculate the star's distance (see An Astronomer's Toolbox III-1). By observing the period, Hubble was able to determine the absolute magnitude of the Cepheid in M31 and to calculate the distance to the star and, therefore, the distance to the nebula containing it.

Straightforward calculations using modern data on the distance of Type I Cepheids demonstrate that Andromeda (M31) is some 2.2 million light-years *beyond* the Milky Way. This proves that M31 is not an open or globular cluster (the traditional nebulae) in our Galaxy but rather an enormous separate stellar system—a separate galaxy. Today, M31 is called the Andromeda Galaxy. It is the only object not part of the Milky Way that can be seen with the naked eye from the Earth's northern hemisphere.

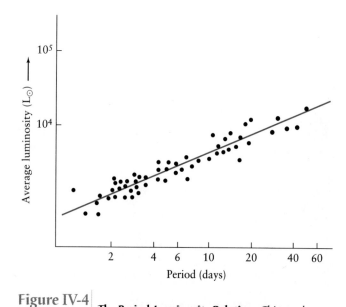

**Figure IV-4**   **The Period-Luminosity Relation**   This graph shows the relationship between the periods and average luminosities of classical (Type I) Cepheid variables. Each black dot represents a Cepheid whose brightness and period have been measured. The red line is the best fit to the data.

their movements away from Earth offer us clues to the structure and origin of the universe.

Quasars and active galaxies, exceptionally powerful sources of radiation, are examined in Chapter 16. We learn of their structure and the causes of their emissions. In Chapter 17 we consider the beginning, evolution, and possible endings of the whole universe. Because this subject involves the most distant observations, it entails the greatest uncertainty of all the topics we have covered. Finally, in Chapter 18 we take a brief look at the possible existence of life elsewhere in the universe.

## WHAT DID YOU THINK?

1 *Are all the stars located in the Milky Way?* While all the stars we can see without the aid of telescopes are in the Milky Way, most stars in the universe are in other galaxies.

2 *How many galaxies exist?* There are an estimated 50 billion galaxies.

## Key Words

galaxy, 342        Milky Way Galaxy, 342        nebula (*plural* nebulae), 343        Shapley–Curtis debate, 343

## Key Ideas

* A century ago, astronomers were divided on whether all stars and nebulae are part of the Milky Way Galaxy.

* The Shapley–Curtis debate was the first major discussion of whether the Milky Way contains all the stars in the universe.

* Cepheid variable stars are important in determining the distance to other galaxies.

* Edwin Hubble first determined that there are other galaxies far outside the Milky Way.

## Review Questions

1 What was the Shapley–Curtis debate all about? Was a winner declared at the end of the debate? Whose ideas turned out to be correct?

2 How did Edwin Hubble prove that Andromeda is not a nebula in our Milky Way Galaxy?

## What If ...

3 The billions of what we now call galaxies, with their myriad stars, were actually as close as the stars in the Milky Way? What would be different about the night sky? Can you think of any other things that would be different under such conditions?

## Observing Projects

4 On a clear night, locate and observe the Andromeda Galaxy. It is "up" in the early evening between August and March in the northern hemisphere. If you are not conversant with the constellations, you can find it first with your Starry Night software and print a star map to take outside with you. Or, you can find it by first locating the **W** made by five bright stars in Cassiopeia. Also locate Polaris, the North Star. Polaris is a medium-bright star found by following the line from the two stars in the Big Dipper farthest from the handle upward until you reach a medium-bright star. Draw a mental line from Polaris to the second star on the right side of Cassiopeia, if you see it as the letter W, or the second star on the left of Cassiopeia, if you see it as the letter M. Keep moving south along that line one-half as far again, and you should see a fuzzy "star," which is Andromeda. Describe what you see. Examine Andromeda through binoculars and again through the largest telescope available. Describe what you see.

# AN ASTRONOMER'S ALMANAC

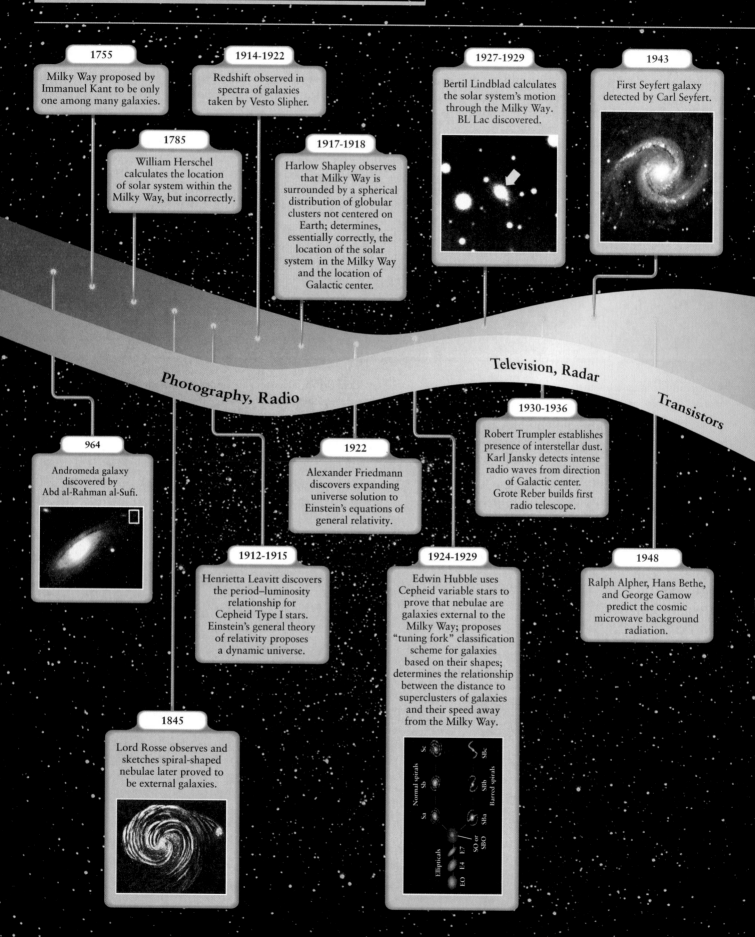

**1755**
Milky Way proposed by Immanuel Kant to be only one among many galaxies.

**1914-1922**
Redshift observed in spectra of galaxies taken by Vesto Slipher.

**1927-1929**
Bertil Lindblad calculates the solar system's motion through the Milky Way. BL Lac discovered.

**1943**
First Seyfert galaxy detected by Carl Seyfert.

**1785**
William Herschel calculates the location of solar system within the Milky Way, but incorrectly.

**1917-1918**
Harlow Shapley observes that Milky Way is surrounded by a spherical distribution of globular clusters not centered on Earth; determines, essentially correctly, the location of the solar system in the Milky Way and the location of Galactic center.

Photography, Radio

Television, Radar

Transistors

**964**
Andromeda galaxy discovered by Abd al-Rahman al-Sufi.

**1922**
Alexander Friedmann discovers expanding universe solution to Einstein's equations of general relativity.

**1930-1936**
Robert Trumpler establishes presence of interstellar dust. Karl Jansky detects intense radio waves from direction of Galactic center. Grote Reber builds first radio telescope.

**1912-1915**
Henrietta Leavitt discovers the period–luminosity relationship for Cepheid Type I stars. Einstein's general theory of relativity proposes a dynamic universe.

**1924-1929**
Edwin Hubble uses Cepheid variable stars to prove that nebulae are galaxies external to the Milky Way; proposes "tuning fork" classification scheme for galaxies based on their shapes; determines the relationship between the distance to superclusters of galaxies and their speed away from the Milky Way.

**1948**
Ralph Alpher, Hans Bethe, and George Gamow predict the cosmic microwave background radiation.

**1845**
Lord Rosse observes and sketches spiral-shaped nebulae later proved to be external galaxies.

# THE UNIVERSE

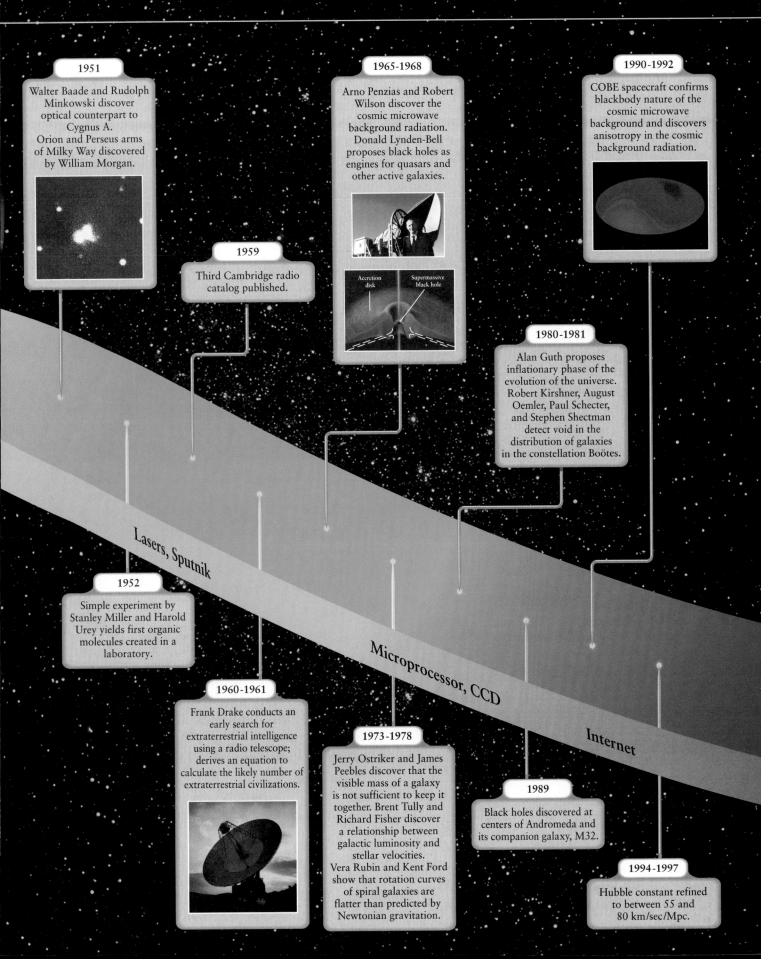

**1951**

Walter Baade and Rudolph Minkowski discover optical counterpart to Cygnus A.
Orion and Perseus arms of Milky Way discovered by William Morgan.

**1959**

Third Cambridge radio catalog published.

**1965-1968**

Arno Penzias and Robert Wilson discover the cosmic microwave background radiation. Donald Lynden-Bell proposes black holes as engines for quasars and other active galaxies.

Accretion disk    Supermassive black hole

**1990-1992**

COBE spacecraft confirms blackbody nature of the cosmic microwave background and discovers anisotropy in the cosmic background radiation.

**1980-1981**

Alan Guth proposes inflationary phase of the evolution of the universe. Robert Kirshner, August Oemler, Paul Schecter, and Stephen Shectman detect void in the distribution of galaxies in the constellation Boötes.

Lasers, Sputnik

**1952**

Simple experiment by Stanley Miller and Harold Urey yields first organic molecules created in a laboratory.

Microprocessor, CCD

Internet

**1960-1961**

Frank Drake conducts an early search for extraterrestrial intelligence using a radio telescope; derives an equation to calculate the likely number of extraterrestrial civilizations.

**1973-1978**

Jerry Ostriker and James Peebles discover that the visible mass of a galaxy is not sufficient to keep it together. Brent Tully and Richard Fisher discover a relationship between galactic luminosity and stellar velocities.
Vera Rubin and Kent Ford show that rotation curves of spiral galaxies are flatter than predicted by Newtonian gravitation.

**1989**

Black holes discovered at centers of Andromeda and its companion galaxy, M32.

**1994-1997**

Hubble constant refined to between 55 and 80 km/sec/Mpc.

# 14

RIVUXG

**Our Galaxy**

This wide-angle photograph spans half the Milky Way as seen from the equatorial latitudes. The Northern Cross is at the left, and the Southern Cross is at the right. The center of the Galaxy is in the constellation Sagittarius, in the middle of this photograph. The dark lines and blotches are caused by hundreds of interstellar clouds of gas and dust that obscure the light from background stars, rather than by a lack of stars.

# The Milky Way Galaxy

## *In this chapter you will discover*

\* the properties of our Milky Way Galaxy

\* Earth's location in the Milky Way

\* how interstellar gas and dust enable star formation to continue in our Galaxy

\* that observations reveal the presence of significant mass in the Milky Way that astronomers have yet to identify

\* that a black hole exists at the center of our Galaxy

## WHAT DO YOU THINK?

1. Where in the Milky Way is the solar system located?

2. Is the Sun moving through the Milky Way and, if so, how fast?

3. How many stars does the Milky Way Galaxy contain?

Like the curtain in a theater, the Milky Way raises our expectations. What breathtaking natural displays lie beyond the translucent glow of stars and gas? The Milky Way stretches all the way around the sky in a continuous band that is almost perpendicular to the plane of the ecliptic. Galileo, the first person to look at the Milky Way through a telescope, immediately discovered that it contains countless dim stars. How far do they extend and what exists in the space between them? Today, we know that most of the stars in the Galaxy lie in an enormous, rotating, disk-shaped assemblage, very much like the spiral nebulae that Lord Rosse first sketched more than 150 years ago. To discover this, astronomers first had to find our place in the Milky Way.

## THE STRUCTURE OF OUR GALAXY

Because the Milky Way completely encircles us, astronomers long ago suspected that the Sun and all the stars in the sky are part of it as well. In the 1780s, William Herschel took the first steps toward mapping its structure. He attempted to deduce the Sun's location in the Galaxy by counting the number of stars in 683 regions of the sky. He reasoned that the greatest density of stars should be seen toward the Galaxy's center and a lesser density seen toward the edge. However, Herschel found roughly the same density of stars all along the Milky Way. He therefore concluded that we are at the center of the Galaxy.

Herschel was wrong: The Earth has no privileged place in the Milky Way. The Sun is 28,000 ly (8600 pc) from the Galaxy's center, the **galactic nucleus.** Herschel's physical understanding of the cosmos was so incomplete that he misinterpreted his observations and thus came to an incorrect conclusion.

## 14-1 Interstellar dust hides the true extent of the Milky Way

While studying star clusters in the 1930s, R. J. Trumpler discovered the reason for Herschel's mistake. Herschel did not know about interstellar gas and dust, which affected his counts of the stars. Trumpler noticed that remote star clusters appear dimmer than would be expected just from their distance alone. Something must be blocking starlight on its way toward the Earth. He concluded that interstellar space is not a perfect vacuum. Instead, it contains dust that absorbs light from distant stars. Great patches of this dust are clearly visible in wide-angle photographs, such as the figure that opens this chapter. Like the stars, this dust is concentrated in the plane of the Galaxy.

This interstellar dust completely obscures from view visible light emanating from the center of our Galaxy. Any visual photons from there are absorbed or scattered before they reach us. Therefore, Herschel was seeing only nearby stars, and he measured apparent magnitudes that were dimmer than they would have been had there been no interstellar dust. Without adjusting for the effects of the dust, he concluded the stars were farther away than they really are. He also had no idea of either the enormous size of the Galaxy or the vast number of stars concentrated around the galactic center and invisible to us.

**Insight into science   A little knowledge** Incomplete information often leads to incorrect interpretation of data and, therefore, to incorrect conclusions. Herschel's lack of knowledge about the matter in interstellar space prevented him from correctly interpreting the distribution of stars surrounding the Earth and, thus, led to an inaccurate position for the Sun within the Galaxy.

Because interstellar dust is concentrated in the plane of the Galaxy, the absorption of starlight is strongest in those parts of the sky covered by the Milky Way. Above or below the plane of the Galaxy, our view is relatively unobscured. Knowledge of our true position in the Galaxy eventually came from observations of globular clusters in these unobscured portions of the sky.

As we saw in Build Your Foundation IV, Henrietta Leavitt published her studies of Type I Cepheids in 1912. Ever since, these variable stars have been important tools in determining distances to the stars. Shortly after Leavitt's discovery, Harlow Shapley, who later debated the existence of other galaxies with Heber Curtis, turned his attention to the family of pulsating stars known as RR Lyrae variables, which are quite similar to Cepheid variables. RR Lyrae

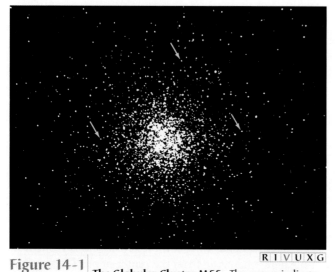

R I V U X G

**Figure 14-1**    **The Globular Cluster M55**    The arrows indicate three RR Lyrae variable stars in this globular cluster located in the constellation Sagittarius. From the average apparent magnitude (as seen in this photograph) and the average absolute magnitudes of these stars (known to be roughly 100 L$_\odot$), astronomers have deduced that the distance to this cluster is 20,000 ly.

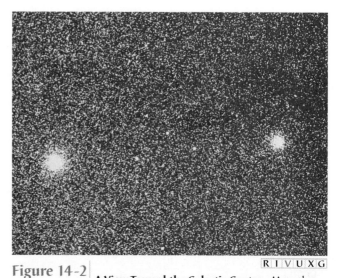

R I V U X G

**Figure 14-2**    **A View Toward the Galactic Center**    More than a million stars fill this view, which covers a relatively clear window just 4° south of the galactic nucleus in Sagittarius. Note the two prominent globular clusters. Although most regions of the sky toward Sagittarius are thick with dust, there is very little obscuring matter in this tiny section of the sky.

variables are commonly found in globular clusters (Figure 14-1). We saw in Chapter 11 that RR Lyrae variable stars also have a period–luminosity relationship. Therefore, they, too, can be used to determine distances.

Shapley used the period–luminosity relationship to determine the distances to the then-known 93 globular clusters in the sky. From their directions and distances, he mapped out the distribution of these clusters in three-dimensional space. By 1915, Shapley had noticed a peculiar property of globular clusters. Most are located only in one-half of the sky, widely scattered around the constellation Sagittarius. Figure 14-2 shows two globular clusters in this part of the sky.

By 1917, Shapley had discovered that the globular clusters are located in a spherical distribution centered not on the Earth, but about a point in the Milky Way toward Sagittarius. Shapley then made a bold conjecture: *The globular clusters orbit the center of the Milky Way. They therefore outline the true size and extent of the Galaxy.* His pioneering research gave strong evidence that the Earth is not at the center of the Galaxy.

## 14-2  Radio observations help map the galactic disk

Decades after Shapley deduced the location of the Galaxy's center from the distribution of globular clusters, astronomers finally measured the distance from the Sun to the very center of our Galaxy. They did this by detecting radio

waves from the galactic nucleus, right through the interstellar gas and dust.

Because of their long wavelengths, radio waves easily penetrate the interstellar medium without being scattered or absorbed. Radio observations of gas clouds orbiting the galactic center confirm that we are located 28,000 ly from it. (In the same way, the *Magellan* spacecraft used radio waves to pierce the veil of clouds around Venus. Astronomers used the information thus provided to create a detailed map of that planet's surface.)

The distance to the galactic nucleus establishes a scale from which the dimensions of other galactic features can be determined. The galactic nucleus is surrounded by a flattened sphere of stars, called the **nuclear bulge,** that is about 20,000 ly in diameter. Swirling out from the nuclear bulge in the plane of the disk are several **spiral arms.** The visible **disk** of our Galaxy is about 100,000 ly in diameter and about 2000 ly thick (Figure 14-3). The spherical distribution of globular clusters defines the **halo** of the Galaxy. If we could view our Galaxy edge-on from a great distance, it would probably look like NGC 4565, shown in Figure 14-4.

The various components of the Galaxy overlap and interpenetrate each other. For example, the disk slices through the nuclear bulge. Nevertheless, these components have maintained their separate identities, chemistries, and motions for billions of years. This happens in large measure because space is so large compared to the sizes of the stars, clouds, and other objects in it that individual objects in different components rarely collide with each other.

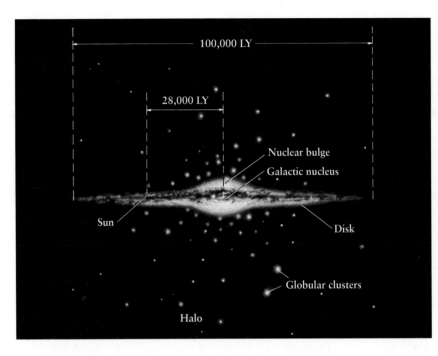

Radio observations also reveal details about the concentrations of gas and dust in the Galaxy's spiral arms. Because hydrogen is by far the most abundant element in the universe, astronomers looked for radio emissions from concentrations of hydrogen gas in the disk of the Galaxy. This technique is useful for detecting gas devoid of the carbon monoxide (see Chapter 11) that usually highlights interstellar regions. Unfortunately, the major transitions of electrons in the hydrogen atom (see Figure 4-16) produce photons at ultraviolet and visible wavelengths that do not penetrate the interstellar medium. How can radio telescopes detect all this hydrogen? The answer lies in atomic physics.

In addition to mass and charge, particles such as protons and electrons possess a tiny amount of angular momentum commonly called **spin.** An electron or a proton can be crudely visualized as a tiny spinning sphere. According to the laws of quantum mechanics, the electron and proton in a hydrogen atom can only spin in either parallel or opposite directions (Figure 14-5); they can have no other spin orientations. If the electron in a hydrogen atom flips from one orientation to the other, the atom must gain or lose a tiny amount of energy. In particular, when flipping from parallel to opposite spins, the atom emits a low-energy radio photon whose wavelength is 21 cm. In 1951 a team of astronomers succeeded in detecting the faint hiss of 21-cm radio static from spin flips of interstellar hydrogen.

The detection of **21-cm radio radiation** was a major breakthrough in mapping the galactic disk. To see why, suppose that you aim your radio telescope across the Galaxy as sketched in Figure 14-6. Your radio receiver picks up 21-cm emission from hydrogen clouds at points 1, 2, 3, and 4 (point *S* is the location of the Sun). However, the radio waves from these various clouds are Doppler shifted (see Section 4-7) by slightly different amounts, because they are moving at different speeds as they travel around the Galaxy. These various Doppler shifts smear the 21-cm

R I V U X G

**Figure 14-4** **Edge-on View of a Spiral Galaxy** If we could view the Milky Way edge-on from a great distance, it would look like spiral galaxy NGC 4565 in the constellation of Coma Berenices. A thin layer of dust and gas is clearly visible in the plane of the galaxy. Also note the reddish color of the nuclear bulge.

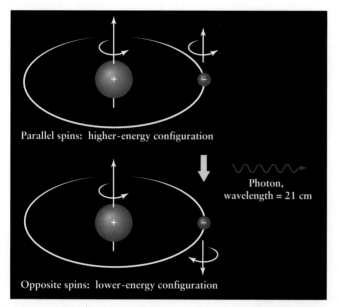

Figure 14-5 **Electron Spin and the Hydrogen Atom** Due to their spin, electrons and protons are both tiny magnets. When an electron and the proton it orbits are spinning in the same direction, their energy is higher than when they are spinning in opposite directions. When the electron flips from the higher-energy to the lower-energy configuration, the atom loses a tiny amount of energy that is radiated as a radio photon with a wavelength of 21 cm.

Figure 14-7 **A Map of the Galaxy** This map, based on radio telescope surveys of 21-cm radiation, shows the distribution of hydrogen gas in a face-on view of the Galaxy. This view hints at spiral structure. The galactic nucleus is marked with a dot surrounded by a circle. Details in the large, blank, wedge-shaped region toward the bottom of the map are unknown, because gas in this part of the sky is moving perpendicular to our line of sight and thus does not exhibit a detectable Doppler shift.

radiation over a range of wavelengths. Because these radio waves from gas clouds in different parts of the Galaxy arrive at our radio telescopes with slightly different wavelengths, it is possible to identify which radio signals come from which gas clouds and thus to produce a map of the Galaxy, such as that shown in Figure 14-7.

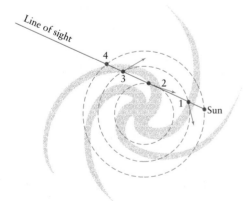

Figure 14-6 **A Technique for Mapping the Galaxy** Hydrogen clouds at different locations along our line of sight are moving around the center of the Galaxy at different speeds. Radio waves from the various gas clouds therefore exhibit slightly different Doppler shifts, permitting astronomers to sort out the gas clouds and map the Galaxy.

Our map reveals numerous arched lanes of neutral hydrogen gas but gives only a vague hint of spiral structure. We need even more information from space in order to improve our understanding of the Galaxy's spiral arms. Note that photographs of other galaxies (Figure 14-8) show spiral arms outlined by "spiral tracers"—bright, population I stars and emission nebulae. As we saw in Chapter 11, these features indicate active star formation. Thus, another useful way to chart the spiral structure of our Galaxy is to map the locations of star-forming complexes marked by H II regions, giant molecular clouds, and massive, hot, young stars in OB associations.

Dust absorption limits the range of visual observations in the plane of the Galaxy to less than 10,000 ly from the Earth. Nevertheless, there are enough OB associations and H II regions visible in the sky to plot the spiral arms near the Sun. Furthermore, as we saw in Chapter 11 (recall Figure 11-4), carbon monoxide is a good tracer of molecular clouds. Radio observations of this molecule have recently been used to chart remote regions of the Galaxy. Taken together, all these observations indicate that our Galaxy has at least four major spiral arms and several short arm segments (Figure 14-9a). Recent observations also suggest that a bar of stars and gas may cross the nuclear bulge. If so, the Milky Way resembles the galaxy in Figure 14-9b more than it does Figure 14-9a.

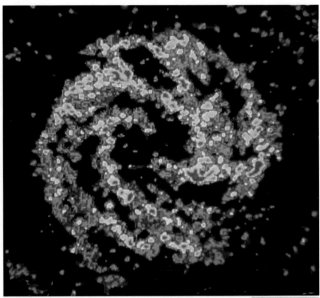

R I V U X G
a

R I V U X G
b

**Figure 14-8** **Two Views of a Spiral Galaxy** The galaxy M83 is in the southern
constellation of Centaurus, about 12 million light-years from Earth. **(a)** At visible
wavelengths, spiral arms are clearly illuminated by young stars and glowing H II regions.
**(b)** A radio view at 21-cm wavelength shows the emission from neutral hydrogen gas.
Note that the spiral arms are more clearly demarcated by visible hot stars and H II regions
than by 21-cm radio emission.

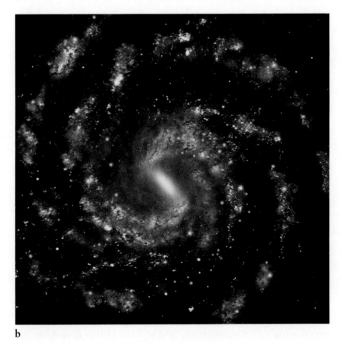

a

b

**Figure 14-9** **Our Galaxy Seen Face-on** **(a)** Our Galaxy has four major spiral arms
and several shorter arm segments. The Sun is located near the short Orion arm, between
two major spiral arms. The Galaxy's diameter is about 100,000 ly, and the Sun is about
28,000 ly from the galactic center. **(b)** If the Milky Way has a bar of stars crossing its
nuclear bulge, as some recent observations suggest, then it probably looks more like this.

The Sun is located near a relatively short arm segment called the Orion arm, which includes the Orion Nebula (see Figure 11-12) and neighboring sites of vigorous star formation in that constellation. Two major spiral arms border either side of the Sun's position. On the side toward the galactic center is the Sagittarius arm, which stargazers in the northern hemisphere see in the summer when they look at the portion of the Milky Way stretching across Scorpius and Sagittarius (see the figure that opens this chapter). Directed away from the galactic center is the Perseus arm, which is visible in the northern hemisphere in the winter. The remaining two major spiral arms are usually referred to as the Centaurus arm and the Cygnus arm, neither of which can be seen at visible wavelengths because of obscuring dust in the interstellar medium.

## 14-3 The Galaxy is rotating

Just as the orbital motion of the planets keeps them from falling into the Sun, the motion of the stars and interstellar clouds around the galactic center keeps these bodies apart. If the stars and clouds in our Galaxy were not in relative motion, their mutual gravitational forces would have caused them to fall into one massive lump billions of years ago. In that case, our Galaxy would not have spiral arms and we would not be here. However, just as detecting the positions of the stars and clouds has been difficult, so, too, has been measuring the orbital motion of the stars and gas, collectively called the *galactic rotation*.

Radio observations of 21-cm radiation from hydrogen gas provide important clues about our Galaxy's rotation. By measuring Doppler shifts, astronomers can determine the speed of objects toward or away from us across the Galaxy. These observations clearly indicate that our Galaxy does not rotate like a rigid body (Earth or Mars, for example) but rather exhibits the same *differential rotation* as the Sun and Jupiter. That is, stars at different distances from the galactic center orbit the Galaxy at different rates.

The Sun itself must be moving around the center of the Galaxy. Otherwise, the solar system would be pulled by gravity into the heart of the Milky Way. Observations show that the Sun has a nearly circular orbit around the center of the Galaxy. Because of the stars' differential rotation in the Galaxy, the Sun is like a car on a circular freeway with the fast lane on one side and the slow lane on the other. As sketched in Figure 14-10, stars in the fast lane are passing the Sun and thus appear from our vantage point to be moving in one direction, while stars in the slow lane are being overtaken by the Sun and therefore appear to be moving in the opposite direction. This is like the retrograde motion of the planets discussed in Chapter 2.

Unfortunately, like the 21-cm observations, studying the motion of nearby stars and gas reveals only how fast they are moving relative to the Sun. To get a complete picture of the Galaxy's rotation, we must find out how fast the Sun itself is orbiting the center of the Galaxy. The Swedish astronomer Bertil Lindblad proposed a method of computing this speed. He noted that not all the stars in the sky move in the orderly pattern shown in Figure 14-10. Globular clusters in the halo of our Galaxy do not participate in the general rotation of the Galaxy. Their combined velocities around the center of the Galaxy must average to zero or else they would drift, en masse, relative to the rest of the Galaxy. Using the motions of globular clusters as a reference, astronomers have calculated that the Sun orbits the galactic center at a speed of 230 km/s, or about 828,000 km (half a million miles) per hour.

Given the Sun's speed and its distance from the galactic center, astronomers can calculate the Sun's orbital period. Traveling at 828,000 km per hour, our Sun takes about 230 million years to complete one trip around the Galaxy. This result demonstrates the vastness of the Milky Way. When last the solar system passed our present location in the Galaxy, early dinosaurs of the Triassic period ruled the Earth.

By combining the true speed of the Sun with the relative speed of the stars around us, as measured by radio astronomers, we can determine the actual orbital speeds

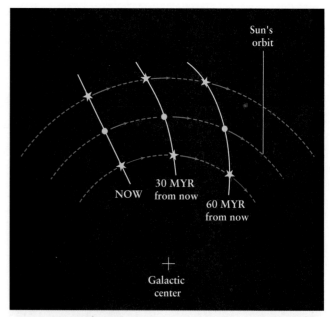

**Figure 14-10**  **Differential Rotation of the Galaxy**  Stars at different distances from the galactic center have different angular speeds. It takes stars farther from the center longer to go around the Galaxy than it does stars closer to the center. As a result, stars closer to the Galaxy's center than the Sun are overtaking the solar system, while stars farther from the center are lagging behind us.

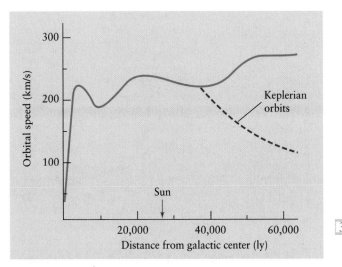

**Figure 14-11** **The Galaxy's Rotation Curve** The blue curve shows the orbital speeds of stars and gas in the Galaxy out to a distance of 60,000 ly from the galactic center. The dashed red curve labeled Keplerian orbits indicates how this orbital speed should decline beyond the confines of most of the Galaxy's visible mass. Because the data (blue curve) do not show any such decline, there apparently is an abundance of invisible dark matter that extends to great distances from the galactic center. This additional mass gives the outer stars higher speeds than they would have otherwise.

of the stars. This computation gives us the **rotation curve** of the Galaxy, a graph showing the orbital speeds of stars and interstellar clouds at various distances from the center of the Milky Way (Figure 14-11).

Knowing the Sun's velocity around the Galaxy from the rotation curve, we can use Kepler's third law to estimate the mass of the Galaxy. Putting in the numbers, we obtain a mass of about $1.1 \times 10^{11}$ M$_\odot$, and that is only the mass *inside* the Sun's orbit around the Galaxy. What could lie beyond?

## MYSTERIES AT THE GALACTIC FRINGES AND NUCLEUS

Looking at the Milky Way extending all around the Earth, we know that stars and other matter extend beyond the Sun's orbit. Unfortunately, because this matter does not affect the Sun's motion, its mass cannot be calculated from Newton's law of gravitation. In recent years, astronomers have been astonished to discover just how much matter apparently lies beyond the orbit of the Sun. They have also found clues to an astounding amount of mass within our Galaxy, the nature of which remains a mystery.

### 14-4 Most of the matter in the Galaxy has not yet been identified

According to Kepler's third law, the farther a star or cloud is from the center of the Milky Way, the slower it should be moving, just as the orbital speeds of the planets decrease with increasing distance from the Sun (see the dashed red line in Figure 14-11). In reality, however, galactic orbital speeds continue to *climb* well beyond the visible edge of the galactic disk. This means there must be more gravitational force from the Galaxy acting on the distant stars and clouds than we can see or have taken into account.

These observations lead astronomers to suspect that a surprising amount of matter must therefore lie beyond the Sun's orbit in the Galaxy. Nearly 90% of the mass of our Galaxy has yet to be located. If we include the unknown mass, the Milky Way's total mass could exceed $1 \times 10^{12}$ M$_\odot$. Accounting for the dark matter, this total mass implies the existence of roughly 200 billion stars in the Galaxy.

To make matters even more mysterious, much of this outlying matter is **dark matter**: It does not show up on photographs. Some astronomers suspect that this dark matter is spherically distributed all around the Galaxy along with the globular clusters in a massive halo. A fraction of the dark matter is composed of neutrinos. The nature of the rest of the dark matter—whether black holes, gas, super Jupiters, brown dwarfs, or something far more exotic—has yet to be determined. The unidentified matter in our Galaxy is sometimes called the **missing mass.** This is a poor name because the matter is not missing; we just do not yet know its location or nature.

Some searches for the dark matter use the fact from general relativity that matter causes passing light to change direction (see Figure 13-2). The Galaxy's dark matter may contain relatively massive objects, like brown dwarfs or black holes. If so, this dark matter will cause the light from more distant stars to change direction as the dark objects pass between those stars and the Earth (Figure 14-12a). Indeed, as a dark object moves directly between a more distant star and the Earth, the star's light is focused toward us, causing the star to brighten. Called **microlensing**, this brightening of distant stars for several weeks has been observed (Figure 14-12b). Microlensing observations indicate that only a small fraction of the Galaxy's dark matter is in the form of these compact, massive objects.

### 14-5 The galactic nucleus is also still poorly understood

The nucleus of our Galaxy is an active, crowded place. If you lived on a planet near the galactic center, you could see a million stars as bright as Sirius, the brightest star in our

a

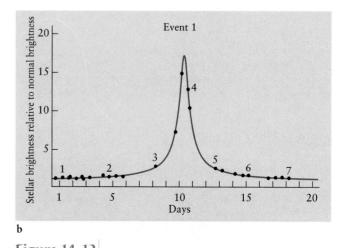

b

**Figure 14-12** | **Microlensing by Dark Matter in the Galactic Halo** (a) Gravitational fields cause light to change direction. A brown dwarf or black hole in the Galaxy's halo passing between the Earth and a more distant star will focus the starlight in our direction, making distant objects appear brighter than they do normally. (b) The light curve of the gravitational microlensing of light from a star in the Galaxy's nuclear bulge by an intervening object. The nature of the intervening object is not yet known.

own night sky. The total intensity of starlight from all those nearby stars would be equivalent to 200 of our full Moons. Night would never really fall. Stranger still, the galactic center turns out to be the site of intense activity.

Because of the absorption of light by interstellar dust at visible wavelengths, most of our knowledge about the center of the Galaxy comes from infrared and radio observations. Figure 14-13 shows three infrared views looking toward the nucleus of the Galaxy. Figure 14-13a is a wide-angle view covering a 50° segment of the Milky Way through Sagittarius and Scorpius. The prominent band across this image is a thin layer of dust in the plane of the Galaxy. The numerous knots and blobs along the dust layer are interstellar clouds heated by young O and B stars. Figure 14-13b is an IRAS view of the galactic center. Numerous streamers of dust (in blue) surround it. The strongest infrared emission (in white) comes from **Sagittarius A,** which is a grouping of several powerful sources of radio waves. One of these sources, called Sagittarius A* (pronounced A-star), is believed to be the galactic nucleus. Figure 14-13c shows stars within 1 ly of Sagittarius A*, with resolution of 0.02 ly.

Radio observations give a different picture of the center of our Galaxy. In 1960, Doppler shift measurements of 21-cm radiation revealed two enormous arms of hydrogen. One arm, which is located between Earth and the galactic nucleus, is approaching us at a speed of 53 km/s. The other arm, on the far side of the galactic nucleus, is receding from us at a rate of 135 km/s. The total amount of hydrogen in these expanding arms is at least several million solar

masses. Something quite extraordinary must have happened about 10 million years ago to expel such an enormous amount of gas from the central region of the Galaxy.

In addition to 21-cm radiation from neutral hydrogen gas, astronomers have also detected radio noise coming from the galactic center. This radio emission, which is produced by high-speed electrons spiraling around a magnetic field, is called **synchrotron radiation.** Despite its small size, Sagittarius A is one of the brightest sources of synchrotron radiation in the entire sky.

Some of the most detailed radio images of the galactic nucleus come from the Very Large Array (VLA). Figure 14-14a is a wide-angle view of Sagittarius A, covering an area about 250 ly across. Huge filaments perpendicular to the plane of the Galaxy stretch 200 ly northward of the galactic disk, then abruptly arch southward toward Sagittarius A. The orderly arrangement of these filaments suggests that a magnetic field may be controlling the distribution and flow of ionized gas, just like magnetic fields on the Sun funnel such gas to create solar prominences.

If the center of our Galaxy isn't active and bizarre enough, recent gamma-ray observations reveal positrons being ejected from that region. (Recall that positrons have positive electrical charges but are otherwise identical to electrons.) The source of these positrons is presently unknown. Positrons can be detected because when a positron and an electron collide, they annihilate each other and release their energy as gamma rays with well-defined wavelengths. These special gamma rays have been observed emanating from the galactic center.

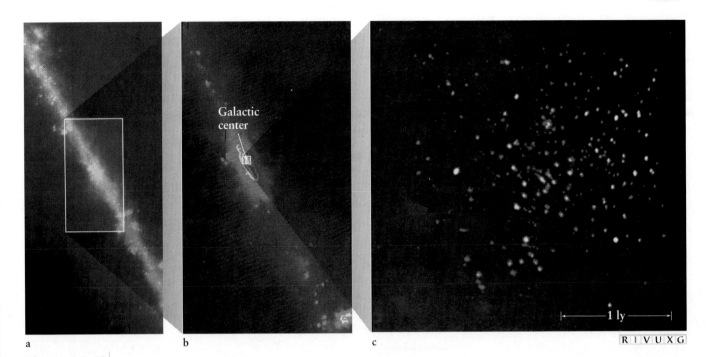

**Figure 14-13** **The Galactic Center** **(a)** This wide-angle view at infrared wavelengths shows a 50° segment of the Milky Way centered on the nucleus of the Galaxy. Black represents the dimmest regions of infrared emission, with blue the next dimmest, followed by yellow and red; white represents the strongest emission. The prominent band across this photograph is a layer of dust in the plane of the Galaxy. Numerous knots and blobs along the plane of the Galaxy are interstellar clouds of gas and dust heated by nearby stars. **(b)** This close-up infrared view of the galactic center covers the area outlined by the white rectangle in (a). **(c)** This infrared image shows about 300 of the brightest stars less than 1 ly from Sagittarius A\*, which is at the center of the picture. The distribution of stars and their observed motions around the galactic center imply a very high density (about a million solar masses per cubic light-year) of less luminous stars.

**Figure 14-14** **Three Views of the Galactic Nucleus** Images (a) and (b) show the appearance of the center of our Galaxy at radio wavelengths. The strongest radio emission is shown in red, weaker emission being colored green through blue. **(a)** This view covers an area of the sky about the same size as the full Moon, corresponding to a diameter of 250 ly. The galactic nucleus is toward the lower right, at the center of the strongest emission. **(b)** This high-resolution view shows details of the galactic center covering an area 30 ly across. The pinwheellike structure is centered on Sagittarius A\*. **(c)** An infrared image of the suspected location (+) of the black hole in the Galaxy's center.

The inner core of Sagittarius A, shown in Figure 14-14b, covers an area about 30 ly across. A pinwheellike feature surrounds Sagittarius A* at the center of this view. One arm of this pinwheel is part of a ring of gas and dust orbiting the galactic center. Something must be holding this high-speed gas in such tight orbit about the galactic nucleus. Using Kepler's third law, astronomers calculate that $2.5 \times 10^6$ $M_\odot$ is needed to prevent this gas from flying off into interstellar space. The observed broadening of spectral lines suggests that an object with the mass of two and a half million Suns in a volume only the size of the solar system is concentrated at Sagittarius A*.

Most astronomers believe that the object is a supermassive black hole (Figure 14-14c). As we saw in Chapter 13, extraordinary activity is also occurring in the nuclei of many other galaxies, implying the presence of supermassive black holes at their centers as well. Astronomers are actively studying these regions in an effort to understand the complex, intriguing activities that are happening there.

## WHAT DID YOU THINK?

1 *Where in the Milky Way is the solar system located?* The solar system is near the Orion spiral arm, about 28,000 ly from the center of the Galaxy.

2 *Is the Sun moving through the Milky Way and, if so, how fast?* The Sun orbits the center of the Milky Way Galaxy at a speed of 828,000 km per hour.

3 *How many stars are in the Milky Way Galaxy?* The Milky Way has about 200 billion stars.

## Key Words

dark matter (missing mass), 355
disk (of a galaxy), 350
galactic nucleus, 349
halo (of a galaxy), 350

microlensing, 355
nuclear bulge, 350
rotation curve (of a galaxy), 355
Sagittarius A, 356

spin (of an electron or proton), 351
spiral arm, 350
synchrotron radiation, 356
21-cm radio radiation, 351

## Key Ideas

### The Structure of Our Galaxy

* Our Galaxy has a disk about 100,000 ly in diameter and about 2000 ly thick, with a high concentration of interstellar dust and gas. It contains around 200 billion stars.

* Interstellar dust obscures our view into the plane of the galactic disk at visual wavelengths. Despite the intervening interstellar dust, hydrogen clouds can be detected by the 21-cm radio waves emitted by changes in the relative spins of electrons and protons.

* The center, or nucleus, of the Milky Way is surrounded by a flattened sphere of stars called the nuclear bulge. The entire Galaxy is surrounded by a spherical distribution of globular clusters called the halo of the Galaxy.

* Young OB associations, H II regions, and molecular clouds in the galactic disk outline huge spiral arms where stars are forming.

* The Sun is located about 28,000 ly from the galactic nucleus, between two major spiral arms. The Sun moves in its orbit at a speed of about 828,000 km per hour and takes about 230 million years to complete one orbit about the center of the Galaxy.

### Mysteries at the Galactic Fringes and Nucleus

* From studies of the rotation of the Galaxy, astronomers estimate that its total mass is about $1 \times 10^{12}$ $M_\odot$. Much of this mass is presently undetectable.

* The galactic nucleus has been studied at infrared and radio wavelengths, which pass readily through intervening interstellar dust and H II regions that illuminate the spiral arms. These observations have revealed many details of the galactic nucleus, but much remains unexplained.

* A supermassive black hole of about $2.5 \times 10^6$ $M_\odot$ apparently exists in the galactic nucleus.

## Review Questions

The answers to all computational problems, which are preceded by an asterisk (*), appear at the end of the book.

1 Why do you suppose that, as seen in the northern hemisphere, the Milky Way is far more prominent in July than in December? Your Starry Night software may be useful in answering this question.

2 What observations led Harlow Shapley to conclude that we are not at the center of the Galaxy?

3 Explain why globular clusters spend more time in the galactic halo than in the plane of the disk, even though their eccentric orbits take them across the disk of the Galaxy.

4 How do hydrogen atoms generate 21-cm radiation? What do astronomers learn about our Galaxy from observations of that radiation?

5 Why do astronomers believe that vast quantities of dark matter surround our Galaxy?

6 What is synchrotron radiation?

7 Why are there no massive O and B stars in globular clusters?

8 How would you estimate the total number of stars in the Galaxy?

9 What evidence suggests that a supermassive black hole might be located at the center of our Galaxy?

* 10 Approximately how many times has the solar system orbited the center of the Galaxy since the Sun and planets were formed?

## Advanced Questions

11 Why don't astronomers detect 21-cm radiation from the hydrogen in giant molecular clouds?

12 Describe the rotation curve you would get if the Galaxy rotated like a rigid body.

13 Compare the apparent distribution of open clusters, which contain young stars, with the distribution of globular clusters relative to the Milky Way. Can you think why open clusters were originally referred to as galactic clusters?

* 14 The visible disk of the Galaxy is about 100,000 ly in diameter and 2000 ly thick. If about five supernovae explode in the Galaxy each century, how often on the average would you expect to see a supernova within 1000 ly of the Sun?

15 Give reasons for the rapid rise in the Galaxy's rotation curve (see Figure 14-11) at distances close to the galactic center.

## Discussion Question

16 What observations would you propose to determine the nature of the hidden mass in our Galaxy's halo?

## What If ...

17 The solar system was located in a globular cluster 28,000 ly above the plane of the Galaxy's disk? Make a drawing of what the Milky Way might look like from that location.

18 The solar system were located at the edge of the galactic disk? How would the Milky Way appear to us?

## Observing Projects

19 Observe the Milky Way on a clear, moonless night with a high-quality pair of binoculars. During the summer months, look toward the galactic center in Sagittarius and follow the Milky Way arching across the sky through the constellations of Aquila and Cygnus. In the winter, follow the Milky Way as it sweeps across such recognizable constellations as Canis Major, Auriga, and Perseus. Note the details in the structure of the Milky Way. For example, toward the galactic center, you should be able to see a few stretches of the Milky Way that clearly display dark patches and mottling where clouds of interstellar gas and dust have blocked the light from background stars. Enjoy the spectacle.

20 To determine the angle between the equatorial coordinate system and the Milky Way, turn on your Starry Night software and turn off daylight, horizon, and planets. Set the "location"

at 90° north (if you are in the northern hemisphere) or 90° south (if you are in the southern hemisphere). Turn on the ecliptic coordinates and the coordinate numbers (use settings/options/equatorial grid/number grid). Make sure that the Milky Way is visible (you can change its color and brightness, if necessary). Next, drag the screen around until you see the Milky Way crossing the celestial equator. Put the celestial equator in the middle of the screen and either print the screen or directly measure the angle between the celestial equator and roughly the middle of the Milky Way (some estimation is necessary here). What is the right ascension of the crossing between the Milky Way and the celestial equator? Now rotate the celestial sphere by 12 h around the celestial equator. Is the Milky Way again crossing the celestial equator? Measure the angle between the two. How does it compare with the previous measurement you made? Explain.

R I V U X G

**A Spiral Galaxy**
Andromeda, a beautiful spiral galaxy, is the only galaxy visible to the naked eye from the Earth's northern hemisphere. Without a telescope, it appears to be a fuzzy blob in the constellation of Andromeda. Located only 2/3 Mpc (2.2 Mly) from us, Andromeda is gravitationally bound to the Milky Way, and it covers an area in the sky roughly 5 times as large as the full Moon.

# Galaxies

## In this chapter you will discover

* how Edwin Hubble categorized galaxies by their shapes

* why some galaxies have spiral arms

* that galaxies are found in clusters and may be surrounded by halos of dark matter

* how some galaxies devour others in dramatic collisions

* the apparent large-scale structure of the universe and its proposed rate of expansion

## WHAT DO YOU THINK?

1. Do all galaxies have spiral arms?

2. Are most of the stars in a spiral galaxy located in its arms?

3. Are galaxies isolated objects?

4. Are all other galaxies moving away from the Milky Way?

O ff in a galactic suburb, far from the center of the action, the Sun is just one of several hundred billion stars in the Milky Way. And our Galaxy is but one of an estimated 50 billion galaxies in the universe. By studying galaxies, we enter the realm of truly cosmic systems. While nearby stars are separated from us by light-years (trillions of miles), neighboring galaxies are hundreds of thousands of times farther away, and the most distant galaxies are billions of light-years from Earth.

Recall from Chapter 10 that astronomers classify stars using the Hertzsprung-Russell diagram to help them understand stellar properties. For precisely the same reason, astronomers classify galaxies by various physical properties. We will see that it is truly remarkable how few overall shapes galaxies have. This classification scheme is where we begin our study of the large-scale structure of the universe.

## TYPES OF GALAXIES

Unlike the Milky Way, most galaxies do not have spiral arms. However, despite their incredible number, there is surprising consistency in their overall shapes. Edwin Hubble began cataloging their appearance in the 1920s, after his measurements of Cepheid variables proved that they lie far outside the Milky Way. The **Hubble classification** of galaxies—spirals, barred spirals, ellipticals, irregulars, and their subclasses—is still used today.

### 15-1 The winding of a spiral galaxy's arms is correlated to the size of its nuclear bulge

**Spiral galaxies** are characterized by a nuclear bulge and by arched lanes of stars and glowing interstellar clouds, which appear as spiral arms. The closest spiral galaxy to the Milky Way is the Andromeda Galaxy, M31, pictured at the opening of this chapter. Andromeda is visible to the naked eye as a fuzzy blob in the skies of the northern hemisphere.

Hubble noted that how tightly the spiral arms are wound varies among different spiral galaxies. Further, the size of nuclear bulges varies. Hubble also observed that these two variations are correlated: The tighter the spiral, the larger the nuclear bulge.

As shown in Figure 15-1, spirals with tightly wound spiral arms (and fat nuclear bulges) are called *Sa* (for spiral type *a*) *galaxies*. Those with moderately wound spiral arms (and a moderate nuclear bulge) are *Sb galaxies*. Finally, loosely wound spirals (with tiny nuclear bulges) are

| Sa | NGC 1357 | Sb | M81 | Sc | NGC 4321 |

R I V U X G

**Figure 15-1** **Various Spiral Galaxies (Face-on Views)** Edwin Hubble classified spiral galaxies according to the tightness of the spiral arms and the size of the nuclear bulge. Sa galaxies have the largest nuclear bulges and the most tightly wound spiral arms, while Sc galaxies have the smallest nuclear bulges and the least tightly wound arms.

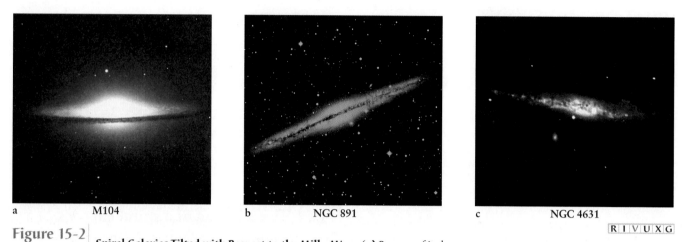

a    M104          b    NGC 891          c    NGC 4631

R I V U X G

**Figure 15-2**  **Spiral Galaxies Tilted with Respect to the Milky Way**  (a) Because of its large nuclear bulge, this galaxy is classified as an Sa. If we could see it face-on, the spiral arms would be tightly wound around a voluminous bulge. (b) Note the smaller nuclear bulge in this Sb galaxy. (c) At visible wavelengths, interstellar dust obscures the relatively insignificant nuclear bulge of this Sc galaxy.

*Sc galaxies.* A typical spiral galaxy contains an estimated 100 billion stars and measures nearly $10^5$ ly in diameter.

Because not all spiral galaxies are oriented face-on to the Earth, their spiral arms are not always evident. However, we can still classify many of them just from observing the size of their central bulges. For example, M104 (Figure 15-2a) has a huge central bulge. It must therefore be an Sa galaxy with tightly wound arms. An Sb galaxy (Figure 15-2b) has a smaller central bulge. The tiny central bulge

of an Sc galaxy (Figure 15-2c) is hardly noticeable at all in this galaxy, which is steeply tilted toward our view.

Besides the degree of winding, the overall appearance of individual spiral arms varies from galaxy to galaxy. In some galaxies, called **flocculent spirals** (from the word meaning "fleecy"), the spiral arms are broad, fuzzy, chaotic, and poorly defined (Figure 15-3a). Other galaxies, called **grand-design spirals**, exhibit beautiful arching arms outlined by brilliant H II regions and OB associations. In

a    M33          b    M74          R I V U X G

**Figure 15-3**  **Variety in Spiral Arms**  The differences in spiral galaxies suggest that at least two mechanisms create spiral arms. (a) This galaxy has the fuzzy, poorly defined spiral arms hypothesized to be created by self-propagating star formation. (b) This galaxy has the thin, well-defined spiral arms hypothesized to be created by a spiral density wave.

these latter galaxies, the spiral arms are thin, delicate, graceful, and well-defined (Figure 15-3b).

## 15-2 Self-propagating star formation and spiral density waves produce spiral arms

For many years the very existence of spiral arms confounded astronomers. They cannot simply come about from the orbital motions of stars. As shown in Figure 14-11, most stars and interstellar gas and dust orbit a galaxy's center at roughly the same linear (straight-line) velocity. Consequently, stars farther from the galactic center take longer to complete their orbits than stars closer in, because they have farther to travel. As a result, the spiral arms should eventually "wind up," from the inside out, wrapping themselves tightly around the nucleus, much more tightly than

is seen in any galaxy (Figure 15-4). Indeed, after a few galactic rotations, the spiral structure should disappear altogether. Why, then, do we observe spiral arms in our Galaxy and many others? This *winding dilemma* suggests that we must look for physical processes, rather than just the motions of stars, to generate spiral arms.

Imagine first a dense interstellar cloud somewhere in the disk of a young galaxy. Stars are just being created in the cloud, and the galaxy does not yet have spiral arms. As hot, massive stars form, their radiation and stellar winds compress the nearby interstellar gas, triggering the formation of *additional* stars. Moreover, massive stars quickly explode as supernovae and produce shock waves, which further compress surrounding gases. As new stars thus trigger the birth of still other stars, the star-forming region grows, a process called *self-propagating star formation*.

The continuing birth of stars also helps account for spiral arms. The galaxy's differential rotation drags the

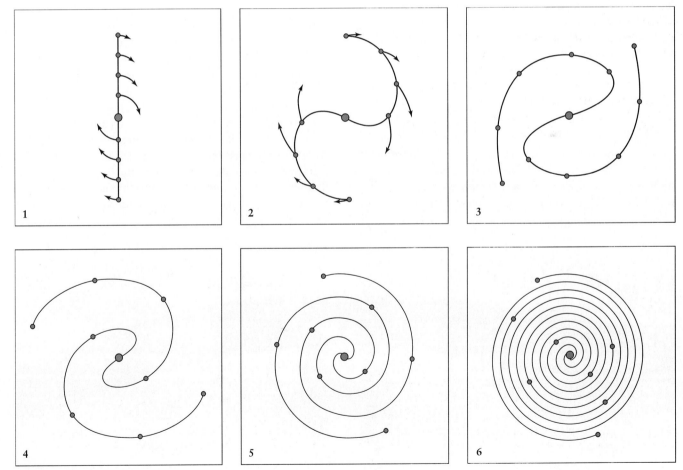

**Figure 15-4** **The Winding Dilemma** The rotation curve of our Galaxy indicates that stars farther from the center take longer to go around than do stars closer in. Therefore, if the spiral arms were created by the motion of stars, eventually the arms would become tightly wound. Such tightening is not observed in our Galaxy or in other galaxies.

inner edges of the young region ahead of the outer edges. As the newly formed stars spread out, they form a spiral arm highlighted by bright O and B stars and glowing nebulae. The high-mass O and B stars explode and then dim before the spiral arms wind up, while newer arms continually develop.

The different types of spiral structures suggest that more than one mechanism gives rise to the creation of a spiral galaxy. Bursts of star formation come and go more or less at random across a galaxy. If self-propagating star formation were the only process leading to spiral arms, bits and pieces of spiral arms would appear only to disappear as the bright, massive stars in them die off. This process creates chaotic spiral arms, such as those observed in flocculent galaxies (see Figure 15-3a), but not the smooth spirals of grand-design galaxies. So, self-propagating star formation cannot be the whole story.

Our search for a second mechanism begins in a pond. If you throw a rock into the water, you create ripples. As you know from experience, the ripples move outward in concentric rings from where the rock struck. But suppose the pond water is rotating when you throw in a rock. *Now* how do the ripples move? Could they be spiral-shaped? According to the astronomer Bertil Lindblad, working in the 1920s, spiral ripples can travel around the disks of some galaxies. Like water in a pond, interstellar gas and dust sustain these ripples, called **spiral density waves.**

Russian scientists in the 1970s tested Lindblad's model. Lacking the computer power to create simulations of galaxies, they threw pebbles into pie pans of water rotating on phonograph turntables! The resulting water wave patterns were indeed spirals. Spiral density waves never wind up. Instead, they orbit in a rigid spiral pattern at about half the speed of the interstellar medium through which they move.

But can we explain why density waves plus rotation leads to spiral galaxies? In the mid-1960s, two American astronomers looked carefully at how galactic spiral density waves traveling through the disk of a galaxy lead to spiral arms. C. C. Lin and Frank Shu argued that density waves cause interstellar gas and dust to pile up temporarily. This occurs because, unlike water waves, which move up and down as they travel along, spiral density waves are compressional waves, just like sound. As a result, they cause interstellar gas and dust to become denser as this material passes through the waves.

Lin and Shu demonstrated that a spiral arm is a galactic traffic jam. Imagine workers painting a line down a busy freeway. The cars normally cruise at 65 mph, but the crew of painters causes a bottleneck. The cars slow down temporarily to avoid hitting other cars and the slowly moving paint truck. As seen from the air, there is a noticeable congestion of cars around the painters (Figure 15-5). An individual car spends only a few moments in the moving traffic jam before resuming its usual speed, but the traffic jam itself lasts all day long. In much the same way, interstellar gas and dust are disturbed relatively briefly, but the density wave maintains well-defined spiral arms for much longer.

The interstellar medium sweeps through the more slowly moving spiral density waves. The waves compress some of this interstellar gas and dust, creating new clouds, which then collapse to form new, metal-rich stars. We see spiral arms because they contain numerous, bright O and B stars and copious quantities of dust and gas that these stars illuminate. The sprawling dust lanes in Figure 15-3b attest to the recent passage of that material through a shock wave.

**Figure 15-5    Compression Wave in Traffic Flow**   When normal traffic flow is slowed down, cars bunch together. In a grand-design galaxy, a density wave moves through the stars and gas. The wave is merely a region of slightly denser matter, which, in turn, creates more gravitational force. This compresses the gas and enhances star formation, which highlights the spiral density wave.

Stars as dim as the Sun contribute virtually nothing to the brightness of spiral arms, which explains why the winding dilemma does not occur in the spiral density wave model: The massive stars highlighting the wave explode before they finish passing through it. The remaining, longer-lived, lower-mass stars, like our Sun, fill the space between the spiral arms without emitting much light. Indeed, as pronounced as spiral arms may appear, only about 5% more stars are found there than are found in between the spiral arms.

It takes an enormous amount of energy to compress interstellar gas and dust. After a billion years or so, even spiral density waves would begin to fade away. Some driving mechanism must keep them going, like throwing more rocks in the pond. The most likely explanation for reinvigorating spiral structure is the passage of a nearby companion galaxy. As this latter galaxy periodically passes close by, its gravitational attraction pulls on the gas, stars, and dust of the spiral galaxy to generate new density waves. Indeed, grand-design galaxies are usually found in the presence of a companion galaxy.

Although most spiral galaxies have two arms, a sizable minority have more. The Milky Way, for example, has at least four spiral arms. Astronomers have not yet established why the numbers of arms vary.

## 15-3 Bars of stars run through the nuclear bulges of barred spiral galaxies

Most astronomers believe that the Milky Way is a spiral galaxy. As we noted in Chapter 14, however, some evidence suggests that it is a **barred spiral**, a spiral galaxy with a bar of stars crossing through the nuclear bulge. The arms in barred spirals extend from the ends of the bar rather than from the nuclear bulge itself. Computer models suggest that bars may arise in galaxies with less dark matter than in unbarred spirals.

Hubble found that the winding of the spiral arms in barred spirals again correlates with the size of the nuclear bulge (Figure 15-6). An *SBa* (for spiral, barred, type a) *galaxy* has a large central bulge (and tightly wound spiral arms). Likewise, a barred spiral with moderately wound spiral arms (and a moderate central bulge) is an *SBb galaxy*, and an *SBc galaxy* has loosely wound spiral arms (and a tiny central bulge). Observations indicate that ordinary spiral galaxies outnumber barred spirals by about 2 to 1.

The motions of the stars in many nearby galaxies have been measured from Doppler shifts of their spectra. In all spiral and barred spiral galaxies discovered to date, the arms trail around behind as the galaxies rotate. They are thus called **trailing-arm spirals.**

## 15-4 Elliptical galaxies display a variety of sizes and masses

**Elliptical galaxies,** named for their distinctive shapes, have no spiral arms. Hubble subdivided elliptical galaxies according to how round or oval they look. The roundest elliptical galaxies are called *E0 galaxies,* while the most elongated are *E7 galaxies.* Elliptical galaxies of intermediate elongation are numbered E1 to E6 (Figure 15-7).

The Hubble scheme classifies galaxies solely by their appearance from our Earth-bound view. But whenever

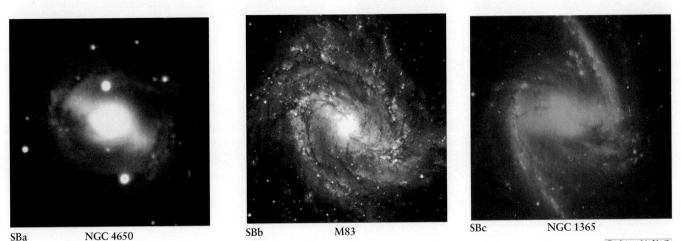

SBa          NGC 4650          SBb          M83          SBc          NGC 1365

R I V U X G

**Figure 15-6**  **Various Barred Spiral Galaxies**  As with spiral galaxies, Edwin Hubble classified barred spirals according to the tightness of their spiral arms (which correlates with the sizes of their nuclear bulges). SBa galaxies have the most tightly wound spirals and largest nuclear bulges, while SBc galaxies have the least tightly wound spirals and the smallest nuclear bulges.

| E0 | M105 | E3 | NGC 4365 | E6 | NGC 3377 |

**Figure 15-7** **Various Elliptical Galaxies** Hubble classified elliptical galaxies according to how round or elongated they appear. An E0 galaxy is round; a very elongated elliptical galaxy is an E7. Three examples are shown here.

R I V U X G

we observe anything in the sky, we are seeing a two-dimensional view of a three-dimensional object. In the case of elliptical galaxies, we observe length and width, but we have no way of knowing anything about the third dimension of depth. What looks like an E0 galaxy (essentially circular) might actually be egg-shaped when viewed from another angle. Conversely, an elongated E7 galaxy might look circular when viewed face-on. Observational evidence suggests that many elliptical galaxies are indeed shaped differently in virtually every direction.

Elliptical galaxies look far less dramatic than their spiral and barred spiral cousins because they contain relatively little interstellar gas and dust. Because stars form in interstellar clouds, few stars should be forming in ellipticals. Observations of spectra confirm the hypothesis that these galaxies contain primarily Population II, low-mass, long-lived stars.

Elliptical galaxies range tremendously in size and mass—from the biggest to the smallest galaxies in the universe. Figure 15-8 shows two **giant elliptical galaxies** that form part of a cluster of galaxies in the constellation Virgo. These enormous giant elliptical galaxies are each about two million light-years in diameter, 20 times the diameter of the Milky Way Galaxy.

Giant ellipticals, containing some 10 trillion solar masses, are rare compared to other types of galaxies, while **dwarf elliptical galaxies** are extremely common. Dwarf ellipticals are only a fraction the size of an average elliptical galaxy and contain so few stars—only a few million—that we see these galaxies as nearly transparent. Because you

**Figure 15-8** **Giant Elliptical Galaxies** The Virgo cluster is a rich, sprawling collection of more than 2000 galaxies about 50 million light-years from Earth. Only the center of this huge cluster appears in this photograph. The two largest galaxies in the cluster are the giant elliptical galaxies M84, on the right, and M86, on the left.

R I V U X G

**Figure 15-9** **A Dwarf Elliptical Galaxy** This nearby E4 galaxy, called Leo I, is about 600,000 light-years from Earth. It is only about 3000 ly in diameter and so sparsely populated with stars that you can see right through its center. It is a satellite of the Milky Way.

R I V U X G

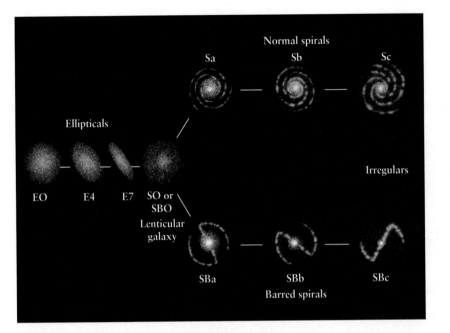

**Figure 15-10 Hubble's Tuning Fork Diagram** Hubble summarized his classification scheme for galaxies with this tuning fork diagram. Elliptical galaxies are classified by how oval they appear, while spirals and barred spirals are classified by the sizes of their central bulges and the correlated winding of their spiral arms. An S0 or SB0 galaxy, also called a lenticular galaxy, is a transitional type between ellipticals and spirals. Irregular galaxies are often positioned between the tines of the fork.

can actually see straight through the center of a dwarf elliptical galaxy and out the other side (Figure 15-9), distant ones are hard to detect. Hence, many more dwarf ellipticals undoubtedly exist than have been identified.

## 15-5 Hubble represented galaxies with different shapes in a tuning fork diagram

Edwin Hubble connected the three regularly shaped types of galaxies—spirals, barred spirals, and ellipticals—in a diagram shaped like a tuning fork (Figure 15-10). According to his scheme, S0 or SB0 galaxies, called **lenticulars** (lens-shaped), are a transition type between ellipticals and the two kinds of spirals. Although they look somewhat like ellipticals, lenticular galaxies have both a central bulge and a disk like spiral galaxies, but they lack spiral arms.

Hubble found some galaxies that cannot be classified as spirals, barred spirals, or ellipticals. He called these **irregular galaxies.** Examples include the Large Magellanic Cloud (LMC) and the Small Magellanic Cloud (SMC), both of which can be seen with the naked eye from southern latitudes and are among the nearest galaxies to the Milky Way. For want of any better scheme, the irregular galaxies are sometimes placed between the ends of the tuning fork tines of the Hubble diagram.

Telescopic views of the Magellanic clouds easily distinguish individual stars (Figures 15-11 and 15-12). Note that the SMC does not exhibit any of the geometric symmetry characteristic of spirals or ellipticals and is therefore a true

irregular. However, the LMC does have a vague barlike structure.

The idea that one type of galaxy may change into another has been in and out of favor with astronomers for

**Figure 15-11 The Large Magellanic Cloud (LMC)** At a distance of only 179,000 ly, this irregular galaxy is the second closest known companion of our Milky Way Galaxy. (The Milky Way's closest known companion, the Sagittarius Dwarf, is shown in Figure 15-17a.) About 62,000 ly across, the LMC spans 22% across the sky, about 44 times the angular size of the full Moon. Note the huge H II region (called the Tarantula Nebula or 30 Doradus) toward the left side of this image. Its diameter of 800 ly and mass of 5 million Suns makes it the largest known H II region.

**Figure 15-12** **The Small Magellanic Cloud (SMC)** The SMC is only slightly farther away from us than the LMC. It is just 26,000 ly across. Because of its sprawling, asymmetrical shape, the SMC is classified as an irregular galaxy. Note that the SMC is rich in young blue stars.

decades. Indeed, this was part of Hubble's motivation in relating the different types of galaxies as he did. As we will see later in this chapter, extensive observations by the Hubble Space Telescope and by ground-based telescopes suggest that interactions between galaxies sometimes lead to changes in their structures. However, there is no evidence that most galaxies change overall shape.

# CLUSTERS AND SUPERCLUSTERS

As you consider the vastness of the universe, it is hard to imagine that structures as huge and apparently isolated as galaxies orbit each other in groups. But they do.

## 15-6 Galaxies are gravitationally bound into clusters and superclusters

Galaxies are not scattered randomly throughout the universe but rather group together in **clusters**. The Fornax cluster (Figure 15-13), so called because it is located in the constellation of Fornax, the Furnace, is typical. Observations of galactic motion suggest that members of a cluster of galaxies are gravitationally bound together: The galaxies orbit each other and occasionally even collide.

Clusters of galaxies are themselves grouped together in huge associations called **superclusters**. A typical supercluster contains dozens of individual clusters spread over a volume 150 million light-years across. In Figure 15-14, which covers many hundreds of square degrees as seen from Earth, note the delicate, filamentary structure of superclusters spread across the sky.

The arrangement of superclusters in space was clarified in the early 1980s, when astronomers began discovering enormous **voids** between superclusters where exceptionally few galaxies are observed. These voids are roughly spherical and measure between 100 million and 400 million light-years in diameter. They may not all be completely empty, however. Observations apparently reveal hydrogen clouds in some of them, while others may be subdivided by strings of dim galaxies. Recent surveys show

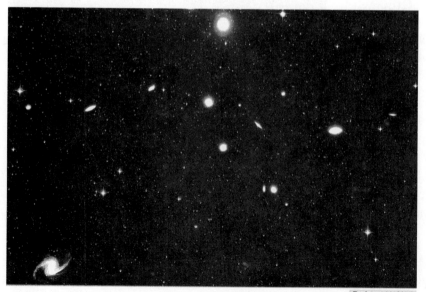

**Figure 15-13** **A Cluster of Galaxies** This group of galaxies, called the Fornax cluster, is about 60 million light-years from Earth. Both elliptical and spiral galaxies are easily identified. The barred spiral galaxy at the lower left is NGC 1365, the largest and most impressive member of the cluster. For a closer view of NGC 1365, see Figure 15-6.

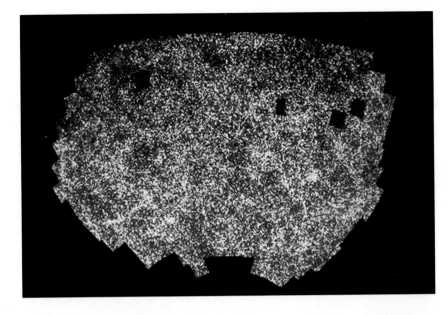

**Figure 15-14** **Large-Scale Distribution of Galaxies** This map shows the distribution of roughly two million galaxies over about 10% of the sky. Each dot is shaded according to the number of galaxies it contains. White dots indicate more than 20 galaxies, blue dots between 1 and 19 galaxies; there are no galaxies in dark blue areas. The clusters form a lacy, filamentary structure. The larger, elongated, bright areas are superclusters and filaments, which generally surround darker voids containing few galaxies.

that most galaxies are distributed on the surfaces between these voids (Figure 15-15). One notable sheet of galaxies, shown as a ribbon in the two-dimensional Figure 15-15, has been dubbed the **Great Wall**. The arc of this "wall of galaxies" runs for some 500 million light-years. Like a sheet of paper seen edge-on, the Great Wall also has depth, extending at least 15 million light-years perpendicular to the plane of Figure 15-15.

The distribution of clusters of galaxies throughout the universe is sometimes said to be "sudsy" because it resembles a collection of giant soap bubbles. Galaxies surround voids in the same way that bubbles are concentrated on the surface of soap film. Many astronomers suspect that this sudsy pattern contains important clues about conditions shortly after the Big Bang that led to the formation of the universe we inhabit.

## 15-7 Clusters of galaxies may appear densely or sparsely populated and regular or irregular in shape

A cluster of galaxies is said to be either **poor** or **rich**, depending on how many galaxies it contains. For example, the Milky Way Galaxy, the Andromeda Galaxy, and the Large and Small Magellanic clouds belong to a poor cluster called the **Local Group**. The Local Group contains some 40 galaxies, over a third of which are dwarf ellipticals. Figure 15-16 shows a map covering most of the Local Group.

Astronomers continue to find new galaxies in the Local Group. Located in the constellation of Sagittarius just 80,000 ly from the center of the Milky Way, the Sagittarius

**Figure 15-15** **A Slice of the Universe** This map shows the locations of 9325 galaxies in two wedge-shaped slices of the universe. The Coma cluster (Figure 15-18) forms part of the so-called Great Wall, a sheet of galaxies 15 million light-years thick and covering an area of 200 by 500 million ly. The Southern Wall is a similar collection of galaxies. To produce this map, a team of astronomers painstakingly measured the redshifts of galaxies out to a distance of nearly 500 million light-years (150 Mpc) from Earth. Note how there are regions almost devoid of galaxies, surrounded by thin regions full of them. This distribution of galaxies is like a slice cut through the soapsuds in a kitchen sink.

**Figure 15-16** **The Local Group**
Our Galaxy belongs to a poor, irregular cluster consisting of about 40 galaxies called the Local Group. This map shows the distribution of the 21 galaxies nearest to the Milky Way. (The galaxies are actually distributed in a three-dimensional space.) The Andromeda Galaxy (M31) is the largest and most massive galaxy in the Local Group. The second largest is the Milky Way itself. M31 and the Milky Way are each surrounded by a dozen satellite galaxies. The recently discovered Sagittarius Dwarf Galaxy (Figure 15-17) is the Milky Way's nearest known neighbor.

Dwarf Galaxy (Figure 15-17a) was only discovered in 1994. Although our closest known galactic neighbor, it had been hidden from view until now by the obscuring gas and dust of our own Galaxy. The Sagittarius Dwarf appears to be coming apart under the tidal forces from our own Galaxy, which will absorb it over the next 100 million years. Another new galaxy in the Local Group, discovered in 1997, is about 3 million light-years from the Milky Way. Named Antlia, it a dwarf galaxy with about one million stars (Figure 15-17b). It is highly likely that other nearby galaxies will be discovered in the coming years.

a

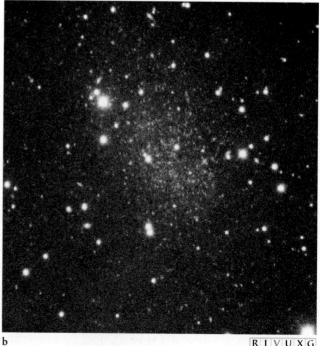

b

**Figure 15-17** **The Nearest and Newest Members of the Local Group** **(a)** The Sagittarius Dwarf is a dwarf elliptical galaxy that lies some 80,000 ly from the Milky Way. The curved lines in this image outline areas of equal brightness and emphasize the structure of the galaxy. Discovered in 1994, the Sagittarius Dwarf is the closest known galaxy to the Milky Way. It is so close that within the next 100 million years or so, the Milky Way will absorb this galaxy. **(b)** Antlia is the most recently discovered member of the Local Group. First detected in 1997, it lies about 3 million ly away, outside the region depicted in Figure 15-16. Antlia contains only about a million stars.

R I V U X G

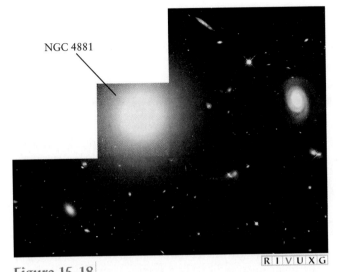

**Figure 15-18**   **The Coma Cluster**   This rich, regular cluster containing thousands of galaxies is about 300 million light-years from Earth. Regular clusters are composed mostly of elliptical and lenticular galaxies and are common sources of X rays. The brightest member of the Coma cluster shown here is the elliptical galaxy NGC 4881.

NGC 4881

R I V U X G

**Figure 15-19**   **The Hercules Cluster**   This irregular cluster, which is about 700 million light-years from Earth, contains a high proportion of spiral galaxies, often associated in pairs and small groups.

R I V U X G

Astronomers also categorize clusters of galaxies according to their shapes. A **regular cluster** is distinctly spherical, with a marked concentration of galaxies at its center. Numerous gravitational interactions over the ages have spread the galaxies into a distinctive spherical distribution. In contrast, galaxies in an **irregular cluster** are more randomly scattered about a sprawling region of the sky.

The nearest example of a rich, regular cluster is the Coma cluster, located about 300 million light-years from us in the constellation of Coma Berenices (Figure 15-18; see also Figure 15-15). Despite its great distance, more than 1000 bright galaxies within it are easily visible from Earth. Certainly the Coma cluster must contain many thousands of dwarf ellipticals soon to be detected—maybe as many as 10,000 galaxies overall.

Rich, regular clusters like the Coma cluster contain mostly elliptical and lenticular galaxies. Only 15% of the Coma cluster's galaxies are spirals and irregulars. Irregular clusters, such as the Virgo cluster and the Hercules cluster (Figure 15-19), display a more even mixture of galaxy types. Two-thirds of the 200 brightest galaxies in the Hercules cluster are spirals, nearly a fifth are ellipticals, and the rest are irregular.

## 15-8 Galaxies in a cluster can collide and combine

Occasionally, two galaxies in a cluster or two galaxies from different, nearby clusters pass through one another

(Figure 15-20). There is so much space between stars that the probability of two stars crashing into each other is extremely small. On the other hand, the galaxies' huge clouds of interstellar gas and dust are so large that they do collide, slamming into each other and producing

R I V U X G

**Figure 15-20**   **The Cartwheel Galaxy**   This ring-shaped assemblage 500 million light-years from Earth is the likely result of one galaxy, probably the blue-white one on the right, having passed through the middle of the larger one, shown above it. Astronomers suspect that the passage created a circular density wave in the Cartwheel that stimulated a burst of star formation, creating many bright blue and white stars.

strong shock waves. The colliding interstellar clouds are stopped in their tracks.

Collisions can merge galaxies together—or hurl their stars far into space. A violent collision can strip both galaxies of their interstellar gas and dust while the stars in each one keep right on going. The violence of the collision heats the gas stripped from these galaxies to extremely high temperatures. This process may be a major source of the hot **intergalactic gas** often observed in rich, regular clusters.

In a less violent collision between two galaxies or in a near miss, the compressed interstellar gas may cool sufficiently over time to allow new stars to form. Such collisions can thus stimulate prolific star formation, which may account for the **starburst galaxies** that blaze with the light of numerous newborn stars. These galaxies are characterized by bright centers surrounded by clouds of warm interstellar dust, indicating a recent, vigorous episode of star birth (Figure 15-21). The warm dust is so abundant that starburst galaxies are among the most luminous objects in the universe at infrared wavelengths.

The starburst galaxy M82 in Figure 15-21 is one member of a nearby cluster of about a dozen galaxies that includes the beautiful spiral galaxy M81 and a fainter companion called NGC 3077 (Figure 15-22a). A radio survey of that region of the sky revealed enormous streams of

**Figure 15-21**    R I V U X G    **A Starburst Galaxy** Prolific star formation is occurring at the center of this galaxy, called M82 or NGC 3034, located about 12 million light-years from Earth in the constellation Ursa Major. This activity was probably triggered by gravitational interactions with neighboring galaxies. Note the turbulent appearance of the interstellar dust around the galaxy's center.

a    R I V U X G        b    R I V U X G

**Figure 15-22**    **The M81 Group** The starburst galaxy M82 is in a nearby cluster of about a dozen galaxies, including the spectacular spiral M81. Several of the galaxies in this cluster are connected by streamers of hydrogen gas. **(a)** This photograph shows the three brightest galaxies at visual wavelengths. **(b)** This radio image, created from data taken by the Very Large Array, shows the streamers of hydrogen gas that connect the bright galaxies and also several dim ones, seen as regions of bright orange here.

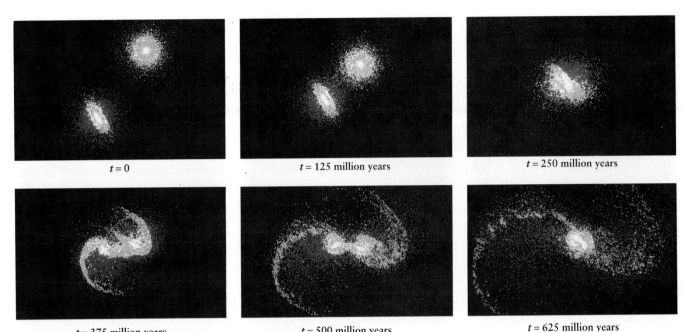

| | | |
|---|---|---|
| $t = 0$ | $t = 125$ million years | $t = 250$ million years |
| $t = 375$ million years | $t = 500$ million years | $t = 625$ million years |

**Figure 15-23** **A Simulated Collision Between Two Galaxies** This computer simulation shows the collision and merging of two galaxies accompanied by an ejection of stars into intergalactic space. The frames progress at 125 Myr intervals. Stars in the disk of each galaxy are colored blue, while stars in their central bulges are yellow. Red indicates dark matter that surrounds each galaxy. Compare the bottom frames here with Figure 15-24.

hydrogen gas connecting these and other, smaller galaxies in that cluster (Figure 15-22b). The loops and twists in these streamers suggest that the three galaxies have had several close encounters over the ages.

The effects of the gravitational interactions between colliding galaxies can also hurl thousands of stars out into intergalactic space along huge, arching streams. These antenna-like shapes have been dramatically illustrated in computer simulations like the one shown in Figure 15-23. Note the similarity between the ending frames of this simulation and the colliding galaxies shown in Figure 15-24.

In a rich cluster, astronomers suspect that many near misses between galaxies must occur. If galaxies are surrounded by extended halos of dim stars, these near misses could strip the galaxies of these outlying stars. In this way, a loosely dispersed sea of dim stars might come to populate the space between galaxies in a cluster. One of the projects assigned to the Hubble Space Telescope is to search for these dim stars in extended halos and in intergalactic space.

Similarly, after galaxies collide, some interior stars are flung far and wide, scattering material into intergalactic space. Such isolated stars have been observed by the Hubble Space Telescope. However, other stars slow down, often causing the remnants of the galaxies to merge. Several dramatic examples of **galactic mergers** have been discovered (Figure 15-25). Astronomers also speak of **galactic**

**cannibalism,** which occurs when a large galaxy captures and "devours" a smaller one. Cannibalism differs from mergers in that the dining galaxy is significantly bigger

R I V U X G

**Figure 15-24** **Colliding Galactic Antennae** Pairs of colliding galaxies often exhibit long "antennae" of stars ejected by the collision, as shown at left. This particular system, called NGC 4038–4039, is located 63 Mly away in the southern constellation Corvus. **Inset:** The collision of the two galaxies has compressed a vast amount of interstellar gas and dust, triggering a firestorm of star formation. This is shown in the numerous red H II regions and blue clusters of hot young stars.

**Figure 15-25**  **Merging Galaxies**  This contorted object in the constellation of Ophiuchus is the result of two spiral galaxies in the process of merging. The widespread blue area reveals that the collision between the two galaxies has triggered an immense burst of star formation.

than its dinner, whereas merging galaxies are about the same size.

Many astronomers suspect that giant ellipticals are the product of galactic mergers and cannibalism, including spiral-spiral collisions. As we have seen, giant galaxies typically occupy the centers of rich clusters. Once formed, giant ellipticals can continue to grow because smaller galaxies are often located around them (see Figure 15-8). As they pass through the extended halo of a giant elliptical, these smaller galaxies slow down and are eventually consumed by the larger galaxy.

Figure 15-26, a computer simulation, shows a large, disk-shaped galaxy devouring a small satellite galaxy. The large galaxy consists of 90% stars (in blue) and 10% gas (in white) by mass. It is surrounded by a halo of dark matter having a mass about 3.3 times that of the disk. The satellite galaxy, which has a tenth of the mass of the large galaxy, contains only stars (in orange). Initially, the satellite is in circular orbit about the large galaxy. Note that spiral arms appear in the large galaxy as the collision proceeds. Two billion years elapse as the satellite spirals in toward the core of the large galaxy. Although much material is stripped from the satellite, most of its stars plunge into the nucleus of the large galaxy.

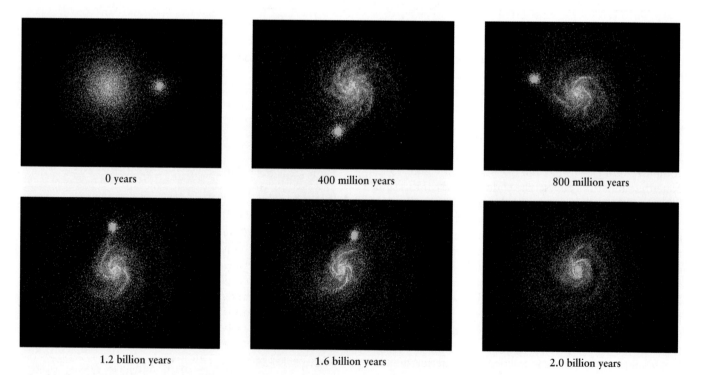

| 0 years | 400 million years | 800 million years |

| 1.2 billion years | 1.6 billion years | 2.0 billion years |

**Figure 15-26**  **Simulated Galactic Cannibalism**  This computer simulation shows a small galaxy (stars in yellow) being devoured by a larger, disk-shaped galaxy (stars in blue, gas in white). The pictures display progress at 400-million-year intervals. Note how spiral arms are induced in the disk galaxy by its interaction with the satellite galaxy.

## 15-9 Galactic halos may account for some dark matter in the universe

What prevents the rapidly moving galaxies in clusters and superclusters from wandering away from each other? There must be sufficient matter to provide the necessary gravitational force to bind galaxies together in clusters and clusters into superclusters. However, *no cluster or supercluster of galaxies contains enough visible matter to stay bound together.* A lot of nonluminous (dark) matter must be scattered about each cluster of galaxies, or else the galaxies would have long ago wandered apart in random directions, and the clusters would no longer exist. Recall that a similar problem arose in Chapter 14 regarding dark matter in individual galaxies. Analyses demonstrate that the total mass needed to bind a typical rich cluster is 10 times greater than the mass of material that shows up on visible-light images.

Astronomers using X-ray telescopes have discovered the nature of some of the remaining dark matter. Satellite observations have revealed X rays pouring from the space between galaxies in rich clusters. This radiation is emitted by substantial amounts of hot intergalactic gas at temperatures between 10 million and 100 million Kelvin. The mass of this hot intergalactic medium is typically as great as the combined mass of all the visible galaxies in a rich cluster.

Extensive evidence supports the idea of extended halos surrounding galaxies. Many galaxies have rotation curves similar to that of our Milky Way (recall Figure 14-11). These rotation curves remain remarkably flat out to surprisingly great distances from the galaxies' centers. For example, Figure 15-27 shows the rotation curves of four spiral galaxies. In all cases, the orbital speed is fairly constant, even out where stars and nebulae are too dim and widely scattered for reliable measurements. According to Kepler's third law, we should see a decline in orbital speed toward the outer portions of a galaxy. The fact that in most cases we do not implies that we have still not detected the true edge of these and many similar galaxies. Astronomers conclude that a considerable amount of dark matter must extend well beyond the visible portion of a galaxy's disk.

## SUPERCLUSTERS IN MOTION

Whenever an astronomer finds an object in the sky, one of the first tasks is to determine its composition. As we saw in Chapters 3 and 4, that means attaching a spectrograph to a telescope and recording the object's spectrum. As long ago as 1914, V. M. Slipher, working at the Lowell Observatory in Arizona, took spectra of "spiral nebulae." He was surprised to discover that the spectral lines of 11 of the 15 spiral nebulae he studied were substantially redshifted. These

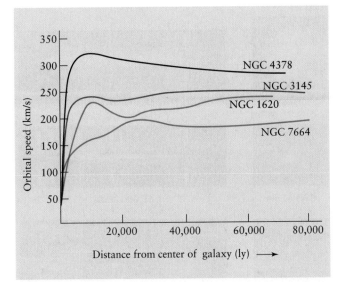

**Figure 15-27** **The Rotation Curves of Four Spiral Galaxies**
This graph shows how the orbital speed of material in the disks of four spiral galaxies varies with the distance from the center of each galaxy. If most of each galaxy's mass was concentrated near the center of the galaxy, these curves would fall off at large distances. But these and many other galaxies have flat rotation curves that do not fall off. This indicates the presence of extended halos of dark matter.

redshifts indicate that they are moving away from us at significant speeds. This marked dominance of redshifts was presented by Heber D. Curtis in the Shapley-Curtis debate as evidence that the "spiral nebulae" could not be part of our Milky Way Galaxy. It also left scientists with a major puzzle: Is the entire universe expanding?

## 15-10 The redshifts of superclusters indicate that the universe is expanding

During the 1920s, Edwin Hubble and Milton Humason recorded the spectra of many galaxies with the 100-in. telescope on Mount Wilson, confirming that many galaxies are rapidly receding from the Milky Way. Using the Doppler effect (recall Figure 4-17), Hubble calculated the speed at which each galaxy is moving away from us.

Using techniques such as the brightness of Cepheid variables (Section IV-2), Hubble could also estimate the distances to a number of these galaxies. He found a direct correlation between the distance to a galaxy and the size of its redshift: *Galaxies in distant superclusters are moving away from us more rapidly than galaxies in nearby superclusters.* Figure 15-28 shows this correlation for five elliptical galaxies. Such recessional motion pervades the universe and is now called the **Hubble flow.**

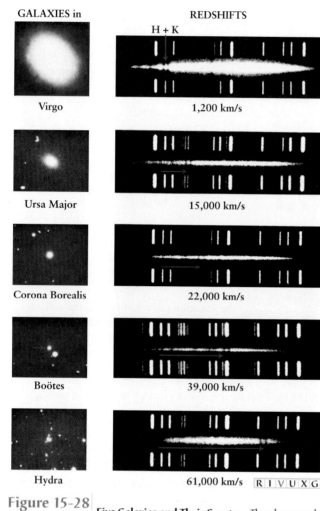

GALAXIES in          REDSHIFTS

Virgo                1,200 km/s

Ursa Major           15,000 km/s

Corona Borealis      22,000 km/s

Boötes               39,000 km/s

Hydra                61,000 km/s

**Figure 15-28** **Five Galaxies and Their Spectra** The photographs of these five elliptical galaxies were all taken at the same magnification. They are labeled according to the constellation in which each galaxy is located. The spectrum of each galaxy is the hazy band between the comparison spectra at the top and bottom of each plate. In all five cases, the so-called H and K lines of calcium are seen. The recessional velocity (calculated from the Doppler shifts of the H and K lines) appears below each spectrum. Note that the fainter—and thus more distant—a galaxy is, the greater is its redshift.

When Hubble plotted these data on a graph of distance versus speed, he found that the points lie nearly along a straight line. Figure 15-29 is a modern version of Hubble's 1929 graph. The relationship between the distances to galaxies and their recessional motion is most easily stated as a formula called the **Hubble law:**

$$\text{Recessional velocity} = H_0 \times \text{distance}$$

$H_0$ is a constant, commonly called the **Hubble constant.**

The relationship between the distances to galaxies and their redshifts is one of the most important astronomical

discoveries of the twentieth century. It tells us that we are living in an expanding universe. The Hubble law reveals the speed of the expanding universe. An Astronomer's Toolbox 15-1 shows how the Hubble law works. The Hubble law does not apply to the galaxies in the Local Group or to the other clusters in our local supercluster, because these galaxies and clusters are all bound together by mutual gravitational attraction. Hubble flow implies only that the superclusters of galaxies are moving away from each other.

**Insight into science** **Power in numbers** Much of the power of science comes from expressing systematic data in mathematical form. By themselves, the spectra of galaxies give a qualitative picture of their motion, and the pattern of recessional velocities suggests a deeper insight into nature. When the data are at last converted into the equation called the Hubble law, it becomes clear that the universe is expanding, and we can begin to look for the reason behind this expansion.

The exact value of the Hubble constant is a topic of active research and heated debate among astronomers. For this book, we use roughly the average of all present data:

$$H_0 = 75 \text{ km/s/Mpc}$$

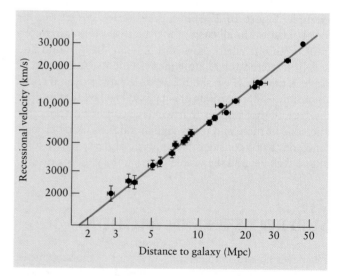

**Figure 15-29** **The Hubble Law** The distances and recessional velocities of 20 spiral galaxies are plotted on this graph. The short horizontal and vertical lines around each point, called error bars, represent uncertainties in the distances and velocities, respectively. The straight line is the "best fit" for the data. This linear relationship between distance and speed is called the Hubble law.

# AN ASTRONOMER'S TOOLBOX 15-1

## The Hubble Law

The Hubble law describes our expanding universe. It is a simple formula:

$$v = H_0 \times d$$

where $v$ is the recessional velocity, $d$ is the distance, and $H_0$ is the Hubble constant. It is the formula for the straight line displayed in Figure 15-29, and the Hubble constant is the slope of this line.

As Figure 15-29 shows, we usually measure a galaxy's velocity in kilometers per second and its distance in megaparsecs. Different techniques for estimating the distances to remote galaxies currently lead to differing estimates for $H_0$, ranging from 50 km/s/Mpc (read "50 kilometers per second per megaparsec") to 88 km/s/Mpc.

*Example:* A galaxy 1 Mpc from us is moving away at a rate of

$$v = (75 \text{ km/s/Mpc}) \times (1 \text{ Mpc}) = 75 \text{ km/s}$$

A galaxy 2 Mpc away is therefore receding at 150 km/s, and so on. A galaxy located 100 million parsecs from Earth rushes away from us with a speed of 7500 km/s! Uncertainty in the value of $H_0$ means a very great uncertainty in the distances of remote galaxies.

*Compare!* While a supercluster of galaxies 10 Mpc away recedes at 750 km/s, the solar system orbits the center of our Galaxy at a comparable 230 km/s.

*Self-Test:* At what speed is the galaxy observed to be 8 billion light-years away moving from us?

---

To determine the Hubble constant, astronomers must measure the redshifts and distances to many galaxies. Although redshift measurements can be quite precise, it is very difficult to measure the distances to remote galaxies accurately. Indeed, conflicting measurements of distance are the main reason that the value of $H_0$ is so controversial.

Recall from Build Your Foundation III that we can always find the distance to an object if we know both its apparent and absolute magnitudes. Astronomers use the term **standard candle** to denote any object whose absolute magnitude is known. Cepheid variables, the brightest supergiants, globular clusters, and supernovae are all useful as standard candles, because astronomers know their absolute magnitudes. To determine the distance to a galaxy, an astronomer must measure the apparent magnitude of one or more of these standard candles in that galaxy. As soon as both the apparent and absolute magnitudes are known, the distance to the galaxy can be easily calculated.

The distances to nearby galaxies can be determined by fairly reliable methods. For example, with the Hubble Space Telescope, Cepheid variable stars can now be seen out to 200 Mly from Earth. The distances to galaxies in this nearby volume of space can thus be determined from the period-luminosity law.

Beyond 200 Mly, even the brightest Cepheid variables, which have absolute magnitudes of about −6, are not visible with current technology. Astronomers then turn to more luminous stars, such as the brightest red and blue supergiants, which can be seen out to distances of 500 million and 800 million light-years, respectively. Thus, out to these limits, we can determine the distances to galaxies from the apparent magnitudes of these luminous supergiants.

When individual stars are no longer discernible, astronomers turn to the light from entire star clusters and gas clouds. The brightest globular clusters can be seen out to 1.3 billion light-years from Earth, and the brightest H II regions out to 3 billion light-years. From the apparent brightness of these clusters and nebulae, distances to remote galaxies have been estimated that suggest a value for the Hubble constant of 55 km/s/Mpc.

A key method to measure the distance to the most remote galaxies is to measure the brightness of supernova explosions within them. The brightest supernovae reach an absolute magnitude of −19 at the peak of their outbursts (Figure 15-30). Using this method, astronomers in 1998 measured a supernova nearly 8 billion light-years away.

In the 1970s, the astronomers Brent Tully and Richard Fisher developed another method for determining distances. They discovered that the width of the hydrogen 21-cm emission line of a spiral galaxy (discussed in Chapter 14) is related to the galaxy's absolute magnitude: *The broader the line, the brighter the galaxy.* This correlation is called the **Tully-Fisher relation**. By measuring the width of the 21-cm line, we can determine a galaxy's absolute magnitude. Combining this with observations of the galaxy's apparent magnitude allows us to calculate its distance from us, just as we did with spectroscopic parallax (see Section 10-6).

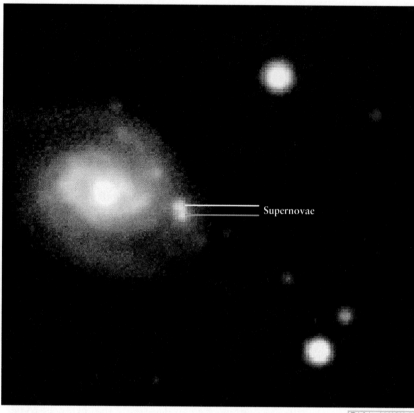

Supernovae

R I V U X G

**Figure 15-30** **Two Supernovae in NGC 664**
In 1997 the rare occurrence of two supernovae in the same galaxy was observed in the spiral galaxy NGC 664, located about 300 Mly (90 Mpc) from Earth. Supernovae observed in remote galaxies are important standard candles used by astronomers to determine the distances to these faraway objects. The two supernovae overlap each other as shown. The upper, yellow-orange supernova was observed to occur two months before the hotter, blue one, which was observed to occur less than two weeks before this image was made and had not yet achieved maximum brightness.

To understand the Tully-Fisher relation, we begin by noting that more massive galaxies rotate faster than less massive ones. Tully and Fisher then made the natural assumption that brighter galaxies are more massive than dimmer ones. The Tully-Fisher relation works like this: As a galaxy rotates, some of its stars move toward us and the light from these stars is blueshifted. Other stars are moving away and the light from the receding stars is redshifted. Taken together, the motions of the stars in a rotating galaxy spread out the 21-cm emission line. (The wider the emission line, then, the faster the galaxy rotates, and so the more massive it must be.)

Because line widths can be measured quite accurately, astronomers can use the Tully-Fisher relation to determine the luminosities (absolute magnitudes) of spiral galaxies and thus their distances. Measurements of 21-cm-line widths for distant, receding galaxies yields a Hubble constant of 88 km/s/Mpc. This is considered by most astronomers to be the upper limit to its possible value.

The major obstacle in pinning down the Hubble constant is that the farther we look into space, the fewer standard candles we have to work with. Consequently, it becomes less possible to check our estimates of distances. Unfortunately, remote galaxies are precisely the objects whose distances we need to determine the Hubble constant accurately.

By refining the standard candles through more observations with instruments like the Hipparcos satellite; the Hubble Space Telescope (HST); and large, new, ground-based instruments such as the Keck telescopes, astronomers hope to determine once and for all the true value for $H_0$. This advance will contribute significantly to our understanding of the structure and evolution of galaxies and of the universe.

## WHAT DID YOU THINK?

1  *Do all galaxies have spiral arms?* No. Galaxies may be either spiral, barred spiral, elliptical, or irregular. Only spirals and barred spirals have arms.

2  *Are most of the stars in a spiral galaxy located in its arms?* No. The spiral arms contain only 5% more stars than the regions between the arms.

3  *Are galaxies isolated objects?* No. Galaxies are grouped in clusters, and clusters are grouped in superclusters.

4  *Are all other galaxies moving away from the Milky Way?* All galaxies except those in our local supercluster are receding from us.

# Key Words

barred spiral galaxy, 365

cluster (of galaxies), 368

dwarf elliptical galaxy, 366

elliptical galaxy, 365

flocculent spiral galaxy, 362

galactic cannibalism, 373

galactic merger, 373

giant elliptical galaxy, 366

grand-design spiral galaxy, 362

Great Wall, 369

Hubble classification, 361

Hubble constant, 376

Hubble flow, 375

Hubble law, 376

intergalactic gas, 372

irregular cluster (of galaxies), 371

irregular galaxy, 367

lenticular galaxy, 367

Local Group, 369

poor cluster (of galaxies), 369

regular cluster (of galaxies), 371

rich cluster (of galaxies), 369

spiral density wave, 364

spiral galaxy, 361

standard candle, 377

starburst galaxy, 372

supercluster (of galaxies), 368

trailing-arm spiral, 365

Tully-Fisher relation, 377

void, 368

# Key Ideas

## Types of Galaxies

* The Hubble classification system groups galaxies into four major types: spiral, barred spiral, elliptical, and irregular.

* The arms of spiral and barred spiral galaxies are sites of active star formation.

* According to the theory of self-propagating star formation, spiral arms of flocculent galaxies are caused by the births and deaths of stars over extended regions of a galaxy. Differential rotation of a galaxy stretches the star-forming regions into elongated arches of stars and nebulae we see as spiral arms.

* According to the spiral density wave theory, spiral arms of grand-design galaxies are caused by density waves. The gravitational field of a spiral density wave compresses the interstellar clouds through which it passes, thereby triggering the formation of stars, including OB associations, which highlight the arms.

* Elliptical galaxies contain much less interstellar gas and dust than do spiral galaxies; little star formation is occurring in elliptical galaxies.

## Clusters and Superclusters

* Galaxies group into clusters rather than being randomly scattered through the universe.

* A rich cluster contains hundreds or even thousands of galaxies; a poor cluster may contain only a few dozen. A regular cluster has a nearly spherical shape with a central concentration of galaxies; in an irregular cluster, the distribution of galaxies is asymmetrical.

* Our Galaxy is a member of a poor, irregular cluster called the Local Group.

* Rich, regular clusters contain mostly elliptical and lenticular galaxies; irregular clusters contain more spiral and irregular galaxies. Giant elliptical galaxies are often found near the centers of rich clusters.

* The observable mass of a cluster of galaxies is not large enough to account for the observed motions of the galaxies; a large amount of unobserved mass must be present between the galaxies.

* Hot intergalactic gases emit X rays in rich clusters.

* When two galaxies collide, their stars initially pass each other but their interstellar gas and dust collide violently, either stripping the gas and dust from the galaxies or triggering prolific star formation. The gravitational effects of a galactic collision can cast stars out of their galaxies into intergalactic space.

* Galactic mergers occur; a large galaxy in a rich cluster may grow steadily through galactic cannibalism, perhaps producing a giant elliptical galaxy.

## Superclusters in Motion

* A simple linear relationship exists between the distance from the Earth to galaxies in other superclusters and the redshifts of those galaxies (a measure of the speed at which they are receding from us). This relationship is the Hubble law, recessional velocity $= H_0 \times$ distance, where $H_0$ is the Hubble constant.

* Astronomers use standard candles—Cepheid variables, the brightest supergiants, globular clusters, H II regions, supernovae in a galaxy—and the Tully-Fisher relation to estimate intergalactic distances. Because of difficulties in measuring the distances to remote galaxies, the value of the Hubble constant, $H_0$, is not known with certainty.

# Review Questions

1 What is the Hubble classification scheme? Which category includes the biggest galaxies? Into which category do the smallest galaxies fall? Which type of galaxy is the most common?

2 In which types of galaxies are new stars most likely forming? Describe the observational evidence that supports your answer.

3 What is the difference between a flocculent spiral galaxy and a grand-design spiral galaxy?

4 Briefly describe how the theory of self-propagating star formation accounts for the existence of spiral arms in some spiral galaxies.

5 Briefly describe how the spiral density wave theory accounts for the existence of spiral arms in some spiral galaxies.

6 How is it possible that galaxies in our Local Group still remain to be discovered? In what part of the sky are these galaxies located? What sorts of observations might reveal these galaxies?

7 Can any galaxies besides our own be seen with the naked eye? If so, which ones can you name?

8 What is the difference between a rich cluster of galaxies and a poor one? What is the difference between a regular cluster of galaxies and an irregular one?

9 How can a collision between galaxies produce a starburst galaxy?

10 Why do astronomers believe that considerable quantities of dark matter must exist in clusters of galaxies?

11 Explain why the dark matter in galaxy clusters cannot be neutral hydrogen.

12 What is the Hubble law?

13 Some galaxies in the Local Group exhibit blueshifted spectral lines. Why aren't these blueshifts violations of the Hubble law?

14 What is a standard candle? Why are standard candles important to astronomers trying to measure the Hubble constant?

15 What kinds of stars would you expect to find populating space between galaxies in a cluster?

## Advanced Questions

The answers to all computational problems, which are preceded by an asterisk (*), appear at the end of the book.

*16 Suppose a spectrum of a distant galaxy showed that its redshift corresponds to a speed of 22,000 km/s. How far away is the galaxy?

*17 A cluster of galaxies in the southern constellation of Pavo (the Peacock) is located 100 Mpc from Earth. How fast, on

average, are galaxies in this cluster receding from us? Why do different galaxies in the cluster show different velocities?

18 How might you determine what fraction of a galaxy's redshift is caused by the galaxy's orbital motion about the center of mass of its cluster?

## Discussion Questions

19 Discuss the advantages and disadvantages of using the various standard candles to determine extragalactic distances.

20 Discuss whether the various Hubble types of galaxies actually represent some sort of evolutionary sequence.

21 Discuss the sorts of phenomena that can occur when galaxies collide. Do you think that such collisions can change the Hubble type of a galaxy? Explain.

22 From what you know about stellar evolution, the interstellar medium, and the spiral density wave theory, explain the appearance and structure of the spiral arms of grand-design spiral galaxies.

## What If ...

23 The solar system was located in an active star-forming region in a spiral arm, rather than on the edge of the Orion arm? How would the solar system be different?

24 The Milky Way collided with the Andromeda Galaxy? What would we experience here on Earth? (This merger is actually expected to occur in the distant future.)

## Observing Projects

25 Using a telescope with an aperture of at least 30 cm (12 in.) in a dark location, observe as many of the spiral galaxies listed in the following table as you can. You can locate them and print out their locations on a star chart using your Starry Night

software. Many of these galaxies are members of the Virgo cluster, which can best be seen from the northern hemisphere during the spring. Because all galaxies are quite faint, be sure to schedule your observations for a moonless night. The best view is obtained when a galaxy is near the meridian. While at the eyepiece, make a sketch of what you see. Can you distinguish any spiral structure? After completing your observations, compare your sketches with photographs found in such popular books as *Galaxies* by Timothy Ferris (Sierra Club Books, 1980) and *The Color Atlas of the Galaxies* by James Wray (Cambridge University Press, 1988).

| Spiral galaxy | R.A. | Decl. | Hubble type |
|---|---|---|---|
| M31 (NGC 224) | 0$^h$ 42.7$^m$ | +41° 16′ | Sb |
| M58 (NGC 4579) | 12  37.7 | +11  49 | Sb |
| M61 (NGC 4303) | 12  21.9 | + 4  28 | Sc |
| M63 (NGC 5055) | 13  15.8 | +42  02 | Sb |
| M64 (NGC 4826) | 12  56.7 | +21  41 | Sb |
| M74 (NGC 628) | 1  36.7 | +15  47 | Sc |
| M83 (NGC 5236) | 13  37.0 | –29  52 | Sc |
| M88 (NGC 4501) | 12  32.0 | +14  25 | Sb |
| M90 (NGC 4569) | 12  36.8 | +13  10 | Sb |
| M94 (NGC 4736) | 12  50.9 | +41  07 | Sb |
| M98 (NGC 4192) | 12  13.8 | +14  54 | Sb |
| M99 (NGC 4254) | 12  18.8 | +14  25 | Sc |
| M100 (NGC 4321) | 12  22.9 | +15  49 | Sc |
| M101 (NGC 5457) | 14  03.2 | +54  21 | Sc |
| M104 (NGC 4594) | 12  40.0 | –11  37 | Sa |
| M108 (NGC 3556) | 11  11.5 | +55  40 | Sc |

**26** Using a telescope with an aperture of at least 30 cm (12 in.) in a dark location, observe as many of the elliptical galaxies listed in the following table as you can. You can locate them and print out their locations on a star chart using your Starry Night software. Six of these galaxies are in the Virgo cluster, which is located in the northern hemisphere's evening sky from March through June. The accompanying photograph, covering a 2° × 5° swath of the cluster, may help you find some of these galaxies. As in the previous Observing Project, be sure to schedule your

observations for a moonless night, when the galaxies you wish to observe will be near the meridian. Do these elliptical galaxies differ in appearance from spiral galaxies?

| Elliptical galaxy | R.A. | Decl. | Hubble type |
|---|---|---|---|
| M49 (NGC 4472) | 12$^h$ 29.8$^m$ | + 8° 00′ | E4 |
| M59 (NGC 4621) | 12  42.0 | +11  39 | E3 |
| M60 (NGC 4649) | 12  43.7 | +11  33 | E1 |
| M84 (NGC 4374) | 12  25.1 | +12  53 | E1 |
| M86 (NGC 4406) | 12  26.2 | +12  57 | E3 |
| M89 (NGC 4552) | 12  35.7 | +12  33 | E0 |
| M110 (NGC 205) | 00  40.4 | +41  41 | E6 |

**27** Using a telescope with an aperture of at least 30 cm (12 in.), observe as many of the interacting galaxies listed in the table below as you can. You can locate them and print out their locations on a star chart using your Starry Night software. As in the previous Observing Projects, be sure to schedule your observations for a moonless night when the galaxies you wish to observe will be near the meridian. While at the eyepiece, make a sketch of what you see. Can you distinguish hints of interplay among the galaxies? After completing your observations, compare your sketches with photographs found in popular books, such as *Galaxies* by Timothy Ferris (Sierra Club Books, 1980).

| Interacting galaxies | R.A. | Decl. |
|---|---|---|
| M51 (NGC 5194) | 13$^h$ 29.9$^m$ | +47° 12′ |
| NGC 5195 | 13  30.0 | +47  16 |
| M65 (NGC 3623) | 11  18.9 | +13  05 |
| M66 (NGC 3627) | 11  20.2 | +12  59 |
| M81 (NGC 3031) | 9  55.6 | +69  04 |
| M82 (NGC 3034) | 9  55.8 | +69  41 |
| M95 (NGC 3351) | 10  44.0 | +11  42 |
| M96 (NGC 3368) | 10  46.8 | +11  49 |
| M105 (NGC 3379) | 10  47.8 | +12  35 |

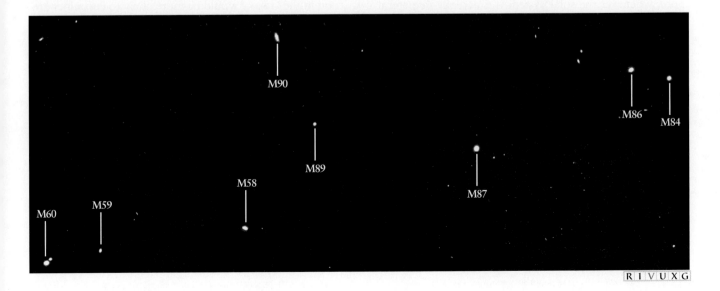

R I V U X G

# The Solar System Were Located Closer to the Center of the Galaxy?

It is a clear, moonless night. Stars by the thousands twinkle serenely against the ebony darkness of space. Overhead the soft white span of the Milky Way catches your attention, and you try to see individual stars in its glowing haze.

Most of the 6000 stars visible throughout the year are within 300 ly of Earth. The rest of the Galaxy's 200 billion or so stars are too dim or too obscured by interstellar gas and dust to be easily observed. What if the solar system were one-third of its present distance from the center of our Galaxy? At that distance we would still be in the realm of the spiral arms, extremely close to the nuclear bulge, and none of the stars we see now would be visible.

**Population Explosion** Out in the galactic suburbs, where we live today, there is about one star per 300 cubic light-years. The displaced Earth would be surrounded by nearly 5 times as many stars, or one per 60 cubic light-years. And the solar system would pass much more frequently through the dust-rich spiral arms of the Galaxy. Scattering starlight, this interstellar matter would glow as diaphanous wisps throughout our nighttime sky. Furthermore, whenever the solar system was actually in a spiral arm, several nearby stars would be of the high-mass, high-luminosity variety. The combined light from these stars and the shimmering clouds would be so great that for millions of years at a time night would never fall.

**Evolution Interruption** Recall from Chapter 12 that high-mass stars evolve much more rapidly than average-mass stars like the Sun. If we were closer to the center of the Milky Way and frequently passing through the spiral arms of the Galaxy, massive stars would explode near us much more frequently than they do today. Those nearby supernovae, occurring within 50 ly of Earth, would deposit lethal radiation, damage life, and cause mass extinctions. As a result, the direction of evolution would change more frequently.

At Earth's current location, the closest star, Proxima Centauri, is more than 3 ly away. Would Earth be in danger of colliding with a star if the solar system was closer to the center of the Galaxy? Several stars would certainly be much closer than Proxima Centauri is now. However, stars are so small compared to the vastness of a galaxy that the likelihood of a collision would still approach nil.

Much more likely would be the passage of a star so close to the Sun that the Earth's orbit would be disturbed. If the Earth's orbit became more elliptical, say, the change of seasons would be noticeably affected. With the seasonal effect of the Earth's tilt compounded by greater changes in distance between the Earth and Sun, one hemisphere of the Earth would suffer much more extreme temperatures than it does today, while the other would see less variation. This would affect the evolution of life, of course, and the distribution of life-forms on the planet.

**Close Encounters** Earth-evolved life at our new location in the Galaxy would be more likely to encounter any sentient beings on other planets nearby. Today, after nearly a century of broadcasting radio and television signals, a nearly 200-ly-diameter sphere of space centered on the Earth is filled with such signals. There are about 6300 stars in that sphere. At our new location there would be 31,500 stars in the same volume and the probability of many more stars with life-supporting planets orbiting them. Perhaps one of the reasons that life on Earth has been able to evolve so long unaware of other intelligent life is because the solar system exists near the fringe of the Galaxy.

1776 BATSE Gamma-Ray Bursts

+90

+180

−180

−90

R I V U X G

**The Most Powerful Known Bursts** Gamma-ray bursters have been observed everywhere in the sky, indicating that, unlike X-ray bursters, they do not originate in the disk of the Milky Way Galaxy. This map of the entire sky "unfolded" onto the page shows 1776 bursts detected by the Compton Gamma-Ray Observatory. The colors indicate the wavelengths from shortest to longest; they are coded red, orange, yellow, yellow-green, green, light blue, dark blue, violet.

# Quasars and Active Galaxies

## *In this chapter you will discover*

✴ distant, luminous quasars

✴ the unusual spectra and small volumes of quasars

✴ other bright objects, called active galaxies

✴ supermassive black holes that probably serve as central engines for quasars and active galaxies

✴ how observations from the central regions of these objects have helped astronomers devise theories about their power

✴ gamma-ray bursters

## WHAT DO YOU THINK?

1  What do quasars look like?

2  What powers a quasar?

The Sun releases far more energy than anything on Earth, and its brightness is reassuring. The energy emitted into space by O- and B-type stars is truly spectacular, and the momentary energy output of supernovae boggles the mind. But all of these are as nothing compared to the power of quasars, active galaxies, and gamma-ray bursters.

Quasars emit more energy each *second* than the Sun does in 200 years. And they continue to do so for millions, even billions, of years. Something remarkable must exist deep within to make them so luminous. Our knowledge of these objects began in an amateur astronomer's backyard.

## QUASARS

Grote Reber built the first radio telescope in 1936 in his backyard in Illinois, opening the realm of nonvisual astronomy. By 1944, Reber had detected strong radio emissions from sources in the constellations of Sagittarius, Cassiopeia, and Cygnus. Two of these sources, Sagittarius A (Sgr A) and Cassiopeia A (Cas A), happen to be in our Galaxy. The first is the galactic nucleus (see Chapter 14) and the second is a supernova remnant (see Chapter 12). However, Reber's third source, called Cygnus A (Cyg A), proved hard to categorize (Figure 16-1). The mystery only deepened in 1951, when Walter Baade and Rudolph Minkowski, using the 200-in. optical telescope on Mount Palomar, discovered a strange-looking galaxy at the same position (Figure 16-1 inset).

The galaxy associated with Cyg A is very dim. Nevertheless, Baade and Minkowski managed to photograph its spectrum. They detected a redshift corresponding to a speed of 14,000 km/s. According to the Hubble law, this speed indicates that Cyg A lies 690 million light-years (211 Mpc) from Earth!

Because it is one of the brightest radio sources in the sky, the enormous distance to Cyg A astounded astronomers. Although barely visible through the giant optical

R I V U X G

**Figure 16-1**  **Cygnus A (3C 405)**  This radio image was produced from observations made at the Very Large Array. Most of the radio emission from Cygnus A comes from the radio lobes located on either side of the peculiar galaxy seen in the inset. The two radio lobes each extend about 160,000 ly from the optical galaxy, and each contains a brilliant, condensed region of radio emission. **Inset:** At the heart of this system of gas lies a strange-looking galaxy, which has a redshift corresponding to a recessional speed of 6% of the speed of light. According to the Hubble law, this corresponds to a distance of about 690 million light-years from Earth. Because Cygnus A is one of the brightest radio sources in the sky, this remote galaxy's energy output must be enormous.

R I V U X G

telescope at Palomar, Cyg A's radio waves can be picked up by amateur astronomers with backyard equipment. Its energy output must therefore be colossal. In fact, Cyg A shines with a radio luminosity $10^7$ times as bright as that of an ordinary galaxy, such as M31 in Andromeda, yet is far more distant. The object creating the Cyg A radio emissions has to be something extraordinary.

## 16-1  Quasars look like stars but have huge redshifts

Cygnus A is not the only powerful radio source in the far-distant sky. Starting in the 1950s, radio astronomers busily made long lists of radio sources. One of the most famous lists, the *Third Cambridge Catalogue*, was published in 1959. (The first two catalogues produced by this British team were filled with inaccuracies.) Even today, astronomers often refer to its 471 radio sources by their "3C numbers." Cyg A, for example, is designated 3C 405, because it is the 405th source on the Cambridge list. Because of its extraordinary luminosity, astronomers were eager to learn whether any other sources in the 3C catalog had similar properties.

One interesting case was 3C 48. In 1960, Allan Sandage used the Palomar telescope to discover a "star" at the location of this radio source (Figure 16-2). Recall that stars

**Figure 16-3**   **The Quasar 3C 273**  This greatly enlarged radio view shows the starlike object associated with the radio source 3C 273. Note the luminous jet on one side. By 1963 astronomers determined that the redshift of this quasar is so great that, according to the Hubble law, it is nearly 2 billion light-years from Earth.

R I V U X G

are blackbodies whose peak intensity is typically visible light and whose radio emission is much less intense. Because ordinary stars are not strong sources of radio emission, 3C 48 had to be something unusual. Indeed, its spectrum showed a series of emission lines that, initially, no one could identify. Although 3C 48 was clearly an oddball, many astronomers thought it was just another strange star in our Galaxy.

Another such "star," called 3C 273, was discovered in 1962. Like 3C 48, this object and the luminous "jet" of bright gas found protruding from one side of it emit a series of bright spectral lines that no one could then identify (Figure 16-3). These spectral lines are brighter than the background radiation at other wavelengths, called the *continuum.*

A breakthrough finally came in 1963, when Maarten Schmidt at the California Institute of Technology found that four of the brightest spectral lines of 3C 273 are positioned relative to one another, just as are the four familiar spectral lines of hydrogen. However, these emission lines from 3C 273 are found at much longer wavelengths than the usual wavelengths of the hydrogen lines. The strong emission lines mean that something unusual is heating the gas. Schmidt reasoned that they *are* hydrogen lines, but subjected to a substantial redshift.

Spectra for stars in our Galaxy exhibit comparatively small Doppler shifts, because these stars cannot move

R I V U X G

**Figure 16-2**   **The Quasar 3C 48**  For several years astronomers erroneously believed that this object (arrow) is simply a peculiar, nearby star that happens to emit radio waves. Actually, the redshift of this starlike object is so great that, according to the Hubble law, it must be roughly 4 billion light-years away.

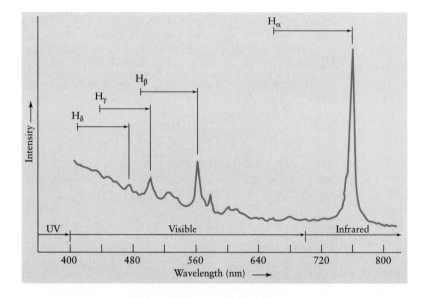

**Figure 16-4** **The Spectrum of 3C 273** Four bright emission lines caused by hydrogen dominate the spectrum of 3C 273. The arrows indicate how far these spectral lines are redshifted from their usual wavelengths.

extremely fast relative to the Sun without soon escaping from the Galaxy. Schmidt thus concluded that 3C 273 is not a nearby star after all. Pursuing this conclusion, he promptly found that its redshift corresponds to a speed of almost 15% the speed of light. According to the Hubble law, this huge redshift implies the incredible distance to 3C 273 of roughly 2 billion light-years (Bly).

Figure 16-4 shows the spectrum of 3C 273. Remember from Chapter 4 that a spectrograph records energy intensity at different wavelengths. The emission lines appear as peaks, and absorption lines appear as valleys. The emission lines are caused by excited gas atoms emitting radiation at specific wavelengths.

Inspired by Schmidt's success, astronomers looked again at the spectral lines of 3C 48. The redshift corresponds to a velocity of nearly one-third the speed of light. Therefore, 3C 48 must be nearly twice as far away as

3C 273, or about 4 Bly from Earth, assuming that the Hubble constant, $H_0$, is 75 km/s/Mpc. As we found in the previous chapter, the value of the Hubble constant is not known with certainty. If it is higher, these sources are nearer to us than the distances given in this chapter; if it is lower, they are farther away.

Because of their starlike appearances and strong emissions, 3C 48 and 3C 273 were dubbed **quasi-stellar radio sources,** a term soon shortened to **quasars.** Quasars need not specifically be *radio* sources; most, in fact, are not. But the name *quasar* has stuck, and thousands have been discovered since the pioneering days of the early 1960s. Quasars, also called **quasi-stellar objects (QSOs),** look like stars but have incredibly higher energy outputs.

Virtually all quasars have enormous redshifts, in some cases corresponding to speeds greater than 90% of the speed of light. Figure 16-5 shows the spectrum of a quasar

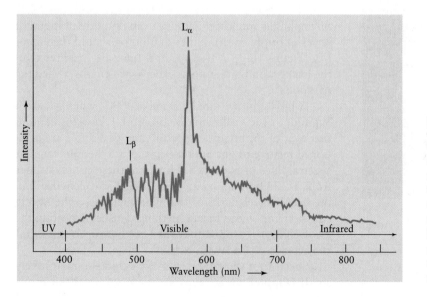

**Figure 16-5** **The Spectrum of a High-Redshift Quasar** The light from this quasar, known as PKS 2000-330, is so highly redshifted that spectral lines normally in the far-ultraviolet ($L_\alpha$ and $L_\beta$) can be seen at visible wavelengths. Note the many deep absorption lines on the short-wavelength side of $L_\alpha$. These lines, collectively called the "Lyman-alpha forest," are believed to be created by remote clouds of gas along our line of sight to the quasar. Hydrogen in these clouds absorbs photons from the quasar at wavelengths less redshifted than the quasar's $L_\alpha$ line.

whose redshift corresponds to 92% of the speed of light. From the Hubble law, it follows that the distances to these high-redshift quasars are typically in the range of 10 billion to 13 billion light-years—their light has taken that long to reach Earth. *When we look at these quasars, we are seeing objects as they existed when the universe was very young.*

## 16-2 A quasar emits a huge amount of energy from a small volume

Galaxies are big and bright. A typical large galaxy, like our own Milky Way, contains hundreds of billions of stars and shines with the luminosity of 10 billion Suns. The most gigantic and most luminous galaxies, such as the giant ellipticals, are only 10 times brighter. But beyond 8 billion light-years from Earth, even the brightest galaxies are too faint to be easily detected. Most ordinary galaxies are too dim to be detected at half that distance. If quasars can be seen up to 13 billion light-years away, they must be far more luminous than galaxies. Indeed, a typical quasar is 100 times brighter than our Milky Way. Stranger still, quasars fluctuate in brightness.

In the mid-1960s, several astronomers discovered evidence of these fluctuations: Some newly identified quasars had actually been photographed in the past. One photograph of 3C 273, for example, dates back to 1887. By carefully examining these old images, astronomers could see that *quasars occasionally flare up.* As Figure 16-6 shows for another quasar, 3C 279, energy from prominent outbursts reached Earth around 1937 and 1943. During these outbursts, the luminosity of 3C 279 increased by a factor of at least 25. Because of the enormous distance to this quasar, during those peak periods it must have been shining at least 10,000 times brighter than the entire Milky Way.

Because quasars fluctuate in brightness, astronomers have been able to place strict upper limits on their size. As we saw in Chapter 13, an object cannot vary in brightness faster than the time it takes light to travel across that ob-

ject. For example, an object that is 1 ly in diameter cannot vary in brightness with a period of less than 1 year.

Many quasars vary in brightness over only a few months, weeks, or days. In fact, X-ray observations reveal large variations in as little as 3 hours! This rapid flickering means that the source of the quasar's energy must be quite small by galactic standards. The energy-emitting region of a typical quasar—the "powerhouse" that blazes with the luminosity of 100 galaxies—is less than 1 light-day in diameter. Something must be producing the luminosity of 100 galaxies from a volume with approximately the same diameter as our solar system.

## 16-3 Active galaxies bridge the energy gap between ordinary galaxies and quasars

Some astronomers in the 1960s could not accept that quasars have such huge energy output. They argued that quasars are closer and therefore emit much less energy. Something else, they insisted, must be causing their massive redshifts. No such mechanism has been found, and most quasars are still believed to be at great distances from us. In recent years, however, astronomers have discovered objects called **active galaxies,** with luminosities between those of ordinary galaxies and remote quasars.

Active galaxies are exotic. Some have unusually bright, starlike nuclei; others have strong emission lines in their spectra; still others are highly variable. Some have jets and beams of radiation emanating from their cores like the emissions observed for 3C 273, and most of these objects are more luminous than ordinary galaxies. Some, called **peculiar galaxies** (denoted "pec"), appear to be blowing themselves apart. Any of the Hubble classes of galaxies (see Chapter 15) can be peculiar.

Carl Seyfert at the Mount Wilson Observatory discovered the first active galaxies in 1943 while surveying spiral galaxies. Now called **Seyfert galaxies,** these spirals reveal

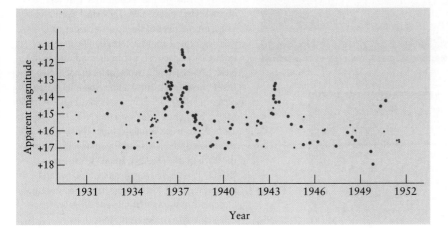

**Figure 16-6** **The Brightness of 3C 279** This graph shows variations in the brightness of the quasar 3C 279. Note the large outburst observed in 1937. The data were obtained by carefully examining old photographic plates in the files of the Harvard College Observatory.

**Figure 16-7** **The Seyfert Galaxy NGC 1566** This Sc galaxy is a Seyfert galaxy some 50 Mly (16 Mpc) from Earth in the southern constellation Dorado (the Goldfish). The nucleus of this galaxy is a strong source of radiation whose spectrum shows emission lines of highly ionized atoms.

exceptionally bright, starlike nuclei and strong emission lines in their spectra. For example, the rich spectrum of NGC 4151 has many prominent emission lines. Some of these lines are produced by iron atoms with a dozen or more electrons stripped away, indicating that NGC 4151 contains some extremely hot gas. Seyfert galaxies also vary

in brightness. For example, at times the magnitude of NGC 4151 changes over the course of a few days.

Another example of a Seyfert galaxy is NGC 1566, shown in Figure 16-7. At infrared wavelengths this galaxy shines with the brilliance of $10^{11}$ Suns. This extraordinary luminosity has been observed to vary by as much as $7 \times 10^9 \ L_\odot$ over only a few weeks. In other words, the infrared power output of the nucleus of NGC 1566 rises and falls by an amount nearly equal to the total luminosity of our entire Galaxy.

Many more Seyfert galaxies have been discovered in recent years. Approximately 10% of the most luminous galaxies in the sky are Seyfert galaxies. Some of the brightest Seyfert galaxies shine as brightly as faint quasars, which leads most astronomers to suspect that the nuclei of Seyfert galaxies are, in fact, low-luminosity quasars.

Some Seyfert galaxies show vestiges of violent explosions in their nuclei. Filaments of gas tens of thousands of light-years long protrude from the nucleus of NGC 1275 in all directions (Figure 16-8). Spectroscopic studies indicate that this gas is being blasted away from the galaxy's nucleus at 3000 km/s. In 1977, Vera Rubin and her colleagues reported observations demonstrating that NGC 1275 actually consists of two colliding galaxies, a spiral and an elliptical. As we saw in Chapter 15, a collision or close encounter between two galaxies can eject matter into intergalactic space. More and more of this intergalactic medium is being discovered.

**Figure 16-8** **The Active Galaxy NGC 1275 (3C 84)** This ground-based image (**far left**) is of the Seyfert galaxy NGC 1275, the largest and brightest member of the Perseus cluster of galaxies located about 250 million light-years from Earth. Spectroscopic studies indicate that NGC 1275 is actually two galaxies in collision. **Inset (near left):** This Hubble Space Telescope image at visible wavelengths shows about 50 small, bright, blue objects surrounding NGC 1275. They are believed to be young globular clusters formed as a result of the collision. The small white spot is the nucleus of the galaxy. **Inset (bottom):** NGC 1275 is also a strong source of X rays and radio waves. This X-ray image from the Einstein Observatory shows that most of the galaxy's X-ray emission comes from its nucleus.

Other active galaxies are the **BL Lacertae objects,** or **blazars.** The name BL Lacertae objects comes from their prototype, BL Lacertae (BL Lac) in the constellation of Lacerta (the Lizard). BL Lac (Figure 16-9) was discovered in 1929, when it was mistaken for a variable star, largely because its brightness varies by a factor of 15 in only a few months. BL Lac's most intriguing characteristic is a *totally featureless spectrum,* exhibiting neither absorption nor emission lines. A BL Lac object is thought to be an elliptical galaxy with a bright starlike center, much as a Seyfert galaxy is a spiral galaxy with a quasarlike center.

## 16-4 Active galaxies lie at the center of double radio sources

The high-speed ejection of matter from active galaxies is seen best at nonvisible wavelengths. For example, Figure 16-10 is a combined optical and radio photograph of the galaxy NGC 5128 in the southern constellation of Centaurus. An unusually broad dust lane stretches across the galaxy. NGC 5128 is classified as an "E0 (pec)." This galaxy, also called Centaurus A, is one of the brightest sources of radio waves in the sky. In part, the brightness of Centaurus A at radio wavelengths comes from its proximity to Earth, only 13 million light-years away, and it was one of the first radio sources discovered when radio telescopes were erected in Australia. As shown superimposed

**Figure 16-9** **BL Lacertae** This photograph shows fuzz around BL Lacertae. The redshift of this fuzz indicates that BL Lacertae is about 900 Mly (280 Mpc) from the Earth. BL Lac objects appear to be giant elliptical galaxies with bright starlike nuclei, much as Seyfert galaxies are spiral galaxies with quasarlike nuclei. BL Lac objects contain much less gas and dust than Seyfert galaxies.

on the visible image in Figure 16-10, radio waves from NGC 5128 pour from two regions, called **radio lobes,** on either side of the galaxy's dust lane. Detailed observations of Centaurus A reveal both radio and X-ray (Figure 16-10 inset) jets originating in the galaxy's nucleus. Particles and energy stream out of the galaxy's nucleus toward the radio

**Figure 16-10** **The Peculiar Galaxy NGC 5128 (Centaurus A)** This extraordinary galaxy is located in the constellation of Centaurus, roughly 13 million light-years from Earth. At visible wavelengths a dust lane crosses the face of the galaxy. Superimposed on this visible image, a false-color radio image shows that vast quantities of radio radiation pour from matter ejected from the galaxy perpendicular to the dust lane. **Inset:** This X-ray image from the Einstein Observatory shows that NGC 5128 has a bright X-ray nucleus. An X-ray jet protrudes from the nucleus along a direction perpendicular to the galaxy's dust lane.

lobes at nearly the speed of light. Such galaxies, with two radio lobes, are now called **double radio sources**.

By 1970 radio astronomers had discovered dozens of other double radio sources. An active galaxy, usually resembling a giant elliptical, is often found between the two radio lobes. The visible galaxy associated with Cyg A is also located between two radio lobes (see Figure 16-1). At radio wavelengths, double radio sources are among the brightest objects in the universe.

All double radio sources seem to have a central "engine" that ejects charged particles and magnetic fields outward along two oppositely directed jets. After traveling many thousands or even millions of light-years, this material slows down, allowing ejected electrons and magnetic fields to produce the radio radiation that we detect. This type of radio emission, called *synchrotron radiation* (see Section 14-5), occurs whenever energetic electrons move in a spiral within a magnetic field. The radio waves that come from the lobes of a double radio source have all the characteristics of synchrotron radiation.

The idea that a double radio source emits jets of particles is supported by the existence of **head-tail sources**, so named because each such source appears to have a region of concentrated radio emission (the head) with two tails of gas streaming from it in opposite directions and then sweeping back. NGC 1265, an active elliptical galaxy in the Perseus cluster of galaxies, shows these properties. NGC 1265 is known to be moving at a high speed (2500 km/s) relative to the Perseus cluster as a whole. In a radio map (Figure 16-11), its radio emission has a distinctly windswept appearance.

**Figure 16-11**    R I V U X G

**The Head-Tail Source NGC 1265** This active elliptical galaxy is moving at a high speed through the intergalactic medium. Because of this motion, the two tail jets trail the galaxy at its head, giving this radio source a distinctly windswept appearance.

Particles ejected in the two jets from NGC 1265 are deflected by the galaxy's passage through the sparse intergalactic medium. The head-tail source leaves behind a trail of particles, like the trail of smoke pouring from a rapidly moving steam train. But what is an active galaxy's locomotive, its "central engine"? What gives rise to its enormous jets?

## SUPERMASSIVE CENTRAL ENGINES

As long ago as 1968, the British astronomer Donald Lynden-Bell suggested how quasars and active galaxies could produce such enormous amounts of energy from such small volumes. Bell argued that the gravitational field of a supermassive black hole (see Chapter 13) could be the central engine.

### 16-5  Supermassive black holes apparently lurk at the centers of most galaxies

As we saw in Chapter 13, confirming the existence of black holes is difficult. At best, we can only rule out sources other than black holes to explain the data. Traces of a supermassive black hole's gravitational attraction have been noted in a growing number of galaxies, including M31, M32, M87, M104, and the Milky Way. Black holes may well lie at the hearts of most other galaxies.

The Andromeda Galaxy (M31) is the largest, most massive galaxy in the Local Group (see the figure that opens Chapter 15). In the mid-1980s, several astronomers made careful spectroscopic observations of the core of M31. Using measurements of Doppler shifts, they determined that stars within 50 ly of the galactic core are orbiting the nucleus at exceptionally high speeds, which suggests that a massive object is located at the galaxy's center. Without the gravity of such an object to keep the stars in their high-speed orbits, they would have escaped from the core region long ago. From such observations, astronomers estimate the mass of the central object in M31 to be about 50 million solar masses. That much matter confined to such a small volume strongly suggests the existence of a supermassive black hole.

Located near M31 is a small elliptical galaxy called M32 (Figure 16-12). High-resolution spectroscopy indicates that stars close to the center of M32 are also orbiting this galaxy's nucleus at unusually high speeds, which could be explained by the presence of a supermassive black hole there. Furthermore, a picture taken by the Hubble Space Telescope (Figure 16-12 inset) shows that the concentration of stars at the core of M32 is truly remarkable. The density of stars there is more than 100 million times great-

**Figure 16-12** **The Elliptical Galaxy M32** This small galaxy is a satellite of M31, a portion of which is seen at the left of this wide-angle photograph. Both galaxies are roughly 2.2 million light-years from Earth. **Inset:** This high-resolution image from the Hubble Space Telescope shows the center of M32. Note the concentration of stars at the nucleus of the galaxy. The area covered in this view is only 175 ly across.

er than the density of stars in the Sun's neighborhood. The concentration of stars and their high speeds strongly support the belief that a supermassive black hole exists at the center of M32.

Astronomers have uncovered evidence of supermassive black holes in other, more remote galaxies (Figure 16-13). John Kormendy used a 3.6-m telescope on Mauna Kea to examine the core of M104 spectroscopically. Once again, high-speed gas orbiting the galaxy's nucleus was found. These observations suggest that the center of this galaxy is dominated by a supermassive black hole containing a billion solar masses. Similar spectroscopic observations of the edge-on S0 galaxy NGC 3151 also reveal evidence for a billion-solar-mass black hole at its core.

Recent observations with the Hubble Space Telescope show distinct evidence for a supermassive black hole at the center of the giant elliptical galaxy M87. Located some 50 million light-years from Earth, M87 is an active galaxy that has long been recognized as unusual. In 1918, Heber Curtis at the Lick Observatory reported that the center of M87 has "a curious straight ray ... apparently connected with the nucleus by a thin line of matter." Figure 16-14a shows M87 and several neighboring galaxies; M87's nucleus and jet are buried in that galaxy's glare. Figure 16-14b is a radio image showing the gas emission from M87, which is a powerful source of radio waves and X rays.

The Hubble Space Telescope image of M87 in Figure 16-14c shows an exceptionally bright, starlike nucleus and

**Figure 16-13** **The Sombrero Galaxy (M104)** This spiral galaxy in Virgo is nearly edge-on to our Earth-based view. Spectroscopic observations indicate that a billion-solar-mass black hole is located at the galaxy's center.

a    R I V U X G    b    R I V U X G    c    R I V U X G

**Figure 16-14**   **The Giant Elliptical Galaxy M87**   M87 is located near the center of the
sprawling, rich Virgo cluster, which is about 50 million light-years from Earth. **(a)** This long exposure
shows the extent of M87 and some smaller, neighboring galaxies. The numerous fuzzy spots that
surround M87 are globular clusters; each contains about a million stars. **(b)** This radio image shows
the intergalactic gas emitted by M87. These gas clouds dwarf the giant galaxy, the bright spot near the
center of the picture. **(c)** This Hubble Space Telescope image of M87 shows the gas disk in its nucleus.
M87's extraordinarily bright nucleus and the gas jets result from a 3-billion-solar-mass black hole,
whose gravity causes huge amounts of gas and an enormous number of stars to crowd around it.

a surrounding disk of gas with trailing spiral arms. To produce this fiery glow, stars must be packed so tightly at the
center of M87 that their density is at least 300 times
greater than that normally found at the centers of giant
ellipticals. The motions of the stars in this central cluster-
ing support the hypothesis that a black hole with a mass of
nearly 3 billion Suns resides at the center of M87.

## 16-6  Jets of matter ejected from around a black hole may explain quasars and active galaxies

If supermassive black holes lie at the centers of galaxies,
the gravitational energy associated with them could well
give rise to quasars and active galactic nuclei. Here is one
scenario many astronomers find convincing: Consider a
black hole at the center of a galaxy. Because the centers of
galaxies are congested places, any black hole there must
capture a massive accretion disk of gas and dust. According to Kepler's third law, the inner regions of such a disk
orbit the most rapidly. The inner material therefore constantly rubs against the more slowly moving gases in the
outer regions, heating them up. As energized gases spiral
toward the black hole, they release their energy violently,
giving the hole its brilliant luminosity.

The crowding of infalling matter would also explain
the jets of gas we observe streaming out from active galaxies. In our scenario, some of the infalling material is
swallowed up by the black hole, but not all. Rather, the

pressures in the accretion disk become so great that before much of the gas crosses the event horizon, it becomes
so compressed that its pressure causes the gas to eject violently, perpendicular to the accretion disk where the hot
gas experiences the least resistance. The result is two oppositely directed beams, depicted in Figure 16-15.

In 1995 the Hubble Space Telescope took a picture of
the giant elliptical galaxy NGC 4261 that some astronomers interpret as showing an accretion disk about 800 ly
in diameter orbiting a supermassive black hole. Indeed, the
speed of the gas and dust in the disk indicates that it is held
in orbit by a 1.2 billion-solar-mass object. The inset in
Figure 16-16 is a Hubble Space Telescope image of the
galaxy's nucleus. The dark, horizontal oval in the picture is
our oblique view of the accretion disk, because ground-
based radio and optical observations show a double-lobed
structure (Figure 16-16) that is bisected by the oval.

What confines the ejected matter to narrow jets?
Theory proposes that the ejecta is initially forced into narrow jets by magnetic fields. The gas still falling toward the
black hole then prevents escaping matter from spreading.
To see why, consider water squirting out of the nozzle of an
ordinary garden hose. The stream of water broadens and
the spray fans out through a wide angle in the air. If the
nozzle is placed in a swimming pool, however, the stream
of water does not fan out as much. Similarly, as the two
jets of hot gas leave the vicinity of the black hole, they must
blast their way through the gas that is still crowding inward. Passage through this material causes the jets to become extremely narrow, concentrated beams. This "rocket

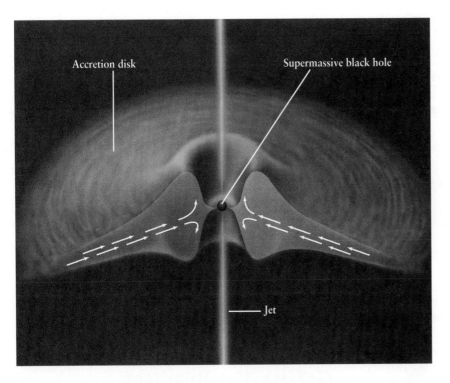

**Figure 16-15** **A Supermassive Black Hole as the Central Engine** The energy output of an active galaxy or a quasar may involve an extremely massive black hole that captures matter from its surroundings. In the scenario depicted here, the inflow of material through an accretion disk is redirected to produce two powerful jets of particles traveling at nearly the speed of light.

nozzle" may explain not only double radio sources but also the jets and beams we see protruding from active galaxies and some quasars.

The main difference between double radio sources, quasars, and BL Lac objects is apparently just the angle at which the central engine is viewed. As Figure 16-17 shows, an observer sees a double radio source when the accretion disk is viewed nearly edge-on, because the jets are nearly in the plane of the sky. At a steeper angle, the observer sees a quasar. If one of the jets is aimed almost directly at Earth, a BL Lac object is seen.

**Insight into science  Occam's razor revisited**
It is possible to create *separate* theories for quasars, BL Lac objects, and radio galaxies. Scientists prefer a comprehensive single theory that explains all three, because it incorporates common properties where only unconnected phenomena existed before. This search for a simple, more powerful explanation is the central engine of progress in science.

R I V U X G

**Figure 16-16** **The Core of an Active Galaxy** The giant elliptical galaxy NGC 4261 is a double radio source located in the Virgo cluster, about 100 million light-years from Earth. **(Far left)** An optical photograph of the galaxy (white) is combined with a radio image (orange and yellow) to show both the visible galaxy, which does not emit much radio energy, and its jets, which do. **Inset:** This Hubble Space Telescope image of the nucleus of NGC 4261 shows a disk of gas and dust about 800 ly (250 pc) in diameter.

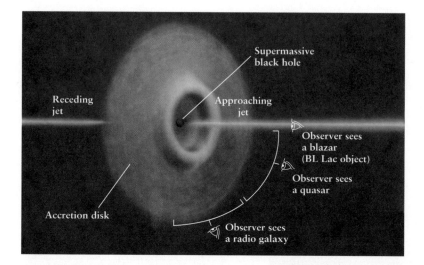

Figure 16-17 **The Orientation of the Central Engine and Its Jets** Double radio sources, quasars, and BL Lacertae objects may be the same type of object viewed from different directions. If one of the jets is aimed almost directly at the Earth, we see a BL Lac object. If the jet is somewhat tilted to our line of sight, we see a quasar. If the jets are nearly perpendicular to our line of sight, we see a double radio source.

The light emitted by distant quasars is subject to Einstein's general theory of relativity. Like the light from stars behind the Sun or behind the dark matter in our Galaxy, the light from distant objects such as quasars is therefore deflected as it travels past other galaxies toward us. This distortion of light, or **gravitational lensing**, can lead to our receiving several images of the distant quasar or galaxy. If the background object is exactly behind an intervening galaxy, the light should actually be focused as a ring, called an **Einstein ring**, rather than as several separate images. Figure 16-18a shows four images of a quasar in a configuration called an **Einstein cross**. Figure 16-18b shows arcs that are part of an Einstein ring.

# GAMMA-RAY BURSTERS

Supermassive black holes are not the only source of extremely powerful emissions from the remote reaches of the universe. Astronomers have also discovered pulses of

a                                    R I V U X G

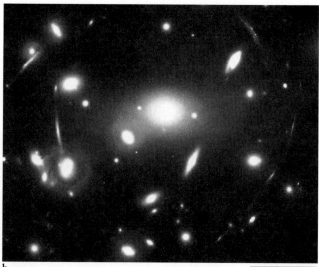

b                                    R I V U X G

**Figure 16-18** **An Einstein Cross and an Einstein Ring** **(a)** This image from the Hubble Space Telescope shows the gravitational lensing of a quasar in the constellation of Pegasus. The quasar, about 8 billion light-years from Earth, is seen as four separate images surrounding a galaxy that is only 400 million light-years away. The diffuse image at the center of this Einstein cross is the core of the intervening galaxy. **(b)** In this image from the Hubble Space Telescope, the quasar that is being lensed is behind the intervening galaxies. The quasar light arrives here as arcs of a ring, called an Einstein ring, after having passed the galaxies. The physical effect that creates these multiple images is a large-scale version of microlensing, depicted in Figure 14-12.

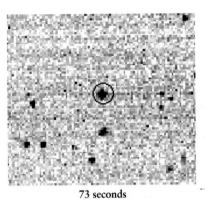

22 seconds          48 seconds          73 seconds

R I V U X G

**Figure 16-19**    **Gamma-Ray Bursts** In 1999 astronomers photographed the visible-light counterpart of a 100-second gamma-ray burst some 9 billion light-years away in the constellation Boötes. The times indicated are after the burst began.

gamma-ray energy from equally distant, yet smaller, sources. These short-lived sources have astonishing energy output.

## 16-7  Gamma-ray bursters are among the most powerful explosions in the known universe

Because atomic blasts emit pulses of gamma rays, the U.S. military put the gamma-ray-detecting *Vela* satellites in orbit around the Earth to monitor illegal nuclear explosions in the 1960s. In 1973, astronomers were told that these satellites were detecting bursts of gamma rays from objects in space. This unexpected news stunned the astronomical community, in part because there was no known mechanism to emit such gamma rays and in part because the energies involved in these blasts are staggering. While the origins of **gamma-ray bursters** are still being debated, most theories are focusing on gamma-ray emission from a black hole swallowing a neutron star or pairs of colliding neutron stars. Calculations confirm that such interactions should indeed lead to rapid, intense bursts of gamma rays.

Unlike X-ray bursters, gamma-ray bursters typically emit energy for less than 100 seconds. And, as this chapter's opening photo shows, they are located all over the celestial sphere. More than 2000 gamma-ray bursters have been observed, and more are being discovered, at a rate of about one per day.

Observations of the spectra of visible light that has accompanied several bursts (Figure 16-19) reveal that the light has passed through intergalactic gas clouds, indicating that the bursters typically originate more than half the distance to the edge of the observable universe. They are billions of light-years away. Taking their distance into account, a typical gamma-ray burster emits as much energy in 100 seconds as the Sun will emit over its entire 10-billion-year lifetime.

### WHAT DID YOU THINK?

1  *What do quasars look like?* They look like stars, but they spew out much more energy than any star.

2  *What powers a quasar?* A quasar is believed to be powered by a supermassive black hole at the center of a galaxy.

## Key Words

## Key Ideas

✳ The refinement of radio astronomy in the late 1940s opened the realm of nonvisual astronomy.

### Quasars

✳ A quasar, or quasi-stellar radio source, is an object that looks like a star but has a huge redshift. This redshift corresponds to an extreme distance from the Earth, according to the Hubble law.

✳ To be seen from Earth, a quasar must be very luminous, typically about 100 times brighter than an ordinary galaxy. Relatively rapid fluctuations in the brightnesses of some quasars indicate that they cannot be much larger than the diameter of our solar system.

✳ An active galaxy is an extremely luminous galaxy that has one or more unusual features: an unusually bright, starlike nucleus; strong emission lines in its spectrum; rapid variations in luminosity; or jets or beams of radiation emanating from its core.

✳ An active galaxy with a bright, starlike nucleus and strong emission lines in its spectrum is categorized as a Seyfert galaxy.

✳ BL Lacertae (BL Lac) objects (also called blazars) have bright nuclei whose cores show relatively rapid variations in luminosity.

✳ Most double radio sources contain an active galaxy located between two characteristic radio lobes. A head–tail radio source shows evidence of jets of high-speed particles emerging from an active galaxy.

✳ Most quasars are probably very distant active galaxies.

### Supermassive Central Engines

✳ Spectroscopic observations indicate huge concentrations of matter at the centers of certain galaxies.

✳ The strong energy emission from quasars, active galaxies, and double radio sources may be produced as matter falls toward a supermassive black hole at the center.

✳ Some matter spiraling in toward a supermassive black hole would be squeezed into two oppositely directed beams that carry particles and energy into intergalactic space.

### Gamma-Ray Bursters

✳ Gamma-ray bursters are events believed to be caused by the collisions of dense objects, such as neutron stars or black holes. They are observed to be billions of light-years away from Earth.

✳ Typical gamma-ray bursters occur for a few tens of seconds and emit more energy than the Sun will radiate over its entire 10-billion-year lifetime.

## Review Questions

1 Suppose you suspected a certain object in the sky to be a quasar. What sort of observations would you perform to confirm your hypothesis?

2 Explain why astronomers do not use any of the standard candles described in Chapter 15 to determine the distances to quasars.

3 Explain how the rate of variability of a source of light can be used to place an upper limit on the size of the source.

4 What is an active galaxy? List the different kinds of active galaxies. How do they differ from one another?

5 Why do astronomers believe that the energy-producing region of a quasar is very small?

6 How is synchrotron radiation produced?

7 What evidence indicates that quasars are extremely distant active galaxies?

8 What is a double radio source?

9 How does SS433 (discussed in Chapter 12) compare with a typical double radio source?

10 What is a supermassive black hole? What observational evidence suggests that supermassive black holes might be located at the centers of certain galaxies?

11 Why do many astronomers believe that the engine at the center of a quasar is a supermassive black hole surrounded by an accretion disk?

12 How does the orientation of the jets emanating from the center of a galaxy relative to our line of sight relate to the type of active galaxy that we observe?

## Advanced Questions

13  In the 1960s some astronomers suggested that quasars might be compact objects ejected at high speeds from the centers of nearby ordinary galaxies. Why does the absence of blueshifted quasars disprove this hypothesis?

14  When quasars were first discovered, many astronomers were optimistic that these extremely luminous objects could be used to probe distant regions of the universe. For example, it was hoped that quasars would provide high-redshift data from which the Hubble constant could be accurately determined. Why have these hopes not been realized?

## Discussion Questions

15  Explore the belief that quasars, double radio sources, and giant elliptical galaxies represent an evolutionary sequence.

16  Some quasars show several sets of absorption lines whose redshifts are less than the redshift of the quasars' emission lines. For example, the quasar PKS 0237-23 has five sets of absorption lines, all with redshifts somewhat less than the redshift of the quasar's emission lines. Propose an explanation for these sets of absorption lines.

## What If ...

17  Earth passed through a jet emitted by a radio galaxy? What would we see and what might happen to the Earth?

R I V U X G

**Structure of the Early Universe**
This microwave map of the sky, produced from data taken by NASA's Cosmic Background Explorer (COBE), shows temperature variations in the cosmic microwave background. Pink regions are about 0.0003 K warmer than the average temperature of 2.73 K; blue regions are about 0.0003 K cooler than the average. These tiny temperature fluctuations date back to the earliest moments of the universe and are related to the large-scale structure of the universe today. The radiation detected to make this map is from a time 300,000 years after the Big Bang.

# Cosmology

## In this chapter you will discover

* cosmology, which seeks to explain how the universe began and whether it will end

* the best theory we have for the origin of the universe—the Big Bang

* how astronomers trace the emergence of matter and the formation of galaxies

* how astronomers explain the overall structure of the universe

* how our understanding of the fate of the universe is in flux

# WHAT DO YOU THINK?

1  What is the universe?

2  Did the universe have a beginning?

3  Will the universe last forever?

People have speculated about the origins of the universe since before the dawn of recorded time, but until modern science and technology were developed over the past two centuries, these theories were untestable and, therefore, matters of faith. Today, astronomers have developed scientific theories of the evolution of the universe along with means of *testing* these theories. Only one theory proposed so far has survived the scrutiny of the scientific process—the Big Bang. With the aid of this theory, we can now describe scientifically how the universe has evolved from about $10^{-43}$ s after it began up to today, and we are closing in on understanding both its very first moment and its fate.

# THE BIG BANG

1  The **universe** consists of all matter, energy, and spacetime. **Cosmology** is the study of the origin, structure, and evolution of the universe. As a science, cosmology got off the ground in 1915, when Einstein published his general theory of relativity. To his surprise and dismay, the equations predicted that the universe is not static: It should either be expanding or contracting.

## 17-1 The expansion rate of the universe enables us to estimate the age of the universe and explain the cosmological redshift

The prediction of a changing universe flew in the face of the belief in an infinite, static universe, a concept promoted by Isaac Newton more than two centuries earlier. Newton believed that each star is fixed in place and held there under the influence of a uniform gravitational pull from every part of the sky. If the stars were not uniformly distributed, he argued, one region would have more mass than another. The denser region's gravity would then attract other stars, causing them to clump together further. Because he did not observe this clumping, Newton concluded that the mass of the universe must be distributed uniformly. He also concluded that without any gravitational sources of change, the universe must be infinitely old and should continue to exist forever without undergoing any major changes in structure.

Newtonian cosmology and prevailing theology made Einstein doubt the implications of his own theory, and so he missed the opportunity to propose that we live in a changing universe. Instead, he adjusted his elegant equations to yield a static cosmos. His did this by adding a repulsive (outward-pushing) term called the **cosmological constant** to his equations so that gravity's attractive force would be counterbalanced and the universe would be static. After observations revealed an expanding universe, Einstein said that adding the cosmological constant was the biggest blunder of his career. While observations revealed that the value of the cosmological constant Einstein used was indeed much too large, observations since 1997 suggest that a weaker, repulsive *cosmological force* throughout the universe may exist, consistent with a smaller value of the cosmological constant. We discuss this further in Section 17-9.

Edwin Hubble is usually credited with the discovery that we live in an **expanding universe**, that is, with discovering the motion of superclusters of galaxies away from one another. As we saw in Chapter 15, Hubble discovered the linear relationship between the distances to galaxies in other superclusters and the redshifts of those galaxies' spectral lines. In other words, the farther a galaxy is from Earth, the greater its redshift. This implies that remote galaxies are moving away from us at speeds proportional to their distances from us (review Figure 15-29).

A redshift caused by the expansion of the universe is properly called a **cosmological redshift**. Doppler shifts are caused by an object's motion through spacetime, whereas the cosmological redshift is caused by the expansion of spacetime itself. Except for the most distant galaxies and quasars, the Doppler shift and the cosmological redshift predict the same relationship between redshift and motion. That is why Hubble was justified in using the Doppler equation to calculate the recession of galaxies.

Because spacetime is expanding, photons we observe from galaxies in other superclusters are all redshifted. Imagine a photon coming toward us from a distant galaxy. As the photon travels through space, space is expanding, and thus the photon's wavelength stretches. When the photon reaches our eyes, we see a drawn-out wavelength, which is the redshift. The longer the photon's journey, the more its wavelength will have stretched. Therefore, astronomers observe larger redshifts in photons from distant galaxies than in photons from nearby galaxies.

How can the universe be expanding if it includes all spacetime? The universe is expanding because spacetime itself expands. According to general relativity, spacetime is not rigid. As it expands, the distance between superclusters of galaxies grows, including the distances of other superclusters from our own. Bear in mind that gravity keeps individual galaxies, clusters, and superclusters the same size; the expansion just acts to increase the separation between superclusters.

Hubble's law gives us a way to calculate the age of the universe. Imagine watching a movie of two superclusters receding from each other. If we then run the film backward, time runs in reverse, and we observe the superclusters approaching each other. Assuming the superclusters approach at a constant rate, the time it would take for them to collide is the time since the Big Bang. We can use a simple equation to make our first estimate of the age of the universe:

$$\frac{\text{Time since the}}{\text{Big Bang}} = \frac{\text{separation distance}}{\text{recessional velocity}}$$

Recall that Hubble found the relationship

$$\text{Recessional velocity} = H_0 \times \text{separation distance}$$

which we can rewrite as

$$H_0 = \frac{\text{recessional velocity}}{\text{separation distance}}$$

Comparing the first and last equations, we see that Hubble's constant is exactly the inverse of the time since the universe began. Using a Hubble constant of 75 km/s/Mpc (see An Astronomer's Toolbox 17-1), we find:

$$1/H_0 = 1/75 \text{ km/s/Mpc} = 13 \text{ billion years}$$

Hubble's constant will almost surely turn out to be lower. Recall from Chapter 15 that some observations place it at 55 km/s/Mpc, which implies an age for the universe of 18 billion years. Because observations of Hubble's constant remain uncertain, we will continue to use the common estimate of 75 km/s/Mpc.

## 17-2 The universe probably originated in an explosion—the Big Bang

Observations of redshifts indicate that the universe has been expanding for billions of years, so, in the past the density of matter was much greater than it is at present. If

## AN ASTRONOMER'S TOOLBOX 17-1

### $H_0$ and the Age of the Universe

The Hubble constant may not look like a measure of time. All we need, however, is a little practice with unit conversion. The trick is to remove all distance units from $H_0$ and then invert it. Recall that we are using $H_0 = 75$ km/s/Mpc, which has different units of distance in the numerator and denominator. As we saw in Chapter 15, this mix of units makes it easy to determine how fast a galaxy is receding once its distance is known.

*Example:* A galaxy 10 Mpc away is receding at

$$75 \text{ km/s/Mpc} \times 10 \text{ Mpc} = 750 \text{ km/s}$$

We know that

$$1 \text{ pc} = 3.09 \times 10^{13} \text{ km}$$

and so, 1 Mpc = $3.09 \times 10^{19}$ km. Because there are $3.156 \times 10^7$ seconds in a year, we get the conversion factors we need:

$$H_0 = \frac{1}{1/75 \text{ km/s/Mpc}} \times (3.09 \times 10^{19} \text{ km/Mpc})$$

$$\times \frac{1}{3.156 \times 10^7 \text{ s/yr}} = 13 \times 10^9 \text{ yr}$$

If the universe had always been expanding at the same speed, then $1/H_0$ would be the age of the universe. However, the universe has been continually slowing down due to the gravitational attraction between superclusters of galaxies. Detailed calculations show that the age of the universe is $2/3 \times 1/H_0$. Using the 13-billion-year estimate, the universe is about 9 billion years old.

*Self-Test:* Why is this a problem?

*Compare!* Decades of observations by many astronomers have confirmed that all distant galaxies are moving away from the Earth. Not only that, but *the Hubble law relating velocities and distances is the same in all directions.* That is, the Hubble constant, $H_0$, is the same no matter where you aim your telescope. The fact that the recession rates are the same in all directions is a condition called **isotropy.** The universe is isotropic: Its global expansion is the same in every direction.

The general expansion away from us suggests that the Earth is at the center of the universe. After all, where else could we be if all the distant galaxies are moving away from us? In fact, the answer is that we could be practically anywhere in the universe! To understand why, see Discover for Yourself: The Expanding Universe.

## DISCOVER FOR YOURSELF | The Expanding Universe

To understand the universe, scientists build models, mathematical representations of the world about us. To develop such pictures, it often helps to begin with a simple analogy. Because the expansion of the universe is hard to visualize, imagine for a moment a chocolate chip cake baking in an oven. As the cake rises, the chocolate chips move farther apart. They do not get larger, nor do they move through the batter. Each chocolate chip remains at rest in its own little bit of cake and is carried along as the cake spreads out.

Now, think of each chocolate chip as a supercluster of galaxies and the batter between the chocolate chips as the rest of spacetime. As the universe expands, the distance between widely separated superclusters of galaxies grows larger and larger. The expansion of the universe *is* the expansion of spacetime.

Suppose you were on one particular chocolate chip, say the chocolate chip labeled 3 at the left in the figure. You would see chocolate chips 2 and 4 moving away from you with equal velocities. Every 10 minutes chocolate chips 2 and 4 get 1 inch farther away as the cake expands, as shown on the right. In that same time interval, chocolate chips 1 and 5 move twice as far from you as chocolate chips 2 and 4. To go twice as far away in the same time, chocolate chips 1 and 5 must be moving twice as fast as the closer chocolate chips 2 and 4. This is exactly what Hubble's law of cosmological expansion (recessional velocity = $H_0 \times$ distance) says: *Double the distance and you double the velocity.*

To see why we need not occupy the center of the universe, put yourself on chocolate chip 4. You see chocolate chips 3 and 5 moving away at the same rate—in fact, they recede at the same rate as you saw chocolate chips 2 and 4 move away when you were on chocolate chip 3. From chocolate chip 4, chocolate chips 1 and 6 are moving away twice as fast as chocolate chips 3 and 5. From any chocolate chip you will see all the other chocolate chips moving away. Similarly, no matter where we are in the universe, we see all the other superclusters receding from us.

**The Expanding Chocolate Chip Cake Analogy** The expanding universe can be compared to a rising chocolate chip cake. Just as all the chocolate chips move apart as the cake rises, all the superclusters of galaxies recede from each other as the universe expands.

we press far enough into the distant past, all the matter in the universe once must have been concentrated to a cosmic singularity—a state of almost inconceivable density. Based on Hubble's observations of an expanding universe, George Gamow in the 1940s proposed that the universe began in a colossal explosion. Ironically, Sir Fred Hoyle, who named this explosion the **Big Bang** in 1950, is one of the very few astrophysicists who do not believe that the Big Bang occurred.

Indeed, observations that the universe is expanding are insufficient evidence for accepting the Big Bang theory. That is why a plausible alternative called the *steady state theory* was proposed in 1948 by Hoyle, Herman Bondi, and Thomas Gold. In this theory the universe is continuously expanding, and, as it expands, new matter is created that takes the place of the receding matter. This new matter then forms new galaxies. Where this new matter comes from is unknown, but the laws of physics do not necessarily exclude its creation. So, why do virtually all astronomers now accept the Big Bang theory rather than the steady state? The answer lies in other observations that are consistent with the former but not with the latter.

Shortly after World War II, astrophysicists Ralph Alpher and George Gamow proposed that the universe immediately following the Big Bang was an incredibly hot black body—far in excess of $10^{12}$ Kelvin. It must therefore have been filled with high-energy, short-wavelength, gamma-ray radiation. As the universe expanded, the cosmological redshift stretched the wavelengths of the radiation created during the Big Bang so that most of the original gamma rays are now radio waves. This means that the energy packed into each unit volume of the universe has decreased, which is another way to say that the universe has cooled.

Calculations indicated that if the universe really was initially "hot," the remnants of that energy should still fill all space today, giving it a temperature only a few Kelvin above absolute zero. This radiation was called the **cosmic microwave background,** because the peak of its blackbody spectrum should lie in the microwave part of the radio spectrum. In the early 1960s, Robert Dicke, P. J. E. Peebles, and their colleagues at Princeton University began designing a microwave radio telescope to detect it.

Meanwhile, just a few miles from Princeton, two physicists had already found the cosmic background radiation. Arno Penzias and Robert Wilson of Bell Laboratories were working on a new horn antenna designed to relay telephone calls to Earth-orbiting communications satellites (Figure 17-1). Penzias and Wilson were deeply puzzled. No matter where in the sky they pointed their antenna, they detected a faint background noise. Even the careful removal of heat-generating pigeon droppings from the antenna failed to eliminate the noise completely. Thanks to a colleague, they soon learned of the then-theoretical cosmic microwave background and the work of Dicke and Peebles in trying to locate it. Communicating with the Princeton astronomers, Penzias and Wilson presented their finding and thus were able to claim the first detection.

Our most accurate cosmic microwave background measurements to date come from the Cosmic Background Explorer (COBE) satellite, which operated between 1989 and 1994 (Figure 17-2). Data from COBE's spectrometer shown in Figure 17-3 demonstrate that this ancient radia-

**Figure 17-2** **The Cosmic Background Explorer (COBE)** This satellite, which operated from 1989 to 1994, measured the spectrum and angular distribution of the cosmic microwave background over a wavelength range of 1 μm to 1 cm. COBE (pronounced "co-bee") found local temperature variations across the sky but no overall deviation from a perfect blackbody spectrum.

tion indeed had the spectrum of a blackbody, with a temperature of 2.73 K. Cosmic background radiation, also commonly called the *3-degree background radiation*, is required by the Big Bang theory but is neither required nor

**Figure 17-1** **The Bell Labs Horn Antenna** This Bell Laboratories horn antenna at Holmdel, New Jersey, was used by Arno Penzias (on the right) and Robert Wilson in 1965 to detect the cosmic microwave background.

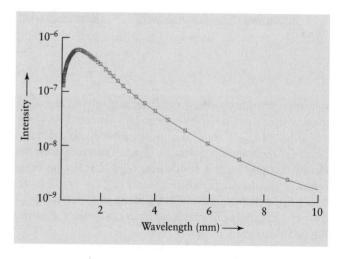

**Figure 17-3** **The Spectrum of the Cosmic Microwave Background** The little squares on this graph are COBE's measurements of the brightness of the cosmic microwave background plotted against wavelength. The data fall along a blackbody curve for 2.73 K to a remarkably high degree of accuracy. The peak of the curve occurs at a wavelength of 1.1 mm, in accordance with Wien's law.

even consistent with the steady-state theory, or any other existing cosmological theory for that matter.

Like the expansion of the universe, the cosmic microwave background is almost perfectly isotropic: Its intensity is nearly the same in every direction in the sky. Figure 17-4 shows a map of the microwave sky. It is slightly warmer than average in the direction of the constellation Leo and slightly cooler in the opposite direction, toward Aquarius. The tiny temperature variation depicted in Figure 17-4 results from the Earth's overall motion through the cosmos. We are moving through the background radiation because of our rotation and orbit around the Sun and the solar system's motion around the Milky Way. More important, the Milky Way, and even the Local Group of galaxies, are themselves in motion (Figure 17-5). If the Earth were exactly at rest with respect to the microwave background, the average intensity of radiation would be the same in all directions.

The observed temperature differences shown in Figure 17-4 mean that the solar system is at present moving toward Leo at a speed of 390 km/s. Taking into account the known velocity of the Sun around the center of our Galaxy, astronomers calculate that the entire Milky Way Galaxy is moving relative to the cosmic microwave background at 600 km/s—some 1.3 million miles per hour—in the general direction of the Centaurus cluster. The gravitational attraction of four nearby clusters of galaxies and an enormous supercluster called the *Shapley concentration* are believed to be pulling us in that direction.

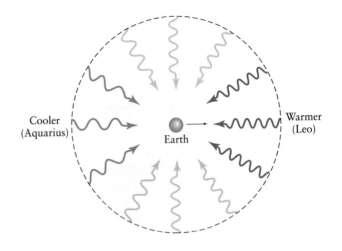

**Figure 17-5**   **Our Motion Through the Microwave Background**   Because of the Doppler effect, we detect shorter wavelengths in the microwave background and a higher temperature of radiation in that part of the sky toward which we are moving. This part of the sky is the area shown in red in Figure 17-4. In the opposite part of the sky, shown in blue in Figure 17-4, the microwave radiation has longer wavelengths and a cooler temperature.

## 17-3 A period of vigorous inflation followed the Big Bang

Detecting the cosmic background radiation provided compelling support for the Big Bang theory, but its near isotropy was, until recently, a major problem. Because light takes time to travel to Earth, *the farther we look into space, the farther back in time we are seeing.* But if the universe has a finite age of around 13 billion years, we cannot see anything farther than 13 billion light-years away. At that distance, the objects we see are just being formed. Therefore, all that we can see from the Earth is contained in an enormous sphere (Figure 17-6). The boundary of this sphere is called the **cosmic particle horizon.** This sphere, with a radius of 13 billion light-years (assuming a 13-billion-year-old universe), is our entire observable universe.

Light from all the luminous objects in this volume has had time to reach the Earth, but light from objects beyond the particle horizon has not yet arrived here, and so we are unaware of objects in these regions. With our technology we can practically see light from this horizon. We are thus almost able to see back to the time before galaxies and stars first formed.

There are regions of space within our cosmic particle horizon that are more than 13 billion ly apart, and so they do not yet know of each other's existence. As we saw from the COBE data, the temperature of the universe is virtually

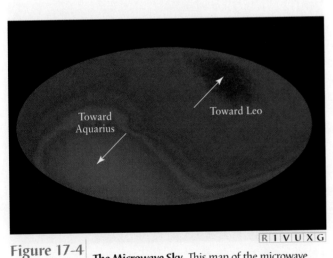

**Figure 17-4**   **The Microwave Sky** This map of the microwave sky was produced from data taken by instruments on board COBE. The galactic center is in the middle of the map, and the plane of the Milky Way runs horizontally across the map. Color indicates temperature— red is warm, blue is cool. The temperature variation across the sky is caused by the Earth's motion through the microwave background. The variation is quite small, only 0.0033 K above the average radiation temperature of 2.726 K.

We cannot yet
see this region of
the universe

Farthest we
can see now

Present cosmic
particle horizon

R I V U X G

**Figure 17-6** **The Observable Universe** This diagram shows how we can only see a part of the entire universe. As time passes, this volume grows, meaning that light from more distant galaxies reaches us. The galaxies we see at the farthest reaches of our telescopes' resolving power are as they were within two billion years after the Big Bang (see inset). These galaxies, formed at the same time as the Milky Way, appear young because the light from their beginnings is just now reaching us. The radius of the cosmic particle horizon is equal to the distance that light has traveled since the Big Bang. Because the Big Bang occurred about 13 billion years ago, the cosmic particle horizon today is about 13 billion light-years away in all directions. **Inset:** This image of the *Hubble Deep Field* shows more than 1000 galaxies located at various distances between us and the cosmic particle horizon. These galaxies are the dots drawn in the wedge-shaped region of the diagram.

the same in all directions. But until recently, astronomers could not explain why these disparate regions share the same temperatures (Figure 17-7). This problem of explaining how such disconnected regions could have the same temperature in a universe that should theoretically have a wide range of temperatures is called the **isotropy problem** or the **horizon problem.**

In the early 1980s, Alan Guth offered a remarkable solution to this long-standing dilemma. Guth mathematically analyzed the suggestion that the universe experienced extremely rapid expansion during the first $10^{-24}$ s of its existence (Figure 17-8). During this brief **inflationary epoch,** the universe ballooned outward in all directions to become many billions of times its previous size. In essence, a repulsive force built up and literally inflated the entire universe.

Inflation thus enlarged an initially tiny region of space so much that it became our entire universe. Because this tiny region had time to achieve a single temperature before inflation, the cosmic background radiation throughout it is still uniform today. Thus, when we examine microwaves from opposite directions in the sky, we are seeing radiation from regions of the universe that were originally in intimate contact. There may be many other "universes" that we cannot detect, each at a different temperature.

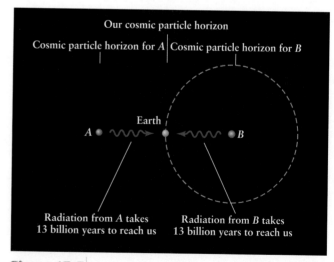

Our cosmic particle horizon

Cosmic particle horizon for *A* | Cosmic particle horizon for *B*

Earth

*A*                              *B*

Radiation from *A* takes
13 billion years to reach us

Radiation from *B* takes
13 billion years to reach us

**Figure 17-7** **The Isotropy Problem** The temperature of radiation in the cosmic microwave background is almost exactly the same from all parts of the sky. This indicates that all parts of the cosmic fireball, from which this radiation comes, must have had almost exactly the same temperature. But regions *A* and *B*, which now lie on our cosmic particle horizon and from which we receive radiation, are so far apart that they should never have been in communication over the lifetime of the universe. Why, then, are they at the same temperature?

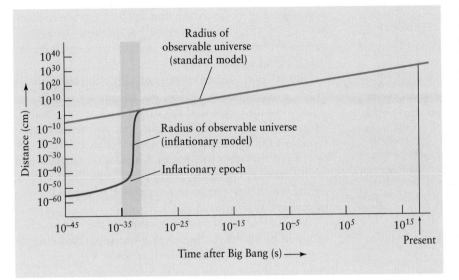

**Figure 17-8** | **Size of the Universe With and Without Inflation** According to the inflationary model, the universe expanded by a factor of about $10^{50}$ shortly after the Big Bang. This sudden growth in the size of the observable universe occurred during a very brief interval, as indicated by the vertical shaded area on the graph. For comparison, the projected size of the early universe without inflation is also shown.

# A Brief History of the Creation of Matter

The composition of the universe during its first few minutes of existence was profoundly different from today. To understand why, we need to expand briefly on the nature of the four known physical forces: gravity, electromagnetism, and the strong and weak nuclear forces. (As noted earlier in this chapter, there may be a fifth force, the repulsive force associated with the expansion of the universe. Little is known of its properties, and thus we will present the observations related to it and their implications later in the chapter.)

## 17-4 All physical forces in nature were initially unified

We are all familiar with the force of gravity. While it is the weakest of all four forces, it is the only force whose effect acts over long distances and is always attractive. The electromagnetic force, which holds electrons in orbit about the nuclei in atoms, also acts over long distances. But over large volumes of space, the net effects of electromagnetism cancel each other, because a negative electric charge exists for every positive charge and a south magnetic pole exists for every north magnetic pole. Therefore, the attractive force of gravity dominates the universe at astronomical distances.

In contrast to gravity and electromagnetism, the strong and the weak nuclear forces are both extremely short range, because their influences extend only over distances less than about $10^{-15}$ m. The **strong nuclear force** holds protons and neutrons together inside the nuclei of atoms.

Without this force, nuclei would disintegrate because of the electromagnetic repulsion of the positively charged protons. Thus, the strong nuclear force overpowers the electromagnetic force inside nuclei.

The **weak nuclear force** is at work in certain kinds of radioactive decay, such as the transformation of a neutron into a proton. Protons and neutrons are composed of more basic particles called **quarks**. A proton is composed of two "up" quarks and one "down" quark, whereas a neutron is made of two "down" quarks and one "up" quark. The weak nuclear force is at work whenever a quark changes from one variety to another. For example, in radioactive decay, a neutron transforms into a proton, and one of the neutron's "down" quarks changes into an "up" quark.

To examine details of the physical forces, scientists use particle accelerators that hurl high-speed electrons and protons at targets. In such experiments, physicists find that as the particles' speeds approach the speed of light, the different forces begin to behave the same way. In fact, during experiments at the CERN accelerator in Europe in the 1980s, particles were slammed together with such violence that the electromagnetic force and the weak nuclear force had equal strength. It seems possible that all four forces would have the same strength if particles were slammed together with energy trillions of times greater than that achievable in the CERN accelerator.

Physicists have no hope of building accelerators that powerful. However, immediately after the Big Bang, the universe was so hot that particles in it had just such energies. The earliest moments of the universe, as predicted by the Big Bang model, thus provide a laboratory in which scientists can explore some of the most elegant ideas in physics. Figure 17-9 summarizes the connections between particle physics and cosmology. Figure 17-10 shows that during the beginning moments of the universe (from 0 to

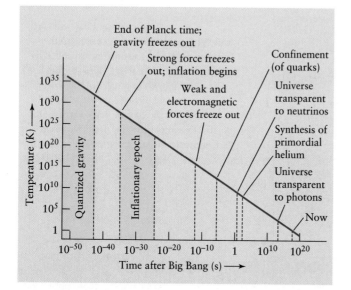

**Figure 17-9** **The Early History of the Universe** Current theory holds that, as the universe cooled, the four forces "froze out" of their initial unified state. The inflationary epoch lasted from $10^{-35}$ s to $10^{-24}$ s after the Big Bang. Neutrons and protons froze out one-millionth of a second after the Big Bang. The universe became transparent to light (that is, photons decoupled from matter) when the universe was 300,000 years old.

$10^{-43}$ s, called the **Planck time**), the four forces are believed to have been unified—only one physical force existed in nature.

By the end of the Planck time, the energy of particles in the universe had declined to the extent that gravity was no longer unified with the other three forces. Most astron-

**Figure 17-10** **Unification of the Four Forces** The strength of the four physical forces depends on the speed or energy with which particles interact. As shown in this figure, the higher the temperature of the universe, the more the forces resemble each other. Also included here is the age of the universe at which the various forces were equal.

omers therefore hypothesize that at $10^{-43}$ s, when the temperature of the universe was $10^{32}$ K, gravity "froze out" (separated from) the otherwise unified hot soup of energy that filled all space.

By $10^{-35}$ s, calculations indicate that the temperature of the universe fell to $10^{27}$ K, and the energy of particles declined to the extent that the strong nuclear force was no longer unified with the electromagnetic and weak nuclear forces. At this point, the strong nuclear force emerged, distinct from the other two forces. Calculations suggest that the inflationary epoch lasted from $10^{-35}$ s to about $10^{-24}$ s, during which time the universe increased in size enormously, perhaps by a factor of $10^{50}$.

At $10^{-12}$ s, when the temperature of the universe had dropped to $10^{15}$ K, a final freeze-out separated electromagnetism from the weak nuclear force. From that moment on, all four forces interacted with particles essentially as they do today.

The next significant event is calculated at $10^{-6}$ s, when the universe's temperature was $10^{13}$ K. Prior to this moment, particles collided so violently that individual protons and neutrons could not exist, because they were constantly being fragmented into quarks. After this moment, appropriately called **confinement**, quarks could finally stick together to form individual protons and neutrons.

## 17-5 During the first second, most of the matter and antimatter in the universe annihilated each other

Early in the first second, equations show that the universe was so hot that photons possessed incredibly high energies. These energies were high enough to create matter and antimatter, according to Einstein's equation $E = mc^2$. As we saw in Chapter 9, to make a particle of mass $m$, you need an amount of energy $E$ at least as great as $mc^2$, where $c$ is the speed of light.

The creation of matter from energy is routinely observed in laboratory experiments involving high-energy gamma rays. When a highly energetic photon collides with a second photon, both vanish, and matter appears in their place, as sketched in Figure 17-11. This process, called **pair production**, always creates one ordinary particle and one so-called antiparticle. For example, if one of the particles is an electron, the other is an antielectron or positron (Figure 17-12, and recall from Chapter 13 that similar pairs of particles are also believed to be produced near a black hole).

There is nothing mysterious about antimatter except that it has an opposite electric charge to matter and has the ability to annihilate ordinary matter. An antielectron is an ordinary electron except that it has a positive charge. Theory predicts that at times earlier than 1 s in the life of the universe, pair production occurred continually. In

**Figure 17-11** **Pair Production and Annihilation** A particle and an antiparticle can be created when high-energy photons collide. Conversely, a particle and an antiparticle can annihilate each other and emit energy in the form of gamma rays.

other words, shortly after the Big Bang all space was chock full of protons, neutrons, electrons, and their antiparticles, all immersed in a phenomenally hot bath of high-energy photons.

After that first second, pair production ceased. As the universe expanded, the temperature declined until the photons could no longer create pairs of particles and antiparticles. Particles and antiparticles often collided, annihilating each other and converting their masses back into high-energy photons. But nothing could replenish the supply of particles and antiparticles.

If particles and antiparticles are always created or destroyed in pairs, why aren't they found in equal numbers in the universe? For every proton created, there should be an antiproton; for every electron there should be an antielectron. And yet, we observe very few antiparticles in the universe today. Furthermore, if all particles had annihilated their antiparticles in the early universe, no matter would be left at all.

Obviously this did not happen. Physicists therefore theorize that a *symmetry breaking* occurred. This means that the number of particles was slightly greater than the number of antiparticles. For every billion antiprotons, perhaps a billion plus one protons formed; for every billion antielectrons, a billion plus one electrons formed. As far as theorists can tell, virtually all the antiparticles created in the first second of time annihilated normal particles shortly thereafter. The remaining particles in the universe were the slight excess of normal matter created during symmetry breaking.

As the universe expanded and cooled further, the remaining protons and neutrons began colliding and fusing together. For the most part, they were quickly separated again by the gamma rays in which they were bathed. However, during the first 3 minutes, enough fusions occurred that were not subsequently unfused to create most of the helium and the trace element lithium that exist today. Hydrogen, helium, and lithium are the three lowest-mass elements. After a few minutes, the universe was too cool to

R I V U X G

**Figure 17-12**  **Creation of Matter from Energy** This photograph shows the conversion of a gamma ray into matter inside a bubble chamber, a device filled with liquid hydrogen that is designed to make the path of a charged particle visible as a long row of tiny bubbles. The path of the gamma ray is not visible because photons are electrically neutral. Near the bottom of the photograph, the energy carried by the gamma ray is converted into an electron and an antielectron. Because of a magnetic field surrounding the bubble chamber, the electron is deflected to the right while the antielectron (positron) veers toward the left.

allow fusion to create more massive elements, such as carbon and oxygen. All elements other than the three lowest-mass ones formed later as a result of stellar evolution.

> 🚀 **Insight into science**  **Share information**  While the evolution of the universe is a truly large-scale event, the activity in the first 3 minutes is nevertheless intimately connected to the creation and evolution of elementary particles, the smallest objects in the universe. Physicists who study elementary particles and astrophysicists who study cosmology now regularly hold meetings to compare notes and connect their fields.

## 17-6 With the formation of neutral atoms, matter came to dominate the universe

Theory holds that for the first 300,000 years, photons prevented neutral atoms from forming. Electrons were ejected as quickly as they came to orbit a nucleus. That early time, called the **radiation-dominated universe,** lasted until the universe was so large that the cosmological redshift had decreased the photons' energies to the point where they could no longer ionize hydrogen. Thereafter, the behavior of matter was dominated by gravitational forces. We live in this **matter-dominated universe.**

The transition from a radiation-dominated universe to a matter-dominated universe, called **decoupling,** occurred about 300,000 years after the Big Bang. We can use Wien's law to calculate the temperature of the cosmic background radiation at this time. The 3000 K temperature at decoupling corresponds to a peak microwave wavelength of 0.001 mm. The temperature history of the universe is graphed in Figure 17-13.

Although the universe is dominated by matter today, the number of photons left over from the Big Bang is immense. From the physics of blackbody radiation, astronomers calculate that there are now 400 million cosmic background microwave photons in every cubic meter of space. In contrast, if all the visible matter in the universe were uniformly spread throughout space, there would be roughly one hydrogen atom in every 3 cubic meters of space. In other words, photons outnumber atoms by more than a billion to one.

Thus, in the first 300,000 years, according to this model, the universe was completely filled with a shimmering expanse of high-energy photons colliding vigorously with protons and electrons. This state of matter, called a *plasma,* is opaque to all wavelengths, just as the glowing gases inside a fluorescent light bulb are opaque. The term **primordial fireball** describes the universe during this time.

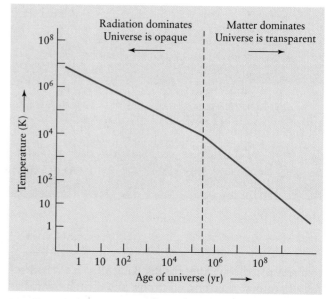

**Figure 17-13** | **The Declining Temperature of the Universe**  As the universe expands in the model, the photons in the radiation background become increasingly redshifted and the temperature of the radiation field falls. Roughly 300,000 years after the Big Bang, when the temperature fell below 3000 K, hydrogen atoms formed and the universe became transparent.

After 300,000 years, when the temperature of the radiation field fell below 3000 K, photons no longer had enough energy to keep protons and electrons apart. Protons and electrons everywhere began combining to form hydrogen atoms. Because hydrogen gas is clear, the universe suddenly became transparent! The photons that just a few seconds earlier had vigorously collided with charged particles could now stream unimpeded across space. Today we see these same photons as the cosmic microwave background.

This dramatic moment, when the universe transformed from being opaque to transparent, is referred to as the **era of recombination.** Because the universe was opaque in its first 300,000 years, we cannot see any further into the past than the era of recombination. The microwave background contains the most ancient photons we expect to ever be able to observe. This microwave background is today a ghostly relic of the former dazzling splendor of the universe.

The Big Bang theory explains the existence of space-time, matter, and energy. It accounts for the hydrogen and most of the helium that exist today. But how do we explain the existence of superclusters of galaxies, clusters of galaxies, and individual galaxies?

To explain such large-scale structures, astrophysicists turn from the physics of gravity to the physics of quantum mechanics (Chapter 4). Whereas gravity is most influential over large scales, quantum mechanics explains the behavior of tiny, elementary particles. Quantum mechanics also predicts the existence of tiny differences from place to place in the density of energy and matter in the early universe. Inflation then expanded these tiny density variations to the colossal sizes of superclusters of galaxies.

These clumps throughout the inflated universe must have initially been only small density enhancements for the microwave background to appear as uniform as it does now. Part of the mission of the COBE satellite was to find these variations, and find them it did! They are only about 1 part in 10,000 denser than the average density of matter in the universe (see the figure that opens this chapter). But these small variations are all that is needed to explain the large-scale structure of the universe.

Wherever the universe had a tiny, slightly denser volume of matter after inflation, the gravity there would have kept all the mass in that region bound together. This material would then have formed one supercluster. A major open question about the early universe is whether the stars formed before the galaxies formed, whether the stars formed as these larger structures were forming, or whether the stars formed after the galaxies were established.

## 17-7 Galaxies formed from huge clouds of primordial gas

Virtually all the matter in the early universe consisted of clouds of hydrogen and helium. Based on observations of the most ancient, metal-poor stars that astronomers have

seen and on the prevailing model of stellar evolution, the galaxies are believed to have begun forming 11 billion to 13 billion years ago. Some galaxies are believed to have begun coalescing from smaller clusters of stars that formed in the pregalactic clouds of gas less than a billion years after the Big Bang. Astronomers have observed such clusters of stars more than 11 billion light-years away.

By observing remote galaxies, astronomers have discovered that most galaxies initially emitted energy we associate with quasars and other active galaxies. This implies that most young galaxies have supermassive black holes at their centers. Such black holes may have attracted mass, enhancing the clumps of gas and stars, if they already existed, thereby creating many of the galaxies.

Observations reveal that galaxies were bluer and brighter in the past than they are today. These changes in color and brightness suggest an overabundance of young, bright, hot, massive stars in newly formed galaxies (Figure 17-14a). As galaxies age, these blue O and B stars become supergiants and eventually die off. Therefore, galaxies grow somewhat redder and dimmer. This is especially true of elliptical galaxies, which appear to form nearly all their stars in one vigorous burst of activity that lasts for about a billion years (Figure 17-14b).

In contrast, spiral galaxies have been forming stars for at least the past 10 billion years, although at a gradually decreasing rate. There is still plenty of interstellar hydrogen in the disks of spiral galaxies like our Milky Way to fuel star formation today. That is why O and B stars still highlight their spiral arms. Figure 17-14b compares the rates at

**Figure 17-14** **Stellar Birth Rates** **(a)** Based on observations of star formation at very early times in the universe and also at times from 5 billion years after the universe formed and onward, this figure shows our present knowledge of the overall star-formation history of the universe. Astronomers hope to fill the data gap between 2 billion years and 5 billion years shortly. They expect it to show a peak of star formation activity. **(b)** Most of the stars in an elliptical galaxy are created in a brief burst of star formation when the galaxy is very young. In spiral galaxies, stars form at a more leisurely pace that extends over billions of years.

which spiral and elliptical galaxies form stars. Evidence suggests that, overall, star formation peaked at about 15 times its present rate when the universe was about one-third its present age.

## 17-8 Star formation may determine a galaxy's initial structure

Imagine a developing galaxy, called a *protogalaxy,* forming all at once from a single cloud of gas. Theory proposes that the rate of star formation determines whether this protogalaxy becomes a spiral or an elliptical. If stars form slowly enough, then the gas surrounding them has plenty of time to settle by collision with other infalling gas into a flattened disk, just like the early solar system. Star formation continues because the protogalactic disk contains an ample supply of hydrogen, and a spiral or lenticular galaxy is created. If, however, the initial stellar birthrate is high in the protogalaxy, the theory predicts that virtually all pregalactic gas is used up in the creation of stars before a disk can form. In this case, an elliptical galaxy is created. Figure 17-15 depicts these contrasting scenarios.

Not all galaxies have maintained their initial structure over the life of the universe. As we discussed in Chapter 15, some galaxies collide, and these collisions can change a galaxy's structure from spiral, for example, to elliptical. The debate over galactic evolution remains difficult and often controversial. For instance, why wasn't gas in the early universe all transformed into objects a million times bigger or smaller than galaxies?

Even more troublesome is the puzzle of dark matter (see Sections 14-4 and 15-9). The observable stars, gas, and dust in a galaxy account for only about 10% to 20% of its mass. Other than massive neutrinos, we have very little idea of what the remaining 80% to 90% looks like, how it is distributed in space, or what it is made of. With this level of ignorance, any discussion of galactic evolution must be woefully incomplete. Add in the great distances involved, and it is easy to see why observing the early universe is among the greatest challenges facing astronomers.

## THE FATE OF THE UNIVERSE

We now turn to the future. Will the universe last forever? Or will it someday stop expanding and collapse?

## 17-9 The average density of matter is one factor that determines the future of the universe

During the 1920s, Alexandre Friedmann in Russia, Georges Lemaître in France, Willem de Sitter in the Netherlands, and, of course, Einstein himself applied the general theory of relativity to cosmology. They found that gravity slows

a

b

**Figure 17-15** **The Creation of Spiral and Elliptical Galaxies** A galaxy begins as a huge cloud of primordial gas that collapses gravitationally. **(a)** If the rate of star birth is low, then much of the gas collapses to form a disk, and a spiral galaxy is created. **(b)** If the rate of star birth is high, then the gas is converted into stars before a disk can form, resulting in an elliptical galaxy.

the expansion of the cosmos. In 1998 astronomers announced evidence (see Section 17-1) that there may also be a repulsive cosmological force acting throughout the universe. Calculations show that this force remains constant as the universe expands. If it exists, it is much weaker than the attractive gravitational force between superclusters used to be. But as the superclusters moved farther and farther apart, their mutual gravitational attractions decreased, while the cosmological force remained constant in strength. Therefore, a time may have been reached when this repulsive force began dominating the universe, causing it to start expanding faster and faster.

To understand the possible fate of the universe, consider a cannonball shot straight upward from the surface of the Earth. The Earth's gravitational force slows the ball's ascent. If the cannonball's speed upward is less than the escape velocity from the Earth's gravity (about 11 km/s), it will fall back to Earth. If the cannonball's speed equals the escape velocity, it will just barely escape falling back to Earth and come to a stop an infinite distance away. And if the ball's speed exceeds the escape velocity, it can easily leave the Earth, despite the relentless pull of the Earth's gravity.

One key factor in determining the fate of the universe is the amount of mass it contains. Is there enough mass to create sufficient gravitational attraction between superclusters to overcome both the energy supplied by the Big Bang and the proposed repulsive cosmological force? If so, the expansion will someday stop, and everything will collapse upon itself and ultimately end in a **Big Crunch** in the distant future. If not, the universe will expand forever into the so-called **Big Chill**. Let's explore the gravitational attraction issue.

If the average density of matter throughout space is sufficiently low, gravity will be too weak to stop the universe's expansion. In that case, the universe will continue to expand forever, like the cannonball that escapes from the Earth's gravitational clutches. The superclusters of galaxies will continue to rush away from each other forever. Under such circumstances, astronomers say that the universe is **unbound.**

Conversely, if the average density of matter across space is sufficiently high, then gravity will eventually halt its expansion. Like the cannonball that falls back to Earth, the universe will begin contracting. In this case, gravity will pull the superclusters of galaxies back toward each other. If this is the case, the universe is **bound.**

If the cosmological force does not exist and the average density of matter is at the *critical density*, the superclusters will just barely manage to keep moving away from each other forever under the influence of their mutual gravity, slowing to a stop an infinite time in the future. This hypothesis presents a special case of an unbound universe. Estimates of the critical density depend on the Hubble

constant. Calculations based on a Hubble constant of 75 km/s/Mpc yield a critical density of $2.4 \times 10^{-26}$ kg/m$^3$, which is equivalent to a density of about 14 hydrogen atoms per cubic meter of space.

If the universe has critical density and the cosmological force is present, then the repulsive force will eventually dominate and ensure that the superclusters move apart forever. Present-day estimates of the density suggest that the universe is unbound. The average density of luminous matter that we see in the sky seems to be about $5 \times 10^{-28}$ kg/m$^3$, which is about one-fiftieth of the critical density. However, as discussed in Chapters 14 and 15, we can see only a small fraction of all the matter in the universe. The rest remains to be discovered. Some physicists argue that this dark matter is probably not composed of particles like protons, neutrons, or electrons; otherwise, we would have already detected it. Instead, they believe that the dark matter is vastly different from anything we have ever encountered.

## 17-10 The expansion of the universe is related to the cosmological redshift

The changing speed of the universe as it expands shows up as a deviation from the straight-line relationship predicted by the Hubble law. Consider galaxies several billion light-years from Earth. The light from these galaxies has taken billions of years to get to our telescopes, so redshift measurements reveal how fast the universe was expanding billions of years ago. Because the universe was expanding at a different speed in the past than it is today, our data deviate slightly from the straight-line Hubble law.

Figure 17-16 displays the relationship between the changing speed of the universe and the Hubble law for great distances. Astronomers describe this relationship with the **deceleration parameter,** designated $q_0$. Appropriately, $q_0 = 0$ corresponds to expansion forever at a constant rate. This is possible only if the universe is completely empty, and thus there is no gravity to slow down the expansion and no cosmological force to speed it up. As sketched in Figure 17-16, if $q_0$ is 0, the universe will expand forever, with no deceleration at all.

If $q_0$ is between 0 and ½, the universe is unbound and will continue to expand forever. Such a universe contains matter at less than the critical density. If $q_0$ is greater than ½, the universe is bound, because the density of matter exceeds the critical density. If $q_0 = ½$, matter is at the critical density. Ignoring the cosmological force, such a universe just barely manages to expand forever.

Allowing for the existence of the cosmological force, $q_0$ could be negative if the universe has less than the critical density. This is because the cosmological force eventually

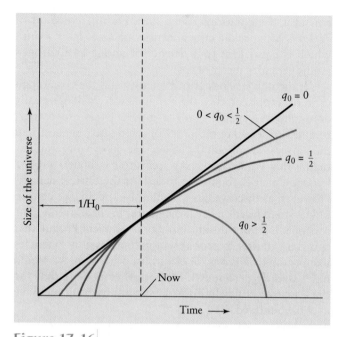

**Figure 17-16**    **Deceleration and the Size of the Universe**
This graph shows the possible evolution of the universe. If $q_0$ is less than ½, the universe is unbound and will expand forever. The universe is marginally bound if $q_0 = $ ½. If $q_0$ is greater than ½, the universe is bound and will someday collapse.

gram. If the data points fall above the $q_0 = $ ½ line, the universe is bound. If the data points fall below the $q_0 = $ ½ line, the universe is unbound.

Unfortunately, such observations are extremely difficult to make. The change in the universe's expansion rate is so slight that galaxies nearer than a billion light-years offer no help in determining $q_0$. And beyond a billion light-years, uncertainties cloud distance measurements. For example, Figure 17-17 shows data obtained by Allan Sandage with the 200-in. telescope at the Mount Palomar Observatory. Note how the data points are scattered about the $q_0 = $ ½ line. This scatter prevents us from determining conclusively whether the universe is bound or unbound. Nevertheless, because the data lies close to the $q_0 = $ ½ line, many astronomers think that the universe is close to being marginally bound.

Since the 1960s, team after team of astronomers have reported various values for $q_0$, some slightly larger than ½ and some slightly smaller. Because $q_0 = $ ½ is the special case separating a bound cosmological model from an unbound model, the predicted fate of the universe swung back and forth. Then, in 1998, the question took an unexpected turn when two groups reported that $q_0$ is *negative*, meaning that distant parts of the universe are accelerating outward, rather than decelerating. They made this assertion by observing Type Ia supernovae in distant galaxies (Figure 17-18).

dominates over gravity, causing the superclusters to accelerate away from each other. In that case, $q_0$ should properly be called the *acceleration parameter*.

Now, $q_0 = $ ½ is a truly special case. Calculations show that immediately after the Planck time, the slightest deviation from the precise critical density would have mushroomed the universe rapidly. If its density had been even slightly less than the critical density near the beginning of time, the universe would have doubled its diameter every $10^{-35}$ s, and the matter in it would have become so spread out that supercluster-sized gas clouds could never form.

If, on the other hand, its density had been slightly greater than the critical density, the expansion would have been much slower than actually occurred, and the universe would have quickly stopped expanding and collapsed in a Big Crunch. In other words, immediately after the Big Bang, the fate of the universe hung in the balance, like a pencil teetering on its point: The tiniest deviation from critical density would have rapidly propelled the universe away from $q_0 = $ ½.

Ever since Edwin Hubble discovered that the universe is expanding, astronomers have struggled to determine the deceleration parameter. In theory, it should be possible to determine $q_0$ by measuring the redshifts and distances of many remote galaxies and then plotting the data on a dia-

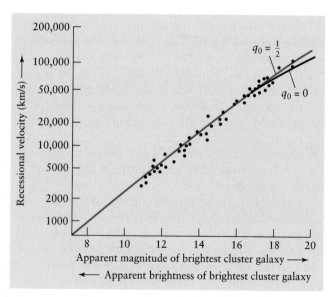

**Figure 17-17**    **The Hubble Diagram** This graph shows the Hubble law extended to include extremely remote galaxies. The magnitude of the brightest galaxy in a cluster is directly correlated with the distance to the cluster. If the data points fall between the curves marked $q_0 = 0$ and $q_0 = $ ½, then the universe is unbound. If the data points fall above the curve marked $q_0 = $ ½, then the universe is bound.

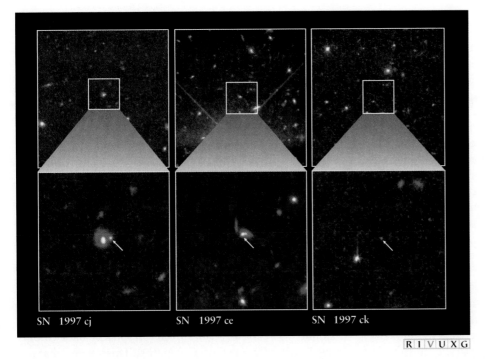

SN 1997 cj        SN 1997 ce        SN 1997 ck

R I V U X G

**Figure 17-18** **Distant Supernovae** These Hubble Space Telescope images show three distant supernovae whose light curves were used to determine the distance to the galaxies they are in. From left to right they are: SN 1997cj in Ursa Major, SN 1997ce in Lynx, and SN 1997ck in Hercules. SN 1997ck is the most distant supernova discovered to date, erupting about 7.7 billion years ago. The distances determined by these supernovae are greater than the distances determined from the cosmological redshift of the spectra from their galaxies. This supports the notion that an outward (cosmological) force is acting over vast distances in the universe.

The light curves (Section 12-8) for Type Ia supernovae (explosions of white dwarfs in binary star systems) are very well known. Furthermore, studies show that supernovae in nearby galaxies behave similarly to those observed in the Milky Way, regardless of their environment. Because they are so bright, such supernovae are excellent standard candles for objects billions of light-years away. Assuming that supernovae in distant galaxies also behave similarly to those in our Galaxy, the light curves of supernovae in distant galaxies reveal that these galaxies are farther away than predicted by the Hubble law. This means that distant galaxies are moving outward faster than they would if gravity from other superclusters was the only force acting on them. This can only happen if there is a cosmological force and the universe has insufficient mass to close itself. These results have profound implications for the entire study of astronomy, and hence they are being carefully scrutinized by many astronomers and astrophysicists around the world.

## 17-11 The average density of matter also affects the shape of the universe

Our present understanding of the shape of the universe is undergoing new scrutiny. It is based in part on the general theory of relativity, which predicts that gravity curves the fabric of spacetime. It is based in part on studies of different possible ways universes can be shaped, a study called *topology*. We consider the two possible situations: Our universe might have no regions that loop around to connect to other regions or, like a donut, it may have such loops. We consider each case separately.

If the universe has no bridges or loops connecting different parts, then there are three possibilities for the shape of space. Imagine shining two powerful laser beams out into space. Suppose that we can align these two beams so that they are perfectly parallel as they leave the Earth. Suppose, further, that nothing gets in the way of these two beams: We can follow them for billions of light-years across the universe, across the spacetime whose shape we wish to determine. The three possibilities for the paths and, hence, the shape of the universe, are:

1   Two beams of light remain perfectly parallel, even after traversing billions of light-years. In this case, space is not curved. That is, the universe has *zero curvature* and space is *flat*, and it extends without limit. If there is no cosmological force, this is the $q_0 = \frac{1}{2}$ universe. As we have seen, such a universe will just barely last forever. If there is a cosmological force, then a flat universe is also destined to last forever, eventually increasing in size at an accelerating rate.

2   Two beams of light gradually get closer and closer together as they move across the universe, eventually intersecting at some enormous distance from Earth. In this case, space is positively curved. Recall that lines of longitude on the Earth's surface are parallel at the equator but intersect at the poles. Thus, the shape of the universe is like the surface of a sphere: Space is spherical and the universe has *positive curvature*.

It is easiest to understand such a universe by visualizing a two-dimensional version of the universe as being the surface of a balloon on whose surface you must remain. In such a universe you could, in principle, keep going in a straight line (which is a curve on the balloon's surface) and eventually end up back where you started. In this case, $q_0 > \frac{1}{2}$. Like the surface of a balloon, a three-dimensional universe with positive curvature has no outside nor center and it is said to be **closed.** Without a cosmological force, a closed universe would expand to a finite size and then collapse back to the Big Crunch. With the cosmological force, a closed universe could expand forever, if that force is sufficiently strong. Otherwise, it would collapse.

3   Two beams of light gradually diverge farther and farther apart as they move across the universe. In this case, the universe has negative curvature. A horse's saddle is a good example of a negatively curved or hyperbolic surface. Parallel lines drawn on a saddle always diverge. Thus, in a negatively curved universe, we would describe space as *hyperbolic.* In this case $q_0 < \frac{1}{2}$. A hyperbolic universe extends without limit and is called **open.**

If the cosmological force does exist and is repulsive, as it now appears to be, such a universe will expand forever. On the other hand, an attractive cosmological force could actually cause a hyperbolic universe to eventually stop expanding and then contract.

Figure 17-19 illustrates these cases with two-dimensional analogies: a sphere, a plane, and a saddle. Of course, real space is three-dimensional. Both the flat and the hyperbolic universes are open and will therefore grow larger forever.

The second alternative for the overall structure of the universe is that it has one or more handles that connect different parts. If the universe has this property, we say it is *multiply connected.* It is then possible for the universe to be closed, but flat or hyperbolic. This is most easily visualized for a flat, two dimensional universe that is curved into the shape of a donut.

Imagine a flat version of our Galaxy (recall the balloon universe and the galaxies on it) located someplace on the donut's surface. Because of the different curves of the donut, you can draw lines from our Galaxy back to ourselves along many different paths. This means that we should be able to see the light from many other galaxies and other objects in space coming at us from different di-

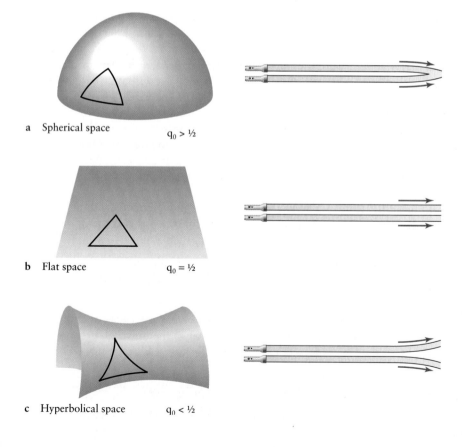

a   Spherical space          $q_0 > \frac{1}{2}$

b   Flat space               $q_0 = \frac{1}{2}$

c   Hyperbolical space       $q_0 < \frac{1}{2}$

**Figure 17-19** | **The Geometry of the Universe**  The shape of spacetime (represented here as two-dimensional for ease of visualization) is determined by the matter contained in the universe. The curvature is either **(a)** positive, **(b)** zero, or **(c)** negative, depending on whether the average density throughout space is greater than, equal to, or less than the critical density.

rections. In other words, we would see our Galaxy, and many other galaxies, in different places in the sky. Because light from these images travels different distances around the universe, we would be able to see images of the Milky Way at various earlier times.

To determine whether the universe is multiply connected, astronomers have begun studying the cosmic microwave background, to see if it shows any repetitive patterns indicating the same light coming from different directions. (It turns out that this is easier than looking for multiple images of galaxies, which could also, in principle, be done.) A multiply connected universe could have the accelerating property seen by the distant galaxies and yet have no outside. It would be a finite, ever-expanding universe.

## 17-12 Inflation accounts for the apparent flatness of the universe

Because observations reveal that $q_0$ is today approximately ½, the density of the universe immediately after the Big Bang must have been equal to the critical density to an incredibly high degree of precision. Calculations demonstrate that, in order for $q_0$ to be even roughly ½ today, density right after the Big Bang must have been equal to the critical density to better than 50 decimal places! What could have happened immediately after the Planck time to ensure density critical to such astounding accuracy? Because critical density means that space is flat, this enigma is called the **flatness problem.**

Inflation accounts for the flatness of the observable universe. The newborn universe might still have begun with a highly curved shape, because its mass might have strongly curved spacetime, just as the mass of a black hole curves spacetime today. Nonetheless, inflation expanded the universe so much that the portion we can see today looks quite flat. The observable universe, which extends outward from Earth at least 13 billion light-years in all di-

rections, is such a tiny fraction of the entire inflated universe that any overall curvature is virtually undetectable. To see why, think about a small portion of the Earth's surface, such as a playing field. For all practical purposes, it is impossible to detect the Earth's curvature over such a small area, so the field looks very flat. Like such a field, the segment of spacetime within Earth's cosmic particle horizon looks flat.

So what is the fate of the universe? There is growing, although by no means incontrovertible, evidence that the universe is unbound, possibly hyperbolic and, hence, it will last forever.

It is interesting that the amount of matter that has been identified is not thousands, millions, or trillions of times too little to close the universe and make it recollapse. Rather, the known mass is a few percent of the critical density. The average density of the universe has a crucial bearing on whether the current inflationary model is correct: This model of the Big Bang *requires* that the universe have exactly the critical density, so that $q_0 = ½$. The possible existence of the cosmological force will require that inflation and, indeed, much of our understanding of cosmic evolution be reconsidered. The story of cosmology is still being written.

## WHAT DID YOU THINK?

[1] *What is the universe?* It is all matter, energy, and spacetime.

[2] *Did the universe have a beginning?* Yes, it probably occurred between 13 billion and 18 billion years ago in the Big Bang.

[3] *Will the universe last forever?* Current observations support the belief that the universe will last forever.

## Key Words

## Key Ideas

### The Big Bang

✶ Most astronomers believe that the universe began as an exceedingly dense cosmic singularity that expanded explosively in an event called the Big Bang. The Hubble law describes the ongoing expansion of the universe and the rate at which superclusters of galaxies move apart.

✶ Depending on the actual value of Hubble's constant, the observable universe extends between 13 billion and 18 billion light-years in every direction from the Earth to what is called the cosmic particle horizon. We cannot see any objects that may exist beyond the cosmic particle horizon because light from these objects has not had enough time to reach us.

✶ According to the theory of inflation, matter originally near our location in the universe spread throughout a volume of the universe so large that we cannot yet observe much of it. The observable universe today is thus a growing volume of space containing matter and radiation that was in close contact with our matter and radiation during the first instant after the Big Bang. This explains the isotropic appearance of the universe.

### A Brief Physical History of Matter

✶ Four basic forces—gravity, electromagnetism, the strong nuclear force, and the weak nuclear force—explain the interactions observed in the universe.

✶ According to current theory, all four forces were identical just after the Big Bang. At the end of the Planck time (about $10^{-43}$ s after the Big Bang), gravity "froze out" to become a separate force. A short time later, the strong nuclear force became a distinct force. A final "freeze-out" separated the electromagnetic force from the weak nuclear force.

✶ Theory holds that before the Planck time, the universe was so dense that known laws of physics do not properly describe the behavior of spacetime, matter, and energy.

✶ During the first 300,000 years of the universe, astronomers believe that matter and energy formed an opaque plasma called the primordial fireball. Cosmic microwave background radiation is the greatly redshifted remnant of the universe as it existed about 300,000 years after the Big Bang.

✶ By 300,000 years after the Big Bang, spacetime expansion caused the temperature of the universe to fall below 3000 K,

enabling protons and electrons to combine to form hydrogen atoms. In the current model, this event is called the era of recombination. The universe became transparent during the era of recombination, meaning that the microwave background radiation contains the oldest photons in the universe.

✶ The universe today is matter-dominated, but during its first 300,000 years the universe was radiation-dominated.

✶ Clusters of galaxies and individual galaxies formed from pieces of enormous hydrogen and helium clouds, each of which became a separate supercluster of galaxies.

✶ All the superclusters of galaxies are moving away from each other.

### The Fate of the Universe

✶ The average density of matter in the universe determines the curvature of space and the ultimate fate of the universe.

✶ According to current models, if the average density of matter in the universe is greater than the critical density, then space is spherical (with positive curvature), the deceleration parameter $q_0$ has a value greater than ½, the universe is bound, and it will ultimately collapse.

✶ If the average density of matter in the universe is less than the critical density, then space is hyperbolic (with negative curvature), $q_0$ has a value less than ½, and the universe is unbound and will continue to expand forever.

✶ If the average density of matter in the universe is exactly equal to the critical density, then space is flat (with zero curvature), $q_0$ is exactly equal to ½, and the universe is marginally bound, and expansion will just barely continue forever.

✶ The observations that the universe is nearly flat and that the cosmic microwave background is almost perfectly isotropic may be the result of a brief period of very rapid expansion (the inflationary epoch) in the very early universe.

✶ There is initial evidence of a repulsive, cosmological force that is causing superclusters to accelerate outward. If confirmed, this means that the universe has less than the critical density of mass and that it will expand forever.

## Review Questions

1 What does it mean when astronomers say that we live in an expanding universe?

2 Explain the difference between a Doppler redshift and a cosmological redshift. Why is it incorrect to think of the

redshifts of remote galaxies and quasars as being a result of the Doppler effect?

3 In what ways are the fate of the universe, the shape of the universe, and the average density of the universe related?

4  Suppose that the universe will expand forever. What will eventually become of the microwave background radiation?

5  What does it mean to say that the universe is matter-dominated? When was the universe radiation-dominated? How did radiation domination show itself?

6  Explain the difference between an electron and an antielectron.

7  Where do astronomers believe most of the photons in the cosmic microwave background come from?

8  Describe an example of each of the four basic physical forces in the universe.

9  What is the observational evidence for (a) the Big Bang, (b) the inflationary epoch, and (c) the confinement of quarks?

## Advanced Questions

The answers to all computational problems, which are preceded by an asterisk (*), appear at the end of the book.

*10  How old would a $q_0 = \frac{1}{2}$ universe be if $H_0 = 60$ km/s/Mpc?

11  Explain why the detection of cosmic microwave background radiation is a major blow to the steady state theory.

12  The separation of nuclei by energetic photons in the early universe was caused by the same mechanism as what other transformation covered earlier in this book?

## Discussion Questions

13  Suppose we lived in a radiation-dominated universe. How would such a universe be different from what we now observe? How would we be different?

14  Discuss the theological implications of the idea that we cannot use science to tell us what existed before the Big Bang.

15  Assume that the cosmological force exists. Explain why gravitation dominated the behavior of the early universe, but why the cosmological force could eventually determine the fate of the universe.

## What If ...

16  The universe were destined to collaspe, and that collapse were under way? What would be different in space and on Earth under those conditions?

17  Our solar system formed very early in the life of our Galaxy, say when the universe was just two billion years old? What would be different in space and on Earth?

# The Search for Extraterrestrial Life

## *In this chapter you will discover*

✳ how scientists search for life that may exist elsewhere in the universe

✳ how scientists estimate how many planets orbiting other stars could support complex life

✳ the requirements scientists believe are necessary for life to exist

✳ the instruments used to search for life beyond our own solar system—and the results of those searches

✳ how we are trying to communicate with extraterrestrial life

**The Prevalence of Life** This painting, entitled *DNA Embraces the Planets*, artistically expresses the suspicion of many scientists that carbon-based life may be a common phenomenon in the universe. Other scientists argue, however, that we may be unique and no intelligent alien civilizations exist. In either case, humanity has the clear mandate to preserve and protect the abundance of life-forms with which we share our planet.

## WHAT DO YOU THINK?

1  How do astronomers search for extraterrestrial intelligence?

2  Have astronomers located any extraterrestrial civilizations?

The heavens inspire us to contemplation. We want to look beyond ourselves, to the formation of the Earth, the nature of the stars, and even the creation of the universe. Perhaps no subject is as compelling, however, as life elsewhere in the universe. Are we alone? What are the chances that we might someday make contact with an alien civilization?

There are no firm answers yet to these questions. We saw in Chapter 6 what may be fossilized bacterial life from Mars (see Figure 6-27). Life may also have evolved on Jupiter's moon Europa. But these locations are unsuitable for the evolution of complex, self-aware life, like ourselves. Despite all the alleged UFO sightings and alleged personal encounters, no one has produced evidence convincing to scientists that an intelligent extraterrestrial lifeform has *ever* visited Earth. Yet the search for extraterrestrial life continues, both in our solar system and beyond.

## 18-1  Our planet is special to us, but is not unique in its possibilities for life

As you have seen throughout this book, one of the great lessons of modern astronomy is that the circumstances that created the Earth are quite typical. We inhabit one of nine planets orbiting an ordinary star—just one among hundreds of billions of stars in the Milky Way Galaxy, which is just one of tens of billions of galaxies throughout the universe. We have already found more than a dozen planets orbiting other stars in our Galaxy, along with disks of gas and dust surrounding newly forming stars and similar to the one from which astronomers believe the Earth formed (see Section II-6). Furthermore, we have found water, essential for life as we know it, in many places in the solar system and beyond. The other elements necessary to make life as we know it are also ubiquitous in our Galaxy. Is it possible that we are also biologically commonplace? The question is at the heart of the search for extraterrestrial intelligence, or **SETI.**

All terrestrial life is based on the unique properties of the element carbon. Carbon atoms form chemical bonds that can combine to form especially long, complex molecules. These molecules can be further linked in elaborate chains, lattices, and fibers. No wonder carbon-based compounds, called **organic molecules,** are the stuff of life. The variety and stability of living organisms depend on the complex, self-regulating chemical reactions of organic molecules.

Carbon is also among the most abundant elements in the universe. This is exceptionally fortunate, because it is the only element suitable to serve as the backbone of life. Carbon is one of only five elements, along with boron (B), nitrogen (N), silicon (Si), and phosphorus (P), capable of bonding with three or more other atoms. Three is the minimum number of bonds necessary to allow for the complexity and diversity of biological molecules. Indeed, biologists have already identified more than six million carbon compounds. Carbon is unique among elements that can bond to three other atoms, because its bonds are flexible yet strong. Silicon-silicon bonds come apart under the slightest disturbance, in comparison, while silicon-oxygen bonds are so strong they create rocks. The other three elements that might serve as the backbone of life—boron, nitrogen, and phosphorous—have similar bonding problems.

Other constituents of organic molecules, such as hydrogen, oxygen, and sulfur, are also plentiful. In interstellar clouds, carbon atoms have combined with other elements to produce an impressive variety of organic compounds. Since the 1960s, radio astronomers have detected telltale microwave emission lines from interstellar clouds that help identify dozens of these carbon-based compounds. Examples include ethyl alcohol ($CH_3CH_2OH$), formaldehyde ($H_2CO$), methyl cyanoacetylene ($CH_3C_3N$), and acetaldehyde ($CH_3CHO$). The unique versatility and abundance of carbon suggest that extraterrestrial biology would also be based on organic chemistry.

Further evidence of extraterrestrial organic molecules comes from meteorites called carbonaceous chondrites (Figure 18-1). Many of the carbonaceous chondrites that have fallen to Earth contain a variety of organic substances. Some scientists wonder whether these organic molecules infiltrated the meteorites after they landed here. As noted in Chapter 8, carbonaceous chondrites are ancient meteorites dating from the formation of the solar system. If they contained organic molecules while in space, it seems reasonable to conclude that even from their earliest days, the planets have been continually bombarded with organic compounds. In 1997, NASA launched the *Stardust* mission to rendezvous with the comet Wild 2 in January 2004, collect dust and other debris from the comet, and return to Earth in 2006. Scientists hope to study this material for organic compounds uncontaminated by exposure to the Earth's atmosphere and surface.

Interstellar space is not the only source of organic material. In a classic experiment performed in 1952, the American chemists Stanley Miller and Harold Urey demonstrated that simple chemicals can combine to form prebiological compounds. Hoping to capture conditions on a primitive Earth, Miller and Urey combined hydrogen,

**Figure 18-1** **A Carbonaceous Chondrite** Carbonaceous chondrites are ancient meteorites that date back to the formation of the solar system. Chemical analysis of newly fallen specimens discloses that they are rich in organic compounds (arrow).

ammonia, methane, and water vapor in a closed container. They then subjected their mixture to an electric arc to simulate lightning bolts. At the end of a week, the inside of the container had become coated with a reddish-brown substance rich in compounds essential to life.

Actually, Earth's primordial atmosphere was more likely to have contained carbon dioxide, nitrogen, and hydrogen, along with water vapor "outgassed" from volcanoes and deposited by comet impacts. Modern versions of the Miller-Urey experiment have therefore used these common gases (Figure 18-2), and, once again, organic compounds are produced. Could these compounds be forming elsewhere in the universe today?

Scientists have not yet created life in a test tube. Biologists have yet to figure out, among other things, how simple organic molecules gathered into cells and developed systems for self-replication.

Moreover, widespread abundance of organic precursors does not guarantee that life is commonplace throughout the universe. If a planet is too close to a star, the temperature and ultraviolet radiation levels will be too high for life to evolve. If the planet is too far away, it will be too cold. If a planet's environment is hostile, life may either never get started or quickly become extinct. If the star is too massive, it will explode before life on the planets orbiting it has a chance to evolve. If the star is too small, the planet would have to be very close to it to be warm enough for life, but then tidal forces from the star would lock the planet in synchronous rotation, making most of its surface uninhabitable.

Nevertheless, many chemical components of life easily arise under conditions like those on the primordial Earth. It therefore seems reasonable to suppose that life could

have originated on suitably located Earthlike planets. Recall from Build Your Foundation II that astronomers have begun discovering extrasolar planets, although no terrestrial ones orbiting Sunlike stars have been found—yet.

## 18-2 The Drake equation predicts how many civilizations are in the Milky Way

Just how many planets are likely to harbor complex life throughout our Galaxy? The first person to tackle this question quantitatively was Frank Drake. Drake proposed that the number of technologically advanced civilizations in the Galaxy (designated by the letter $N$) could be estimated by the **Drake equation**:

$$N = R_* f_p n_e f_l f_i f_c L$$

**Figure 18-2** **The Miller-Urey Experiment Updated** Modern versions of this classic experiment prove that numerous organic compounds important to life can be synthesized from gases that were present in Earth's primordial atmosphere. This experiment supports the hypothesis that life on Earth arose as a result of ordinary chemical reactions.

in which

$R_*$ = rate at which solar-type stars form in the Galaxy

$f_p$ = fraction of stars that have planets

$n_e$ = number of planets per star system suitable for life

$f_l$ = fraction of those habitable planets on which life actually arises

$f_i$ = fraction of those life-forms that evolve into intelligent species

$f_c$ = fraction of those species that develop adequate technology and then choose to send messages out into space

$L$ = lifetime of that technologically advanced civilization

The Drake equation expresses the number of extraterrestrial civilizations quantitatively as a product of terms, some of which can be estimated from what we know about stars and stellar evolution. For instance, thanks to new evidence for extrasolar planets, astronomers can now hope to determine the first two terms, $R_*$ and $f_p$ by observation. We should probably exclude stars larger than about 1.5 $M_\odot$, because they have main-sequence lifetimes shorter than the time it took for intelligent life to develop here on Earth—some 3.8 to 4.0 billion years. If that is typical of the time needed to evolve higher life-forms, then a massive star probably becomes a giant or even explodes as a supernova before self-aware creatures can evolve on any of its planets.

Although low-mass stars have much longer lifetimes, they, too, seem unsuited for life because they are so cool. Only planets very near a low-mass star would be sufficiently warm for water to be a liquid, an essential ingredient for life to evolve as we know it. And, as mentioned earlier, a planet that close would become tidally coupled to the star, developing synchronous rotation. One side would have continuous daylight, while the other would be in perpetual, frigid darkness.

This leaves main-sequence stars like the Sun, those with spectral types between F5 and M0. Based on statistical studies of star formation in the Milky Way, astronomers calculate that roughly one of these Sunlike stars forms in the Galaxy each year, yielding a value of $R_* \approx 1$ per year.

We learned in Build Your Foundation II that the planets in our solar system formed as a natural consequence of the birth of the Sun. We have also seen evidence that similar processes of planetary formation may be commonplace around isolated stars. Many astronomers therefore give $f_p$ a value of 1, meaning they believe it likely that most Sunlike stars have planets.

Unfortunately, the rest of the terms in the Drake equation are very uncertain. Let's work with some hypothetical values. The chances that a planetary system has an Earthlike world are not known. Were we to consider our own solar system as representative, we could put $n_e$ at 1. Let's be more conservative, however, and suppose that 1 in 10 solar-type stars is orbited by a habitable planet, making $n_e = 0.1$.

From what we know about the evolution of life on Earth, we might assume that, given appropriate conditions, the development of life is a certainty, which would make $f_l = 1$. This is, of course, an area of intense interest to biologists. For the sake of argument, we might also assume that evolution naturally leads to the development of intelligence (a conjecture that is hotly debated) and also make $f_i = 1$. It is anyone's guess as to whether these intelligent extraterrestrial beings would attempt communication with other civilizations in the Galaxy, but if we assume they all would, $f_c$ would also be put at 1.

The last variable, $L$, the longevity of technological civilization, is the most uncertain of all—it cannot be tested! Looking at our own example, we see a planet whose atmosphere and oceans are increasingly polluted, potentially destroying the food chain. When we add in how close we have come to destroying ourselves with weapons of mass destruction, it may be that we humans are among the lucky few technological civilizations to squeak through its first years. In other words, $L$ may be as short as 100 years. Putting all these numbers together, we arrive at

$$N = 1 \times 1 \times 0.1 \times 1 \times 1 \times 1 \times 100 = 10$$

Therefore, out of the hundreds of billions of stars in the Galaxy, there may be only ten civilizations technologically advanced enough to communicate with us.

Of course, this is just an estimate. A wide range of values have been proposed for the terms in the Drake equation, and these various numbers produce vastly different estimates of $N$. Some scientists argue that there is exactly one advanced civilization in the Galaxy and that we are it. Others speculate that there may be tens of millions of planets inhabited by intelligent creatures. We just don't know.

## 18-3  SETI uses radio telescopes

How might we ascertain whether extraterrestrial civilizations exist, given the tremendous distances that separate us from other stars and the huge number of stars with the potential to support life-bearing planets? Looking for individual extrasolar planets is technically demanding and time consuming. As we saw in Section II-6, it demands careful observation of data available only in the past 20 years. Furthermore, the discovery of a planet does not imply that it harbors life. Astronomers therefore need a way to bypass the daunting search for planets, and, starting in the 1950s and 1960s, they identified a promising approach to searching directly for extraterrestrial intelligence. They look for radio transmissions from distant civilizations.

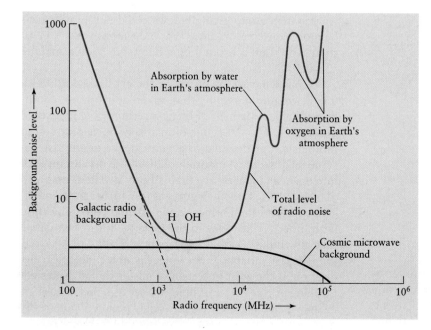

Figure 18-3 **The Water Hole** The so-called water hole is a range of radio frequencies from about 1000 to 10,000 megahertz (MHz) that happens to have relatively little cosmic noise. Some scientists suggest that this noise-free region would be well suited for interstellar communication.

As we have seen, radio waves can travel immense distances without being significantly altered by the interstellar medium. Because they penetrate gas and dust, radio waves are a logical choice for interstellar communication. Even if alien civilizations are not trying to communicate with us, they probably use radio waves in their own technology. However, where should we look in order to maximize the odds of detecting alien signals?

We could try random frequencies in the hopes of hearing something—anything. But the number of directions to look and frequencies to check are both staggering. We need to find some way to improve the odds. It turns out that some radio wavelengths travel farther without being absorbed by interstellar gas and dust, or without competing with noise from other sources, than do other wavelengths. SETI pioneer Bernard Oliver pointed out that a range of relatively clear frequencies exists in the neighborhood of the microwave emission lines of hydrogen and the hydroxyl radical (OH) (Figure 18-3). This region of the microwave part of the radio spectrum is called the **water hole**, a humorous reference to the H and OH lines being so close together. If extraterrestrial beings *are* purposefully sending messages into space, it seems reasonable that they would choose to transmit on these frequencies.

Alternatively, aliens might reason that other civilizations are already listening at certain scientifically important frequencies. For example, astronomers who study the distribution of hydrogen around the Galaxy have their radio telescopes tuned to 21-cm radiation (Chapter 14). Alien civilizations might therefore try to send signals toward us at this wavelength.

Even if only a few alien civilizations are scattered across the Galaxy, we have the technology to detect radio transmissions from them. One of the first searches began in 1960, when Frank Drake used a radio telescope at the National Radio Astronomy Observatory in West Virginia to "listen" to two Sunlike stars, τ (tau) Ceti and ε (epsilon) Eridani, but without success.

SETI searches use radio telescopes along with sophisticated electronic equipment and powerful computers to conduct both a targeted search and an all-sky survey. Most of the SETI programs in operation today are privately funded. Several are actively listening and analyzing enormous volumes of data. Indeed, there is so much data to analyze that one SETI project, called SETI@home, has enlisted the help of home computer users. These SETI researchers provide actual data from radio telescopes (Figure 18-4) to be analyzed, along with a data analysis program. While the computer's screensaver is on, the program runs, the data is analyzed, and the results are sent back to the researchers to be combined with data from other home computer users. The processed data is replaced with fresh data to be analyzed. Another SETI program plans to use an array of small, inexpensive TV satellite antennas to look for signals from space.

Occasionally, a SETI project does detect powerful or unusual signals from space. However, because none of these signals is ever repeated, SETI researchers do not believe they have yet discovered any extraterrestrial civilization. Despite at least 40 unsuccessful searches to date using radio telescopes around the world, the search for extraterrestrial intelligence continues.

The detection of a message from an alien civilization would be one of the greatest events in human history. Such a message could dramatically change the course of civilization through the sharing of scientific information or an awakening of social or humanistic enlightenment. In only a few years, our industry and social structure could advance centuries into the future. The technological and humanistic enlightenment would touch every person on Earth.

## 18-4 Humans have been sending signals into space for more than a century

We have been doing more than just listening passively for voices from the cosmos. Astronomers have intentionally broadcast well-focused signals through radio telescopes toward likely star systems. For more than a century, we humans have been transmitting other messages about ourselves, too. It all began with the first radio broadcast. Space within 100 ly of the solar system is now filled with signals from radio and television shows. If other advanced civilizations fall within that sphere, they could very well be listening to radio programs or watching television shows from decades past, trying to make sense of our species.

Launched in 1977, the two *Voyager* spacecraft are now making their way into interstellar space, carrying a more pedestrian message. Each has a plaque describing who and where we are, as well as phonograph records (have you ever seen one?) with human voices (Figure 18-5). Because space is so vast and the spacecraft are so

**Figure 18-4** **A Deep-Space Tracking Antenna** This dish-shaped antenna, located in the Mojave Desert in California, was originally built by NASA to track interplanetary spacecraft. It was recently used in an all-sky survey, along with an antenna at the Arecibo Observatory in Puerto Rico, to search for extraterrestrial intelligence. In 1996 an antenna in Canberra, Australia, joined the network.

a

b

**Figure 18-5** **Human Memorabilia on *Voyager*** The *Voyager 1* and *Voyager 2* spacecraft, now in interstellar space, each carries a plaque (**a**) with images of a man and a woman, as well as a phonograph record (**b**). There are also instructions for playing the record, which contains information about our biology, our technology, and our knowledge base. Each record also contains the sounds of children's voices. It is remotely possible that another race might someday discover the spacecraft.

small, the likelihood of their being discovered is truly remote. And yet millions, perhaps billions, of years from now, one of the spacecraft just might reach another race of intelligent creatures. From its path and the information on board, those creatures might be able to determine where the *Voyager* spacecraft came from. If their travel budgets are sufficiently large, they might even decide to come and visit us. At the least, they could send us radio messages. Will our descendants be here to receive them?

## WHAT DID YOU THINK?

1  *How do astronomers search for extraterrestrial intelligence?* They search for radio signals from other advanced civilizations.

2  *Have astronomers located any extraterrestrial civilizations?* No extraterrestrial civilizations have yet been discovered.

## Key Words

## Key Ideas

✳  The chemical building blocks of life exist throughout the Milky Way Galaxy. Organic molecules have been discovered in interstellar clouds, in some meteorites, and in comets.

✳  The Drake equation is used to estimate the number of technologically advanced civilizations in the Galaxy whose radio transmissions we might discover.

✳  Astronomers are using radio telescopes to search for signals from other self-aware life in the Galaxy. This effort is called the search for extraterrestrial intelligence, or SETI. SETI is primarily done at frequencies where radio waves pass most easily through the interstellar medium. So far, these searches have not detected any life off Earth.

✳  Everyday radio and television transmissions from Earth, along with intentional broadcasts into space, may be detected by other life-forms.

## Discussion Questions

1  What information would you have put on the record sent out on the *Voyager* spacecraft?

2  What arguments could you make against sending messages via radio or spacecraft into interstellar space?

3  Try using the Drake equation with values that you find reasonable. How many civilizations do you estimate there might be in our Galaxy?

4  Discuss the possible biological effects of our probes visiting other life-sustaining worlds. What social effects might probes from other worlds have on us?

5  List some of the pros and cons in the argument that alien spacecraft have visited the Earth.

6  Why would most of a planet locked in synchronous rotation with its star be uninhabitable?

## What If ...

7  Our planet was the moon of a much larger, Jupiterlike planet orbiting the Sun at 1 AU? What might the Earth and life inhabiting it be like?

# Life Had Begun on an Older Earth?

Our view of the universe is hardly more than a snapshot in time. The universe evolves, and Earth's surface, the eclipses, and the stars in the sky appear the way they do only because we experience them at this unique time. What would the universe look like if evolutionary processes on Earth had taken twice as long to produce creatures who were aware of their own existence? What if humans had evolved 9 billion years after the solar system formed, rather than after only 4.5 billion years?

**A Different Earth** As you have learned, our planet's crust is composed of tectonic plates floating on Earth's mantle. The continents are still drifting, and 4.5 billion years from now their distribution will be completely different. No pattern of land that we recognize today will remain intact. The Mediterranean Sea will be gone, and the Atlantic Ocean may be as wide as the Pacific Ocean is today, while the Pacific would probably be much narrower. The islands of the Pacific, from Japan south to Australia, may be a single connected unit. New islands will grow over hot spots in the mantle, just as the Hawaiian Islands do today. The Andes and the Himalayas will have tried to rise higher, only to come tumbling down due to weathering.

Another major change is the length of a day on an older Earth. It will be almost twice as long as our day, and the Moon will be about 1.5 times farther away, appearing correspondingly smaller in the sky and taking 1.8 times as long to orbit the planet. Both the longer day and more distant Moon will result from the gravitational interplay between ocean tides and the Moon. As tidal waters rub against the solid Earth, this friction slows the Earth's rotation and currently causes the day to grow 0.002 seconds longer each century. The energy lost as Earth spins down must be retained in the Earth-Moon system, so the Moon must gain energy in its orbit around Earth. This increase in energy currently forces the Moon to move farther away from Earth at a rate of 3.8 centimeters per year.

Because the Moon will appear smaller in the sky, "humans" on the older Earth will never see a total eclipse of the Sun. The best they will see is an annular eclipse, with a broad ring of the Sun showing around the edges of the Moon. Total lunar eclipses will still occur because the older Earth's shadow will remain wider than the Moon.

**Changes in the Night Sky** Presently, Polaris appears almost exactly over Earth's north rotation pole. But citizens of the older Earth will not see a pole star for two reasons. First, Earth's rotation axis precesses, or wobbles, by 23.5° over a period of 25,800 years, making the pole wander around the sky. It is just by chance that in recent centuries there has been a bright star to mark the north pole. It is unlikely that such a bright star would mark the pole when humans evolve on the older Earth. Second, many bright stars that could be near the pole in the future may no longer be visible, having used up their nuclear fuels in the next 4.5 billion years and exploded.

Perhaps new bright stars will form that will take turns as future polar guide stars as Earth's rotation axis precesses, much as Polaris does now and Thuban in Draco did in the past. Asterisms (patterns of stars), such as the Big Dipper or Orion, will all be different, too, because many stars that exist in our sky now will have exploded within the next 4.5 billion years, while the others will all move relative to each other.

Older Earth's astronomers will, like Edwin Hubble, discover that distant galaxies appear to move away from us because of the expansion of the universe. However, the Hubble constant is not at all constant. When those astronomers measure the speed of distant galaxies moving away from us, they will get a very different number for the Hubble constant.

**A Brighter Sun** The Sun in the time of the older Earth will still have nearly the same surface temperature and yellowish hue that it has now, but it will appear about twice as bright because it will be about 60% bigger. The extra 4.5 billion years will have begun to take their toll on the Sun's nuclear fuel supply.

The Sun has fused hydrogen into helium throughout its present lifetime of 4.5 billion years, using up less than half of the available hydrogen in its core. In another 4.5 billion years, a total of 80% to 90% of the available hydrogen in the core will have been converted into helium. While the Sun's giant phase would still be a billion years off, the humans of the older Earth would face changes to the planet and the solar system as the Sun's output increased. These changes, such as a runaway greenhouse effect in which much of Earth's surface water becomes vaporized, could potentially force those humans to devise a way to quickly leave the planet and colonize elsewhere in the Galaxy.

# APPENDIX

## Tables of Data

### Table A-1
### The Planets: Orbital Data

| Planet | Semimajor axis (AU) | Semimajor axis (10⁶ km) | Sidereal period (yr) | Sidereal period (d) | Synodic period (d) | Mean orbital speed (km/s) | Orbital eccentricity | Inclination of orbit to ecliptic (°) |
|---|---|---|---|---|---|---|---|---|
| Mercury | 0.3871 | 57.9 | 0.2408 | 87.97 | 115.88 | 47.9 | 0.206 | 7.00 |
| Venus | 0.7233 | 108.2 | 0.6152 | 224.70 | 583.92 | 35.0 | 0.007 | 3.39 |
| Earth | 1.0000 | 149.6 | 1.0000 | 365.26 | ——— | 29.8 | 0.017 | 0.00 |
| Mars | 1.5237 | 227.9 | 1.8809 | 686.98 | 779.94 | 24.1 | 0.093 | 1.85 |
| (Ceres) | 2.7656 | 413.7 | 4.603 | 1,681.3 | 466.6 | 17.9 | 0.097 | 10.61 |
| Jupiter | 5.2028 | 778.3 | 11.862 | 4,332.7 | 398.9 | 13.1 | 0.048 | 1.31 |
| Saturn | 9.5388 | 1427.0 | 29.458 | 10,760 | 378.1 | 9.6 | 0.056 | 2.49 |
| Uranus | 19.1914 | 2871.0 | 84.01 | 30,685 | 369.7 | 6.8 | 0.046 | 0.77 |
| Neptune | 30.0611 | 4497.1 | 164.79 | 60,082 | 367.5 | 5.4 | 0.010 | 1.77 |
| Pluto | 39.5294 | 5913.5 | 248.5 | 90,767 | 366.7 | 4.7 | 0.248 | 17.15 |

### Table A-2
### The Planets: Physical Data

| Planet | Equatorial diameter (km) | Equatorial diameter (Earth = 1) | Mass (kg) | Mass (Earth = 1) | Average density (kg/m³) | Rotation period* (days) | Inclination of equator to orbit (°) | Surface gravity (Earth = 1) | Albedo | Escape speed (km/s) |
|---|---|---|---|---|---|---|---|---|---|---|
| Mercury | 4,879 | 0.383 | $3.302 \times 10^{23}$ | 0.055 | 5430 | 58.646 | 2 (?) | 0.39 | 0.106 | 4.3 |
| Venus | 12,104 | 0.949 | $4.869 \times 10^{24}$ | 0.815 | 5240 | $243.01^{R}$ | 177.3 | 0.91 | 0.65 | 10.4 |
| Earth | 12,756 | 1.000 | $5.974 \times 10^{24}$ | 1.000 | 5515 | 1.000 | 23.45 | 1.000 | 0.37 | 11.2 |
| Mars | 6,794 | 0.533 | $6.419 \times 10^{23}$ | 0.107 | 3940 | 1.026 | 25.19 | 0.38 | 0.15 | 5.0 |
| Jupiter | 142,984 | 11.209 | $1.899 \times 10^{27}$ | 317.83 | 1330 | 0.414 | 3.12 | 2.5 | 0.52 | 59.5 |
| Saturn | 120,536 | 9.449 | $5.685 \times 10^{26}$ | 95.16 | 700 | 0.444 | 26.73 | 1.1 | 0.47 | 35.5 |
| Uranus | 51,118 | 4.007 | $8.662 \times 10^{25}$ | 14.50 | 1300 | $0.718^{R}$ | 97.86 | 0.90 | 0.50 | 21.3 |
| Neptune | 49,528 | 3.883 | $1.028 \times 10^{26}$ | 17.204 | 1640 | 0.671 | 29.56 | 1.1 | 0.5 | 23.5 |
| Pluto | 2,300 | 0.18 | $1.3 \times 10^{22}$ | 0.002 | 2030 | $6.387^{R}$ | 118 | 0.07 | 0.5 | 1.3 |

* For Jupiter, Saturn, Uranus, and Neptune, the internal rotation period is given. A superscript R means that the rotation is retrograde (opposite the planet's orbital motion).

### Table A-3
### Satellites of the Planets

| Planet | Satellite | Discoverers | Average distance from center of planet (km) | Orbital (sidereal) period* (days) | Orbital eccentricity | Diameter of satellite (km) | Mass (kg) |
|---|---|---|---|---|---|---|---|
| EARTH | Moon | — | 384,400 | 27.322 | 0.0549 | 3476 | $7.348 \times 10^{22}$ |
| MARS | Phobos | Hall (1877) | 9,380 | 0.319 | 0.01 | $28 \times 23 \times 20$ | $1.1 \times 10^{16}$ |
| | Deimos | Hall (1877) | 23,460 | 1.263 | 0.00 | $16 \times 12 \times 10$ | $1.8 \times 10^{15}$ |
| JUPITER | Metis | Synott (1979) | 127,960 | 0.2948 | 0.00 | 40 | $1 \times 10^{17}$ |
| | Adrastea | Jewitt et al. (1979) | 128,980 | 0.2983 | 0 (?) | $24 \times 16 \times 20$ | $1.9 \times 10^{16}$ |
| | Amalthea | Barnard (1892) | 181,300 | 0.4981 | 0.00 | $270 \times 200 \times 155$ | $7.2 \times 10^{18}$ |
| | Thebe | Synott (1979) | 221,900 | 0.6745 | 0.01 | 100 | $8 \times 10^{17}$ |
| | Io | Galileo (1610) | 421,600 | 1.769 | 0.00 | 3630 | $8.94 \times 10^{22}$ |
| | Europa | Galileo (1610) | 670,900 | 3.551 | 0.01 | 3138 | $4.92 \times 10^{22}$ |
| | Ganymede | Galileo (1610) | 1,070,000 | 7.155 | 0.00 | 5262 | $1.48 \times 10^{23}$ |
| | Callisto | Galileo (1610) | 1,883,000 | 16.689 | 0.01 | 4800 | $1.08 \times 10^{23}$ |
| | Leda | Kowal (1974) | 11,094,000 | 238.72 | 0.15 | 16 | $6 \times 10^{15}$ |
| | Himalia | Perrine (1904) | 11,480,000 | 250.57 | 0.16 | 180 | $1 \times 10^{19}$ |
| | Lysithea | Nicholson (1938) | 11,720,000 | 259.22 | 0.11 | 40 | $8 \times 10^{16}$ |
| | Elara | Perrine (1905) | 11,737,000 | 259.65 | 0.21 | 80 | $8 \times 10^{17}$ |
| | Ananke | Nicholson (1951) | 21,200,000 | $631^R$ | 0.17 | 30 | $4 \times 10^{16}$ |
| | Carme | Nicholson (1938) | 22,600,000 | $692^R$ | 0.21 | 44 | $1 \times 10^{17}$ |
| | Pasiphae | Melotte (1908) | 23,500,000 | $735^R$ | 0.38 | 70 | $2 \times 10^{17}$ |
| | Sinope | Nicholson (1914) | 23,700,000 | $758^R$ | 0.28 | 40 | $8 \times 10^{16}$ |
| SATURN | Pan | Showlater (1990) | 133,570 | 0.573 | 0.00 | 20 | (?) |
| | Atlas | Terrile (1980) | 137,640 | 0.602 | 0 (?) | $40 \times 30 \times 30$ | (?) |
| | Prometheus | Collins et al. (1980) | 139,350 | 0.613 | 0.00 | $140 \times 80 \times 100$ | $3 \times 10^{17}$ |
| | Pandora | Collins et al. (1980) | 141,700 | 0.629 | 0.00 | $110 \times 70 \times 100$ | $2 \times 10^{17}$ |
| | Epimetheus | Walker (1966) | 151,422 | 0.694 | 0.01 | $140 \times 100 \times 100$ | $6 \times 10^{17}$ |
| | Janus | Dolfuss (1966) | 151,472 | 0.695 | 0.01 | $220 \times 160 \times 200$ | $2 \times 10^{18}$ |
| | Mimas | Herschel (1789) | 185,520 | 0.942 | 0.02 | 390 | $3.8 \times 10^{19}$ |
| | Enceladus | Herschel (1789) | 238,020 | 1.370 | 0.00 | 500 | $7.3 \times 10^{19}$ |
| | Tethys | Cassini (1684) | 294,660 | 1.888 | 0.00 | 1050 | $6.2 \times 10^{20}$ |
| | Calypso | Smith et al. (1980) | 294,660 | 1.888 | 0 (?) | $30 \times 20 \times 25$ | (?) |
| | Telesto | Smith et al. (1980) | 294,660 | 1.888 | 0 (?) | 24 | (?) |
| | Dione | Cassini (1684) | 377,400 | 2.737 | 0.00 | 1120 | $1.1 \times 10^{21}$ |
| | Helene | Laques et al. (1980) | 377,400 | 2.737 | 0.01 | $40 \times 30 \times 30$ | (?) |
| | Rhea | Cassini (1672) | 527,040 | 4.518 | 0.00 | 1530 | $2.3 \times 10^{21}$ |
| | Titan | Huygens (1655) | 1,221,850 | 15.945 | 0.03 | 5150 | $1.4 \times 10^{23}$ |
| | Hyperion | Bond (1848) | 1,481,000 | 21.277 | 0.10 | $410 \times 260 \times 220$ | $2 \times 10^{19}$ |
| | Iapetus | Cassini (1671) | 3,561,300 | 79.331 | 0.03 | 1440 | $1.6 \times 10^{21}$ |
| | Phoebe | Pickering (1898) | 12,952,000 | $550.48^R$ | 0.16 | 220 | $4 \times 10^{18}$ |

**Table A-3(continued)**
**Satellites of the Planets**

| Planet | Satellite | Discoverers | Average distance from center of planet (km) | Orbital (sidereal) period* (days) | Orbital eccentricity | Diameter of satellite (km) | Mass (kg) |
|---|---|---|---|---|---|---|---|
| URANUS | Cordelia | Voyager 2 (1986) | 49,750 | 0.335 | 0 (?) | 40 | (?) |
| | Ophelia | Voyager 2 (1986) | 53,760 | 0.376 | 0 (?) | 30 | (?) |
| | Bianca | Voyager 2 (1986) | 59,160 | 0.435 | 0 (?) | 40 | (?) |
| | Cressida | Voyager 2 (1986) | 61,770 | 0.464 | 0 (?) | 70 | (?) |
| | Desdemona | Voyager 2 (1986) | 62,660 | 0.474 | 0 (?) | 60 | (?) |
| | Juliet | Voyager 2 (1986) | 64,360 | 0.493 | 0 (?) | 80 | (?) |
| | Portia | Voyager 2 (1986) | 66,100 | 0.513 | 0 (?) | 110 | (?) |
| | Rosalind | Voyager 2 (1986) | 69,930 | 0.558 | 0 (?) | 60 | (?) |
| | Belinda | Voyager 2 (1986) | 75,260 | 0.624 | 0 (?) | 70 | (?) |
| | Puck | Voyager 2 (1986) | 86,010 | 0.762 | 0 (?) | 150 | (?) |
| | Miranda | Kuiper (1948) | 129,780 | 1.414 | 0.00 | 470 | $6.6 \times 10^{19}$ |
| | Ariel | Lassell (1851) | 191,240 | 2.520 | 0.00 | 1160 | $1.4 \times 10^{21}$ |
| | Umbriel | Lassell (1851) | 265,790 | 4.144 | 0.00 | 1170 | $1.2 \times 10^{21}$ |
| | Titania | Herschel (1787) | 435,840 | 8.706 | 0.00 | 1580 | $3.5 \times 10^{21}$ |
| | Oberon | Herschel (1787) | 582,600 | 13.463 | 0.00 | 1520 | $3.0 \times 10^{21}$ |
| | Caliban | Gladman, Nicholson, Burns, Kavelaars (1997) | 5,753,400 | $415^R$ | 0.16 | 80 | (?) |
| | Sycorax | Gladman, Nicholson, Burns, Kavelaars (1997) | 8,000,000 | $415^R$ | 0.52 | 160 | (?) |
| NEPTUNE | Naiad | Voyager 2 (1989) | 48,230 | 0.296 | 0 (?) | 60 | (?) |
| | Thalassa | Voyager 2 (1989) | 50,070 | 0.312 | 0 (?) | 80 | (?) |
| | Despina | Voyager 2 (1989) | 52,530 | 0.333 | 0 (?) | 150 | (?) |
| | Galatea | Voyager 2 (1989) | 61,950 | 0.429 | 0 (?) | 180 | (?) |
| | Larissa | Voyager 2 (1989) | 73,550 | 0.554 | 0 (?) | 190 | (?) |
| | Proteus | Voyager 2 (1989) | 117,640 | 1.121 | 0 (?) | 415 | (?) |
| | Triton | Lassell (1846) | 354,800 | $5.877^R$ | 0.00 | 2700 | $2.15 \times 10^{22}$ |
| | Nereid | Kuiper (1949) | 5,513,400 | 359.2 | 0.75 | 340 | (?) |
| PLUTO | Charon | Christy (1978) | 19,640 | 6.387 | 0.00 | 1190 | $1.9 \times 10^{21}$ |

* A superscript R means that the satellite orbits in a retrograde direction (opposite to the planet's rotation).

## Table A-4
### The Nearest Stars

| Name* | Parallax (arcsec) | Distance (light-years) | Spectral type | Radial velocity** (km/s) | Proper motion (arcsec/year) | Apparent visual magnitude | Absolute visual magnitude | Luminosity (Sun = 1) |
|---|---|---|---|---|---|---|---|---|
| Sun | | | G2 V | | | −26.7 | +4.83 | 1.00 |
| Proxima Centauri | 0.772 | 4.22 | M5.5 V | −22 | 3.853 | +11.01 | +15.45 | $8.2 \times 10^{-4}$ |
| Alpha Centauri A | 0.742 | 4.40 | G2 V | −25 | 3.710 | −0.01 | +4.34 | 1.77 |
| Alpha Centauri B | 0.742 | 4.40 | K0 V | −21 | 3.724 | +1.35 | +5.70 | 0.55 |
| Barnard's Star | 0.549 | 5.94 | M4 V | −111 | 10.358 | +9.54 | +13.24 | $3.6 \times 10^{-3}$ |
| Wolf 359 | 0.418 | 7.80 | M6 V | +13 | 4.689 | +13.45 | +16.6 | $3.5 \times 10^{-4}$ |
| Lalande 21185 | 0.392 | 8.32 | M2 V | −84 | 4.802 | +7.49 | +10.46 | 0.023 |
| L 726-8 A | 0.381 | 8.56 | M5.5 V | +29 | 3.366 | +12.41 | +15.3 | $9.4 \times 10^{-4}$ |
| L 726-8 B | 0.381 | 8.56 | M6 V | +32 | 3.366 | +13.2 | +16.1 | $5.6 \times 10^{-4}$ |
| Sirius A | 0.379 | 8.61 | A1 V | −9 | 1.339 | −1.44 | +1.45 | 26.1 |
| Sirius B | 0.379 | 8.61 | white dwarf | −9 | 1.339 | +8.44 | +11.3 | $2.4 \times 10^{-3}$ |
| Ross 154 | 0.336 | 9.71 | M3.5 V | −12 | 0.666 | +10.37 | +13.00 | $4.1 \times 10^{-3}$ |
| Ross 248 | 0.316 | 10.32 | M5.5 V | −78 | 1.626 | +12.29 | +14.8 | $1.5 \times 10^{-3}$ |
| Epsilon Eridani | 0.311 | 10.49 | K2 V | +17 | 0.977 | +3.72 | +6.18 | 0.40 |
| Lacaille 9352 | 0.304 | 10.73 | M1.5 V | +10 | 6.896 | +7.35 | +9.76 | 0.051 |
| Ross 128 | 0.300 | 10.87 | M4 V | −31 | 1.361 | +11.12 | +13.50 | $2.9 \times 10^{-3}$ |
| L 789-6 | 0.294 | 11.09 | M5 V | −60 | 3.259 | +12.33 | +14.7 | $1.3 \times 10^{-3}$ |
| 61 Cygni A | 0.287 | 11.36 | K5 V | −65 | 5.281 | +5.20 | +7.49 | 0.16 |
| 61 Cygni B | 0.285 | 11.44 | K7 V | −64 | 5.172 | +6.05 | +8.33 | 0.095 |
| Procyon A | 0.286 | 11.40 | F5 IV–V | −4 | 1.259 | +0.40 | +2.68 | 7.73 |
| Procyon B | 0.286 | 11.40 | white dwarf | −4 | 1.259 | +10.7 | +13.0 | $5.5 \times 10^{-4}$ |
| BD +59° 1915 A | 0.281 | 11.61 | M3 V | −1 | 2.238 | +8.94 | +11.18 | 0.020 |
| BD +59° 1915 B | 0.281 | 11.61 | M3.5 V | +1 | 2.313 | +9.70 | +11.97 | 0.010 |
| Groombridge 34 A | 0.280 | 11.65 | M1.5 V | +12 | 2.918 | +8.09 | +10.33 | 0.030 |
| Groombridge 34 B | 0.280 | 11.65 | M3.5 V | +11 | 2.918 | +11.07 | +13.3 | $3.1 \times 10^{-3}$ |
| Epsilon Indi | 0.276 | 11.82 | K5 V | −40 | 4.704 | +4.69 | +6.89 | 0.27 |
| GJ 1111 | 0.276 | 11.82 | M6.5 V | −5 | 1.288 | +14.79 | +17.0 | $2.7 \times 10^{-4}$ |
| Tau Ceti | 0.274 | 11.90 | G8 V | −17 | 1.922 | +3.49 | +5.68 | 0.62 |
| GJ 1061 | 0.270 | 12.08 | M5.5 V | −20 | 0.836 | +13.03 | +15.2 | $1.0 \times 10^{-3}$ |
| L 725-32 | 0.269 | 12.12 | M4.5 V | +28 | 1.372 | +12.10 | +14.25 | $1.7 \times 10^{-3}$ |
| BD +05° 1668 | 0.263 | 12.40 | M3.5 V | +18 | 3.738 | +9.84 | +11.94 | 0.011 |
| Kapteyn's star | 0.255 | 12.79 | M0 V | +246 | 8.670 | +8.86 | +10.89 | 0.013 |
| Lacaille 8760 | 0.253 | 12.89 | M0 V | +28 | 3.455 | +6.69 | +8.71 | 0.094 |
| Krüger 60 A | 0.250 | 13.05 | M3 V | −33 | 0.990 | +9.85 | +11.9 | 0.010 |
| Krüger 60 B | 0.250 | 13.05 | M4 V | −32 | 0.990 | +11.3 | +13.3 | $3.4 \times 10^{-3}$ |

* Stars that are components of binary systems are labeled A and B.
** A positive radial velocity means the star is receding; a negative radial velocity means the star is approaching.

Compiled from the Hipparcos General Catalogue and from data reported by the Research Consortium on Nearby Stars. The table lists all known stars within 4.00 parsecs (13.05 light-years).

### Table A-5
### The Visually Brightest Stars

| Name | Designation | Distance (light-years) | Spectral type | Radial velocity* (km/s) | Proper motion (arcsec/year) | Apparent visual magnitude | Apparent visual brightness** (Sirius = 1) | Absolute visual magnitude | Luminosity (Sun = 1) |
|---|---|---|---|---|---|---|---|---|---|
| Sirius A | α CMa A | 8.61 | A1 V | −9 | 1.339 | −1.44 | 1.000 | +1.45 | 26.1 |
| Canopus | α Car | 313 | F0 I | +21 | 0.031 | −0.62 | 0.470 | −5.53 | $1.4 \times 10^4$ |
| Arcturus | α Boo | 36.7 | K2 III | −5 | 2.279 | −0.05 | 0.278 | −0.31 | 190 |
| Rigel Kentaurus | α Cen A | 4.4 | G2 V | −25 | 3.71 | −0.01 | 0.268 | +4.34 | 1.77 |
| Vega | α Lyr | 25.3 | A0 V | −14 | 0.035 | +0.03 | 0.258 | +0.58 | 61.9 |
| Capella | α Aur | 42.2 | G8 III | +30 | 0.434 | +0.08 | 0.247 | −0.48 | 180 |
| Rigel | β Ori A | 773 | B8 Ia | +21 | 0.002 | +0.18 | 0.225 | −6.69 | $7.0 \times 10^5$ |
| Procyon | α CMi A | 11.4 | F5 IV–V | −4 | 1.259 | +0.4 | 0.184 | +2.68 | 7.73 |
| Achernar | α Eri | 144 | B3 IV | +19 | 0.097 | +0.45 | 0.175 | −2.77 | 5250 |
| Betelgeuse | α Ori | 427 | M2 Iab | +21 | 0.029 | +0.45 | 0.175 | −5.14 | $4.1 \times 10^4$ |
| Hadar | β Cen | 525 | B1 II | −12 | 0.042 | +0.61 | 0.151 | −5.42 | $8.6 \times 10^4$ |
| Altair | α Aql | 16.8 | A7 IV–V | −26 | 0.661 | +0.76 | 0.132 | +2.2 | 11.8 |
| Aldebaran | α Tau A | 65.1 | K5 III | +54 | 0.199 | +0.87 | 0.119 | −0.63 | 370 |
| Spica | α Vir | 262 | B1 V | +1 | 0.053 | +0.98 | 0.108 | −3.55 | $2.5 \times 10^4$ |
| Antares | α Sco A | 604 | M1 Ib | −3 | 0.025 | +1.06 | 0.100 | −5.28 | $3.7 \times 10^4$ |
| Pollux | β Gem | 33.7 | K0 III | +3 | 0.627 | +1.16 | 0.091 | +1.09 | 46.6 |
| Fomalhaut | α PsA | 25.1 | A3 V | +7 | 0.368 | +1.17 | 0.090 | +1.74 | 18.9 |
| Deneb | α Cyg | 3230 | A2 Ia | −5 | 0.002 | +1.25 | 0.084 | −8.73 | $3.2 \times 10^5$ |
| Mimosa | β Cru | 353 | B0.5 III | +20 | 0.05 | +1.25 | 0.084 | −3.92 | $3.4 \times 10^4$ |
| Regulus | α Leo A | 77.5 | B7 V | +4 | 0.249 | +1.36 | 0.076 | −0.52 | 331 |

Data in this table was compiled from the Hipparcos General Catalogue.

* A positive radial velocity means the star is receding; a negative radial velocity means the star is approaching.

** This is the ratio of the star's apparent brightness to that of Sirius, the brightest star in the night sky.

Note: Acrux, or α Cru (the brightest star in Crux, the Southern Cross), appears to the naked eye as a star of apparent magnitude +0.87, the same as Aldebaran, but it does not appear in this table because Acrux is actually a binary star system. The blue-white component stars of this binary system have apparent magnitudes of +1.4 and +1.9, and thus they are dimmer than any of the stars listed here.

## Table A-6
### Some Important Astronomical Quantities

| | |
|---|---|
| Astronomical unit: | 1 AU = $1.496 \times 10^{11}$ m |
| Light-year: | 1 ly = $9.461 \times 10^{15}$ m = 63,240 AU |
| Parsec: | 1 pc = $3.086 \times 10^{16}$ m = 3.262 ly |
| Solar mass: | 1 $M_\odot$ = $1.989 \times 10^{30}$ kg |
| Solar radius: | 1 $R_\odot$ = $6.960 \times 10^{8}$ m |
| Solar luminosity: | 1 $L_\odot$ = $3.827 \times 10^{26}$ W |
| Earth's mass: | 1 $M_\oplus$ = $5.974 \times 10^{24}$ kg |
| Earth's equatorial radius: | 1 $R_\oplus$ = $6.378 \times 10^{6}$ m |
| Moon's mass: | 1 $M_\oplus$ = $7.348 \times 10^{22}$ kg |
| Moon's equatorial radius: | 1 $R_\oplus$ = $1.738 \times 10^{6}$ m |

## Table A-7
### Some Important Physical Constants

| | |
|---|---|
| Speed of light: | $c = 2.998 \times 10^{8}$ m/s |
| Gravitational constant: | $G = 6.668 \times 10^{-11}$ N m$^2$ kg$^{-2}$ |
| Planck constant: | $h = 6.625 \times 10^{-34}$ J s |
| | $= 4.136 \times 10^{-15}$ eV s |
| Boltzmann constant: | $k = 1.380 \times 10^{-23}$ J K$^{-1}$ |
| | $= 8.617 \times 10^{-5}$ eV K$^{-1}$ |
| Stefan–Boltzmann constant: | $\sigma = 5.669 \times 10^{-8}$ W m$^{-2}$ K$^{-4}$ |
| Mass of electron: | $m_e = 9.108 \times 10^{-31}$ kg |
| Mass of $^1$H atom: | $m_H = 1.673 \times 10^{-27}$ kg |

# AN ASTRONOMER'S TOOLBOX A-1

## Temperature Scales

Three temperature scales are in common use. Throughout most of the world, temperatures are expressed in degrees Celsius (°C), named in honor of the Swedish astronomer Anders Celsius, who proposed it in 1742. The Celsius temperature scale (also known as the "centigrade scale") is based on the behavior of water, which freezes at 0°C and boils at 100° C at sea level on Earth.

Scientists usually prefer the Kelvin scale, named after the British physicist Lord Kelvin (William Thomson), who made many important contributions to our knowledge about heat and temperature. On the Kelvin temperature scale, water freezes at 273 K and boils at 373 K. Note that we do not use the degree symbol with the Kelvin temperature scale.

Because water must be heated to 100 K or 100°C to go from its freezing point to its boiling point, you can see that the size of a Kelvin is the same as the size of a degree Celsius. When considering temperature *changes,* measurements in Kelvins and in degrees Celsius lead to the same number.

A temperature expressed in Kelvins is always equal to the temperature in degrees Celsius plus 273. Scientists prefer the Kelvin scale because it is closely related to the physical meaning of temperature. All substances are made of atoms, which are very tiny (a typical atom has a diameter of about $10^{-10}$ m) and constantly in motion. The temperature of a substance is directly related to the average speed of its atoms. If something is hot, its atoms are moving at high speeds. If a substance is cold, its atoms are moving much more slowly.

The coldest possible temperature is the temperature at which atoms move as slowly as possible (they can never quite stop completely). This minimum possible temperature, called absolute zero, is the starting point for the Kelvin scale. Absolute zero is 0 K, or −273°C. Because it is impossible for anything to be colder than 0 K, there are no negative temperatures on the Kelvin scale.

In the United States, many people still use the now-archaic Fahrenheit scale, which expresses temperatures in degrees Fahrenheit (°F). When the German physicist Gabriel Fahrenheit introduced this scale in the early 1700s, he intended 0°F to represent the coldest temperature then achievable (with a mixture of ice and saltwater) and 100°F to represent the temperature of a healthy human body. On the Fahrenheit scale, water freezes at 32°F and boils at 212°F. Because there are 180 degrees Fahrenheit between the freezing and boiling points of water, a degree Fahrenheit is only 100/180 (= 5/9) the size of the other scales.

The following equation converts from degrees Fahrenheit to degrees Celsius:

$$T_C = \frac{5}{9} (T_F - 32)$$

To convert from Celsius to Fahrenheit, a simple rearrangement of terms gives the relationship

$$T_F = \frac{9}{5} T_C + 32$$

where $T_F$ is the temperature in degrees Fahrenheit and $T_C$ is the temperature in degrees Celsius.

*Example:* Consider a typical room temperature of 68°F. Using the second equation, we can convert this measurement to the Celsius scale as follows:

$$T_C = \frac{5}{9} (68 - 32) \approx 20°C$$

To arrive at the Kelvin scale, we simply add 273 degrees to the value in degrees Celsius. Thus, 68°F ≈ 20°C = 293 K

*Compare!* The accompanying figure displays the relationships among these three temperature scales.

Kelvin    Celsius    Fahrenheit

$15.5 \times 10^6$    $15.5 \times 10^6$    $27.9 \times 10^6$ — Sun's core temperature

5800    5527    9981 — Sun's surface temperature

373    100    212 — Boiling point of water
273    0    32 — Freezing point of water

0    −273    −460 — Absolute zero

# GLOSSARY

**A ring** The outermost ring of Saturn visible from Earth; it is located just beyond Cassini's division.

**absolute magnitude** The apparent magnitude that a star would have if it were 10 parsecs from Earth.

**absorption line spectrum** Dark lines superimposed on a continuous spectrum.

**acceleration** A change in the direction or magnitude of a velocity.

**accretion** The gradual accumulation of matter by an astronomical body, usually caused by gravity.

**accretion disk** An orbiting disk of matter spiraling in toward a star or black hole.

**achromatic lens** A compound lens designed to minimize the effect of chromatic aberration.

**active galaxy** A very luminous galaxy, often containing an active galactic nucleus.

**active optics** A system that adjusts a reflecting telescope in response to changes in temperature and shape of the mount; it helps optimize an image.

**adaptive optics** Primary telescope mirrors that are continuously and automatically adjusted to compensate for the distortion of starlight due to the motion of the Earth's atmosphere.

**AGB star** *See* **asymptotic giant branch (AGB)star.**

**albedo** The fraction of sunlight that a planet, asteroid, or satellite scattered directly back into space.

**amino acids** A class of chemical compounds that are the building blocks of proteins.

**angle** The opening between two straight lines that meet at a point.

**angular diameter (angular size)** The arc angle across an object.

**angular resolution** The angular size of the smallest detail of an astronomical object that can be distinguished with a telescope.

**annular eclipse** An eclipse of the Sun in which the Moon is too distant to cover the Sun completely so that a ring of sunlight is seen around the Moon at mideclipse.

**anorthosite** A light-colored rock found throughout the lunar highlands and in some very old mountains on Earth.

**aphelion** The point in its orbit where a planet or other solar system body is farthest from the Sun.

**Apollo asteroid** An asteroid that is sometimes closer to the Sun than is the Earth.

**apparent magnitude** A measure of the brightness of light from a star or other object as seen from Earth.

**arc angle** The measurement of the angle between two objects or two parts of the same object.

**arcminute (arcmin)** One-sixtieth of a degree of arc.

**arcsecond (arcsec)** One-sixtieth of a minute of arc.

**asteroid** Any of the rocky objects larger than a few hundred meters in diameter (and not classified as a planet or moon) that orbits the Sun.

**asteroid belt** A 1½-astronomical-unit-wide region between the orbits of Mars and Jupiter in which most of the asteroids are found.

**astronomical unit (AU)** The average distance between the Earth and the Sun: $1.5 \times 10^8$ kilometers $\approx$ 93 million miles.

**astronomy** The branch of science dealing with objects and phenomena that lie beyond the Earth's atmosphere.

**astrophysics** That part of astronomy dealing with the physics of astronomical objects and phenomena.

**asymptotic giant branch (AGB) star** A red giant star that has completed core helium fusion and has re-expanded for a second time.

**atom** The smallest particle of an element that has the properties characterizing that element.

**atomic number** The number of protons in the nucleus of an atom.

**aurora** (*plural* **aurorae**) Light radiated by atoms and ions in the Earth's upper atmosphere; seen most commonly in the polar regions.

**autumnal equinox** The intersection of the ecliptic and the celestial equator where the Sun crosses the equator moving from north to south.

**average density** The mass of an object divided by its volume.

**B ring** The brightest of the three rings of Saturn visible from Earth; it lies just inside the Cassini division.

**barred spiral galaxy** A spiral galaxy in which the spiral arms begin from the ends of a bar running through the nuclear bulge.

**belt asteroid** An asteroid whose orbit lies in the asteroid belt.

**belts** (of Jupiter) Dark, reddish bands in Jupiter's cloud cover.

**Big Bang** An explosion that took place roughly 15 billion years ago, creating all space, matter, and energy in which the universe emerged.

**Big Chill** Possible endstate of the universe, cooling and expanding forever.

**Big Crunch** Possible end state of the universe, collapsing to a tiny volume.

**binary star** Two stars revolving about each other; a double star.

**birth line** A line on the Hertzsprung-Russell diagram corresponding to where stars with different masses transform from protostars to pre–main-sequence stars.

**BL Lacertae (BL Lac) object** A type of active galaxy; a blazar.

**black hole** An object whose gravity is so strong that the escape velocity from it exceeds the speed of light.

**blackbody** A hypothetical perfect radiator that absorbs and re-emits all radiation falling upon it.

**blackbody curve** The curve obtained when the intensity of radiation from a blackbody at a particular temperature is plotted against wavelength.

**blackbody radiation** Electromagnetic radiation emitted by a blackbody.

**blazar** A BL Lacertae object.

**blueshift** A shift of all spectral features toward shorter wavelengths; the Doppler shift of light from an approaching source.

**Bohr atom** A model of the atom, described by Niels Bohr, in which electrons revolve about the nucleus in various circular orbits.

**bound universe** A universe with enough mass to stop expanding and eventually begin to contract.

**breccia** A rock formed by the sudden amalgamation of various rock fragments under pressure.

**brown dwarf** Any of the planetlike bodies with less than $0.08 \, M_{\odot}$ and more than about 13 $M_{Jupiter}$; such bodies do not have enough mass to sustain fusion in their cores.

**burster** A nonperiodic ray source that emits powerful bursts of X rays or gamma rays.

**C ring** The faint, inner portion of Saturn's main ring system.

**caldera** The crater at the summit of a volcano.

**Callisto** One of the four Galilean satellites of Jupiter.

**capture theory** The idea that the Moon was created at a different location in the solar system and subsequently captured by Earth's gravity.

**carbon fusion** The thermonuclear fusion of carbon nuclei to produce nitrogen.

**carbonaceous chondrites** A class of extremely ancient, carbon-rich meteorites.

**Cassegrain focus** An optical arrangement in a reflecting telescope in which light rays are reflected by a secondary mirror through a hole in the primary mirror.

**Cassini division** A prominent gap between Saturn's A and B rings discovered in 1675 by J. D. Cassini.

**celestial equator** A great circle on the celestial sphere 90° from the celestial poles.

**celestial poles** Points about which the celestial sphere appears to rotate.

**celestial sphere** A hypothetical sphere of very large radius centered on the observer; the apparent sphere of the night sky.

**Celsius scale** *See* **temperature (Celsius).**

**center of mass** The point around which a rigid system is perfectly balanced in a gravitational field; also, the point in space around which mutually orbiting bodies have elliptical orbits.

**Cepheid variable star** One of two types of yellow, supergiant, pulsating stars.

**Cerenkov radiation** Radiation produced by particles traveling through a substance faster than light can.

**Ceres** The largest known asteroid and the first to be discovered.

**Chandrasekhar limit** The maximum mass of a white dwarf, about 1.4 $M_{\odot}$.

**charge-coupled device (CCD)** A type of solid-state silicon wafer designed to detect photons.

**chromatic aberration** An optical property whereby different colors of light passing through a lens are focused at different distances from it.

**chromosphere** The layer in the solar atmosphere between the photosphere and the corona.

**circumpolar stars** All the stars that never set at a given latitude; all the stars between Polaris and the northern horizon.

**close binary** A binary star whose members are separated by a few stellar diameters.

**closed universe** A universe that contains enough matter to cause it to recollapse. It is finite in extent and has no "outside."

**cluster** (of galaxies) A collection of a few hundred to a few thousand galaxies bound by gravity.

**cocreation theory** The theory that the Moon formed simultaneously with the Earth and in orbit around it.

**collision-ejection theory** The theory that the Moon was created by the impact of a planet-sized object with the Earth; presently considered the most plausible theory of the Moon's formation.

**color index** The difference in the magnitudes of a star's brightness measured in two separate wavelength bands.

**color-magnitude diagram** A plot of the surface temperatures (colors) versus the absolute magnitudes of stars.

**coma** (of a comet) The nearly spherical, diffuse gas surrounding the nucleus of a comet near the Sun.

**comet** A small body of ice and dust in orbit about the Sun. While passing near the Sun, a comet's vaporized ices give rise to a coma, tails, and a hydrogen envelope.

**conduction** (thermal) The transfer of heat by passing energy directly from atom to atom.

**configuration** (of a planet) A particular geometric arrangement of the Earth, a planet, and the Sun.

**confinement** The moment shortly after the Big Bang when quarks bound together to form particles like protons and neutrons.

**conic section** The curve of intersection between a circular cone and a plane. This curve can be a circle, ellipse, parabola, or hyperbola.

**conjunction** The alignment of two bodies in the solar system so that they appear in the same part of the sky as seen from Earth.

**conservation of angular momentum** The law of physics stating that the total amount of angular momentum in an isolated system remains constant.

**constellation** Any of the 88 contiguous regions that cover the entire celestial sphere, including all the objects in each region; also, a configuration of stars often named after an object, a person, or an animal.

**contact binary** A close binary system in which both stars fill or overflow their Roche lobes.

**continental drift** The gradual movement of the continents over the surface of the Earth due to plate tectonics.

**continuous spectrum** A spectrum of light over a range of wavelengths without any spectral lines.

**continuum** *See* **continuous spectrum.**

**convection** The transfer of energy by moving currents of fluid or gas containing that energy.

**convective zone** A layer in a star where energy is transported outward by means of convection; also known as the *convective envelope* or *convection zone*.

**core** The central portion of any astronomical object.

**core helium fusion** The fusion of helium to form carbon and oxygen at the center of a star.

**core hydrogen fusion** The fusion of hydrogen to form helium at the center of a star.

**corona** The Sun's outer atmosphere.

**coronagraph** A specially designed telescope with a baffle that blocks out the solar disk so that the corona can be photographed.

**coronal hole** A dark region of the Sun's inner corona as seen at X-ray wavelengths.

**coronal mass ejection** Large volumes of high energy gas released from the Sun's corona.

**cosmic censorship** The belief that the only connection between a black hole and the universe is the black hole's event horizon.

**cosmic microwave background** Photons from every part of the sky with a blackbody spectrum at 2.73 K; the cooled-off radiation from the primordial fireball that originally filled all space.

**cosmic particle horizon** A sphere, centered on the Earth, whose radius equals the distance traveled by light since the Big Bang.

**cosmic ray** High speed particles traveling through space.

**cosmic ray shower** Groups of particles from Earth's atmosphere propelled Earthward by the impact of a cosmic ray.

**cosmological constant** A number sometimes inserted in the equations of general relativity that represents a pressure that opposes gravity throughout the universe.

**cosmological redshift** An increase in wavelength from distant galaxies and quasars caused by the expansion of the universe.

**cosmology** The study of the formation, organization, and evolution of the universe.

**coudé focus** A reflecting telescope in which a series of mirrors direct light to a remote focus away from the moving parts of the telescope.

**crater** A circular depression on a planet or moon caused by the impact of a meteoroid, asteroid, or comet or by a volcano.

**crescent Moon** A lunar phase during which the Moon appears less than half full.

**critical density** The average density throughout the universe for which space is flat and galaxies just barely continue to recede from each other infinitely far into the future.

**critical surface** An imaginary double-teardrop-shaped surface surrounding the stars in a binary that delineates the gravitational domain of each star; Roche lobes.

**crust** The solid surface layer of some astronomical bodies, including the terrestrial planets, the moons, the asteroids, and some stellar remnants.

**dark matter (missing mass)** The as-yet-undetected matter in the universe that is underluminous and probably quite different from ordinary matter.

**dark nebula** A cloud of interstellar gas and dust that obscures the light of more distant stars.

**deceleration parameter** ($q_0$) A number that specifies how fast the expansion of the universe is slowing down.

**declination** The coordinate on the celestial sphere exactly analogous to latitude on Earth; measured north and south of the celestial equator.

**decoupling** The epoch in the early universe when electrons and ions first combined to create stable atoms; the time when electromagnetic radiation ceased to dominate over matter.

**deferent** A fixed circle in the Earth-centered universe along which a smaller circle (an epicycle) moves carrying a planet, the Sun, or Moon.

**degeneracy** The condition in which all the lower energy states for particles (electrons or neutrons) in a gas have been filled, thereby causing the gas to behave differently than an ordinary gas.

**degree (°)** A unit of angular measure or of temperature measure.

**dense core** Any of the regions of interstellar gas clouds that are slightly denser than normal and destined to collapse to form one or a few stars.

**density** The ratio of the mass of an object to its volume.

**density wave** A spiral-shaped compression of the gas and dust in a spiral galaxy.

**density-wave theory** An explanation of spiral arms in galaxies elaborated by C. C. Lin and F. Shu.

**detached binary** A binary system in which the surfaces of both stars are inside their Roche lobes.

**differential rotation** The rotation of a nonrigid object in which parts at different latitudes or different radial distances move at different speeds.

**differentiation** The separation of different kinds of material into different layers inside a planet.

**diffraction grating** *See* **grating.**

**direct motion** The gradual, eastward apparent motion of a planet against the background stars as seen from Earth.

**disk** (of a galaxy) A flattened assemblage of stars, gas, and dust in a spiral galaxy like the Milky Way.

**diurnal motion** Cyclic motion with a one-day period.

**Doppler effect (or Doppler shift)** The change in wavelength of radiation due to relative motion between the source and the observer along the line of sight.

**double-line spectroscopic binary** A spectroscopic binary whose spectrum exhibits spectral lines of both stars.

**double radio source** An extragalactic radio source characterized by two large lobes of radio emission, often located on either side of an active galaxy.

**Drake equation** A mathematical equation used to estimate the number of extraterrestrial civilizations that may exist in our Galaxy.

**dust tail** A comet tail caused by dust particles escaping from the comet's nucleus.

**dwarf elliptical galaxy** A small elliptical galaxy with far fewer stars than a typical galaxy.

**dwarf star** Any star smaller than a giant, such as a main-sequence star or a white dwarf.

**dynamo theory** The generation of a magnetic field by circulating electric charges.

**eccentricity** *See* **orbital eccentricity.**

**eclipse** The blocking of part or all of the light from the Moon by the Earth (lunar eclipse) or from the Sun by the Moon (solar eclipse).

**eclipse path** The track of the tip of the Moon's shadow along the Earth's surface during a total or annular solar eclipse.

**eclipsing binary** A double star system in which stars periodically pass in front of each other as seen from Earth.

**ecliptic** The annual path of the Sun on the celestial sphere; the plane of the Earth's orbit around the Sun.

**Einstein cross** The appearance of four images of the same galaxy or quasar due to gravitational lensing by an intervening galaxy.

**Einstein ring** The circular or arc-shaped image of a distant galaxy or quasar created by gravitational lensing by an intervening galaxy.

**ejecta blanket** The ring of material surrounding a crater that was ejected during the crater-forming impact.

**electromagnetic radiation** Radiation consisting of oscillating electric and magnetic fields, such as gamma rays, X rays, visible light, ultraviolet and infrared radiation, and radio waves.

**electromagnetic spectrum** The entire array or family of electromagnetic radiation.

**electron** A negatively charged subatomic particle usually found in orbit about the nucleus of an atom.

**electron degeneracy pressure** A powerful pressure produced by repulsion of closely packed (degenerate) electrons.

**element** A substance that cannot be decomposed by chemical means into simpler substances.

**ellipse** A closed curve obtained by cutting completely through a circular cone with a plane; the shape of planetary orbits.

**elliptical galaxy** A galaxy with an elliptical shape, little interstellar matter, and no spiral arms.

**elongation** The angle between a planet and the Sun as seen from Earth.

**emission line spectrum** A spectrum that contains only bright emission lines.

**emission nebula** A glowing gaseous nebula whose light comes from fluorescence caused by a nearby star.

**Encke division** A thin gap in Saturn's A ring, possibly first seen by J. F. Encke in 1838.

**energy** The ability to do work.

**energy flux** The amount of energy emitted from each square meter of an object's surface per second.

**energy level** (in an atom) A particular amount of energy possessed by an electron in orbit around a nucleus.

**epicycle** In the Earth-centered universe, a moving circle about which planets revolve.

**equations of stellar structure** A set of relationships that describe the interactions of matter, energy, and gravity inside a star.

**equinox** Either of the two days of the year when the Sun crosses the celestial equator and is therefore directly over the Earth's equator; *see also* **autumnal equinox** *and* **vernal equinox**.

**era of recombination** The time, roughly 300,000 years after the Big Bang, when the universe became transparent.

**ergoregion** The region of space immediately outside the event horizon of a rotating black hole where it is impossible to remain at rest.

**escape velocity** The speed needed by one body to just escape the gravitational attraction of another body and move into free space.

**Europa** One of the Galilean satellites of Jupiter.

**event horizon** The location around a black hole where the escape velocity equals the speed of light; the boundary of a black hole.

**evolutionary track** On the Hertzsprung-Russell diagram, the path followed by a point representing an evolving star.

**excitation** The process of imparting energy to an electron in an atom or ion.

**expanding universe** The motion of the superclusters of galaxies away from each other.

**eyepiece lens** A magnifying lens used to view the image produced at the focus of a telescope.

**F ring** A thin ring just beyond the outer edge of Saturn's main ring system.

**Fahrenheit scale** *See* **temperature (Fahrenheit)**.

**filament** A dark curve seen above the Sun's photosphere that is the top view of a solar prominence.

**first quarter Moon** A phase of the waxing Moon when Earth-based observers see half of the Moon's illuminated hemisphere.

**fission theory** The theory that the Moon formed from matter flung off the Earth because the planet was rotating extremely fast.

**flare** A sudden, temporary outburst of photons and particles from an extended region of the solar surface.

**flatness problem** The quandary associated with the improbable fact that space throughout the universe seems to be essentially flat.

**flocculent spiral galaxy** A spiral galaxy whose spiral arms are broad, fuzzy, and poorly demarcated.

**focal length** The distance from a lens or concave mirror to where converging light rays meet.

**focal plane** The plane at the focal length of a lens or concave mirror on which an extended object is focused.

**focal point** *See* **focus**.

**focus** (of a lens or concave mirror) The place at the focal length where light rays from a point object (that is, one that is too distant or tiny to resolve) are converged by a lens or concave mirror.

**focus** (*plural* **foci**) (of an ellipse) The two points inside an ellipse, the sum of whose distances from any point on the ellipse is constant.

**force** That which can change the momentum of an object.

**frequency** The number of waves that cross a given point per unit time; the number of vibrations per unit time.

**full Moon** A phase of the Moon during which its full daylight hemisphere can be seen from Earth.

**galactic cannibalism** A collision between two galaxies of unequal mass and size in which the smaller galaxy is absorbed by the larger galaxy.

**galactic merger** A collision and subsequent merger of two roughly equal-sized galaxies.

**galactic nucleus** The center of a galaxy; the center of the Milky Way Galaxy.

**galaxy** A large assemblage of stars, gas, and dust bound together by their mutual gravitational attraction.

**Galilean satellite (Galilean moon)** Any one of the four large moons of Jupiter that is visible from Earth through a small telescope.

**gamma ray** The most energetic form of electromagnetic radiation.

**gamma-ray burster** An object that emits a short burst of gamma rays; they are believed to be outside our Galaxy.

**Ganymede** One of the Galilean satellites of Jupiter.

**gas (ion) tail** The relatively straight tail of a comet produced by the solar wind acting on ions in a comet's coma.

**general theory of relativity** A description of spacetime formulated by Einstein explaining how gravity affects the geometry of space and the flow of time.

**geocentric cosmology** The belief that the Earth is at the center of the universe.

**giant elliptical galaxy** A very large, extremely massive elliptical galaxy, usually located near the center of a rich cluster of galaxies.

**giant molecular cloud** A large interstellar cloud of cool gas and dust in a galaxy.

**giant star** A star whose diameter is roughly 10 to 100 times that of the Sun.

**gibbous Moon** A phase of the Moon in which more than half, but not all, of the Moon's daylight hemisphere is visible from Earth.

**globular cluster** A large spherical cluster of gravitationally bound stars usually found in the outlying regions of a galaxy.

**grand-design spiral galaxy** A spiral galaxy whose spiral arms are thin, graceful, and well-defined.

**granulation** The rice-grain–like structure of the solar photosphere due to convection of solar gases.

**granules** Lightly colored convection features about 1000 kilometers in diameter seen constantly in the solar photosphere.

**grating** An optical device consisting of closely spaced lines ruled on a piece of glass that is used like a prism to disperse light into a spectrum.

**gravitation (gravity)** The tendency of all matter to attract all other matter.

**gravitational lensing** The distortion of the appearance of an object by a source of gravity between it and the observer.

**gravitational redshift** The redshift of photons leaving the gravitational field of any massive object, such as a star or black hole.

**gravitational waves (gravitational radiation)** Ripples in the overall geometry of space produced by nonspherical moving objects.

**gravity** *See* gravitation.

**Great Dark Spot** A large, dark, oval-shaped storm that used to be in Neptune's southern hemisphere.

**Great Red Spot** A large, red-orange, oval-shaped storm in Jupiter's southern hemisphere.

**Great Wall** A huge arc of galaxies between two voids in the cosmos.

**greatest elongation** The largest possible angle between the Sun and an inferior planet.

**greenhouse effect** The trapping of infrared radiation near a planet's surface by the planet's atmosphere.

**ground state** The lowest energy level of an atom.

**H I region** A region of neutral hydrogen in interstellar space.

**H II region** A region of ionized hydrogen in interstellar space.

**halo** (of a galaxy) A spherical distribution of globular clusters, isolated stars, and possibly dark matter that surrounds a galaxy.

**Hawking process** The formation of real particles from virtual ones just outside a black hole's event horizon; the means by which black holes evaporate.

**head-tail source** A radio galaxy whose radio emission is deflected from the galaxy.

**heliocentric cosmology** A theory of the formation and evolution of the solar system with the Sun at the center.

**helioseismology** The study of vibrations of the solar surface.

**helium flash** The explosive ignition of helium fusion in the core of a low-mass, giant star.

**helium fusion** The thermonuclear fusion of helium to produce carbon and oxygen.

**helium shell flash** The explosive ignition of helium fusion in a thin shell surrounding the core of a low-mass star.

**Hertzsprung-Russell (H-R) diagram** A plot of the absolute magnitude or luminosity of stars versus their surface temperatures or spectral classes.

**highlands** Heavily cratered, mountainous regions of the lunar surface.

**horizon problem (isotropy problem)** The difficulty in explaining why seemingly disconnected regions of the universe have the same temperature.

**horizontal branch stars** A group of post–helium-flash stars near the main sequence on the Hertzsprung-Russell diagram of a typical globular cluster.

**hot-spot volcanism** The creation of volcanoes on a planet's surface caused by a reservoir of hot magma in the planet's mantle under a thin part of the crust.

**Hubble classification** A system of classifying galaxies according to their appearance into one of four broad categories: spirals, barred spirals, ellipticals, and irregulars.

**Hubble constant ($H_0$)** The still uncertain constant of proportionality in the relation between the recessional velocities of remote galaxies and their distances; the correct value will determine the age of the universe.

**Hubble flow** The recession of the galaxies caused by the expansion of the universe.

**Hubble law** The relationship that states that the redshifts of remote galaxies are directly proportional to their distances from Earth.

**hydrocarbon** A molecule based on hydrogen and carbon.

**hydrogen envelope** An extremely large, tenuous sphere of hydrogen gas surrounding the head of a comet.

**hydrogen fusion (hydrogen burning)** The thermonuclear fusion of hydrogen to produce helium.

**hydrostatic equilibrium** A balance between the weight of a layer in a star and the pressure that supports it.

**hyperbola** An open curve obtained by cutting a cone with a plane.

**impact breccia** A rock consisting of various fragments cemented together by the impact of a meteoroid.

**impact crater** A crater on the surface of a planet or moon produced by the impact of an asteroid, meteoroid, or comet.

**inferior conjunction** The configuration when Mercury or Venus is directly between the Sun and the Earth.

**inflationary epoch** A brief period shortly after the Big Bang during which the scale of the universe increased very rapidly.

**infrared radiation** Electromagnetic radiation of a wavelength longer than visible light but shorter than radio waves.

**instability strip** A region on the Hertzsprung-Russell diagram occupied by pulsating stars.

**interferometry** A method of increasing resolving power by combining electromagnetic radiation obtained by two or more telescopes.

**intergalactic gas** Gas located between the galaxies within a cluster of galaxies.

**interstellar dust** Microscopic solid grains of various compounds in interstellar space.

**interstellar gas** Sparse gas in interstellar space.

**interstellar medium** Interstellar gas and dust.

**inverse-square law** The gravitational attraction between two objects and the apparent brightness of a light source are both inversely proportional to the square of its distance.

**Io** One of the Galilean satellites of Jupiter.

**ion** An atom that has become electrically charged due to the loss or addition of one or more electrons.

**ionization** The process by which an atom loses or gains electrons.

**iron meteorite** A meteorite composed of iron with a small admixture of nickel; also called an *iron*.

**irregular cluster** (of galaxies) An unevenly distributed group of galaxies bound together by their mutual gravitational attraction.

**irregular galaxy** An asymmetrical galaxy having neither spiral arms nor an elliptical shape.

**isotope** Any of several forms of the same chemical element whose nuclei all have the same number of protons but different numbers of neutrons.

**isotropy** The fact that the average number of galaxies at different distances from Earth is the same in all directions; also, the fact that the temperature of the cosmic microwave background is essentially the same in all directions.

**isotropy problem (horizon problem)** The difficulty in explaining why seemingly disconnected regions of the universe have the same temperature.

**Jeans instability** The condition under which gravitational forces overcome thermal forces to cause part of an interstellar cloud to collapse and form stars and planets.

**kelvin** *See* **temperature (Kelvin)**.

**Kepler's laws** Three statements, formulated by Johannes Kepler, that describe the motions of the planets.

**Kerr black hole** Any rotating, uncharged black hole.

**kiloparsec (kpc)** One thousand parsecs; about 3260 light-years.

**Kirchhoff's laws** Three statements formulated by Gustav Kirchhoff describing spectra and spectral analysis.

**Kirkwood gaps** Gaps in the spacing of asteroid orbits discovered by Daniel Kirkwood.

**Kuiper belt** A doughnut-shaped ring of space around the Sun beyond Pluto that contains many frozen comet bodies, some of which are occasionally deflected toward the inner solar system.

**Large Magellanic Cloud (LMC)** An irregular galaxy, companion to the Milky Way.

**last quarter Moon** A phase of the waning Moon when Earth-based observers see half of the Moon's illuminated hemisphere.

**law of equal areas** Kepler's second law.

**law of inertia** The physical law that an object will stay at rest or move at a constant speed in a fixed direction unless acted upon by an outside force.

**lenticular galaxy** A disk-shaped galaxy without spiral arms.

**light** Electromagnetic radiation, which travels in packets called photons.

**light curve** A graph that displays variations in the brightness of a star or other astronomical object over time.

**light-gathering power** A measure of how much light a telescope intercepts and brings to a focus.

**light-year (ly)** The distance that light travels in a vacuum in one year.

**lighthouse model** The explanation that a pulsar pulses by rotating and funneling energy outward via magnetic fields that are not aligned with the rotation axis.

**limb** (of the Sun) The apparent edge of the Sun as seen in the sky.

**limb darkening** The phenomenon whereby the Sun is darker near its limb than near the center of its disk.

**line of nodes** The line along which the plane of the Moon's orbit intersects the plane of the ecliptic.

**liquid metallic hydrogen** A metallike form of hydrogen that is produced under extreme pressure.

**Local Group** The cluster of galaxies of which our own Galaxy is a member.

**long-period comet** A comet that takes tens of thousands of years or more to orbit the Sun once.

**luminosity** The rate at which electromagnetic radiation is emitted from a star or other object.

**luminosity class** The classification of a star of a given spectral type according to its luminosity; the classes are supergiant, bright giant, giant, subgiant, and main sequence.

**lunar** Referring to the Moon.

**lunar eclipse** An eclipse during which the Earth blocks light that would have struck the Moon.

**lunar month** *See* **synodic month**.

**lunar phases** The names given to the apparent shapes of the Moon as seen from Earth.

**Lyman series** A series of spectral lines of hydrogen produced by electron transitions to and from the lowest energy state of the hydrogen atom.

**magnetic dynamo** A theory that explains phenomena of the solar cycle as a result of periodic winding and unwinding of the Sun's magnetic field in the solar atmosphere.

**magnetic field** A region of space near a magnetized body within which magnetic forces can be detected.

**magnetosphere** The region around a planet occupied by its magnetic field.

**magnification (magnifying power)** The number of times larger in angular diameter an object appears through a telescope than when it is seen by the naked eye.

**magnitude** A measure of the amount of light received from a star or other luminous object.

**magnitude scale** The system of denoting magnitudes.

**main sequence** A grouping of stars on the Hertzsprung-Russell diagram extending diagonally across the graph from the hottest, brightest stars to the dimmest, coolest stars.

**main-sequence star** A star, fusing hydrogen to helium in its core, whose surface temperature and luminosity place it on the main sequence on the Hertzsprung-Russell diagram.

**mantle** (of a planet) That portion of a terrestrial planet located between its crust and core.

**mare** (*plural* **maria**) Latin for "sea," a large, relatively crater-free plain on the Moon.

**mare basalt** Dark, solidified lava that covers the lunar maria.

**mass** A measure of the total amount of material in an object.

**mass-luminosity relation** The linear relationship between the masses and luminosities of main-sequence stars.

**matter-dominated universe** A universe in which the radiation field that fills all space is unable to prevent the existence of neutral atoms.

**megaparsec (Mpc)** One million parsecs.

**mesosphere** The layer in the Earth's atmosphere above the stratosphere.

**metal-poor star** A star whose abundance of heavy elements is significantly less than that of the Sun.

**metal-rich star** A star whose abundance of heavy elements is comparable to that of the Sun.

**meteor** The streak of light seen when any space debris vaporizes in the Earth's atmosphere; a "shooting star."

**meteor shower** Frequent meteors that seem to originate from a common point in the sky.

**meteorite** A fragment of space debris that has survived passage through the Earth's atmosphere.

**meteoroid** A small rock in interplanetary space.

**microlensing** The gravitational focusing of light from a distant star by a closer object to give a brighter image of the star.

**Milky Way Galaxy** The galaxy in which our solar system resides.

**minor planet** *See* **asteroid**.

**missing mass** *See* **dark matter**.

**model** A hypothesis that has withstood observational or experimental tests.

**molecule** A bound combination of two or more atoms.

**momentum** A measure of the inertia of an object; an object's mass multiplied by its velocity.

**neap tide** The least change from high to low tide during a day; it occurs during the first and third quarter phases of the Moon.

**nebula** (*plural* **nebulae**) A cloud of interstellar gas and dust.

**neon fusion** The thermonuclear fusion of neon nuclei.

**neutrino** A subatomic particle, with no electric charge and little mass, that is important in many nuclear reactions and in supernovae.

**neutron** A nuclear particle with no electric charge and with a mass nearly equal to that of the proton.

**neutron degeneracy pressure** A powerful pressure produced by degenerate neutrons.

**neutron star** A very compact, dense stellar remnant composed almost entirely of neutrons.

*New General Catalogue* **(NGC)** A catalog of star clusters, nebulae, and galaxies, first published in 1888.

**new Moon** The phase of the Moon when it is nearest the Sun in the sky.

**Newtonian laws of motion** A branch of physics based on Newton's laws of mechanics and gravitation.

**Newtonian reflector** An optical arrangement in a reflecting telescope in which a small, flat mirror reflects converging light rays to a focus on one side of the telescope tube.

**nonthermal radiation** Radiation emitted by charged particles moving through a magnetic field; synchrotron radiation.

**north celestial pole** The location on the celestial sphere directly above the Earth's northern rotation pole.

**northern lights (aurora borealis)** Light radiated by atoms and ions in the Earth's upper atmosphere due to high-energy particles from the Sun and seen mostly in the northern polar regions.

**nova** (*plural* **novae**) A star that experiences a sudden outburst of radiant energy, temporarily increasing its luminosity by a factor of between $10^4$ and $10^6$.

**nuclear** Referring to the nucleus of an atom.

**nuclear bulge** A distribution of stars in the shape of a flattened sphere that surrounds the nucleus of a spiral galaxy like the Milky Way.

**nuclear density** The density of matter in the nucleus of an atom; about $10^{17}$ kilograms per cubic meter $(kg/m^3)$.

**nucleus** (of an atom) The massive part of an atom, composed of protons and neutrons, about which electrons revolve.

**nucleus** (of a comet) A collection of ices and dust that constitute the solid part of a comet.

**nucleus** (of a galaxy) The concentration of stars, gas, and dust, and often a supermassive black hole, at the very center of a galaxy.

**OB association** An unbound group of very young, massive stars predominantly of spectral types O and B.

**OBAFGKM sequence** The sequence of stellar spectral classifications from hottest to coolest stars.

**objective lens** The principal lens of a refracting telescope.

**observable universe** All space that is nearer to us than the distance traveled by light since the time of the Big Bang.

**Occam's razor** The principle of choosing the simplest scientific theory that correctly explains any phenomenon.

**occultation** The eclipsing of an astronomical object other than the Moon or Sun by another astronomical body.

**Oort cloud** A spherical region of the solar system beyond the Kuiper belt where most comets are believed to spend most of their time.

**open cluster** A loose association of young stars in the disk of the galaxy; a galactic cluster.

**open universe** A universe with a hyperbolic shape; lacks the mass necessary to someday stop expanding and recollapse. It will expand forever.

**opposition** The configuration of a planet when it is at an elongation of 180° and thus appears opposite the Sun in the sky.

**optical double** A pair of stars that appear to be near each other but are unbound and at very different distances from Earth.

**optics** The branch of physics dealing with the behavior and properties of light.

**orbit** The path of an object that is moving about a second object.

**orbital eccentricity** A measure between 0 and 1 indicating how close to circular a planet's orbit is (the eccentricity of a circular orbit is 0).

**organic molecule** A carbon-based compound.

**oxygen fusion** The thermonuclear fusion of oxygen nuclei.

**ozone layer** The lower stratosphere, where most of the ozone in the air exists.

**pair production** The creation of a particle and an antiparticle from energetic photons.

**Pallas** The second asteroid to be discovered.

**parabola** An open curve formed by cutting a circular cone at an angle parallel to the sides of the cone.

**parallax** The apparent displacement of an object relative to more distant objects caused by viewing it from different locations.

**parsec (pc)** A unit of distance equal to 3.26 light-years.

**partial eclipse** A lunar or solar eclipse in which the eclipsed object does not appear completely covered.

**Pauli exclusion principle** A principle of quantum mechanics that says that two identical particles cannot simultaneously have the same position and momentum.

**peculiar galaxy** Any Hubble class of galaxy that appears to be blowing apart.

**penumbra** The portion of a shadow in which only part of the light source is covered by the shadow-making body.

**penumbral eclipse** A lunar eclipse in which the Moon passes only through the Earth's penumbra.

**perihelion** The point in its orbit where a planet is nearest the Sun.

**period** The interval of time between successive repetitions of a periodic phenomenon.

**period-luminosity relation** A relationship between the period and average luminosity of a pulsating star.

**periodic table** A listing of the chemical elements according to their properties; invented by D. Mendeleev.

**phase** (of the Moon) The appearance of the Moon at different points in its orbit of the Earth.

**photodisintegration** The breakup of nuclei in the core of a massive star due to the effects of energetic gamma rays.

**photometry** The measurement of light intensities.

**photon** A discrete unit of electromagnetic energy.

**photon pressure** The force per unit area exerted by photons on stellar or interstellar gas.

**photosphere** The region in the solar atmosphere from which most of the visible light escapes into space.

**physics** Basic principles that govern the behavior of physical reality.

**pixel** A contraction of the term "picture element"; usually refers to one square of a grid into which the light-sensitive component of a charge-coupled device is divided.

**plage** A bright spot on the Sun believed to be associated with an emerging magnetic field.

**Planck time** The brief interval of time, about $10^{-43}$ second, immediately after the Big Bang, when all four forces (gravity, electromagnetism, weak, strong) had the same strength.

**Planck's law** A relationship between the energy carried by a photon and its wavelength.

**planetary differentiation** The process early in the life of each planet whereby denser elements sank inward and lighter ones rose.

**planetary nebula** A luminous shell of gas ejected from an old, low-mass star.

**planetesimal** Primordial asteroidlike object from which the planets accreted.

**plasma** A hot, ionized gas.

**plate tectonics** The motions of large segments (plates) of the Earth's surface over the underlying mantle.

**polymer** A long molecule composed of many smaller molecules.

**poor cluster** (of galaxies)  A cluster of galaxies with only a few members.

**Population I star**  A star whose spectrum exhibits spectral lines of many elements heavier than helium; a metal-rich star.

**Population II star**  A star whose spectrum exhibits comparatively few spectral lines of elements heavier than helium; a metal-poor star.

**positron**  An electron with a positive rather than negative electric charge; an antielectron.

**powers of ten**  A convenient method of writing large and small numbers that uses a number between 1 and 10 multiplied by a power of 10.

**pre–main-sequence star**  The stage of star formation just before the main sequence; it involves slow contraction of the young star.

**precession** (of the Earth)  A slow, conical motion of the Earth's axis of rotation caused by the gravitational pull of the Moon and Sun on the Earth's equatorial bulge.

**precession of the equinoxes**  The slow westward motion of the equinoxes along the ecliptic because of the Earth's precession.

**primary mirror**  The large, concave, light-gathering mirror in a reflecting telescope, analogous to the objective lens on a refracting telescope.

**prime focus**  The point in a reflecting telescope where the primary mirror focuses light.

**primordial black hole**  A relatively low-mass black hole hypothetically formed at the beginning of the universe.

**primordial fireball**  The extremely hot gas that filled the universe immediately following the Big Bang.

**prism**  A wedge-shaped piece of glass used to disperse white light into a spectrum.

**prominence**  Flamelike protrusion seen near the limb of the Sun and extending into the solar corona.

**proper motion**  The change in the location of a star on the celestial sphere.

**proton**  A heavy, positively charged nuclear particle.

**protoplanet**  The embryonic stage of a planet when it is growing because of collisions with planetesimals.

**protostar**  The earliest stage of a star's life before fusion commences and when gas is rapidly falling onto it.

**protosun**  The Sun prior to the time when hydrogen fusion began in its core.

**pulsar**  A pulsating source associated with a rapidly rotating neutron star with an off-axis magnetic field.

**quantum gravitation**  A theory being developed in which gravitation is quantized in particles called gravitons, similar to photons, the quanta of electromagnetic radiation.

**quantum mechanics**  The branch of physics dealing with the structure and behavior of atoms and their interactions with each other and with light.

**quark**  A particle that is a building block of the heavy nuclear particles such as protons and neutrons.

**quarter Moon**  A phase of the Moon when it is located 90% from the Sun in the sky.

**quasar** (quasi-stellar radio source)  A starlike object with a very large redshift.

**quasi-stellar object (QSO)**  A quasar.

**radial velocity**  That portion of an object's velocity parallel to the line of sight.

**radial-velocity curve**  A plot showing the variation of radial velocity with time for a binary star or variable star.

**radiation**  Electromagnetic energy; photons.

**radiation (photon) pressure**  The transfer of momentum carried by radiation to an object on which the radiation falls.

**radiation-dominated universe**  The time at the beginning of the universe when the electromagnetic radiation prevented ions and electrons from combining to make neutral atoms.

**radiative zone**  A region inside a star where energy is transported outward by the movement of photons through a gas from a hot location to a cooler one.

**radio astronomy**  The branch of astronomy dealing with observations at radio wavelengths.

**radio galaxy**  A galaxy that emits an unusually large amount of radio waves.

**radio lobes**  Vast regions of radio emission on opposite sides of a radio galaxy.

**radio telescope**  A telescope designed to detect radio waves.

**radio wave**  Long-wavelength electromagnetic radiation.

**radioactivity**  The process whereby certain atomic nuclei naturally decompose by spontaneously emitting particles.

**red giant**  A large, cool star of high luminosity.

**red supergiant**  An extremely large, cool star of high luminosity; a star in the upper right corner of the Hertzsprung-Russell diagram.

**redshift**  The shifting to longer wavelengths of the light from remote galaxies and quasars; the Doppler shift of light from any receding source.

**reflecting telescope (reflector)**  A telescope in which the principal light-gathering component is a concave mirror.

**reflection**  The rebounding of light rays off a smooth surface.

**reflection nebula**  A comparatively dense cloud of gas and dust in interstellar space that is illuminated by a star between it and the Earth.

**refracting telescope (refractor)**  A telescope in which the principal light-gathering component is a lens.

**refraction**  The bending of light rays upon passing from one transparent medium to another.

**regolith** The powdery, lifeless material on the surface of a moon or planet.

**regular cluster** (of galaxies) An evenly distributed group of galaxies bound together by mutual gravitational attraction.

**resolution** The degree to which fine details in an optical image can be distinguished.

**resolving power** A measure of the ability of an optical system to distinguish fine details in the image it produces.

**resonance** The large response of an object to a small periodic gravitational tug from another object.

**retrograde motion** The occasional backward (that is, westward) apparent motion of a planet against the background stars as seen from Earth. Retrograde motion is an optical illusion.

**retrograde rotation** The rotation of a planet opposite to its direction of revolution around the Sun. Only Pluto, Uranus, and Venus have retrograde rotation.

**revolution** The orbit of one body about another.

**rich cluster** (of galaxies) A cluster of galaxies with many members.

**right ascension** The celestial coordinate analogous to longitude on Earth and measured around the celestial equator from the vernal equinox.

**rille** A winding crack or depression in the lunar surface.

**ringlet** Any one of numerous, closely spaced, thin bands of particles in Saturn's ring system.

**Roche lobe** The teardrop-shaped regions around each star in a binary star system inside of which gas is gravitationally bound to that star.

**rotation** The spinning of a body about an axis passing through it.

**rotation curve** (of a galaxy) A graph showing how the orbital speed of material in a galaxy depends on the distance from the galaxy's center.

**RR Lyrae variable star** A type of pulsating star with a period less than one day.

**Sa, Sb, Sc** Categories of spiral galaxies determined by the sizes of their nuclear bulges or how tightly wound their spiral arms are; an Sa galaxy is the most tightly wound.

**Sagittarius A** The strong radio source associated with the nucleus of the Milky Way galaxy.

**satellite** A body that revolves about a larger one.

**SBa, SBb, SBc** Categories of barred spiral galaxies determined by how tightly wound their spiral arms are; SBa galaxies are the most tightly wound.

**scarp** A cliff on Mercury believed to have formed when the planet cooled and shrank.

**Schmidt corrector plate** A specially shaped lens used with spherical mirrors that corrects for spherical aberration and provides an especially wide field of view.

**Schwarzschild black hole** Any nonrotating, uncharged black hole.

**Schwarzschild radius** The distance from the center to the event horizon in any black hole.

**scientific method** The method of doing science based on observation, experimentation, and the formulation of hypotheses (theories) that can be tested.

**scientific notation** The style of writing large and small numbers using powers of ten.

**scientific theory** An idea about the natural world that is subject to testing and refinement.

**seafloor spreading** The process whereby magma upwelling along rifts in the ocean floor causes adjacent segments of the Earth's crust to separate.

**secondary cosmic rays** Particles from the Earth's atmosphere given high speeds Earthward by cosmic rays from space.

**secondary mirror** A relatively small mirror used in reflecting telescopes to guide the light out the side or bottom of the telescope.

**seeing disk** The size that a star appears to have on a photographic or charge-coupled-device image as a result of the changing refraction of the starlight passing through the Earth's atmosphere.

**seismic waves** Vibrations traveling through or around an astronomical body usually associated with earthquakelike phenomena.

**seismograph** A device used to record and measure seismic waves, such as those produced by earthquakes.

**seismology** The study of earthquakes and related phenomena.

**self-propagating star formation** The process whereby the birth of stars in one part of a galaxy stimulates star formation in a neighboring region of that galaxy.

**semidetached binary** A close binary system in which one star fills or is overflowing its Roche lobe.

**semimajor axis** (of an ellipse) Half of the longest dimension of an ellipse.

**SETI** The search for extraterrestrial intelligence.

**Seyfert galaxy** A spiral galaxy with a bright nucleus whose spectrum exhibits emission lines.

**Shapley-Curtis debate** An inconclusive debate between Harlow Shapley and Heber Curtis in 1920 about whether certain nebulae were beyond the Milky Way.

**shell helium fusion** Helium fusion that occurs in a thin shell surrounding the core of a star.

**shell hydrogen fusion** Hydrogen fusion that occurs in a thin shell surrounding the core of a star.

**shepherd satellite** A small satellite whose gravitational tug is responsible for maintaining a sharply defined ring of matter around a planet such as Saturn or Uranus.

**shock wave**  An abrupt, localized region of compressed gas caused by an object traveling through the gas at a speed greater than the speed of sound.

**shooting star**  *See* meteor.

**short-period comet**  A comet that orbits the Sun in the vicinity of the planets, thereby reappearing with tails every 200 years or less.

**sidereal month**  The period of the Moon's revolution about the Earth measured with respect to the Moon's location among the stars.

**sidereal period**  The orbital period of one object about another measured with respect to the stars.

**single-line spectroscopic binary**  A spectroscopic binary whose periodically varying spectrum exhibits the spectral lines of only one of its two stars.

**singularity**  A place of infinite curvature of spacetime in a black hole.

**Small Magellanic Cloud (SMC)**  An irregular galaxy that is a companion to the Milky Way.

**solar corona**  The Sun's outer atmosphere.

**solar cycle**  A 22-year cycle during which the Sun's magnetic field reverses its polarity twice.

**solar day**  From noontime to the next noontime; for Earth it is 24 hours.

**solar eclipse**  An eclipse during which the Moon blocks the Sun.

**solar flare**  A violent eruption on the Sun's surface.

**solar luminosity ($L_\odot$)**  The total energy emitted by the Sun each second.

**solar model**  A set of equations that describe the internal structure and energy generation of the Sun.

**solar nebula**  The cloud of gas and dust from which the Sun and the rest of the solar system formed.

**solar seismology**  The study of the Sun's interior from observations of vibrations of its surface.

**solar system**  The Sun, planets, their satellites, asteroids, comets, and related objects that orbit the Sun.

**solar wind**  A radial flow of particles (mostly electrons and protons) from the Sun.

**solstice**  Either of two points along the ecliptic at which the Sun reaches its maximum distance north or south of the celestial equator.

**south celestial pole**  The location on the celestial sphere directly above the Earth's south rotation pole.

**southern lights (aurora australis)**  Light radiated by atoms and ions in the Earth's upper atmosphere due to high-energy particles from the Sun; seen mostly in the southern polar regions.

**spacetime**  The concept from special relativity that space and time are both essential in describing the position, motion, and action of any object or event.

**special theory of relativity**  A description of mechanics and electromagnetic theory formulated by Einstein according to which measurements of distance, time, and mass are affected by the observer's motion.

**spectral analysis**  The identification of chemicals by the appearance of their spectra.

**spectral lines**  Dark or bright lines at specific wavelengths in a spectrum.

**spectral type**  A classification of stars according to the appearance of their spectra.

**spectrogram**  The photograph of a spectrum.

**spectrograph**  A device for photographing a spectrum.

**spectroscope**  A device for directly viewing a spectrum.

**spectroscopic binary star**  A double star whose binary nature can be deduced from the periodic Doppler shifting of lines in its spectrum.

**spectroscopic parallax**  A method of determining a star's distance from the Earth by measuring its surface temperature, luminosity, and apparent magnitude.

**spectroscopy**  The study of spectra.

**spectrum (*plural* spectra)**  The result of electromagnetic radiation passing through a prism or grating so that different wavelengths are separated.

**speed**  The rate at which an object moves.

**spherical aberration**  An optical property whereby different portions of a spherical lens or spherical, concave mirror have slightly different focal lengths.

**spicule**  A narrow jet of rising gas in the solar chromosphere.

**spin (of an electron or proton)**  A small, well-defined amount of angular momentum possessed by electrons, protons, and other particles.

**spiral arms**  Lanes of interstellar gas, dust, and young stars that wind outward in a plane from the central regions of some galaxies.

**spiral density wave**  A spiral-shaped pressure wave that orbits the disk of a spiral galaxy and induces new star formation.

**spiral galaxy**  A flattened, rotating galaxy with pinwheel-like spiral arms winding outward from the galaxy's nuclear bulge.

**spoke**  A moving dark region of Saturn's rings.

**spring tide**  The greatest daily difference between high tide and low tide, occurring when the Moon is new or full.

**stable Lagrange points**  Locations throughout the solar system where the gravitational forces from the Sun and a planet keep space debris trapped.

**standard candle** An object whose known luminosity can be used to deduce the distance to a galaxy.

**star** A self-luminous sphere of gas.

**starburst galaxy** A galaxy where there is an exceptionally high rate of star formation.

**Stefan-Boltzmann law** A relationship between the temperature of a blackbody and the rate at which it radiates energy.

**stellar evolution** The changes in size, luminosity, temperature, and chemical composition that occur as a star ages.

**stellar model** The result of theoretical calculations that give details of physical conditions inside a star.

**stellar parallax** The apparent shift in a nearby star's position on the celestial sphere resulting from the Earth's orbit around the Sun.

**stellar spectroscopy** The study of the properties of stars encoded in their spectra.

**stony meteorite** A meteorite composed of rock with very little iron; also called a *stone*.

**stony-iron meteorite** A meteorite composed of roughly equal amounts of rock and iron.

**stratosphere** The second layer in the Earth's atmosphere, directly above the troposphere.

**strong nuclear force** The force that binds protons and neutrons together in nuclei.

**subduction zone** A location where colliding tectonic plates cause the Earth's crust to be pulled down into the mantle.

**subgiant** A star whose luminosity is between that of main-sequence stars and normal giants of the same spectral type.

**summer solstice** The point on the ecliptic where the Sun is farthest north of the celestial equator.

**Sun** The star about which the Earth and other planets revolve.

**sunspot** A temporary cool region in the solar photosphere created by protruding magnetic fields.

**sunspot cycle** The semiregular 11-year period with which the number of sunspots fluctuates.

**sunspot maximum** The time during the solar cycle when the number of sunspots is exceptionally high.

**sunspot minimum** The time during the solar cycle when the number of sunspots is exceptionally low.

**supercluster** (of galaxies) A gravitationally bound collection of many clusters of galaxies.

**supergiant** A star of very high luminosity.

**supergranule** A large convective cell in the Sun's chromosphere containing many granules.

**superior conjunction** The configuration when a planet is behind the Sun as seen from Earth.

**supermassive black hole** A black hole whose mass exceeds 1000 solar masses.

**supernova** (*plural* **supernovae**) A stellar outburst during which a star suddenly increases its brightness roughly a millionfold.

**supernova remnant** A nebula left over after a supernova detonates.

**synchronous rotation** The condition when a moon's rotation rate and revolution rate are equal or when a planet's rotation rate equals its moon's revolution rate.

**synchrotron radiation** The radiation emitted by charged particles moving through a magnetic field; nonthermal radiation.

**synodic month (lunar month)** The period of revolution of the Moon with respect to the Sun; the length of one cycle of lunar phases.

**synodic period** The interval between successive occurrences of the same configuration of a planet as seen from Earth.

**T Tauri stars** Young, variable stars associated with interstellar matter that show erratic changes in luminosity.

**tail** (of a comet) Gas and dust particles from a comet's nucleus that have been swept away from the comet's nucleus by the radiation pressure of sunlight and impact of the solar wind.

**telescope** An instrument for viewing remote objects.

**temperature (Celsius)** Temperature measured on a scale where water freezes at 0° and boils at 100°.

**temperature (Fahrenheit)** Temperature measured on a scale where water freezes at 32° and boils at 212°.

**temperature (Kelvin)** Absolute temperature measured in Celsius degree intervals. Water freezes at 273 K and boils at 373 K.

**terminator** The line dividing day and night on the surface of any body orbiting the Sun; the line of sunset or sunrise.

**terrestrial planet** Any of the planets Mercury, Venus, Earth, or Mars; a planet with a composition and density similar to that of Earth.

**theory** A hypothesis that has been demonstrated to describe a range of phenomena accurately.

**thermal energy** The energy associated with the motions of atoms or molecules in a substance.

**thermal equilibrium** A balance between the input and outflow of heat in a system.

**thermonuclear fusion** A reaction in which the nuclei of atoms are fused together at a high temperature.

**thermosphere** A layer high in the Earth's atmosphere above the mesosphere.

**tidal force** A gravitational force whose strength and/or direction varies over a body and thus tends to deform the body.

**time zone** One of 24 divisions of the Earth's surface separated by 15° along lines of constant longitude (with allowances for some political boundaries).

**total eclipse** A solar eclipse during which the Sun is completely hidden by the Moon, or a lunar eclipse during which the Moon is completely immersed in the Earth's umbra.

**trailing-arm spiral** A spiral arm pointing away from the direction of rotation, characteristic of all spiral galaxies.

**transition** (electronic) The change in energy and orbit of an electron around an atom or molecule.

**Trojan asteroid** One of several asteroids at stable Lagrange points that share Jupiter's orbit about the Sun.

**troposphere** The lowest level of the Earth's atmosphere.

**Tully-Fisher relation** A correlation between the width of the 21-centimeter line of a spiral galaxy and its absolute magnitude.

**turbulence** Random motions in a gas or liquid.

**turnoff point** The location of the brightest main-sequence stars on the Hertzsprung-Russell diagram of a globular cluster.

**21-cm radiation** Radio emission from a hydrogen atom caused by the flip of the electron's spin orientation.

**twinkling** The apparent change in a star's brightness, position, or color due to the motion of gases in the Earth's atmosphere.

**Type I Cepheid** Population I Cepheid variable star found in the disks of spiral galaxies.

**Type Ia supernova** A supernova occurring after a white dwarf accretes enough mass from a companion star to exceed the Chandrasekhar limit.

**Type II Cepheid** Population II Cepheid variable star found in elliptical galaxies and in the halos of disk galaxies that is 1.5 magnitudes dimmer than a Type I Cepheid.

**Type II supernova** A supernova occurring after a massive star's core is converted to iron.

**UBV filters** A system of filters that yield stellar magnitudes at different wavelengths in the ultraviolet, blue, and visible (yellow) spectral regions.

**ultraviolet (UV) radiation** Electromagnetic radiation of wavelengths shorter than those of visible light but longer than those of X rays.

**umbra** The central, completely dark portion of a shadow.

**unbound universe** A universe that lacks the mass necessary to stop its expansion; a universe that will expand forever.

**universal constant of gravitation** The constant of proportionality in Newton's law of gravitation, usually denoted G.

**universal law of gravitation** Newton's law of gravitation, which describes how the gravitational force between two bodies depends on their masses and separation.

**universe** All space along with all the matter and radiation in space.

**Van Allen radiation belts** Two flattened, doughnut-shaped regions around the Earth where many charged particles (mostly protons and electrons) are trapped by the Earth's magnetic field.

**variable star** A star whose luminosity varies.

**velocity** A quantity that specifies both direction and speed of an object.

**vernal equinox** The point on the ecliptic where the Sun crosses the celestial equator from south to north.

**very-long-baseline interferometry (VLBI)** A method of connecting widely separated radio telescopes to make observations of very high resolution.

**virtual particle** A particle and its antiparticle, created simultaneously in pairs and which quickly disappear without a trace.

**visual binary star** A double star in which the two components can be resolved through a telescope.

**void** A huge, roughly spherical region of the universe where exceptionally few galaxies are found.

**water hole** The part of the electromagnetic spectrum at a few thousand megahertz where there is very little background noise from space.

**wavelength** The distance between two successive peaks in a wave.

**waning** An adjective that means "decreasing," as in the "waning crescent Moon" or the "waning gibbous Moon."

**waxing** An adjective that means "increasing," as in the "waxing crescent Moon" or the "waxing gibbous Moon."

**weak nuclear force** A nuclear interaction involved in certain kinds of radioactive decay.

**weight** The force with which a body presses down on the surface of the Earth.

**white dwarf** A low-mass stellar remnant that has exhausted all its thermonuclear fuel and contracted to a size roughly equal to the size of the Earth.

**Widmanstätten patterns** Crystalline structure seen inside iron meteorites.

**Wien's law** A relationship between the temperature of a blackbody and the wavelength at which it emits the greatest intensity of radiation.

**winter solstice** The point on the ecliptic where the Sun reaches its greatest distance south of the celestial equator.

**wormhole** A hypothetical connecting passage between black holes and other places in the universe.

**X ray** Electromagnetic radiation whose wavelength is between that of ultraviolet light and gamma rays.

**X-ray burster**  A neutron star in a binary star system that accretes mass, undergoes thermonuclear fusion on its surface, and therefore emits short bursts of X rays.

**year**  The sidereal period of revolution of the Earth about the Sun.

**Zeeman effect**  A splitting or broadening of spectral lines in the presence of a magnetic field.

**zenith**  The point on the celestial sphere directly overhead.

**zero-age main sequence (ZAMS)**  The positions of stars on the Hertzsprung-Russell diagram that have just begun to fuse hydrogen in their cores.

**zodiac**  A band of 13 constellations around the sky through which the Sun appears to move throughout the year.

**zones** (on Jupiter)  Light-colored bands in Jupiter's cloud cover.

# ANSWERS TO COMPUTATION QUESTIONS & SELF-TESTS

## Foundation I
2. **a.** $5.974 \times 10^{24}$ kg
   **b.** $1.989999 \times 10^{30}$ kg
   **c.** $6.9599 \times 10^5$ km
   **d.** $3.1558 \times 10^7$ s
3. **a.** 8.3 pc = 27.1 ly
   **b.** 6.52 ly = 2.0 pc
   **c.** 8450 AU = $1.26 \times 10^{12}$ km
   **d.** $2.7 \times 10^3$ Mpc = $2.7 \times 10^6$ kiloparsecs

## Chapter 1
36. 1 more sidereal month than synodic month
45. $R_\odot \approx 6.53 \times 10^8$ m

## Chapter 2
13. 5.2 square AU in 1995, 26 square AU in 5 years
14. $a$ = 100 AU, maximum distance is almost 200 AU
15. 2.8 yr
16. use $P^2 = a^3$
21. At 2 AU, 1 year $\approx$ 2.8 present years; At 1/2 AU, 1 year $\approx$ .35 present years
22. .01 × present gravitational attraction

## Chapter 3
**Toolbox 3-1 Self test** $\lambda$ = 404 nm, color: violet
4. 9 times more
17. 1/25
18. Palomar gathers $10^6$ times more light than the human eye.
20. **a.** 222x
    **b.** 100x
    **c.** 36x

## Chapter 4
10. 7 1/2 times
11. 238 nm
12. 6520 K
13. approaching, 13 km/s
14. receding, 21 km/s
15. about 90,000 km/s
19. $\lambda \approx$ 500 nm (same as now), 4 × brighter than now

## Chapter 8
18. $5.0 \times 10^{14}$ tons

## Chapter 9
5. next maximum in 2005, next minimum in 2012
16. 1400 kg/m$^3$
17. 4.8%
18. 500 nm = visible light; 58 nm = ultraviolet; 1.9 nm = x-ray

## Foundation III
**Toolbox III-1** Rigel 237 pc.
Parallax angle = $4.2 \times 10^{-3}$ arcsec
10. 25 times brighter
12. 4.3 pc
14. **a.** 9.7 pc,
    **b.** 0.10 arcsec
15. 1585 times brighter

## Chapter 10
**Toolbox 10-1** In $M_1 + M_2 = a^3/P^2$, use $M_1 = 1.0$ M$_\odot$ and ignore $M_2$, which is at most (Jupiter) 1000 times less than $M_1$.

## Chapter 11
25. 2000
26. 200 times longer

## Chapter 12
25. about 7465 years ago

## Chapter 13
**Toolbox 13-1** $10^{12}$ kg
8. 8.9 km, 89 km

## Chapter 14
10. 20 times
14. about once every 750 years

## Chapter 15
**Toolbox 15-1** $1.84 \times 10^5$ km/s
16. 293 Mpc
17. 7500 km/s

## Chapter 17
10. 11 billion years

# FOR FURTHER READING

## Chapter 1

Allen, D., and C. Allen. *Eclipse.* Allen & Unwin, 1987.

Allen, R. *Star Names: Their Lore and Meaning.* Dover, 1963 (originally published 1899).

*Astronomy* magazine. Kalmbach Publishing, published monthly (also on the Web at www.astronomy.com).

Berry, R. *Discover the Stars.* Harmony, 1987.

Burnham, R., Jr. *Burnham's Celestial Handbook,* 3 vols. Dover, 1978.

Chartrand, M. *Skyguide.* Western Publications, 1982.

Gingerich, O. "The Origin of the Zodiac." *Sky & Telescope,* March 1984.

Harrington, P. S., *Eclipse!* Wiley, 1997.

Kaler, J. *The Ever-Changing Sky.* Cambridge University Press, 1996.

Kunitzsch, P. "How We Got Our Arabic Star Names." *Sky & Telescope,* January 1983.

Motz, L., and C. Nathanson. *The Constellations: An Enthusiast's Guide to the Night Sky.* Doubleday, 1988.

Pasachoff, J. M., and D. H. Menzel. *A Field Guide to the Stars and Planets.* Houghton-Mifflin, 1992.

Ridpath, I., and W. Tirion. *The Monthly Sky Guide,* 4th edition. Cambridge University Press, 1987.

Sagan, C. "The Shores of the Cosmic Ocean." In *Cosmos.* Random House, 1980.

*Sky & Telescope* magazine. Sky Publishing, published monthly (also on the Web at www.skypub.com).

Tirion, W. *Sky Atlas 2000.0.* Sky Publishing and Cambridge University Press, 1981.

## Chapter 2

Christianson, G. "Newton's Principia: A Retrospective." *Sky & Telescope,* June 1987.

———. *This Wild Abyss.* Free Press, 1978.

Cohen, I. B. "Newton's Discovery of Gravity." *Scientific American,* March 1981.

Drake, S. "Newton's Apple and Galileo's Dialogue." *Scientific American,* August 1980.

Gingerich, O. "Copernicus and Tycho." *Scientific American,* December 1973.

———. "The Galileo Affair." *Scientific American,* August 1982.

———. "How Galileo Changed the Rules of Science." *Sky & Telescope,* March 1993.

Krupp, E. C. "Observing the Occasion." *Sky & Telescope,* December 1996.

Lerner, L., and E. Gosselin. "Galileo and the Specter of Bruno." *Scientific American,* November 1986.

Wilson, C. "How Did Kepler Discover His First Two Laws?" *Scientific American,* March 1972.

## Chapter 3

Brunier, S. "Temples in the Sky." *Sky & Telescope,* February, June, and December 1993.

Chaisson, E. *The Hubble Wars.* HarperCollins, 1994.

Cole, S. "Astronomy on the Edge: Using the Space Telescope." *Sky & Telescope,* October 1992.

Dyer, A. "Buying the Best Telescope." *Sky & Telescope,* December 1997.

Fugate, R. Q., and W. J. Wild. "Untwinkling the Stars." *Sky & Telescope,* May and June 1994.

Giacconi, R. "The Einstein X-Ray Observatory." *Scientific American,* February 1980.

Harrington, S. "Selecting Your First Telescope." *Mercury,* July/August 1982.

Henbest, N., and M. Marten. *The New Astronomy.* Cambridge University Press, 1983.

Janesick, J., and M. Blouke. "Sky on a Chip: The Fabulous CCD." *Sky & Telescope,* September 1987.

Jastrow, R., and S. Baliunas. "Mount Wilson: America's Observatory." *Sky & Telescope,* March 1993.

Kondo, Y., W. Wamsteker, and D. Stickland. "IUE: 15 Years and Counting." *Sky & Telescope,* September 1993.

Krisciunas, K. *Astronomical Centers of the World.* Cambridge University Press, 1988.

Learner, R. *Astronomy Through the Telescope.* Van Nostrand Reinhold, 1981.

MacRobert, A. "Astronomy with a $5 Telescope." *Sky & Telescope,* April 1990.

Powell, C. "Mirroring the Cosmos." *Scientific American,* November 1991.

Preston, R. *First Light: The Search for the Edge of the Universe.* Atlantic Monthly Press, 1987.

Readhead, A. "Radio Astronomy by Very-Long-Baseline Interferometry." *Scientific American,* June 1982.

Robinson, L. "The Frigid World of IRAS." *Sky & Telescope,* January 1984.

Schorn, R. A. "Listening to the Universe." *Sky & Telescope,* November 1988.

Sinnott, R., and K. Nyren. "The World's Largest Telescopes." *Sky & Telescope,* July 1993.

Sullivan, W. "Radio Astronomy's Golden Anniversary." *Sky & Telescope,* December 1982.

Tucker, W., and K. Tucker. *The Cosmic Inquirers.* Harvard University Press, 1986.

Verschuur, G. *The Invisible Universe Revealed.* Springer-Verlag, 1987.

## Chapter 4

Achinstein, P. *Particles and Waves: Historical Essays in the Philosophy of Science.* Oxford University Press, 1991.

Bova, B. *The Beauty of Light.* Wiley, 1988.

Cline, B. *Men Who Made a New Physics.* Signet, 1965.

Davies, J. K. "The Extreme Ultraviolet: A Promising New Window on the Universe." *Astronomy,* July 1987.

Feynman, R. *QED.* Princeton University Press, 1985.

Field, G. B., and E. J. Chaisson. *The Invisible Universe.* Vintage Books, 1987.

Gingerich, O. "Unlocking the Chemical Secrets of the Cosmos." *Sky & Telescope,* July 1981.

Gribbin, J. *In Search of Schroedinger's Cat.* Bantam, 1984.

Griffin, R. "The Radial-Velocity Revolution." *Sky & Telescope,* September 1989.

Minnaert, M. G. J. *Light and Color in the Outdoors.* Springer-Verlag, 1993.

Snow, C. *The Physicists.* Macmillan, 1981.

Sobel, M. *Light.* University of Chicago Press, 1987.

Trefil, J. S., *From Atoms to Quarks,* rev. ed. Anchor, 1994.

Van Heel, A., and C. Velzel. *What Is Light?* McGraw-Hill, 1968.

Weinberg, S. *Subatomic Particles.* Scientific American Books, 1983.

Weiss, R. J. A. *A Brief History of Light and Those that Lit the Way.* World Scientific, 1996.

## Chapter 5

Akasofu, S.-I. "The Dynamic Aurorae." *Scientific American,* May 1989.

Beatty, J. K. "The Making of a Better Moon." *Sky & Telescope,* December 1986.

Bolt, B. A. *Earthquakes and Geological Discovery.* Scientific American Library, 1993.

Broecker, W. S. *How to Build a Habitable Planet.* Eldigio Press, 1985.

Burnham, R. "Watching the Lunar Phases." *Astronomy,* July 1989.

Cattermole, P., and P. Moore. *The Story of the Earth.* Cambridge University Press, 1985.

Comins, N. F. *What If the Moon Didn't Exist?* HarperCollins, 1993.

Davis, N. *The Aurora Watcher's Handbook.* University of Alaska Press, 1992.

Hartmann, W. "The Moon's Early History." *Astronomy,* September 1976.

Karl, T. R., N. Nicholls, and J. Gregory. "The Coming Climate." *Scientific American,* April 1992.

Kelley, K., ed. *The Home Planet.* Addison-Wesley and Mir, 1988.

Kitt, M. T. "Eight Lunar Wonders." *Astronomy,* March 1989.

———. "One Day at Copernicus Crater." *Astronomy,* September 1988.

Murphy, J. B., and R. D. Nance. "Mountain Building and the Supercontinent Cycle." *Scientific American,* April 1992.

Press, F., and R. Siever. *Understanding the Earth,* 2nd edition. W. H. Freeman, 1998.

Spudis, P. D. *The Once and Future Moon.* Smithsonian Institution Press, 1996.

Taylor, S. R., and S. M. McLennan. "The Evolution of Continental Crust." *Scientific American,* January 1996.

Van Andel, T. *New Views on an Old Planet.* Cambridge University Press, 1985.

## Chapter 6

Bazilevskiy, A. T. "The Planet Next Door." *Sky & Telescope,* April 1989.

Bolt, B. A. *Earthquakes and Geological Discovery.* Scientific American Library, 1993.

Burgess, E. *Venus: An Arrant Twin.* Columbia University Press, 1985.

Burnham, R. "What Makes Venus Go?" *Astronomy,* January 1993.

Chaikin, A. "Four Faces of Mars." *Sky & Telescope,* July 1992.

Chapman, C. R. "Mercury's Heart of Iron." *Astronomy,* November 1988.

Davies, M., et al., eds. *Atlas of Mercury.* NASA SP-423, 1978.

Eicher, D. J. "*Magellan* Scores at Venus " *Astronomy,* January 1991.

Goldman, S. J. "A Sol in the Life on Pathfinder." *Sky & Telescope,* November 1997.

———. "Venus Unveiled." *Sky & Telescope,* March 1992.

Grinspoon, D. H. *Venus Revealed* Addison-Wesley, 1997.

Haberle, R. M. "The Climate of Mars." *Scientific American,* May 1986.

Hartmann, W. "The Significance of the Planet Mercury." *Sky & Telescope,* May 1976.

———. "What's New on Mars?" *Sky & Telescope,* May 1989.

Horowitz, N. H. *To Utopia and Back.* W. H. Freeman, 1986.

Kargle, J. S. "The Rivers of Venus." *Sky & Telescope,* August, 1997.

Kargle, J. S., and R. G. Strom. "Global Climatic Change on Mars." *Scientific American,* November 1996.

McSween, H. Y. "Nor Any Drop to Drink." *Sky & Telescope,* December 1995.

Murray, B., and E. Burgess. *Flight to Mercury.* Columbia University Press, 1977.

Parker, S. "Mars Global Survey: You Ain't Seen Nothin' Yet." *Sky & Telescope,* January 1998.

Petersen, C. C. "Welcome to Mars." *Sky & Telescope,* October 1997.

Powell, C. S. "Venus Revealed." *Scientific American,* January 1992.

Saunders, R. S. "The Surface of Venus." *Scientific American,* December 1990.

Strom, R. *Mercury, the Elusive Planet.* Cambridge University Press, 1987.

———. "Mercury: The Forgotten Planet." *Sky & Telescope,* September 1990.

Toon, O. B. "How Climate Evolved on the Terrestrial Planets." *Scientific American*, February 1988.

## Chapter 7

Beatty, J. "Getting to Know Neptune." *Sky & Telescope*, February 1990.

———. "A Place Called Uranus." *Sky & Telescope*, April 1986.

———. "Report on the *Voyager* Encounters with Saturn." *Sky & Telescope*, January, October, and November 1981.

Beatty, J., and D. H. Levy. "Crashes to Ashes: A Comet's Demise." *Sky & Telescope*, October 1995.

Beebe, R. F. "Queen of the Giant Storms." *Sky & Telescope*, October 1990.

Bennett, J. "The Discovery of Uranus." *Sky & Telescope*, March 1981.

Berry, R. "Neptune Revealed." *Astronomy*, December 1989.

Binzel, R. "Pluto." *Scientific American*, June 1990.

Brown, R., and D. Cruikshank. "The Moons of Uranus, Neptune, and Pluto." *Scientific American*, July 1985.

Burgess, E. *By Jupiter*. Columbia University Press, 1982.

Burnham, R. "The Saturnian Satellites." *Astronomy*, November 1981.

Carroll, M. "Europa: Distant Ocean, Hidden Life?" *Sky & Telescope*, December 1997

Cuzzi, J. N., and L. W. Esposito. "The Rings of Uranus." *Scientific American*, July 1987.

Elliot, J., et al. "Discovering the Rings of Uranus." *Sky & Telescope*, June 1977.

Goldman, S. J. "*Galileo* in Retrospect." *Sky & Telescope*, December 1997.

Ingersoll, A. "Jupiter and Saturn." *Scientific American*, December 1981.

———. "Uranus." *Scientific American*, January 1987.

Kinoshita, J. "Neptune." *Scientific American*, November 1989.

MacRobert, A. M. "Hunting the Moons and Saturn." *Sky & Telescope*, July 1991.

Morrison, D. "The Enigma Called Io." *Sky & Telescope*, March 1985.

———. *Exploring Planetary Worlds*. Scientific American Library, 1993.

———. *Voyages to Saturn*. NASA SP-451, 1982.

Morrison, D., and J. Samz. *Voyage to Jupiter*. NASA SP-439, 1980.

Pollack, J., and J. Cuzzi. "Rings in the Solar System." *Scientific American*, November 1981.

Rogers, J. H. *The Giant Planet Jupiter*. Cambridge University Press, 1995.

Soderblom, L. A. "The Galilean Moons of Jupiter." *Scientific American*, January 1980.

*Voyager at Jupiter*. Special issue of *Astronomy*, May 1979.

Washburn, M. *Distant Encounters: The Exploration of Jupiter and Saturn*. Harcourt Brace Jovanovich, 1983.

## Chapter 8

Aguirre, E. L. "The Great Comet of 1997." *Sky & Telescope*, July 1997.

Alvarez, W., and F. Asaro. "What Caused the Mass Extinction? An Extraterrestrial Impact." *Scientific American*, October 1990.

Beatty, J. K. "An Inside Look at Halley's Comet." *Sky & Telescope*, May 1986.

———. "Killer Crater in the Yucatán?" *Sky & Telescope*, July 1991.

Binzel, R. P., M. A. Barucci, and M. Fulchignoni. "The Origins of the Asteroids." *Scientific American*, October 1991.

Brandt, J. C., and R. D. Chapman. *Rendezvous in Space: The Science of Comets*. W. H. Freeman, 1992.

Cunningham, C. J. "Giuseppe Piazzi and the 'Missing Planet.'" *Sky & Telescope*, September 1992.

Dodd, R. *Thunderstones and Shooting Stars: The Meaning of Meteorites*. Harvard University Press, 1986.

Durda, D. "All in the Family." *Astronomy*, February 1993.

Knacke, R. "Sampling the Stuff of a Comet." *Sky & Telescope*, March 1987.

Kowal, C. *Asteroids*. Ellis Horwood/John Wiley, 1988.

Levy, D. H. "How to Discover a Comet." *Astronomy*, December 1987.

Luu, J. X., and D. C. Jewitt. "The Kuiper Belt." *Scientific American*, May 1996.

Marsden, B. G. "Comet Swift-Tuttle: Does It Threaten Earth?" *Sky & Telescope*, January 1993.

Morrison, D., and C. R. Chapman. "Target Earth: It *Will* Happen." *Sky & Telescope*, March 1990.

Olson, R. "Giotto's Portrait of Halley's Comet." *Scientific American*, July 1979.

Russell, D. "The Mass Extinctions of the Late Mesozoic." *Scientific American*, January 1982.

Sagan, C., and A. Druyan. *Comet*. Random House, 1985.

Weissman, P. "Are Periodic Bombardments Real?" *Sky & Telescope*, March 1990.

———. "Realm of the Comet." *Sky & Telescope*, March 1987.

Whipple, F. "The Black Heart of Comet Halley." *Sky & Telescope*, March 1987.

———. *The Mystery of Comets*. Smithsonian Institution Press, 1985.

———. "The Spin of Comets." *Scientific American*, March 1980.

## Chapter 9

Baliunas, S., and W. Soon. "The Sun-Climate Connection." *Sky & Telescope*, December 1996.

Eddy, J. *A New Sun*. NASA SP-402, 1979.

Foukal, P. "The Variable Sun." *Scientific American*, February 1990.

Frazier, K. *Our Turbulent Sun*. Prentice-Hall, 1983.

Friedman, H. *Sun and Earth*. Scientific American Library, 1986.

Giovanelli, R. *Secrets of the Sun*. Cambridge University Press, 1984.

Kennedy, J. R., "GONG: Probing the Sun's Hidden Heart." *Sky & Telescope*, October 1996.

Lang, K. R. "SOHO Reveals the Secrets of the Sun." *Scientific American*, March 1997 .

———. *Sun, Earth, and Sky*. Spring-Verlag, 1995.

Leibacher, J. W., et al. "Helioseismology." *Scientific American*, September 1985.

Levine, R. "The New Sun." In J. Cornell and P. Gorenstein, eds. *Astronomy from Space: Sputnik to Space Telescope*. MIT Press, 1983.

Mitton, S. *Daytime Star*. Scribner, 1981.

Nichols, R. "Solar Max: 1980–1989." *Sky & Telescope*, December 1989.

Noyes, R. *The Sun, Our Star*. Harvard University Press, 1982.

Robinson, L. J. "The Disquieting Sun: How Big, How Steady?" *Sky & Telescope*, April 1982.

———. "The Sunspot Cycle: Tip of the Iceberg." *Sky & Telescope*, June 1987.

Schaefer, B. E. "Sunspots that Changed the World." *Sky & Telescope*, April 1997.

Taylor, M. "Observing from the South Pole." *Sky & Telescope*, October 1988.

Wallenhorst, S. "Sunspot Numbers and Solar Cycles." *Sky & Telescope*, September 1982.

Wentzel, D. *The Restless Sun*. Smithsonian Institution Press, 1989.

## Chapter 10

Ashbrook, J. "Visual Double Stars for the Amateur." *Sky & Telescope*, November 1980.

Evans, D., et al. "Measuring Diameters of Stars." *Sky & Telescope*, August 1979.

Gingerich, O. "A Search for Russell's Original Diagram." *Sky & Telescope*, July 1982.

Griffin, R. "The Radial-Velocity Revolution." *Sky & Telescope*, September 1989.

Kaler, J. "Origins of the Spectral Sequence." *Sky & Telescope*, February 1986.

———. *Stars*. Scientific American Library, 1992.

———. *Stars and Their Spectra*. Cambridge University Press, 1997.

Mitton, J., and A. MacRobert. "Colored Stars." *Sky & Telescope*, February 1989.

Nielsen, A. "E. Hertzsprung—Measurer of Stars." *Sky & Telescope*, January 1968.

Phillip, A., and L. Green. "Henry N. Russell and the H-R Diagram." *Sky & Telescope*, April and May 1978.

Tomkin, J., and D. Lambert. "The Strange Case of Beta Lyrae." *Sky & Telescope*, October 1987.

Turon, C. "Measuring the Universe: From Hipparchus to *Hipparcos*." *Sky & Telescope*, July 1997.

## Chapter 11

Blitz, L. "Giant Molecular Cloud Complexes in the Galaxy." *Scientific American*, April 1982.

Cohen, M. *In Darkness Born: The Story of Star Formation*. Cambridge University Press, 1988.

Fortier, E. "Touring the Stellar Cycle." *Astronomy*, March 1987.

Hartley, K. "How a Star Is Born." *Astronomy*, December 1989.

Iben, I., and A. V. Tutukov. "The Lives of Stars: From Birth to Death and Beyond." *Sky & Telescope*, December 1997 and January 1998.

Johnson, B. "Red Giant Stars." *Astronomy*, December 1976.

Kaler, J. B. *Cosmic Clouds*. Scientific American Library, 1997.

———. "Cousins of Our Sun: The G Stars." *Sky & Telescope*, November 1986.

———. "Journeys on the H-R Diagram." *Sky & Telescope*, May 1988.

———. "The K Stars: Orange Giants and Dwarfs." *Sky & Telescope*, August 1986.

———. "The M Stars: Supergiants to Dwarfs." *Sky & Telescope*, May 1986.

———. "The Spectacular O Stars." *Sky & Telescope*, November 1987.

———. *Stars*. Scientific American Library, 1992.

———. "The Temperate F Stars." in *Sky & Telescope*, February 1987.

Kippenhahn, R. *100 Billion Suns: The Birth, Life, and Death of Stars*. Basic Books, 1983.

Lada, C. "Energetic Outflows from Young Stars." *Scientific American*, July 1982.

Robinson, L. "Orion's Stellar Nursery." *Sky & Telescope*, November 1982.

Stahler, S. W. "The Early Life of Stars." *Scientific American*, July 1991.

Stahler, S. W., and N. Comins. "The Difficult Birth of Sunlike Stars." *Astronomy*, September 1988.

Verschuur, G. *Interstellar Matters*. Springer-Verlag, 1988.

## Chapter 12

Bethe, H., and G. Brown. "How a Supernova Explodes." *Scientific American*, May 1985.

Clark, D. *Superstars*. McGraw-Hill, 1984.

Davidson, K. "Crisis in Eta Carinae?" *Sky & Telescope*, January 1998.

———. "From Swords to Supernovae." *Sky & Telescope*, November 1997.

Greenstein, G. *Frozen Star*. Freundlich Books, 1984.

Kirschner, R. P. "Supernova 1987A: The First Ten Years." *Sky & Telescope*, February 1997.

Lattimer, J., and A. Burrows. "Neutrinos from Supernova 1987A." *Sky & Telescope*, October 1988.

Margon, B. "The Bizarre Spectrum of SS433." *Scientific American*, October 1980.

Marschall, L. *The Supernova Story*. Plenum, 1988.

Marschall, L., and K. Brecher. "Will Supernova 1987A Shine Again?" *Astronomy*, February 1992.

White, N. "New Wave Pulsars." *Sky & Telescope*, January 1987.

Woosley, S., and T. Weaver. "The Great Supernova of 1987." *Scientific American*, August 1989.

## Chapter 13

Begelman, M., and M. Rees. *Gravity's Fatal Attraction: Black Holes in the Universe*. Scientific American Library, 1996

Chaisson, E. *Relatively Speaking: Relativity, Black Holes, and the Fate of the Universe*. W. W. Norton, 1988.

Croswell, K. "The Best Black Hole in the Galaxy." *Astronomy*, March 1992.

Greenstein, G. *Frozen Star.* Freundlich Books, 1983.

Lasota, J. P. "Unmasking Black Holes." *Scientific American,* May 1999.

McClintock, J. "Do Black Holes Exist?" *Sky & Telescope,* January 1988.

Rees, M. J. "Black Holes in Galactic Centers." *Scientific American,* November 1990.

Thorne, K. *Black Holes and Time Warps.* W. W. Norton, 1994.

Wheeler, J. A. *A Journey Into Gravity and Spacetime.* Scientific American Library, 1990.

## Chapter 14

Bok, B. "A Bigger and Better Milky Way." *Astronomy,* January 1984.

———. "The Milky Way Galaxy." *Scientific American,* March 1981.

Bok, B., and P. Bok. *The Milky Way,* 5th ed. Harvard University Press, 1981.

Geballe, T. "The Central Parsec of the Galaxy." *Scientific American,* July 1979.

Jayawardhana, R. "Destination: Galactic Center." *Sky & Telescope,* June 1995.

Kaufmann, W. J. "Our Galaxy, Part 1." *Mercury,* May/June 1989.

———. "Our Galaxy, Part 2." *Mercury,* July/August 1989.

Palmer, E. "Unveiling the Hidden Milky Way." *Astronomy,* November 1989.

## Chapter 15

Barrow, J. D., and J. Silk. *The Left Hand of Creation.* Basic Books, 1983.

Bothun, G. D. "The Ghostliest Galaxies." *Scientific American,* February 1997.

Christianson, G. E. *Edwin Hubble: Mariner of the Nebulae.* Farrar, Straus, & Giroux, 1995.

De Vaucouleurs, G. "The Distance Scale of the Universe." *Sky & Telescope,* December 1983.

———. "M31's Spiral Shape." *Sky & Telescope,* December 1987.

Elmegreen, D. M., and B. Elmegreen. "What Puts the Spiral in Spiral Galaxy?" *Astronomy,* September 1993.

Ferris, T. *Galaxies.* Stewart, Tabori, and Chang, 1982.

Field, G. "The Hidden Mass in Galaxies." *Mercury,* May/June 1982.

Freeman, M. "Galaxies." *Smithsonian,* January 1989.

Hartley, K. "Elliptical Galaxies Forged by Collision." *Astronomy,* May 1989.

Hodge, P. *Galaxies.* Harvard University Press, 1986.

Jayawardhana, R. "Our Galaxy's Nearest Neighbor." *Sky & Telescope,* May 1998.

Keel, W. "Crashing Galaxies, Cosmic Fireworks." *Sky & Telescope,* January 1989.

Osterbrock, D., R. Brashear, and J. Gwinn. "Young Edwin Hubble." *Mercury,* January/February 1990.

Rubin, V. "Dark Matter in Spiral Galaxies." *Scientific American,* June 1983.

Silk, J. "Formation of the Galaxies." *Sky & Telescope,* December 1986.

Tully, R. "Unscrambling the Local Supercluster." *Sky & Telescope,* June 1982.

West, M. "Galaxy Clusters: Urbanization of the Cosmos." *Sky & Telescope,* January 1997.

Wray, J. *The Color Atlas of Galaxies.* Cambridge University Press, 1988.

## Chapter 16

Blandford, R., et al. "Cosmic Jets." *Scientific American,* May 1982.

Burns, J., and R. Price. "Centaurus A: The Nearest Active Galaxy." *Scientific American,* November 1983.

Courvoisier, T. J.-L., and E. I. Robson. "The Quasar 3C 273." *Scientific American,* June 1991.

Ford, H., and Z. I. Tsvetanov. "Massive Black Holes in the Hearts of Galaxies." *Sky & Telescope,* June 1996.

Leonard, P. J. T., and J. T. Bonnell. "Gamma Rays of Doom." *Sky & Telescope,* February 1998.

McCarthy, P. "Measuring Distances to Remote Galaxies and Quasars." *Mercury,* January/February 1988.

Mood, J. "Star Hopping to a Quasar." *Astronomy,* May 1987.

Preston, R. *First Light.* Atlantic Monthly Press, 1987.

Verschuur, G. *The Invisible Universe Revealed.* Springer-Verlag, 1987.

Wilkes, B. "The Emerging Picture of Quasars." *Astronomy,* December 1991.

## Chapter 17

Davies, P. *The Forces of Nature,* 2nd ed. Cambridge University Press, 1986.

Discus, D., et al. "The Future of the Universe." *Scientific American,* March 1983.

Disney, M. *The Hidden Universe.* Macmillan, 1984.

Dressler, A. "The Large-Scale Streaming of Galaxies." *Scientific American,* September 1987.

Gribbin, J. *In Search of the Big Bang.* Bantam, 1986.

Guth, A. The Inflationary Universe. Addison Wesley, 1997.

Halliwell, J. J. "Quantum Cosmology and the Creation of the Universe." *Scientific American* December 1991.

Harrison, E. *Darkness at Night.* Harvard University Press, 1987.

Hawking, S. *A Brief History of Time.* Bantam, 1988 (also on CD-ROM from W. H. Freeman).

Henbest, N., and H. Couper. *The Guide to the Galaxies.* Cambridge University Press, 1994.

Kanipe, J. "Beyond the Big Bang." *Astronomy,* April 1992.

Krauss, L. M. "Cosmological Antigravity." *Scientific American,* January 1999.

Luminet, J. P., G. D. Starkman, and J. R. Weeks. "Is Space Finite?" *Scientific American,* April 1999.

Mather, J. C., and J. Boslough. *The Very First Light.* Basic Books, 1996.

Monda, R. "Shedding Light on Dark Matter." *Astronomy,* February 1992.

Rees, M. *Before the Beginning.* Helix/Addison-Wesley, 1997.

Riordan, M., and D. N. Schramm. *The Shadows of Creation: Dark Matter and the Structure of the Universe.* W. H. Freeman, 1991.

Rowan-Robinson, M. *Our Universe: An Armchair Guide.* W. H. Freeman, 1990.

Silk, J. *The Big Bang,* 2nd ed., W. H. Freeman, 1989.

———. *A Short History of the Universe.* Scientific American Library, 1994.

Smoot, G., and K. Davidson. *Wrinkles in Time.* William Morrow, 1993.

Weinberg, S. *The First Three Minutes,* updated edition. HarperCollins, 1988.

## Chapter 18

Beatty, J. K. "The New, Improved SETI." *Sky & Telescope,* May 1983.

Broecker, W. S. *How to Build a Habitable Planet.* Eldigio Press, 1985.

Davies, P. *The Cosmic Blueprint.* Simon & Schuster, 1988.

Dawkins, R. *The Blind Watchmaker.* W. W. Norton, 1987.

Feinberg, G., and R. Shapiro. *Life Beyond Earth: The Intelligent Earthling's Guide to Life in the Universe.* Morrow, 1980.

Finney, B., and E. Jones. *Interstellar Migration and the Human Experience.* University of California Press, 1985.

Goldsmith, D., ed. *The Quest for Extraterrestrial Life: A Book of Readings.* University Science Books, 1980.

Goldsmith, D., and T. Owen. *The Search for Life in the Universe.* Benjamin/Cummings, 1980.

Gould, S. J. "The Evolution of Life on the Earth." *Scientific American,* October 1994.

———. *Wonderful Life.* Scribner, 1990.

Harrison, E. *Masks of the Universe.* Macmillan, 1985.

Hart, M., and B. Zuckerman, eds. *Extraterrestrials: Where Are They?* Pergamon Press, 1982.

Horowitz, N. *To Utopia and Back: The Search for Life in the Solar System.* W. H. Freeman, 1986.

Kutter, G. S. *The Universe and Life.* Jones and Bartlett, 1987.

Marx, G., ed. *Bioastronomy: The Next Steps.* Kluwer, 1988.

Papagiannis, M. D. "Bioastronomy: The Search for Extraterrestrial Life." *Sky & Telescope,* June 1984.

Pendelton, Y. J., and J. D. Famer. "Life: A Cosmic Imperative?" *Sky & Telescope,* July 1997.

Reeves, H. *The Hour of Our Delight: Cosmic Evolution, Order, and Complexity.* W. H. Freeman, 1991.

Regis, E. *Extraterrestrials: Science and Alien Intelligence.* Cambridge University Press, 1985.

Rood, R., and J. Trefil. *Are We Alone?* Scribner, 1981.

Sagan, C., and F. Drake. "The Search for Extraterrestrial Intelligence." *Scientific American,* May 1975.

Schorn, R. A. "Extraterrestrial Beings Don't Exist." *Sky & Telescope,* September 1981.

Shklovskii, I. S., and C. Sagan. *Intelligent Life in the Universe.* Holden-Day, 1966.

Wilson, A. C. "The Molecular Basis of Evolution." *Scientific American,* October 1985.

Zuckerman, B., and M. Hart, eds. *Extraterrestrials—Where Are They?,* 2nd edition. Cambridge University Press, 1995.

# ILLUSTRATION CREDITS

Cover image and frontispiece: Painting © 1999 by Don Dixon, Long Beach, California.

**Foundation I** Facing p. 1: NASA. Fig. I-1: Corbis. Fig. I-2: Scala/Art Resource, NASA, NOAO, New Mexico State University, C. Holmes. p. 5: From top to bottom, R. Williams and the Hubble Deep Field Team (STScI) and NASA; AAT; L. Golub, Naval Observatory, IBM Research, NASA; Richard Bickel/Corbis; Scientific American Books. Fig. I-3: top left, Celestron International; top right, L. Golub, Naval Observatory, IBM Research, NASA; bottom left, NASA; bottom right, NOAO. Fig. I-4: top left, NOAO; top center, Observatoire de Haute-Provence, France; top right, NASA; bottom left, C.R. O'Dell, Rice University, John Bally, University of Colorado, Ralph Sutherland, University of Colorado; bottom center, R.J. Dufour, Rice University; bottom right, F.N. Owen and J.J. Puschell/NRAO/AUI.

**Chapter 1** p. 14: Anglo-Australian Observatory. Fig. 1-1a & b: John Sanford/ Astrostock. Fig. 1-8: Frank Zullo/Science Source/ Photo Researchers. Fig. 1-12: Dennis Di Cicco. Fig. 1-18b: Yerkes Observatory. Fig. 1-18c-i: Lick Observatory. Fig. 1-19: Art Wolfe. Fig. 1-24: M. Harms. Fig. 1-25: W. Dellinges. p. 34: NASA. Fig. 1-27: D. di Cicco.

**Chapter 2** p. 41: NASA. Fig. 2-9b: C. Holmes. Fig. 2-11: J. Marling. p. 52: top left, E. Lessing/Art Resource; top right, Art Resource; bottom left, painting by Jean-Leon Huens, courtesy of National Geographic Society; bottom right, National Portrait Gallery, London.

**Chapter 3** p. 58: Richard Wainscoat, Honolulu, Hawaii. Fig. 3-11 left & right: AURA. Fig. 3-12 left & right: AURA. Fig. 3-16: Yerkes Observatory. Fig. 3-21a: The University of Arizona, Steward Observatory. Fig. 3-21b: The University of Arizona, Steward Observatory. Courtesy of Lori Stiles. Fig. 3-22a & b: NASA/ESA. Fig. 3-23 top & bottom: NOAO. Fig. 3-24: NASA. Fig. 3-25 left & right: NASA. Fig. 3-26a & b: Air Force Research Laboratory, Starfire Optical Range, Kirtland AFB, NM. Fig. 3-26c: Reta Beebe (New Mexico State University), D. Gilmore, L. Bergeron (STScI), and NASA. p. 77: Roger Ressmeyer/ Corbis. Fig. 3-27: W.M. Keck Observatory. Courtesy of Richard J. Wainscoat. Fig. 3-28: IRFA, University of Hawaii, and Roger Ressmeyer (c) 1993 Corbis. Fig. 3-29a-c: Patrick Seitzer, NOAO. p. 78: Celestron International. Fig. 3-30: NRAO. Fig. 3-31: NRAO. Fig. 3-32a: NASA. Fig. 3-32b: (c) 1982 Associated Universities, Inc., under contract with the National Science Foundation (VLA observations by Imke de Pater, J.R. Dickel). Fig. 3-33: ESA. Fig. 3-34a: G.R. Carruthers, NRL. Fig. 3-34b: NASA. Fig. 3-34c: R.C. Mitchell, Central

Washington University. p. 85: D. Montrose/ Custom Medical Stock Photo.

**Chapter 4** p. 86: University of Michigan. Fig. 4-4: NOAO. Fig. 4-7: Carnegie Observatories. Fig. 4-13: Carnegie Observatories. p. 100: NOAO. Fig. 4-19 top & bottom: Yerkes Observatory.

**Foundation II** p. 104: NASA/JPL. M. Buie, K. Horne and D. Tholen. Fig. II-1: (c) 1980, 1984 Anglo-Australian Observatory. Fig. II-2: (c) 1980, 1984 Anglo-Australian Observatory. Fig. II-4: STScI, M.J. McCaughrean (MPIA), C.R. O'Dell (Rice University), NASA. Fig. II-6: NASA. Fig. II-10: NASA. Fig. II-11: NASA. Fig. II-12: STScI, A. Schultz (CSC/ STScI) and NASA. Fig. II-13: Paul Butler and Geoff Marcy; graphics by Leigh Ann McConnaughey, Berkeley. Fig. II-14: STScI, Susan Terebey (Extrasolar Research Corp.), and NASA.

**Chapter 5** p. 122: NASA. Fig. 5-1: NASA. Fig. 5-6: Digital image by Peter W. Sloss, NOAA-NESDIS-NGDC. Fig. 5-6 inset left: NASA. Fig. 5-6 inset right: NASA. Fig. 5-9a: Jules Bucher/ Photo Researchers. Fig. 5-11a: L.A. Frank and J.D. Craven, The University of Iowa. Fig. 5-11b: NASA. Fig. 5-11c: S.-I. Akasofu, Geophysical Institute, University of Alaska. Fig. 5-12: Lick Observatory. Fig. 5-13: NASA. Fig. 5-14: NASA. Fig. 5-15: NASA. Fig. 5-16: NASA. Fig. 5-17: Carnegie Observatories. Fig. 5-18: NASA. Fig. 5-19: NASA. Fig. 5-20: NASA. Fig. 5-21: NASA. Fig. 5-22: NASA. Fig. 5-23 inset: Lunar Planetary Institute. Fig. 5-24a-f: W. Benz. Fig. 5-29: Randall L. Ricklefs, McDonald Observatory, University of Texas at Austin.

**Chapter 6** p. 147: NASA/JPL. Fig. 6-1: Astrogeology Team, U.S. Geological Survey. p. 149 NASA, USGS. Fig. 6-2: NASA, UCO/Lick Observatory. Fig. 6-3: NASA. Fig. 6-4: NASA. Fig. 6-5: NASA. Fig. 6-6: NASA. Fig. 6-8: W. Benz, A.G.W. Cameron, and W. Slattery. Fig. 6-9: NASA. Fig. 6-13a: C.M. Pieters and the U.S.S.R. Academy of Sciences. Fig. 6-14: NASA. Fig. 6-15: S.P. Meszaros and NASA. Fig. 6-16: NASA. Fig. 6-17: NASA/JPL. Fig. 6-18: Donald Parker. Fig. 6-19: NASA. Fig. 6-20: NASA, USGS. Fig. 6-21a: NSSDC/NASA. Fig. 6-21b: NSSDC/NASA and Dr. Michael H. Carr. Fig. 6-22: A.S. McEwen, USGS. Fig. 6-23: NASA. Fig. 6-24: NASA. Fig. 6-25 right: NASA. Fig. 6-25 inset: NASA. Fig. 6-26: (c) CNES; Licensed by SPOT image Corporation/ Photo Researchers. Fig. 6-27a: NASA. Fig. 6-27b: NASA/JPL. Fig. 6-28: NASA. Fig. 6-29: NASA. Fig. 6-30: NASA/JPL. Fig. 6-31a & b: NASA.

**Chapter 7** p. 172: J. Trauger (JPL) and NASA. Fig. 7-1: S. Larson. p. 174: Stephen P. Meszaros/ NASA, M. Buie, K. Horne, and D. Tholen. Fig. 7-2: NASA. Fig. 7-3: NASA. Fig. 7-4a & b: NASA. Fig. 7-5: NASA. Fig. 7-8: H.A. Weaver, T.E Smith, STScI and NASA. Fig. 7-9: NASA. Fig. 7-10a: NASA. Fig. 7-11: JPL, NASA.

Fig. 7-12 top: Brown University, JPL, NASA. Fig. 7-12 center: NASA, JPL. Fig. 7-12 bottom: Brown University, JPL, NASA. Fig. 7-13: NASA. Fig. 7-14: NASA. Fig. 7-15: JPL, NASA. Fig. 7-16: NASA. Fig. 7-17: NASA. Fig. 7-18: NASA. Fig. 7-19: Reta Beebe, New Mexico State University; D. Gilmore and L. Bergeron, STScI; and NASA. Fig. 7-22a: (c) 1998 by Calvin J. Hamilton, Columbia, Maryland. Fig. 7-22b: NASA. Fig. 7-23: NASA. Fig. 7-24: NASA. Fig. 7-25: NASA. Fig. 7-26: NASA. Fig. 7-27: Peter H. Smith, University of Arizona Lunar and Planetary Laboratory; NASA. Fig. 7-28: NASA. Fig. 7-29: Erich Karkoschka, University of Arizona, and NASA. Fig. 7-33 inset: NASA. Fig. 7-34: NASA. Fig. 7-35: NASA. Fig. 7-36: Lawrence Sromovsky, University of Wisconsin-Madison, and NASA. Fig. 7-37: NASA. Fig. 7-38: NASA. Fig. 7-39: NASA. Fig. 7-40 left & right: Lick Observatory. Fig. 7-41: Alan Stern, Southwest Research Institute, Marc Buie, Lowell Observatory, NASA/ESA. Fig. 7-42: U.S. Naval Observatory. Fig. 7-43 top left: Alan Stern (SwRI), Marc Buie (Lowell Obs.), NASA, and ESA; top right: R. Albrecht, ESA/ESO Space Telescope European Coordinating Facility, and NASA; Figure 7-43 inset: A. Stern (SwRI), M. Buie (Lowell Obs.), NASA, ESA. p. 203: Photri.

Chapter 8  p. 204: Gary Goodman. Fig. 8-1 left : NASA. Fig. 8-1 center: Lick Observatory. Fig. 8-1 right: STScI. Fig. 8-3: Yerkes Observatory. Fig. 8-5: NASA. Fig. 8-5 inset: NASA. Fig. 8-6: NASA and Johns Hopkins University. Fig. 8-8: NEAR. Fig. 8-9: Jim Scotti, Spacematch on Kitt Peak. Fig. 8-11a & b: Alan Fitzsimmons, Queen's University of Belfast. Fig. 8-12a: Anglo-Australian Observatory. Fig. 8-12b: Max Planck Institut fur Aeronomie. Fig. 8-13 left & right: Johns Hopkins University and Naval Research Laboratory. Fig. 8-14: Hans Vehrenberg. Fig. 8-16: Palomar Observatory. Fig. 8-18: UCO/Lick Observatory. Fig. 8-19: UCO/Lick Observatory. Fig. 8-21: New Mexico State University Observatory. Fig. 8-22: Tony and Daphne Hallas, Astrophotos. Fig. 8-22 inset: A.H. Weaver and P.D. Sellman, STSI/NASA. Fig. 8-23: R.A. Oriti. Fig. 8-24: Meteor Crater Enterprises. Fig. 8-25: Akira Fujii, Chiro Astronomical Observatory, Koriyama, Japan. Fig. 8-26: R.A. Oriti. Fig. 8-27: R.A. Oriti. Fig. 8-28: R.A. Oriti. Fig. 8-29: R.A. Oriti. Fig. 8-30: Chip Clark. Fig. 8-31: SOVFOTO. Fig. 8-32: J.A. Wood. Fig. 8-33: W. Alvarez. Fig. 8-34 inset top left: Frank T. Kyte, University of California, Los Angeles. Fig. 8-34 inset top right: Virgil L. Sharpton, Lunar and Planetary Institute. Fig. 8-34: Digital image by Peter W. Sloss, NOAA-NESDIS-NGDCD.

Chapter 9  p. 227: Stanford Lockheed Institute for Space Research and NASA. Fig. 9-1: Celestron International. Fig. 9-2: Goran Scharmer, Lund Observatory. Fig. 9-3 inset: NOAO. Fig. 9-5: NOAO. Fig. 9-6: R. Christen and M. Christen, Astro-Physics Inc. Fig. 9-7a: NOAO. Fig. 9-7b: National Solar Observatory. Fig. 9-8b: NOAO. Fig. 9-8c: NOAO. Fig. 9-9: Carnegie Observatories. Fig. 9-11a & b: NOAO. Fig. 9-13: Dr. Alan Title/Stanford Lockheed Institute for Space Research and NASA. Fig. 9-14: Sacramento Peak Observatory. Fig. 9-15a: Naval Research Laboratory. Fig. 9-15b: SOHO. Fig. 9-16: Solar and Astrophysics Laboratory of the Lockheed-Martin Advanced Technology Center. Fig. 9-17: NOAO. Fig. 9-19a, b, & c: High Altitude Observatory/Solar Maximum Mission. Fig. 9-22 inset: Super-Kamiokande group, Institute for Cosmic Ray Research, University of Tokyo.

Foundation III  p. 249: Okiro Fujii, L'Astronomia.

Chapter 10  p. 260: NOAO. Fig. 10-2: R. Bell, University of Maryland, and M. Briley, University of Wisconsin at Oshkosh. Fig. 10-7a-g: Navy Prototype Optical Interferometer, Flagstaff, Arizona. Courtesy of Dr. Christian A. Hummel. Fig. 10-10: Lick Observatory.

Chapter 11  p. 276: Anglo-Australian Observatory. Fig. 11-1a: Anglo-Australian Observatory. Fig. 11-2: Royal Observatory, Edinburgh. Fig. 11-2 inset: R.C. Mitchell, Central Washington University. Fig. 11-3: Anglo-Australian Observatory. Fig. 11-4: R. Maddalena, M. Morris, J. Moscowitz, and P. Thaddeus. Fig. 11-5a: Nancy Levenson/NASA. Fig. 11-5b: Jeff Hester, Arizona State University and NASA. Fig. 11-6: Anglo-Australian Observatory. Fig. 11-7a & b: NOAO. Fig. 11-9: S. Kularni, California Institute of Technology; D. Golimowski, Johns Hopkins University; NASA. Fig. 11-21a & b: Lick Observatory. p. 301: John Elk III/ Bruce Coleman.

Chapter 12  p. 302: J. Morse, University of Colorado and NASA. Fig. 12-4a: Anglo-Australian Observatory. Fig. 12-4b: Howard Bons, STScI; Robin Ciardullo, Pennsylvania State University; NASA. Fig. 12-4c: H. Bond, STScI and NASA. Fig. 12-5d: R. Sahai and J. Trauer, JPL, the WFPC-2 Science Team, and NASA. Fig. 12-6: R.B. Minton. Fig. 12-7a & b: UCO/Lick Observatory.

Chapter 13  p. 327: Wood, Ronsaville, Harlin, Inc. Fig. 13-6: J. Kristian, Carnegie Observatories. Fig. 13-7a: Holland Ford, STScI/Johns Hopkins University; Richard Harms, Applied Research Corp.; Zlatan Tsvetanov, Arthur Davidsen, and Gerard Kriss at Johns Hopkins University; Ralph Bohlin and George Hartig at STScI; Linda Dressel and Ajay K. Kochhar at Applied Research Corp. in Landover, MD.; and Bruce Margon from the University of Washington in Seattle. Fig. 13-7b: NASA. Fig. 13-8: R.P. van der Marel, STScI; F.C. van den Bosch, University of Washington; NASA.

Foundation IV  p.340: (c) 1994 Smithsonian Institution/Corbis "Over the Great Wall," digital image opener by Margaret Geller and enhanced by Roger Resselmeyer. Fig. IV-1a: Birr Castle Demesne. Fig. IV-1b: Lund Humphries. Fig. IV-2: NOAO. Fig. IV-3: Dr. Wendy L. Freedman, Observatories of the Carnegie Institution of Washington; NASA.

Chapter 14  p. 348: Dirk Hoppe. Fig. 14-1: Harvard Observatory. Fig. 14-2: Harvard Observatory. Fig. 14-4: U.S. Naval Observatory. Fig. 14-7: G. Westerhout. Fig. 14-8a & b: Anglo-Australian Observatory, VLA, NRAO. Fig. 14-9a & b: Illustration by Dennis Davidson, courtesy of AMNH/Hayden Planetarium.

# INDEX

Page numbers in *italics* indicate illustrations.

NORTHERN HORIZON

EASTERN HORIZON

WESTERN HORIZON

SOUTHERN HORIZON

# THE NIGHT SKY IN MARCH

Chart time (Local Standard Time):

10 pm...First of March
9 pm...Middle of March
8 pm...Last of March

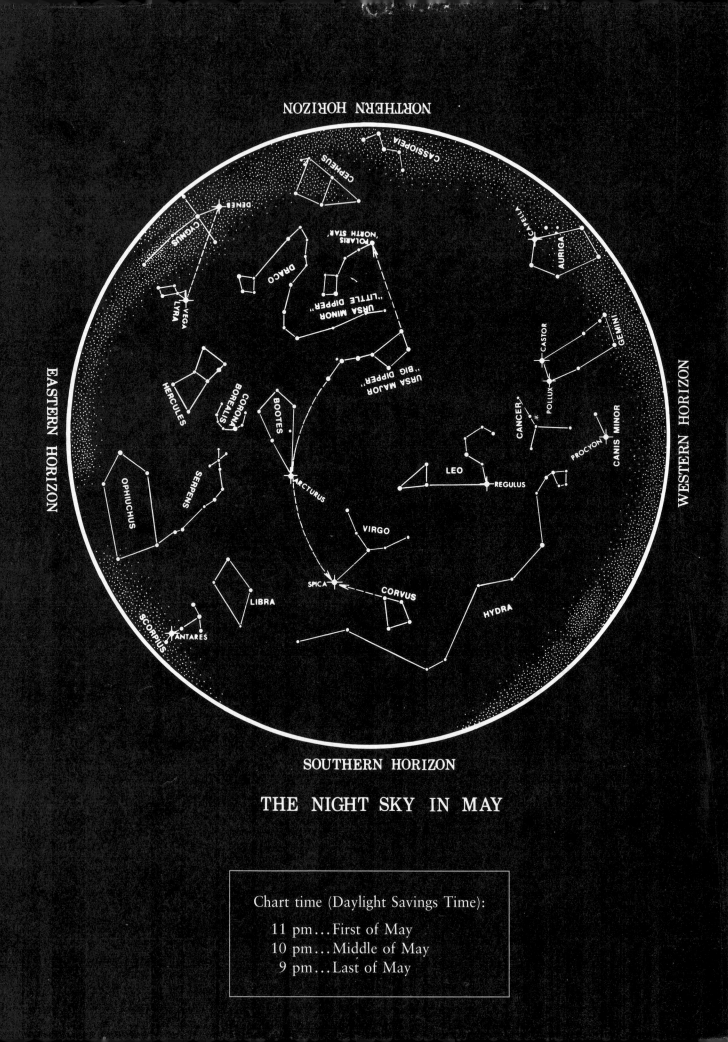

NORTHERN HORIZON

CASSIOPEIA
CEPHEUS
DENEB
CYGNUS
POLARIS 'NORTH STAR'
DRACO
CAPELLA
AURIGA
VEGA
LYRA
URSA MINOR "LITTLE DIPPER"
CASTOR
POLLUX
GEMINI
HERCULES
CORONA BOREALIS
BOOTES
URSA MAJOR "BIG DIPPER"
CANCER
CANIS MINOR
PROCYON
OPHIUCHUS
SERPENS
ARCTURUS
LEO
REGULUS
VIRGO
LIBRA
SPICA
CORVUS
SCORPIUS
ANTARES
HYDRA

EASTERN HORIZON

WESTERN HORIZON

SOUTHERN HORIZON

# THE NIGHT SKY IN MAY

Chart time (Daylight Savings Time):

11 pm...First of May
10 pm...Middle of May
9 pm...Last of May